Vortex Flows and Related Numerical Methods

NATO ASI Series

Advanced Science Institutes Series

A Series presenting the results of activities sponsored by the NATO Science Committee, which aims at the dissemination of advanced scientific and technological knowledge, with a view to strengthening links between scientific communities.

The Series is published by an international board of publishers in conjunction with the NATO Scientific Affairs Division

A Life Sciences **B Physics**	Plenum Publishing Corporation London and New York
C Mathematical and Physical Sciences **D Behavioural and Social Sciences** **E Applied Sciences**	Kluwer Academic Publishers Dordrecht, Boston and London
F Computer and Systems Sciences **G Ecological Sciences** **H Cell Biology** **I Global Environmental Change**	Springer-Verlag Berlin, Heidelberg, New York, London, Paris and Tokyo

NATO-PCO-DATA BASE

The electronic index to the NATO ASI Series provides full bibliographical references (with keywords and/or abstracts) to more than 30000 contributions from international scientists published in all sections of the NATO ASI Series.
Access to the NATO-PCO-DATA BASE is possible in two ways:

– via online FILE 128 (NATO-PCO-DATA BASE) hosted by ESRIN, Via Galileo Galilei, I-00044 Frascati, Italy.

– via CD-ROM "NATO-PCO-DATA BASE" with user-friendly retrieval software in English, French and German (© WTV GmbH and DATAWARE Technologies Inc. 1989).

The CD-ROM can be ordered through any member of the Board of Publishers or through NATO-PCO, Overijse, Belgium.

Series C: Mathematical and Physical Sciences - Vol. 395

Vortex Flows and Related Numerical Methods

edited by

J.T. Beale

Department of Mathematics,
Duke University,
Durham, North Carolina, U.S.A.

G.-H. Cottet

Laboratoire de Modélisation et Calcul,
Université Joseph Fourier,
Grenoble, France

and

S. Huberson

Université du Havre,
Le Havre, France

Kluwer Academic Publishers

Dordrecht / Boston / London

Published in cooperation with NATO Scientific Affairs Division

Proceedings of the NATO Advanced Research Workshop on
Vortex Flows and Related Numerical Methods
Grenoble, France
June 15–19, 1992

A C.I.P. Catalogue record for this book is available from the Library of Congress.

ISBN 0-7923-2250-9

Published by Kluwer Academic Publishers,
P.O. Box 17, 3300 AA Dordrecht, The Netherlands.

Kluwer Academic Publishers incorporates the publishing programmes of
D. Reidel, Martinus Nijhoff, Dr W. Junk and MTP Press.

Sold and distributed in the U.S.A. and Canada
by Kluwer Academic Publishers,
101 Philip Drive, Norwell, MA 02061, U.S.A.

In all other countries, sold and distributed
by Kluwer Academic Publishers Group,
P.O. Box 322, 3300 AH Dordrecht, The Netherlands.

Printed on acid-free paper

Printed in the Netherlands

Table of Contents

PREFACE

The papers presented here are based on talks given at a NATO ARW held in Saint Pierre de Chartreuse, near Grenoble (France) in June 1992. In addition to the editors, the organizers were J.-P. Chollet from Grenoble and A. Zervos from Athenes.

This meeting, following other succesful events of a similar type and in particular the 1990 AMS-SIAM Summer School in Seattle, has proved again the value of bringing together specialists from different fields sharing interest in the computation of vortex flows. The reason is that vortex flows, while being of crucial importance in Fluid Dynamics, raise fundamental mathematical questions and motivate large scale computations with numerical methods adapted to their particular features.

The first part of the book is devoted to recent developments in mathematical (articles 4,5,6) and numerical (1,2,3,7) models for the analysis and the numerical approximation of vortex flows.

The second part is devoted to numerical studies focusing on vortex generation problems around obstacles (9,10) and on related boundary layer (8) and wake (11) models.

The third part concentrates on new developments in the design of Contour Dynamics (14,18) and Vortex methods (12,15,16) and hybrid finite-difference vortex methods (13,17).

In Part IV we have grouped the papers that specifically adress 3 dimensional incompressible flows: vortex sheets (19,22), vortex reconnection and related problems (19,21,23).

Finally Part V is devoted to numerical computations of reacting (24,26) and compressible (25,27) flows.

We would like to thank all the participants and the staff who made this meeting just as enjoyable as it was scientifically rich. We gratefully acknowledge the support from the following agencies (in addition to NATO):
Centre National de la Recherche Scientifique, DRET (French Army), Ministère de l'Education Nationale, IMAG, Cray France, Universités du Havre et de Grenoble and the US Air Force.

The Editors

LOCAL SPECTRAL ANALYSIS OF TURBULENT FLOWS USING WAVELET TRANSFORMS

C. BASDEVANT, V. PERRIER, and T. PHILIPOVITCH
L.M.D.-Ecole Normale Supérieure
24 rue Lhomond
F-75231 Paris Cedex 05

and

M. DO KHAC
D.R.E.T.
26 boulevard Victor
F-00460 Armées
and
O.N.E.R.A.
29 avenue de la Division Leclerc
F-92320 Châtillon

ABSTRACT. Numerical simulations of two-dimensional turbulent flows exhibit strong solitary, long-lived, coherent vortices ; these coherent structures are not taken in account in the *à la* Kolmogorov similarity theory, this brings to a difference between the theoretical predictions of spectral slopes and those coming from numerical experiments. How do these coherent structures contribute to the energy distribution in the enstrophy inertial range?

In order to describe them, we propose in this paper multiscale analysis, localized both in space and scale, namely the continuous wavelet transform and the orthonormal wavelet decomposition. We will discuss here the advantages and the drawbacks of this two kinds of analysis.

After a quick revue of the classical theory of two-dimensional turbulence, we recall the definitions of continuous and discrete wavelet transforms. By applying them on a numerical turbulent field, we test then the robustness of wavelet coefficients when the basic wavelet is changed. Afterwards, wavelet coefficients are used to extract from the flow field spatially localized spectral informations, like local (in space) enstrophy spectrum and enstrophy fluxes. These new quantities will allow the characterization of coherent structures in terms of spatial and spectral behaviours.

1. Two-dimensional turbulence and coherent structures

1.1. THE PHENOMENOLOGICAL THEORY

Incompressible fluid motion in two dimensions is governed by the vorticity equation:

$$\frac{\partial \omega}{\partial t} + J(\psi,\omega) = F(\omega) - D(\omega) \tag{1}$$

J.T. Beale et al. (eds.), Vortex Flows and Related Numerical Methods, 1–26.

Here ω is the vorticity, ψ the stream function, J the Jacobian operator, $F(\omega)$ an external vorticity source and $D(\omega)$ the dissipation term.

In the absence of sources and dissipation ($F = D = 0$), and with suitable boundary conditions such as periodicity or rigid boundaries, the following quantities are quadratic invariants of the dynamics:
Energy:

$$E = -\frac{1}{2} < \psi \, \omega > \tag{2}$$

Enstrophy:

$$Z = \frac{1}{2} < \omega^2 > \tag{3}$$

where the brackets stand for the mean other the flow domain.

Now, from the Fourier transform $\hat{\omega}$ [1] of the vorticity field ω, we can define the enstrophy spectrum $Z(k)$ of the flow by:

$$Z(k) = \frac{1}{2} \int_{|\mathbf{k}|=k} |\hat{\omega}(\mathbf{k})|^2 \, |\mathbf{k}| \, d\theta \tag{4}$$

$Z(k)$ represents the enstrophy integrated in spectral space over an annulus of radius k. The total enstrophy verifies:

$$Z = \int_0^{+\infty} Z(k) \, dk \tag{5}$$

It is linked to the energy spectrum $E(k)$ by the simple relation:
$Z(k) = k^2 E(k)$.

These spectra are of primordial importance in statistical theories of turbulence as they analyze the dynamics of the flow through its non linear interactions between scales identified as Fourier modes. These theories rely on hypotheses of homogeneity and isotropy for the fluid motion. The self similar Kolmogorov theory of three-dimensional turbulence was extended to two-dimensional flows by Kraichnan ([11], [12]) and Leith [13] ; it is based on a double cascade concept:

- Enstrophy is carried from injection scales to small scales where dissipation occurs with a constant transfert rate η through an inertial range where the spectrum follows a power law:

$$Z(k) \approx C \, \eta^{2/3} \, k^{-1} \tag{6}$$

- Energy reaches ever larger scales with a constant transfert rate ϵ in a reverse energy cascade. In this reverse energy cascade, the energy spectrum follows the law :

$$E(k) \approx C' \, \epsilon^{2/3} \, k^{-5/3} \tag{7}$$

[1] $\hat{\omega}(\mathbf{k}) = \int\int_{R^2} \omega(\mathbf{x}) \, e^{-2i\pi \mathbf{k}.\mathbf{x}} \, d^2\mathbf{x}$

1.2. THE PHYSICAL POINT OF VIEW: THE IMPORTANCE OF COHERENT STRUCTURES

Long-term simulations of two-dimensional turbulent flows have been performed by several authors (Basdevant *et al* [3], Mc Williams [15], Benzi [5]) and all led to the result that the enstrophy spectrum does not follow the predicted *à la* Kolmogorov k^{-1} law, but exhibits spectral slopes ranging from k^{-2} to k^{-4}, depending on the nature of the forcing. Indeed, numerical experiments, as well as laboratory experiments (Couder & Basdevant [4]) generate strong solitary, long-lived, coherent vortices, which concentrate most of the enstrophy in a small fraction of physical space. The effect of coherent vortices on the spectral slope cannot be properly reproduced by statistical closure models, which assume a random distribution of phases.

The role of coherent structures in the enstrophy cascade has been simply demonstrated by Babiano *et al* [1] in two numerical experiments:

When the spatial coherence is destroyed in the vorticity field (by resetting at random the phases of all Fourier modes each time-step), the spectrum produced in the stationary regime of the simulation follows the phenomological law in k^{-1}.

Since coherent vortices occupy only a small portion of space and have much stronger amplitudes than the remaining field, Babiano *et al* filter them by clipping the vorticity field at a given level ω_0. They observe that as ω_0 decreases, the clipped enstrophy spectrum converges to a "clean" k^{-1} law. This indicates that the medium that surrounds the coherent vortices in the vorticity field follows a k^{-1} law.

These experiments suggest that the vorticity field can be separate into an active component (coherent structures) responsible for the non linear dynamic of the flow, and a passive component submitted to the cascade process. Several authors (Weiss [23], Mac Williams [15], Elmhaïdi *et al* [8], Zouari *et al* [24]) studied criteria to characterize these two parts of the flow. But we need a tool to determine the energy distribution of coherent vortices in the enstrophy inertial range. Following M. Farge [9], the wavelet transform, which gives informations on the field localized both in space and scale, seems to be the right tool to characterize coherent structures and their contribution to the total enstrophy.

1.3. THE NUMERICAL EXPERIMENT

The flow field analyzed afterwards is taken from a numerical simulation of two-dimensional incompressible Navier-Stokes equations integrated on a doubly periodic square using a pseudo-spectral model of 128^2 points (Basdevant *et al* [3]). The dissipation includes a "linear" friction to dissipate energy at large scale in addition to an iterated Laplacian which modelises the enstrophy dissipation near the cutoff scale. The forcing is defined by keeping the amplitude of a zonal mode at wavenumber 10 constant in time.

The numerical integration was pursued long enough to reach a stationary regime for the enstrophy spectrum. The instantaneous enstrophy spectrum of the stationary final regime is shown on *figure 1*. It is noticed that in opposition to the prediction

(6), the spectral slope is steeper than k^{-1}, in the enstrophy inertial range, above wavenumber 10. In this experiment, the slope is close to k^{-2}.

Figure 2 shows the contour plot of the vorticity field. It exhibits number of strong, isolated, quasi-circular vortices, which have very long lifetimes compared to the more chaotic surrounding structures.

In order to characterize or filter these structures, and to define new fundamental quantities localized in physical and spectral spaces, we will introduce now two-dimensional wavelet transforms.

2. Two-dimensional wavelet analysis

There exist two types of wavelet transforms: the continuous wavelet transform and the discrete wavelet transform. The first one leads to a redundant representation of the signal, it is used for image processing and analysis (astronomy, turbulence). The second one, based on an orthonormal decomposition of the signal, is very convenient for information compression and coding.

In this paper, we will describe and compare the properties of this two different transforms, in two dimensions. We will apply them to the analysis of the turbulent flow described above.

2.1. 2-D CONTINUOUS WAVELET TRANSFORM

The theory of the continuous wavelet transform was first introduced by A. Grossmann and J. Morlet in 1984 [10], and generalized in several dimensions by R. Murenzi [20].

The continuous wavelet transform of a given function f of $L^2(\mathbf{R}^2)$ is defined by:

$$W_f(a,\theta,\mathbf{b}) = \iint_{\mathbf{R}^2} f(\mathbf{x})\ \overline{\psi}_{a,\theta}(\mathbf{x}-\mathbf{b})\ d^2\mathbf{x}, \quad a>0, \theta\in[0,2\pi[, \mathbf{b}\in\mathbf{R}^2 \tag{8}$$

where the family of functions $\psi_{a,\theta}$ is obtained from a single one ψ, called the wavelet mother, by dilation of factor a^{-1} and rotation of angle θ:

$$\psi_{a,\theta}(\mathbf{x}) = \frac{1}{a}\psi(R_\theta(\frac{\mathbf{x}}{a})) \tag{9}$$

As we will see later, this transformation is invertible and isometric, as soon as function ψ verifies the following admissibility condition, that is if ψ verifies:

$$c_\psi = 2\pi \iint_{\mathbf{R}^2} \frac{|\hat{\psi}(\mathbf{k})|^2}{\|\mathbf{k}\|^2}\ d^2\mathbf{k} < +\infty \tag{10}$$

In this paper we shall restrict our analysis to isotropic wavelet mothers, with enough regularity (see section 3.1). The admissibility condition (10) implies that $\hat{\psi}(0)=0$, and the wavelet coefficient $W_f(a,\mathbf{b})$ (there is no more dependance on θ) can be viewed as the filtering of f by a band-pass filter:

$$W_f(a,\mathbf{b}) = a \iint_{\mathbf{R}^2} \hat{f}(\mathbf{k})\ \overline{\hat{\psi}}(a\mathbf{k})\ e^{2i\pi\mathbf{k}.\mathbf{b}}\ d^2\mathbf{k} \tag{11}$$

Here $a\hat{\psi}(a\mathbf{k})$ is the Fourier transform of the wavelet ψ_a, and is numerically supported on a annulus of radius $\sim a^{-1}$, as shown on *figure 3*.

The wavelet transform is isometric and the energy conservation writes:

$$\iint_{\mathbf{R}^2} |f(\mathbf{x})|^2 \, d^2\mathbf{x} \;=\; \frac{1}{c_\psi} \int_0^\infty \frac{da}{a^3} \iint_{\mathbf{R}^2} d^2\mathbf{b} \;\; |W_f(a,\mathbf{b})|^2 \tag{12}$$

The function f can be reconstructed from its wavelet coefficients all over the domain $(a, \mathbf{b})$ by the relation:

$$f(\mathbf{x}) \;=\; \frac{1}{c_\psi} \int_0^\infty \frac{da}{a^3} \iint_{\mathbf{R}^2} d^2\mathbf{b} \;\; W_f(a,\mathbf{b}) \; \psi_a(\mathbf{x}-\mathbf{b}) \tag{13}$$

2.2. 2-D ORTHONORMAL WAVELET TRANSFORM

The theory of orthonormal wavelet bases has been developed since 1985 by Y. Meyer and its group (see [18]). The construction of such bases is provided by the concept of multiresolution analysis, first introduced by S. Mallat [16]. In the two dimensional case, orthonormal wavelet bases on $L^2(\mathbf{R}^2)$ are built from multiresolution analysis of $L^2(\mathbf{R}^2)$ which can be taken as tensor-products of the one of $L^2(\mathbf{R})$. It leads to an orthonormal basis of $L^2(\mathbf{R}^2)$ generated by four families of functions, as follows:

$$\begin{aligned} \varphi_j(x,y) &= \varphi_j(x)\varphi_j(y) \\ \psi_j^{(1)}(x,y) &= \varphi_j(x)\psi_j(y) \\ \psi_j^{(2)}(x,y) &= \psi_j(x)\varphi_j(y) \\ \psi_j^{(3)}(x,y) &= \psi_j(x)\psi_j(y) \end{aligned} \tag{14}$$

where ψ, φ are the generating functions of the one-dimensional multiresolution analysis: ψ is the wavelet mother, φ the scaling function [18].
$\varphi_j(x) = 2^{j/2}\varphi(2^j x)$, $\psi_j(x) = 2^{j/2}\psi(2^j x)$ are the related functions at scale 2^{-j}. Notice that functions ψ and φ are real.

Then, the orthonormal basis of wavelets in two dimensions is generated by three different functions, the 2-D wavelets $\psi^{(1)}$, $\psi^{(2)}$, $\psi^{(3)}$, by mean of dilations of 2^j and translations. Due to orthonormality, a given function f of $L^2(\mathbf{R}^2)$ is decomposed as:

$$f(\mathbf{x}) \;=\; \sum_{j=-\infty}^{+\infty} \sum_{\mathbf{n}\in\mathbf{Z}^2} \sum_{r=1}^{3} c_j^{(r)}(\mathbf{n}) \; \psi_j^{(r)}(\mathbf{x} - 2^{-j}\mathbf{n}) \tag{15}$$

with:

$$c_j^{(r)}(\mathbf{n}) \;=\; \iint_{\mathbf{R}^2} f(\mathbf{x}) \; \psi_j^{(r)}(\mathbf{x} - 2^{-j}\mathbf{n}) \; d^2\mathbf{x} \tag{16}$$

or, with the help of the Fourier transform:

$$c_j^{(r)}(\mathbf{n}) \;=\; 2^{-j} \iint_{\mathbf{R}^2} \hat{f}(\mathbf{k}) \; \overline{\hat{\psi}}^{(r)}(2^j\mathbf{k}) \; e^{2i\pi 2^{-j}\mathbf{n}.\mathbf{k}} \; d^2\mathbf{k} \tag{17}$$

Since ψ_j is a band-pass filter around scale 2^{-j} and φ_j a low-pass filter selecting scales larger than 2^{-j}, the location in Fourier space of wavelets $\psi_j^{(r)}$ is supported by a square annulus of radius $\sim 2^j$ (see *figure 4*).

2.3. PRACTICAL IMPLEMENTATION

Consider the wavelet analysis of a scalar 1-periodic field discretized on a regular grid with $N = 2^p$ points in both directions. Since the field is periodic, we will use periodic wavelets in the continuous and discrete cases and compute wavelet coefficients by the mean of Fast Fourier Transforms.

The largest scale we can define corresponds to all the domaine ; it will be the scale 1 by convention. Other scales (smaller than 1) are discretized logarithmically as follows:

$$a_j = a_0^{-j} \qquad j \geq 0 \tag{18}$$

The choice of a_0 in both case is determined by the reconstruction formula. In the orthogonal case, the multiresolution analysis we choose imposes $a_0 = 2$. For the continuous wavelet transform, the choice of a_0 is constrained by the precision we ask for the reconstruction of the signal through a discretized version of formula (13). Following Daubechies [6], the choice (based on frame theory) $a_0 = 2^{1/4}$ is enough precise for wavelet mothers which are Gaussian derivatives.
The smallest scale we can reach is in both cases $\frac{2}{N} = 2^{1-p}$.

In the continuous case, formula (11) is discretized as follows: for each scale a, we compute:

$$W_f(a, \frac{m}{N}, \frac{n}{N}) = \sum_{p,q=-N/2}^{N/2-1} \hat{f}_{p,q}\ a\overline{\hat{\psi}}(a_p, a_q)\ e^{2i\pi\frac{mq+np}{N}} \tag{19}$$

where $\hat{f}_{p,q}$ stands for the discrete Fourier transform of the discrete values $f(\frac{m}{N}, \frac{n}{N})$, computed by the mean of F.F.T.:

$$\hat{f}_{p,q} = \frac{1}{N^2} \sum_{n,m=0}^{N-1} f(\frac{m}{N}, \frac{n}{N})\ e^{-2i\pi\frac{mq+np}{N}} \tag{20}$$

Since the signal f is periodic, the use of F.F.T. does not induce boundary problems.

The orthonormal wavelet transform of the same field, used in the framework of periodic functions, results without any approximation (see Do Khac *et al* [7]) in the mean of the signal plus wavelet coefficients from scale 1 to scale 2^{1-p}, the scale 2^{-j} being represented by three arrays of $2^j \times 2^j$ coefficients corresponding to the 2D wavelets $\psi^{(1)}$, $\psi^{(2)}$, $\psi^{(3)}$ (*figures 4 & 5*).

The practical implementation of the 2D-periodic decomposition- recomposition algorithm is detailed in [7].

2.4. INTERPRETATION OF WAVELET COEFFICIENTS

We may interprete each wavelet coefficient in term of space and wavenumber information, since it is obtained through the scalar product of the signal against the corresponding wavelet, localized around a particular space and scale position. This localization can be deduced from the localization of the basic wavelet, which in turn depends on its regularity, its number of wanishing moments ; this point will be discussed in section 3.1.

The basic one-dimensional wavelet ψ is localized around a position (b_0, k_0) in physical-Fourier space:

- b_0 is in our examples the center of symmetry of the one-dimensional wavelet ψ ($b_0 = 0$ in the continuous case; in the orthogonal case $b_0 = 1/2$ for the wavelet ψ and $b_0 = 0$ for the scaling function φ).

- k_0 is the mean wavenumber of the one-dimensional wavelet spectrum $\hat{\psi}$ given by the relation :

$$k_0 = \frac{\int_0^\infty k \; |\hat{\psi}(k)| \; dk}{\int_0^\infty \; |\hat{\psi}(k)| \; dk} \tag{21}$$

Then each scale a_j is related to a wavenumber k_j, k_j being the mean wavenumber of the Fourier spectrum $\hat{\psi}_j$ of the wavelet ψ_j:

$$a_j = \frac{k_0}{k_j} \tag{22}$$

Then for the continuous wavelet transform, $W_f(a_j, \mathbf{x})$ represents the contributions of f coming from a point $\mathbf{x}$ belonging to the grid, and from wavenumber k_j.

From the conservation property (12), it appears that, when applied to a given signal f, the quantity $a^{-1}|W_f(a, \mathbf{b})|^2$ is at point $\mathbf{b}$ and scale a (up to a multiplicative constant) the density of energy per unit of area in physical space and unit of wavenumber in spectral space.

As this coefficient is obtained through the scalar product of the signal f against an oscillating wavelet, it may vanish when it happens to be in opposite phase with the signal. These spurious zeroes may be avoided, either by using complex wavelets and analyzing the modulus of the complex wavelet coefficients, or, and it's the way we used, by replacing the energy density of f at scale a and point $\mathbf{b}$ by its mean in physical space over a disk of radius a centered at point $\mathbf{b}$:

$$< a^{-1}|W_f(a, \mathbf{x})|^2 >_{\|\mathbf{x}-\mathbf{b}\| < a}$$

This smoothing procedure, which preserves conservation properties of the wavelet transform, is consistant with Heisenberg uncertainty principle.

For the orthonormal wavelet decomposition (Do Khac *et al* [7]):

– $c_j^1(k, l)$ represents the contributions of f coming from point $(k2^{-j}, (l + 1/2)2^{-j})$ and from wavenumber k_j.

– $c_j^2(k,l)$ represents the contributions of f coming from point $((k+1/2)2^{-j}, l2^{-j})$ and from wavenumber k_j.
– $c_j^3(k,l)$ represents the contributions of f coming from point $((k+1/2)2^{-j}, (l+1/2)2^{-j})$ and from wavenumber k_j.

To be consistant with the continuous wavelet transform (which we use with an isotropic wavelet), we define $O_f(2^{-j}, 2^{-j}(m,n))$, a mean coefficient at scale 2^{-j} and at point $2^{-j}(m+1/2, n+1/2)$:

$$O_f(2^{-j}, 2^{-j}(m,n)) =$$

$$\left(\frac{(c_j^{(1)}(m,n))^2+(c_j^{(1)}(m+1,n))^2}{2} + \frac{(c_j^{(2)}(m,n))^2+(c_j^{(2)}(m,n+1))^2}{2} + (c_j^{(3)}(m,n))^2\right)^{1/2}$$

From the orthonormality of the decomposition we deduce that $O_f(2^{-j}, 2^{-j}(m,n))^2$ is the energy of the signal f integrated in physical space on a square of side 2^{-j}, and integrated in wavenumber space on a wavenumber interval of lenght $\sim 2^j$ (*figure* 4-a).

3. Wavelet analysis of turbulent flow ; robustness of the wavelet transform

In this section we will compare maps of wavelet coefficients obtained with different basic wavelets, in the continuous and discrete cases. As we will see, the wavelet transform is very sensitive to the basic wavelet and we have to be careful in the interpretation of wavelet coefficients.

3.1. WAVELET MOTHERS CHOICE

The required condition for a function to be a basic wavelet is very weak:

- in the continuous case, many functions verify the admissibility condition (10). These functions may be real or complex.

- in the discrete case, one use usually wavelets coming from multiresolution analysis, which leads to real functions.

Our objective here is to link phenomenons observed in physical space with the classical phenomenology defined in term of Fourier modes ; thus the basic wavelet should be a good compromise between space localization and Fourier localization.

The localization in Fourier space asks for a rapid decrease of the wavelet spectrum both at large scales ($k \to 0$) and at small scales ($k \to \infty$). The first condition is met when the wavelet has vanishing moments:

$$\int_{\mathbf{R}} x^l\ \psi(x)\ dx = 0 \quad l = 0,\ 1, \ldots\ m \Rightarrow \hat{\psi}(k) = o(k^m) \quad k \to 0 \tag{23}$$

The last condition requires enough regularity for the wavelet:

$$\psi(x) \in C^p(\mathrm{R}) \Rightarrow \hat{\psi}(k) = O(k^p) \quad k \to \infty \tag{24}$$

In order to compare the orthogonal and the continuous technics, we choose for the continuous case derivatives of isotropic Gaussian, and for the discrete one spline functions (Lemarié [14], Perrier & Basdevant for the periodic case [21]) with the same number of vanishing moments.

Continuous case:

$$\begin{array}{llll} \psi(\mathbf{x}) = -\frac{d^2}{dr^2}e^{-r^2/2} & m=2 & p=\infty & (\text{"}\mathit{Mexicain\ hat}\text{"}) \\ \psi(\mathbf{x}) = \frac{d^4}{dr^4}e^{-r^2/2} & m=4 & p=\infty & \end{array} \tag{25}$$

where $r = \|\mathbf{x}\|$

Discrete case:

$$\begin{array}{lll} Affine\ spline & m=2 & p=0 \\ Cubic\ spline & m=4 & p=2 \end{array} \tag{26}$$

Figures 6 & 7 sketch the physical and Fourier representation of these functions.

The following section shows that the interpretation of wavelet coefficients may be different for $m = 2$ and $m = 4$.

3.2. WAVELET ANALYSIS

Consider now the flow field extracted from the numerical simulation described in §1.3. *Figure 2* shows the contour plot of the vorticity field when the flow has reached a stationary state ; it exhibits large co-rotating as well as counter rotating vortices.

In this section we draw the maps of enstrophy coefficients at a given scale a_j. In the continuous case it corresponds to maps of $|W_\omega(2^{-j/4}, \mathbf{x})|^2$ and for the discrete one to maps of $|O_\omega(2^{-j}, 2^{-j}(m,n))|^2$.

On *figures 8 to 11* are displayed the contour plots of the local entrophy coefficients computed from the continuous and discrete transform at different scales and for two different basic wavelets ($m = 2$ and $m = 4$).

When comparing enstrophy coefficients at different scales we find that vortices are best identified at large and intermediate scales, whereas small scale coefficients are of high magnitude in interacting zones, such as between corotating vortices.

When comparing enstrophy coefficients with different numbers of vanishing moments ($m = 2$, $m = 4$) we can observe that for $m = 2$, as the spectrum of the analyzing wavelet has a broad band, the uncertainty on the wavenumber k_j enlarges, this results for example at small scale in the appareance in the maps of more structures. Conversely if for $m = 4$ wavelet are best localized in Fourier space, we loose some localization in physical space, this results for example in small scales larger in physical space; however from other experiments not reported here, the case $m = 4$ seems to be a good compromise between localization in Fourier and physical space.

When comparing continuous and discrete analysis we find qualitatively the same results except from the oscillatory effect already noticed for continuous wavelet analysis. The comparison would be in fact better if we had used in the figures the smoothed wavelet coefficients defined in section 2.4.

4. Local spectrum

Wavelet analysis separate a field into a sum of contributions localized both in space and scale. When applied to a field possessing a strong spatial variability, it can, much better than Fourier analysis give insights on the intermittency and on the spatial structure of the flow. In the following section we will define the local wavelet spectrum that can be constructed from wavelet coefficients coming from the continuous and discrete wavelet transforms.

4.1. ENSTROPHY FOURIER SPECTRUM

Classical Fourier analysis of incompressible two-dimensional turbulence defines the enstrophy spectrum $Z(k)$ of the flow from the Fourier transform $\hat{\omega}$ of the vorticity field ω by:

$$Z(k) = \frac{1}{2} \int_{|\mathbf{k}|=k} |\hat{\omega}(\mathbf{k})|^2 \, |\mathbf{k}| \, d\theta \ , \quad Z = \int_0^{+\infty} Z(k) \, dk \tag{27}$$

Alternatively from wavelet analysis we may construct a spectrum localized in space and then separate active regions from the non active ones.

4.2. LOCAL ENSTROPHY SPECTRUM DEFINED BY THE MEAN OF CONTINUOUS WAVELET TRANSFORM

Since the continuous wavelet transform is isometric, the total enstrophy of the flow is given from wavelet coefficients all over the domain by:

$$Z = \frac{1}{2c_\psi} \int_0^{\infty} \frac{da}{a^3} \iint_{\mathbb{R}^2} d^2\mathbf{x} \ |W_\omega(a, \mathbf{x})|^2 \tag{28}$$

Remembering that $a = k_0/k$, let us define as the local enstrophy spectrum:

$$\tilde{Z}(k, \mathbf{x}) = \frac{1}{2c_\psi} \frac{k}{k_0^2} |W_\omega(\frac{k_0}{k}, \mathbf{x})|^2 \tag{29}$$

Then the enstrophy repartition between wavenumbers is given by:

$$\tilde{Z}(k) = \iint_{\mathbb{R}^2} \tilde{Z}(k, \mathbf{x}) \, d^2\mathbf{x} \tag{30}$$

and it follows the identity:

$$Z = \int_0^{+\infty} \tilde{Z}(k) \, dk \tag{31}$$

As discussed already in section 2.4, the local spectrum defined by formula (29) may present spurious oscillations due to cancellation effects; we may alternatively define a smoothed spectrum (as Moret-Bailly *et al* for the one-dimensional case [19]) by:

$$\hat{Z}(k,\mathbf{x}) = < \tilde{Z}(k,\mathbf{x'}) >_{\|\mathbf{x}-\mathbf{x'}\|<a=k_0/k} \tag{32}$$

Properties (30) and (31) are still valid for this spectrum.

4.3. LOCAL ENSTROPHY SPECTRUM DEFINED BY THE MEAN OF ORTHOGONAL WAVELET TRANSFORM

Remembering that the wavelet decomposition is orthonormal, the total enstrophy of the flow field expresses as:

$$Z = \frac{1}{2}c_0^2 + \frac{1}{2}\sum_{j=0}^{p-1}\sum_{n,m=0}^{2^j-1}\sum_{r=0}^{3}(c_j^{(r)}(m,n))^2, \tag{33}$$

or using O_ω as defined in §2.4:

$$Z = \frac{1}{2}c_0^2 + \frac{1}{2}\sum_{j=0}^{p-1}\sum_{n,m=0}^{2^j-1} O_\omega(2^{-j},2^{-j}(m,n))^2, \tag{34}$$

where c_0 is the mean value of the vorticity ω, and $c_j^{(r)}(m,n)$ its wavelet coefficients. From these coefficients we may define the local enstrophy spectrum by:

$$\tilde{\tilde{Z}}(k_j,\mathbf{x}_{mn}) = \frac{1}{2}\,O_\omega(2^{-j},2^{-j}(m,n))^2.\frac{2^{2j}}{\Delta k_j} \tag{35}$$

with:

$$\Delta k_j = \sqrt{k_j k_{j+1}} - \sqrt{k_j k_{j-1}} \tag{36}$$

$\mathbf{x}_{mn} = 2^{-j}(m+\frac{1}{2},n+\frac{1}{2})$ and k_j as defined in section 2.4. Notice that, to be compared to the usual Fourier transform, the local spectrum has been defined as a density of enstrophy per unit of area and per unit of wavenumber.

The analogue of the Fourier enstrophy spectrum of the field can be reconstructed by:

$$\tilde{\tilde{Z}}(k_j) = \sum_{m,n=0}^{2^j-1}\tilde{\tilde{Z}}(k_j,\mathbf{x}_{mn}).2^{-2j} \tag{37}$$

and the total enstrophy of the flow field expresses as:

$$Z = \sum_{j=0}^{p-1}\tilde{\tilde{Z}}(k_j)\,\Delta k_j. \tag{38}$$

4.4. NUMERICAL RESULTS

From the local enstrophy spectrum we may study the space variability of the flow by analyzing the spectral slope, that is, as for Fourier spectra, we can determine an exponent m for the spectrum decay, but now using wavelets this exponent is a function of space:

$$Z(k,\mathbf{x}) \sim k^{m(\mathbf{x})} \tag{39}$$

From turbulence theories (Babiano *et al* [1]) we may infer that this exponent should be close to -1 in inter-vortices passively advected regions and larger in active turbulent regions.

Figures 12, 13 & 14 display local enstrophy spectra $Z(k,\mathbf{x})$ for the three particular points $\mathbf{x}$ of the flow field mentionned on *figure 2*: point A corresponds to the center of a strong vortex, point B to a shear zone between two co-rotating vortices and point C to a low energy and low enstrophy area. Spectra have been drawn with orthogonal and continuous wavelet decompositions, in both cas we used four vanishing moments wavelets ($m = 4$). Continuous wavelet spectra displayed are smoothed spectra (formula (32)). These three points characterize three behaviours for the local spectrum:

1. large enstrophy at intermediate scales for point A and B leading to large spectral slopes for the mean spectrum,
2. large enstrophy at small scales for point B which belongs to a shear,
3. enstrophy distributed among all scales for point C which is located outside active regions.

Point B exhibits a spectral slope at small scales close to the k^{-1} predicted from the classical 2D-theory. Conversely point A, corresponding to a strong vortex, exhibits a steeper slope.

Figure 15 shows the mean wavelet spectra $\tilde{Z}(k)$ and $\tilde{\tilde{Z}}(k)$ obtained by spatially averaging all local spectra for the continuous (30), and the discrete transforms (37), compared to the classical Fourier spectrum $Z(k)$ (27). The three spectra would be identical if the basic wavelet were strictly localized in Fourier space.

Notice that the discrepancy is most important at large scale because there the spectrum is a mean value on a few number of degrees of freedom.

5. Enstrophy fluxes

In this section we study enstrophy transfers due to non-linear conservative interactions, the objective being to give evidence of the enstrophy cascade in physical space. In this case the wavelet will provide a new contribution, in the sense that we will be able to define a local (in space) enstrophy or energy transfer, and then see in what part of the flow field the cascades are effective.

5.1. CLASSICAL ENSTROPHY FLUX

The classical enstrophy flux is defined from the Fourier transform of the vorticity and its non-linear derivative. From (1) and (4) the time derivative of the enstrophy spectrum due to non-linear interactions expresses:

$$\frac{dZ(k)}{dt} = Re\left(\int_0^{2\pi} d\theta\ \hat{J}(\omega,\psi)(k)\ \overline{\hat{\omega}}(k)\ k\right) \tag{40}$$

The total enstrophy being conserved by non-linear interaction we have:

$$\frac{dZ}{dt} = \int_0^{+\infty} \frac{dZ(k)}{dt}\ dk \equiv 0 \tag{41}$$

We then define the enstrophy flux through wavenumber k due to non-linear interactions:

$$\Pi_Z(k) = \int_k^{+\infty} \frac{dZ(k')}{dt}\ dk' \tag{42}$$

From (41): $\Pi_Z(0) = 0$

5.2. LOCAL ENSTROPHY FLUX

In a similar way, from (1) and (29) we can obtain the time derivative due to non-linear interactions, of the local enstrophy spectrum:

$$\frac{\partial}{\partial t}\tilde{Z}(k,\mathbf{x}) = \frac{1}{c_\psi}\frac{k}{k_0^2}\ Re\left(W_J(\frac{k_0}{k},\mathbf{x})\ \overline{W_\omega}(\frac{k_0}{k},\mathbf{x})\right) \tag{43}$$

where $W_J(a,\mathbf{x})$ is the wavelet transform of the non linear interactions $J(\omega,\psi)$. Since the continuous wavelet tranform preserves the scalar product, we get:

$$\frac{dZ}{dt} = \int_0^{+\infty} dk \iint_{\mathbf{R}^2} \frac{\partial}{\partial t}\tilde{Z}(k,\mathbf{x})\ d^2\mathbf{x} \equiv 0 \tag{44}$$

We then define the enstrophy flux through wavenumber k, at point $\mathbf{x}$ by:

$$\tilde{\Pi}_Z(k,\mathbf{x}) = \int_k^{+\infty} \frac{\partial}{\partial t}\tilde{Z}(k',\mathbf{x})\ dk' \tag{45}$$

The enstrophy flux through wavenumber k is obtained through integration of the local flux over all the domain:

$$\tilde{\Pi}_Z(k) = \iint_{\mathbf{R}^2} \tilde{\Pi}_Z(k,\mathbf{x})\ d^2\mathbf{x} \tag{46}$$

From previous discussions it should correspond to the classical enstrophy flux:

$$\Pi_Z(k) \sim \tilde{\Pi}_Z(k) \tag{47}$$

Moreover, entrophy conservation imposes:

$$\tilde{\Pi}_Z(0) \equiv 0 \tag{48}$$

Similar definitions can be given in the orthogonal case.

5.3. NUMERICAL RESULTS

On *figure 16* are compared enstrophy fluxes computed with the classical Fourier technic (42) and the continuous wavelet analysis (46) using four vanishing moment wavelets ($m = 4$). The flux obtained by wavelet analysis is clearly consistent with the one obtained by Fourier analysis, however it is smoother. This can be explained by the spectral smoothing induced by the band-pass filtering effect of the wavelets. Alternatively the Fourier flux is much irregular as it is an instantaneous flux, should we have used time averages, the flux would have been more regular and closer to the wavelet flux.

Local enstrophy fluxes will be systematically studied in a subsequent paper. Preliminary results indicate that, as spectra, fluxes behaviours vary strongly in the flow field.

6. Conclusion and perspectives

We described and compared the continuous and discrete wavelet technics for the analysis of two-dimensional field. When applied to turbulent vorticity field, these technics are able to give detailed space-scale informations. Maps of wavelet coefficients, local spectra, local enstrophy fluxes are powerfull tools to separate the different dynamical behaviours of the different space/scales locations of the flow. We noticed however that the wavelet analysis has its own limitations linked to the properties of the wavelet mother. One has to take care of these limitations in the interpretation of the results, specially when looking at enstrophy repartition among scales: the information we get may be biaised by the shape of the spectrum of the basic wavelet.

In a subsequent paper [2] we will apply these technics, following Babiano's works ([8], [24]), to characterize the three components of the flow field which are vortex regions, circulation zones, and turbulent swamps.

References

[1] A. BABIANO, C. BASDEVANT, B. LEGRAS, R. SADOURNY, *Vorticity and passive-scalar dynamics in two-dimensional turbulence*, J. Fluid Mech., **183**, 379-397, 1987.

[2] A. BABIANO, C. BASDEVANT, M. FARGE, V. PERRIER, T. PHILIPOVITCH, in preparation.

[3] C. BASDEVANT, B. LEGRAS, R. SADOURNY, M. BELAN, *A study of barotropic model flows : intermittency waves and predictability*, J. Atmos. Sci. **38**, 2305-2326, 1981.

[4] Y. COUDER, C. BASDEVANT, *Experimental and numerical study of vortex couples in two-dimensional flows*, J. Fluid Mech., **173**, 225-251, 1986.

[5] R. BENZI, G. PALADIN, G. PARIS, A. VULPIANI, *Power spectra in two-dimensional turbulence*, J.Phys.A, 17 35211, 1984.

[6] I. DAUBECHIES, *The wavelet transform, time-frequency localization and signal analysis*, IEEE Transf. on Inform. Theory, **36-5**, 961-1005, 1990.

[7] M. DO KHAC, C. BASDEVANT, V. PERRIER, K. DANG TRAN, *Wavelet analysis of turbulent fields*, 1992, submitted to Physics of Fluids-A.

[8] D. ELMHAÏDI, A. PROVENZALE, A. BABIANO, *Elementary topology of two-dimensional turbulence from a Lagrangian viewpoint and single-particle dispersion*, 1992, submitted to J. of Fluid Mech.

[9] M. FARGE, *Wavelet transforms and their applications to turbulence*, Annual Review of Fluid Mechanics, **40**, 395-457, 1992.

[10] A. GROSSMANN, J. MORLET, *Decomposition of Hardy functions into square integrable wavelets of constant shape* , SIAM J. Math Anal., **15**, 723-736, 1984.

[11] R.H. KRAICHNAN, *Inertial ranges in two-dimensional turbulence*, Phys. Fluid, **10**, 1417-1423, 1967.

[12] R.H. KRAICHNAN, *Inertial range transfer in two-dimensional and three-dimensional turbulence*, J. Fluid Mech., **47**, 525-535, 1971.

[13] C.E. LEITH, *Diffusion approximation for two-dimensional turbulence*, Phys. Fluids, **11**, 671-673.

[14] P.G. LEMARIÉ, *Ondelettes à localisation exponentielle*, Journ. de Math. Pures et Appl.,**67**, 227-236, 1988.

[15] J.C. Mc WILLIAMS, *The emergence of isolated coherent vortices in turbulent flows*, J. Fluid Mech., **146**, 21-43, 1984.

[16] S.G. MALLAT, *A theory for multiresolution signal decomposition: the wavelet representation*, IEEE Trans. Patt. Anal. Mach. Int., 674-693, 1989.

[17] C. MENEVEAU, *Analysis of turbulence in the orthonormal wavelet representation*, Journal Fluid Mech., **232**, 469-520, 1991.

[18] Y. MEYER, *Ondelettes et Opérateurs I*, Ondelettes, Hermann, 1990.

[19] F. MORET-BAILLY, J. LIANDRAT, M.P. CHAUVE, P. TCHAMITCHIAN, *Décomposition en ondelettes d'un signal de vitesse dans une expérience de transition laminaire-turbulent sur un disque tournant*, C.R.Acad.Sci. Paris, Série 2, 1991.

[20] R. MURENZI, *Wavelet transform associated to the n-dimensional Euclidian Group with dilations*, PhD thesis, UCL, Louvain-la-Neuve, 1989.

[21] V. PERRIER, C. BASDEVANT, *Periodical wavelet analysis, a tool for inhomogeneous field investigation - Theory and Algorithms*, La Recherche Aérospatiale, **3**, 53-67, 1989.

[22] T. PHILIPOVITCH, *Analyse, filtrage et étude de la dynamique des structures cohérentes dans les écoulements turbulents bidimensionels*, Thèse de Doctorat de l'Université Paris VII, en préparation.

[23] J. WEISS, *The dynamics of enstrophy transfer in two-dimensional hydrodynamics*, report LJI-TN-121, La Jolla Institute, San Diego, 1981.

[24] N. ZOUARI, D. ELMHAÏDI, A. BABIANO, *Elementary topology of two-dimensional turbulence from a Lagrangian viewpoint and relative dispersion*, 1992, to appear in J. Fluid Mech.

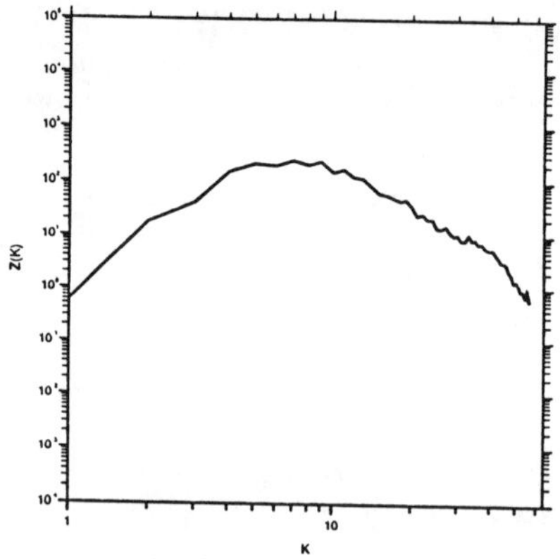

Figure 1: Enstrophy spectrum of the flow within the stationary regime (log-log scale).

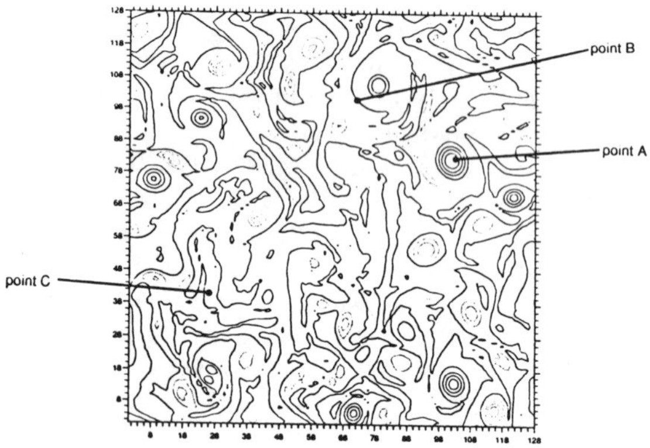

Figure 2: Vorticity field (dot lines for negative values).

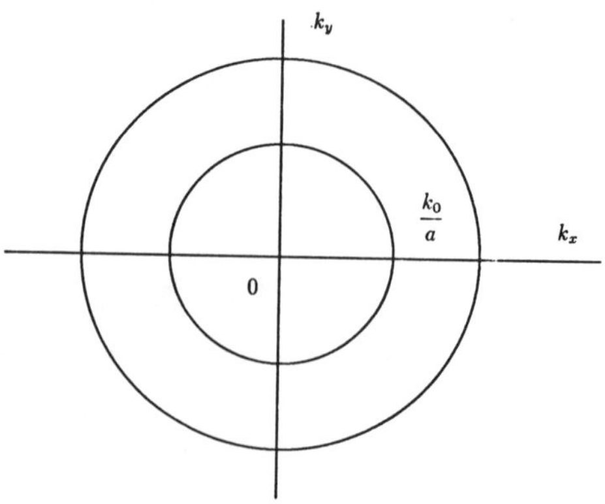

Figure 3: Schematic localization in Fourier space of the support of continuous wavelet basis ψ_a.

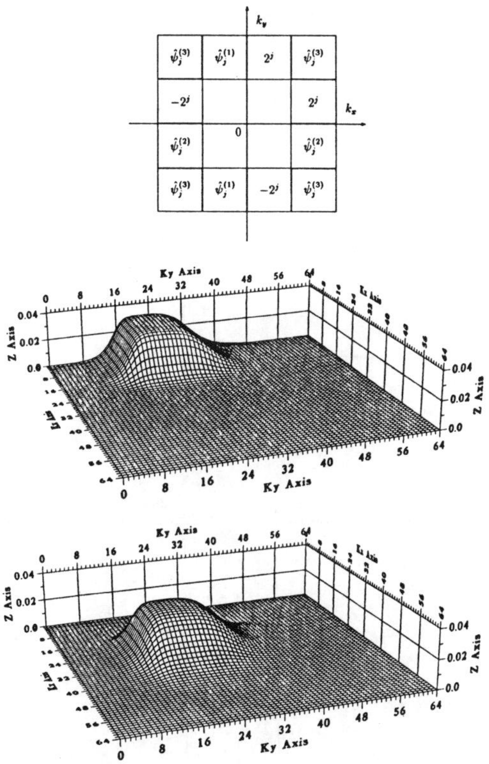

Figure 4:

4-a Schematic localization in Fourier space of the support of orthogonal wavelets basis $\psi_j^{(1)}, \psi_j^{(2)}, \psi_j^{(3)}$.

4-b Modulus of the Fourier coefficients $|\hat{\psi}_5^{(1)}(k_x, k_y)|, k_x > 0, k_y > 0$.

4-c Modulus of the Fourier coefficients $|\hat{\psi}_3^{(3)}(k_x, k_y)|, k_x > 0, k_y > 0$.

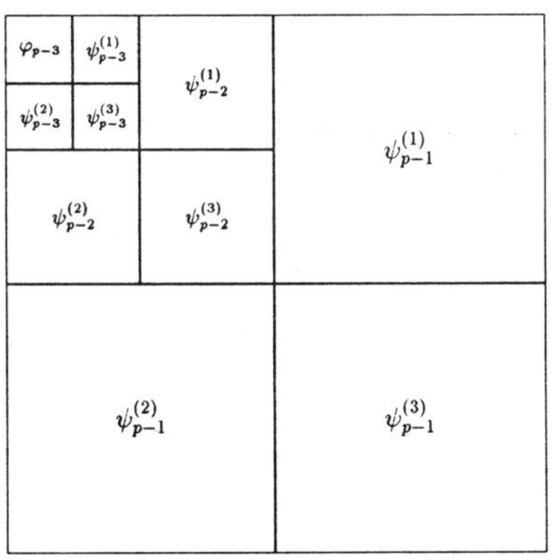

Figure 5: Schematic recursive decomposition of a two-dimensional field. Areas of different squares are proportional to the number of coefficients.

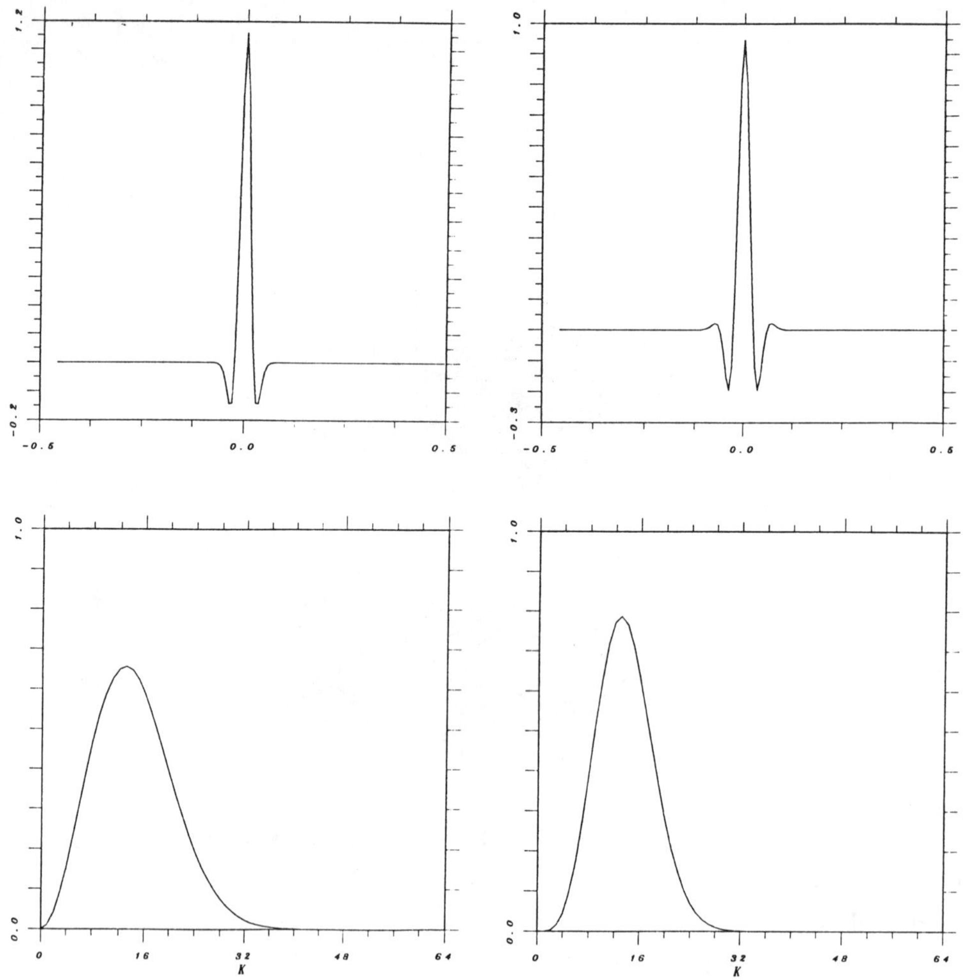

Figure 6: Basic wavelets in the continuous case, the mexican hat and its second derivative.

6-a: One-dimensional section in physical space, $m = 2$.

6-b: One-dimensional section in physical space, $m = 4$.

6-c: One-dimensional section of modulus of Fourier coefficients, $m = 2$.

6-d: One-dimensional section of modulus of Fourier coefficients, $m = 4$.

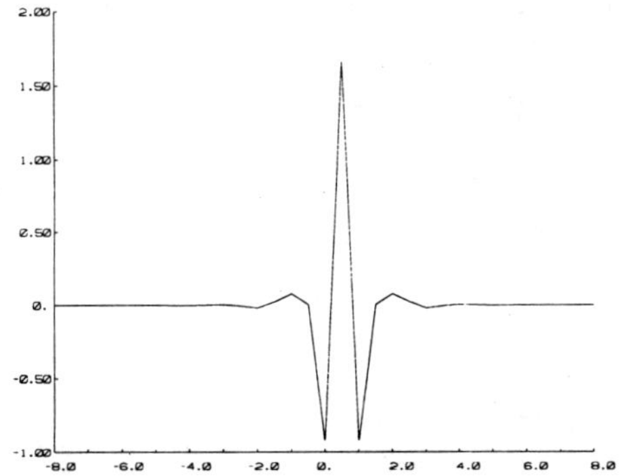
2.00
1.50
1.00
0.50
0.
-0.50
-1.00
-8.0 -6.0 -4.0 -2.0 0. 2.0 4.0 6.0 8.0

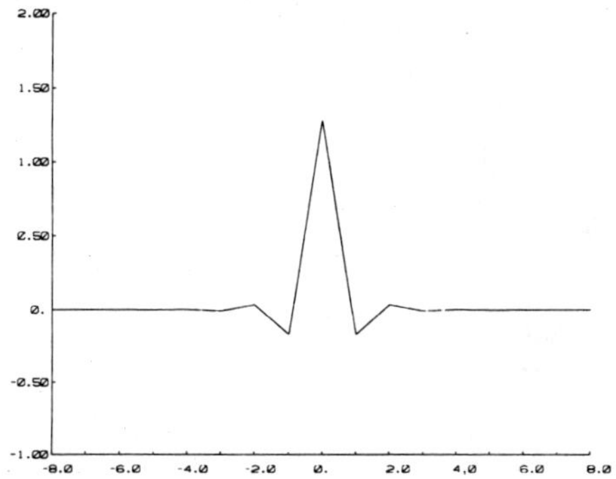
2.00
1.50
1.00
0.50
0.
-0.50
-1.00
-8.0 -6.0 -4.0 -2.0 0. 2.0 4.0 6.0 8.0

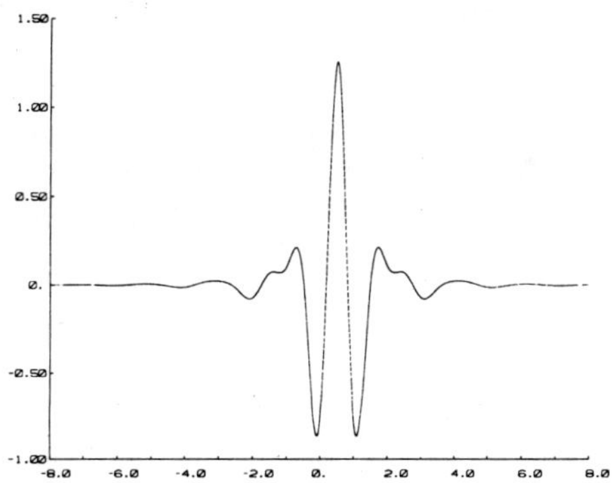
1.50
1.00
0.50
0.
-0.50
-1.00
-8.0 -6.0 -4.0 -2.0 0. 2.0 4.0 6.0 8.0

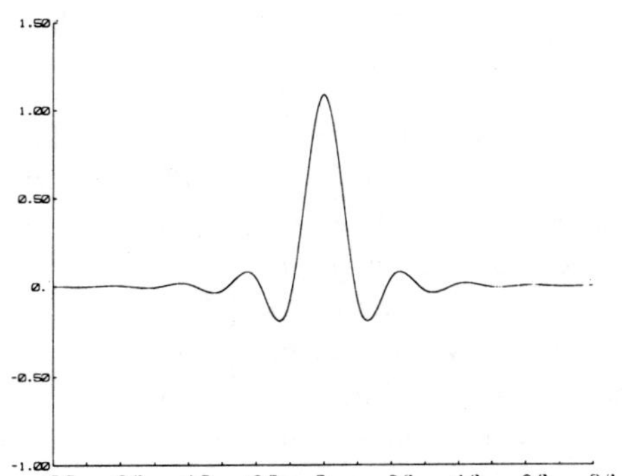
1.50
1.00
0.50
0.
-0.50
-1.00
-8.0 -6.0 -4.0 -2.0 0. 2.0 4.0 6.0 8.0

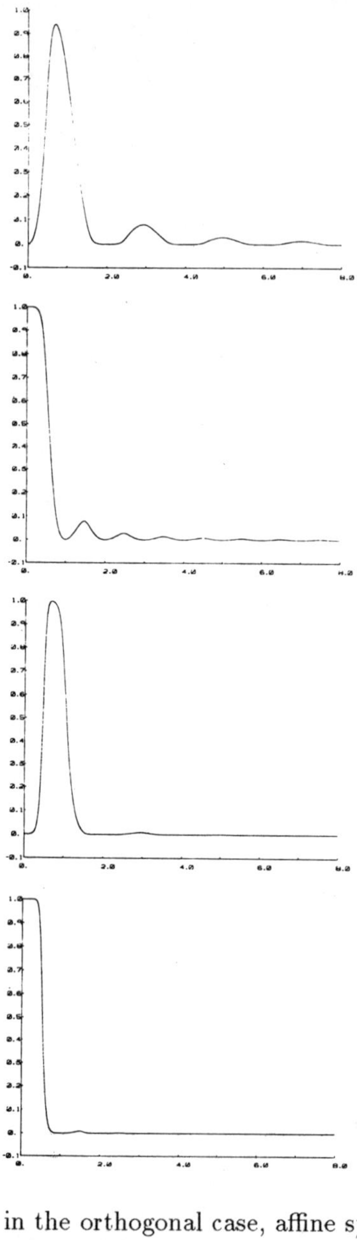

Figure 7: Basic wavelets in the orthogonal case, affine spline and cubic spline.
7-a: The one-dimensional wavelet ψ in physical space, $m = 2$.
7-b: The one-dimensional scaling function ϕ in physical space, $m = 2$.
7-c: Modulus of Fourier coefficients $\hat{\psi}$, $m = 2$.
7-d: Modulus of Fourier coefficients $\hat{\phi}$, $m = 2$.
7-e: The one-dimensional wavelet ψ in physical space, $m = 4$.
7-f: The one-dimensional scaling function ϕ in physical space, $m = 4$.
7-g: Modulus of Fourier coefficients $\hat{\psi}$, $m = 4$.
7-h: Modulus of Fourier coefficients $\hat{\phi}$, $m = 4$.

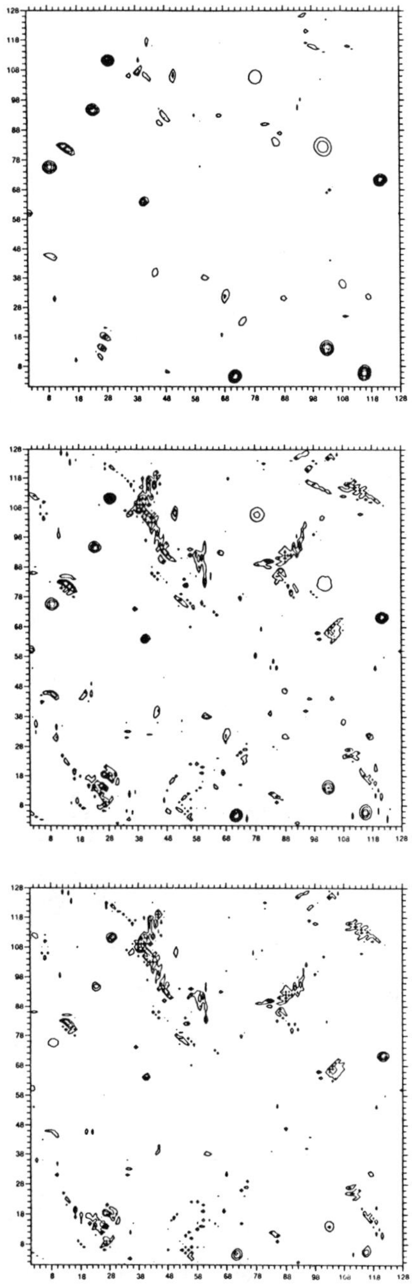

Figure 8: Wavelet enstrophy coefficients for the continuous wavelet analysis, $m = 2$ (Mexican hat):

8-a: $k = 28.46$

8-b: $k = 50.16$

8-c: $k = 87.10$

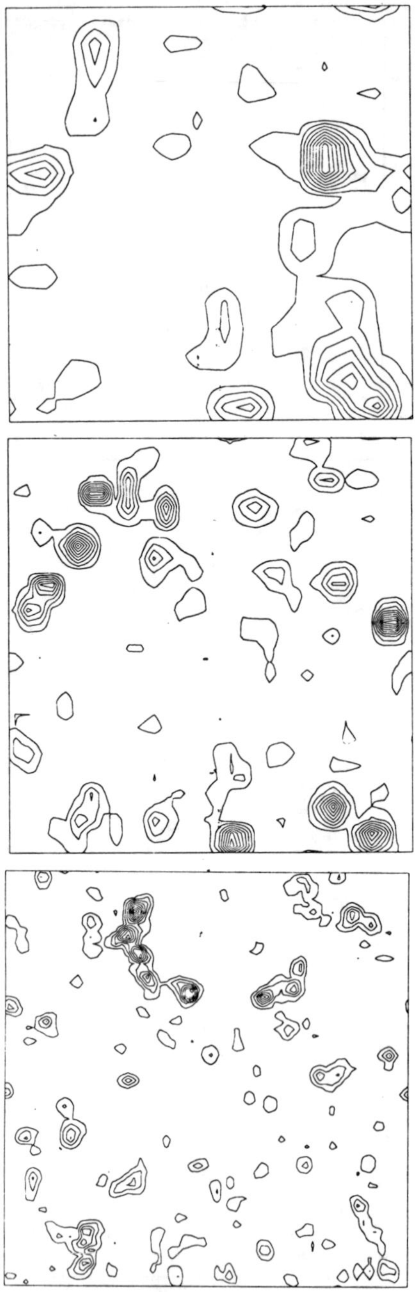

Figure 9: Wavelet enstrophy coefficients for the orthogonal wavelet analysis, $m = 2$ (affine splines):

9-a: $k = 28.46,\ j = 4$
9-b: $k = 50.16,\ j = 5$
9-c: $k = 87.10,\ j = 6$

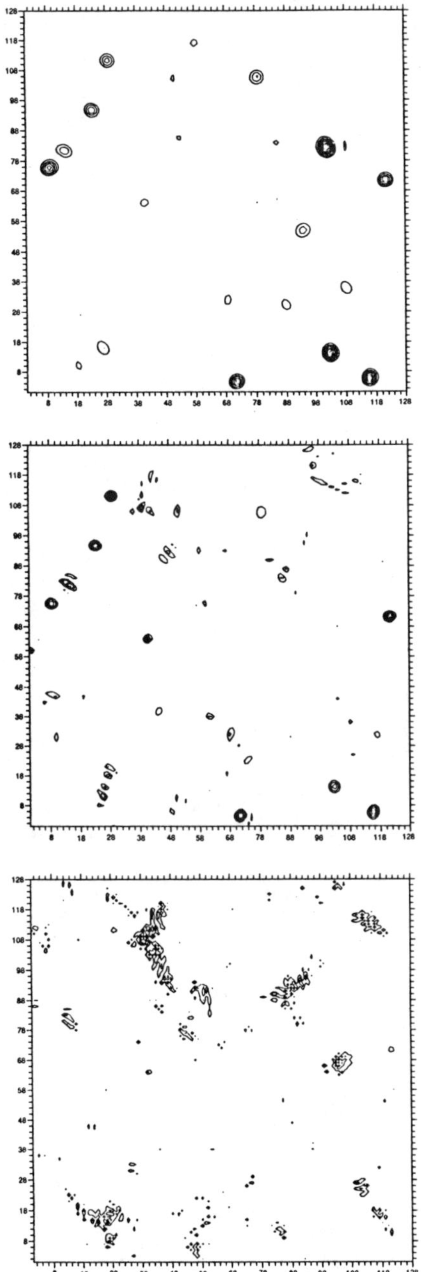

Figure 10: Wavelet enstrophy coefficients for the continuous wavelet analysis, $m = 4$:

10-a: $k = 12.91$
10-b: $k = 25.81$
10-c: $k = 51.59$

Figure 11: Wavelet enstrophy coefficients for the orthogonal wavelet analysis, $m =$ 4 (cubic splines):

11-a: $k = 12.91, \ j = 4$
11-b: $k = 25.81, \ j = 5$
11-c: $k = 51.59, \ j = 6$

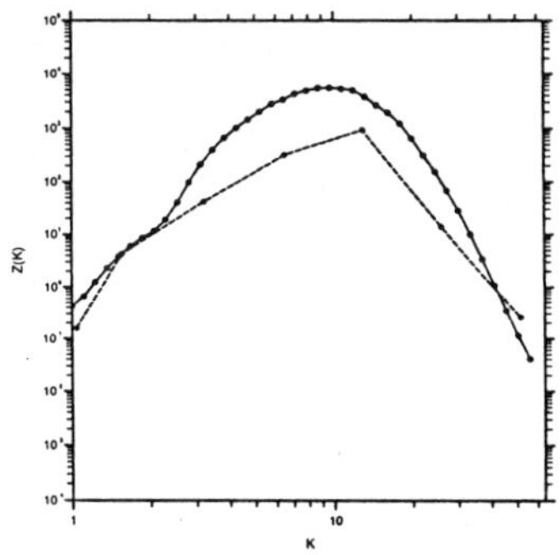

Figure 12: Local enstrophy spectra for point A (see *figure 2*), log-log scale:
Solid line: continuous wavelet analysis.
Dot line: discrete wavelet analysis.

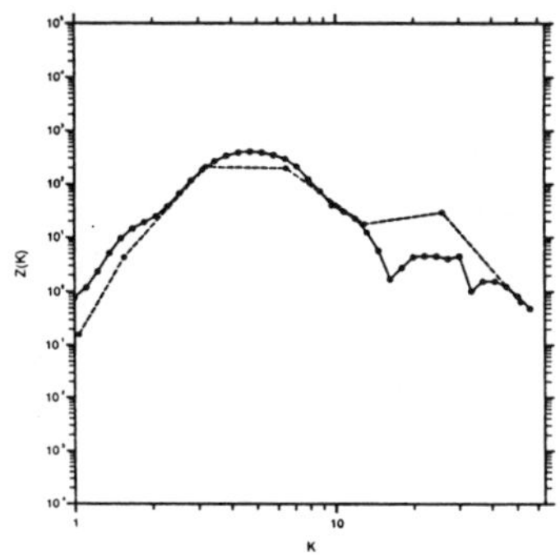

Figure 13: Local enstrophy spectra for point B (see *figure 2*), log-log scale:
Solid line: continuous wavelet analysis.
Dot line: discrete wavelet analysis.

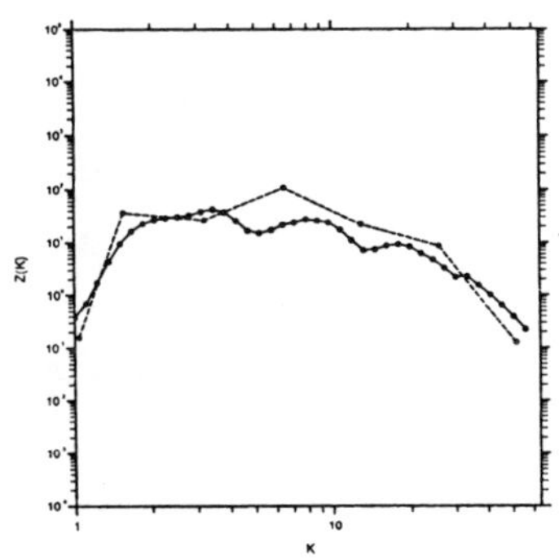

Figure 14: Local enstrophy spectra for point C (see *figure 2*), log-log scale:
Solid line: continuous wavelet analysis.
Dot line: discrete wavelet analysis.

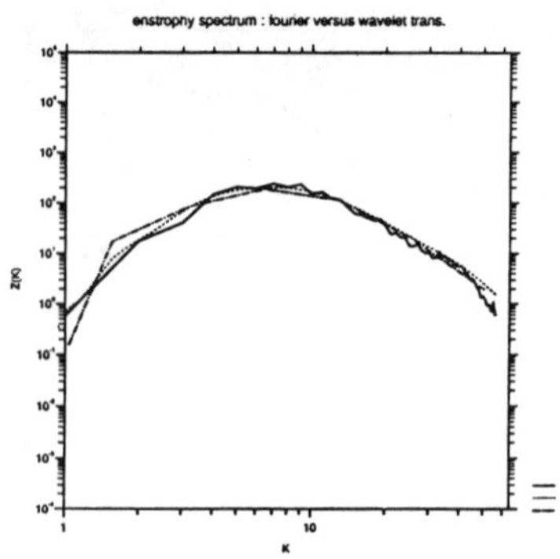

Figure 15: Mean enstrophy spectra, log-log scale:
Solid line: classical Fourier spectrum.
Dot line: continuous wavelet spectrum ($m = 4$).
Interrupted line: discrete wavelet spectrum ($m = 4$).

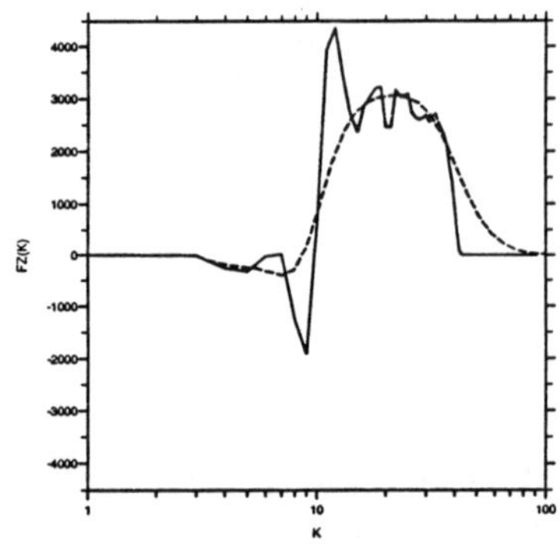

Figure 16: Global enstrophy fluxes computed with the Fourier technic (continuous line) and continuous wavelets (interupted line), log-log scale.

OPERATOR SPLITTING FOR NAVIER-STOKES AND CHORIN-MARSDEN PRODUCT FORMULA

J. THOMAS BEALE
Mathematics Department
Duke University
Durham NC 27706 USA

CLAUDE GREENGARD
IBM Research Division
Thomas J. Watson Research Center
Yorktown Heights NY 10598 USA

and

ENRIQUE THOMANN
Mathematics Department
Oregon State University
Corvallis OR 97331 USA

ABSTRACT. Operator splitting applied to the Navier-Stokes equations, in which alternately the nonlinearity and the diffusion are ignored, requires the accommodation of incompatible initial conditions at each time step. A one-dimensional model problem which can be solved explicitly and in which operator splitting converges is presented. Euler/Stokes and the Chorin-Marsden product formula are discussed. Convergence results are reviewed.

1. Introduction

The Navier-Stokes equations for incompressible fluid flow in a bounded domain describe the evolution of a fluid, by convection and diffusion, subject to the constraints of incompressibility and satisfaction of the appropriate boundary conditions. In an operator splitting approach to the Navier-Stokes equations, Navier-Stokes dynamics are approximated by a sequence of steps in which alternately diffusion and convection are ignored. These steps consist in the solutions, respectively, of the Euler equations and the Stokes equations. The appropriate boundary condition for the Stokes equations is no-slip, while for the Euler equations, the appropriate boundary condition is that the flow be tangential at the boundary. The alternation of boundary conditions is a cause of the difficulty (and interest) of both the analysis of Euler/Stokes splitting and numerical implementation of schemes based on the splitting. Such a fractional step approximation seems natural, in that the Navier-Stokes equations combine these two kinds of evolution having different character,

J.T. Beale et al. (eds.), Vortex Flows and Related Numerical Methods, 27–38.

and each partial step has a special structure not shared by the full Navier-Stokes equations. The Euler/Stokes splitting has recently been shown to converge, in both two and three space dimensions, for arbitrary (smooth) bounded regions. The analysis is reviewed here.

Chorin's random vortex method [6; 7] involves the numerical approximation of the solution of the Euler equations, followed by a generation of vorticity on the boundary, and then a numerical approximation of the solution of the heat equation. The latter two steps can be viewed as an approximation to a solution of the Stokes equations. This scheme has been idealized as the Chorin-Marsden product formula [8].

In the following section, we present a very simple model problem (in one space dimension) which can be explicitly solved and whose operator splitting approximations can be explicitly written down as well. In Section 3, we discuss Euler/Stokes splitting and the Chorin-Marsden product formula. In the case of the upper half-plane, the latter has been shown to converge. This analysis is reviewed in Section 4. The proof of convergence of Euler/Stokes splitting followed earlier results for variants of this splitting. Section 5 contains a review of some of the analysis.

2. Model Problem

As a very simple example of a problem in which operator splitting leads to initial value/boundary value incompatibility at the end of each non-diffusion step, consider the diffusion problem in the interval $]0,\pi[$,

$$\frac{\partial v}{\partial t}(x,t) = \nu \frac{\partial^2 v}{\partial x^2}(x,t) + f(x), \tag{1}$$

$$v(0,t) = v(\pi,t) = 0. \tag{2}$$

For simplicity, we take $v(\cdot,0) \equiv 0$ as the initial condition. Expressing f by its Fourier series,

$$f(x) = \sum_{j=1}^{\infty} \hat{f}_j \sin jx,$$

one finds that the solution v to problem (1,2) is given by

$$v(x,t) = \sum_{j=1}^{\infty} \hat{v}_j(t) \sin jx,$$

where the Fourier coefficients $\hat{v}_j$ are given by

$$\hat{v}_j(t) = \frac{\hat{f}_j}{\nu j^2}\left(1 - e^{-\nu j^2 t}\right).$$

Now consider the operator splitting approximation obtained by iterating solutions over time steps of duration k of the problems

$$\frac{\partial v}{\partial t}(x,t) = f(x)$$

and

$$\frac{\partial v}{\partial t}(x,t) = \nu \frac{\partial^2 v}{\partial x^2}(x,t), \qquad v(0,t) = v(\pi,t) = 0.$$

Let by u^n denote the result after n time steps, $n \geq 0$. Then u^{n+1} has Fourier coefficients

$$\hat{u}_j^{n+1} = \left(\hat{u}_j^n + k\hat{f}_j\right) e^{-\nu j^2 k},$$

and it follows by induction that for all n,

$$\hat{u}_j^n = k\hat{f}_j \left(\frac{1 - e^{-\nu j^2 nk}}{e^{\nu j^2 k} - 1}\right).$$

Set $t = nk$. Then

$$\hat{v}_j(t) - \hat{u}_j^n = k\hat{f}_j(1 - e^{-\nu j^2 t}) \left(\frac{1}{\nu j^2 k} - \frac{1}{e^{\nu j^2 k} - 1}\right).$$

Since the last factor on the right-hand side is bounded independently of ν, j, k, it follows that for some constant C, independent of ν and k,

$$|v(nk) - u^n|_{L^2} \leq Ck.$$

For this example, there are lower rates of convergence in somewhat higher norms, depending on the viscosity. A rate of convergence with constant independent of the viscosity has not been demonstrated for the Euler/Stokes splitting.

3. Euler/Stokes Splitting and the Chorin-Marsden Product Formula

Let Ω be an open subset of $\mathbb{R}^2$ or $\mathbb{R}^3$, with smooth boundary $\partial\Omega$. Denote by N the unit exterior normal vectors to $\partial\Omega$. In the discussion of the two-dimensional case below, the unit tangent vector to $\partial\Omega$ is denoted by τ. We set the coefficient of viscosity to one, since little can be said about the approximations in the limit of small viscosity. For a discussion of the behavior of solutions of the Navier-Stokes equations near boundaries as the viscosity goes to zero, see [14].

The Navier-Stokes equations are given by

$$\begin{aligned} \frac{\partial \mathbf{v}}{\partial t} + (\mathbf{v}\cdot\nabla)\mathbf{v} + \nabla\pi &= \Delta\mathbf{v}, \\ \nabla\cdot\mathbf{v} &= 0, \end{aligned}$$

together with the boundary condition

$$\mathbf{v} = 0 \qquad \text{on } \partial\Omega,$$

and an initial condition

$$\mathbf{v}(\mathbf{x}, 0) = \mathbf{v}_0(\mathbf{x}),$$

where $\mathbf{v}_0$ satisfies the divergence and boundary conditions. Here $\mathbf{v}(\mathbf{x},t)$ is the velocity and π is the pressure. The Euler equations are given by

$$\frac{\partial \mathbf{u}}{\partial t} + (\mathbf{u}\cdot\nabla)\mathbf{u} + \nabla\pi = 0,$$
$$\nabla\cdot\mathbf{u} = 0,$$

together with the boundary condition

$$\mathbf{u}\cdot N = 0 \qquad \text{on } \partial\Omega, \tag{3}$$

and an initial condition

$$\mathbf{u}(\mathbf{x},0) = \mathbf{u}_0(\mathbf{x}),$$

where $\mathbf{u}_0$ satisfies (3). The Stokes equations are given by

$$\frac{\partial \mathbf{u}}{\partial t} + \nabla\pi = \Delta\mathbf{u},$$
$$\nabla\cdot\mathbf{u} = 0,$$

with boundary condition

$$\mathbf{u} = 0 \qquad \text{on } \partial\Omega, \tag{4}$$

and initial data,

$$\mathbf{u}(\mathbf{x},0) = \mathbf{u}_0(\mathbf{x}).$$

The initial state, $\mathbf{u}_0$, can be more general than is suggested by the boundary condition (4). In fact, the appropriate restriction on $\mathbf{u}_0$ is that it satisfy (3) (see [16]).

Euler/Stokes splitting consists in the following procedure. Fix a time step of size k. Given an initial velocity, solve the Euler equations for time k, then solve the Stokes equations over a time step k, with as initial condition the final state of the Euler step. Then repeat these steps. The fact that velocity fields incompatible with the no-slip boundary condition are admissible as initial conditions for the Stokes equations makes it reasonable to consider such an operator splitting. However, even the existence of Euler/Stokes solutions needs to be demonstrated (in three dimensions), since the Euler equations have solutions which are known to exist only for a time interval which depends on the initial data. We return to this question in Section 5.

For the remainder of this section and through the following one, we confine our attention to the two-dimensional case, since there are no convergence results for Chorin-Marsden in three dimensions. The Euler equations admit the following formulation in terms of the vorticity, $\omega = \nabla\times\mathbf{u}$:

$$\frac{\partial \omega}{\partial t} + \mathbf{u}(\omega)\cdot\nabla\omega = 0,$$

with initial condition,

$$\omega(\mathbf{x},0) = \omega_0(\mathbf{x}).$$

Here the velocity field $\mathbf{u}(\omega)$ is defined by

$$\mathbf{u}(\omega) = (-\frac{\partial\psi}{\partial y}, \frac{\partial\psi}{\partial x}), \tag{5}$$

where the stream function ψ satisfies

$$\begin{aligned} \Delta\psi &= -\omega, \\ \psi|_{\partial\Omega} &= 0. \end{aligned}$$

Observe that there is no boundary condition for the vorticity.

The Stokes equations become the heat equation in their vorticity formulation,

$$\frac{\partial\omega}{\partial t} = \Delta\omega. \tag{6}$$

with the non-local boundary condition that the vorticity must satisfy the constraint,

$$\mathbf{u}(\omega) = 0 \qquad \text{on } \partial\Omega.$$

In the random vortex method, the Euler equations are solved approximately during a time step k by the convection of elements of vorticity. At the end of this time step, the velocity $\mathbf{u}(\omega)$ is no longer zero on the boundary, and the vorticity is taken to be given by the sum of its interior value and a (discretized version of a) one-dimensional Dirac distribution on the boundary of magnitude $\mathbf{u}(\omega)(\mathbf{x}) \cdot \tau(\mathbf{x})$ at $\mathbf{x} \in \partial\Omega$. Diffusion over the time step k is then represented by a (stochastic) approximation to the heat equation. (Both Neumann and Dirichlet boundary conditions have been implemented for the heat equation; we confine our attention to the Neumann version.)

Denote by $H(t)(\omega_0)$ the solution at time t of the heat equation (6) with the Neumann boundary condition, $\partial\omega/\partial N = 0$ on $\partial\Omega$. Set

$$\tilde{\Phi}(\omega) = \omega + \Phi(\omega), \tag{7}$$

where

$$\Phi(\omega) = (\mathbf{u}(\omega) \cdot \tau)\delta_{\partial\Omega}.$$

Finally, let $E(t)(\omega_0)$ denote the solution of the Euler equations, with initial data ω_0, at time t. The Chorin-Marsden product formula gives as an approximation to the Navier-Stokes solution at time nk,

$$\omega_n = \left(E(k)H(k)\tilde{\Phi}\right)^n \omega_0. \tag{8}$$

As we have seen, the boundary condition in the vorticity form of the Stokes equations is expressed as a constraint on the vorticity field. In the Chorin-Marsden product formula, the heat equation with Neumann boundary conditions is solved instead. One can also write down a boundary condition which, while non-local, expresses the normal derivative of the vorticity (and thus the vortex flux on the boundary) as a function of the values of ω on the boundary. From multiplication of the Stokes equation by test functions and integration by parts, one can conclude that for solutions, ω, of the Stokes equation,

$$\frac{\partial \omega}{\partial N}(t, \mathbf{x}) = \frac{\partial \phi}{\partial N}(t, \mathbf{x}),$$

where $\phi(\cdot, t)$ is the solution of the Dirichlet problem

$$\Delta \phi = 0$$

$$\phi|_{\partial\Omega} = \omega|_{\partial\Omega}.$$

For further discussion of vorticity boundary conditions, see [2; 9; 13].

4. Convergence of the Chorin-Marsden Product Formula in the Half-Plane

In [5] (referred to henceforth as BP2), discussed later in this section, it was proved that the Chorin-Marsden product formula in the half-plane converges. This work followed an earlier paper, [4] (BP1), in which the production of vorticity on the boundary for Navier-Stokes flow in the upper half-plane was studied.

The Navier-Stokes equations in the region, $\Omega = \{(x, y) : \; y \geq 0\}$, are given in vorticity form by

$$\begin{aligned} \frac{\partial \omega}{\partial t} + \mathbf{u}(\omega) \cdot \nabla \omega &= \Delta \omega, \\ \mathbf{u}(\omega)|_{y=0} &= 0, \end{aligned}$$

where $\mathbf{u}(\omega)$ is defined according to (5). Benfatto & Pulvirenti reformulated the Navier-Stokes equations in BP1 by imposing a homogeneous Neumann boundary condition on the vorticity and by putting in vorticity generation at the boundary as a source term in the differential equation:

$$\frac{\partial \omega}{\partial t} + \mathbf{u}(\omega) \cdot \nabla \omega = \Delta \omega + f(x, t)\delta(y), \tag{9}$$

$$\frac{\partial \omega}{\partial y}|_{y=0} = 0, \tag{10}$$

$$\mathbf{u}(\omega)|_{y=0} = 0, \tag{11}$$

where f is regarded as an unknown. The existence of solutions to this problem is obtained by an iteration process. Given an approximate vorticity, ω_n, let $\mathbf{u}_n = \mathbf{u}(\omega_n)$ be the corresponding velocity field. Consider then the equation

$$\frac{\partial \omega}{\partial t} + \mathbf{u}_n \cdot \nabla \omega = \Delta \omega + f(x, t)\delta(y), \tag{12}$$

together with the conditions (10,11), and initial data $\omega(x, y, 0) = \omega_0(x, y)$. For a given f, let $\omega_{n,f}$ be the solution of the linear convection problem, (12), (10). Imposing the no-slip boundary condition (11) defines an integral equation for f, a solution of which, f_{n+1}, is established by the convergence of its Neumann series. Define $\omega_{n+1} = \omega_{n,f}$ with $f = f_{n+1}$. In this way, one obtains a convergent sequence, ω_n, the limit of which satisfies the Navier Stokes equations.

In their second paper, BP2, Benfatto & Pulvirenti established the convergence of the Chorin-Marsden product formula by comparing this approximation with the solution of the Navier-Stokes equations as obtained in their first paper. A crucial step in this proof is to relate the generation step as defined in the Chorin-Marsden formula in (7), to the singular forcing introduced at the boundary in equation (9). For simplicity in the exposition, we explain this relation for the Stokes equations. Note that the Stokes equations admit a description similar to (9), namely

$$\frac{\partial \omega}{\partial t} = \Delta\omega + f(x,t)\delta(y), \tag{13}$$

with boundary conditions (10), where f is determined by the no-slip condition (11). Also, the Chorin-Marsden product formula alternates a generation step with a solution of the heat equation with Neumann boundary conditions.

Set $S_l^k = (H(k)(\tilde{\Phi}))^l$. Note that formally the generation after a time step k, is of order k. Define $f^{\omega,k}$ by

$$k f^{\omega,k}(x)\,\delta(y) = \Phi(\omega)$$

It follows that for an arbitrary initial vorticity γ,

$$S_l^k(\gamma) = H(kl)(\tilde{\Phi}(\gamma)) + k\sum_{j=1}^{l-1} H(k(l-j))(f_j^{\gamma,k}\,\delta(y)) \tag{14}$$

where $f_j^{\gamma,k}$ denotes the amount that has to be generated at the jth time step and is given by

$$k f_j^{\gamma,k}\,\delta(y) = \Phi\left(H(kj)(\tilde{\Phi}(\gamma)) + \sum_{s=1}^{j-1} kH(k(j-s))(f_s^{\gamma,k}\,\delta(y))\right). \tag{15}$$

The formal limits of these two equations as $k \to 0$ are

$$S_t(\gamma) = H(t)(\tilde{\Phi}(\gamma)) + \int_0^t H(t-s)(f_s^{\gamma}\,\delta(y))ds \tag{16}$$

where f_s^{γ} satisfies

$$\Phi\left(H(t)(\tilde{\Phi}(\gamma)) + \int_0^t H(t-s)(f_s^{\gamma}\,\delta(y))ds\right) = 0. \tag{17}$$

This last equation was introduced and solved in BP1 as the equation defining the singular term in (13).

A similar procedure can be followed for the Navier-Stokes equations and the Chorin-Marsden product formula. In this case the equations (14) and (15) have to be corrected in order to take into account the generation due to the Euler step. Similarly, equations (16) and (17) correspond to the solution of the Navier-Stokes equations as described using equation (9) with boundary conditions (10), and the no-slip constraint (11). It should be remarked that the proof of convergence does not follow from the usual consistency and stability analysis. Instead, it is shown that the approximate solution at time T approximates the solution of the Navier Stokes equation at the same time, within an error of order $k^{1/4}$.

5. Convergence of Euler/Stokes Splitting

Let Ω be a subset of R^n, $n = 2$ or $n = 3$, with smooth boundary $\partial\Omega$. We assume given a sufficiently smooth solution, $\mathbf{v}$, of the Navier-Stokes equations on a time interval $[0, T]$. Let $\mathbf{v}_0 = \mathbf{v}(0)$ be the initial velocity.

Let V^p be the closure in $L^p(\Omega)$ of the space of smooth, divergence-free vector fields of compact support in Ω. The elements satisfy the tangential boundary condition in a weak sense. Let P denote the bounded projection of $L^p(\Omega)$ on V^p (see [11]). Then $A = -P\Delta$, with zero boundary condition, is called the Stokes operator. The Stokes equations can be expressed as an evolution equation in V^p with generator A,

$$\frac{d\mathbf{u}}{dt} + A\mathbf{u} = 0,$$

and we denote the solution of the Stokes equation at time t, with initial condition $\mathbf{u}_0$, by $e^{-tA}\mathbf{u}_0$.

The key property of the Stokes operator which enables operator splitting to converge is its smoothing property; this smoothing compensates for the failure of the initial state of each Stokes step to satisfy the boundary condition. It follows from the analyticity of the Stokes semigroup [12] that for each $s > 0$, there is a constant C such that for all $\mathbf{u} \in V^p$ and $t > 0$,

$$|A^s e^{-tA}\mathbf{u}|_{L^p} \leq Ct^{-s}|\mathbf{u}|_{L^p}.$$

(See [10] for a general discussion of analytic semigroups and fractional powers of their generators and [18] for the special case of the Stokes operator.) From this estimate, one gets a bound on the $H^{2,p}$ norm of $e^{-tA}\mathbf{u}$ in terms of a lower norm of $\mathbf{u}$, namely $H^{r,p}$ with $0 \leq r < 1/p$,

$$|e^{-tA}\mathbf{u}|_{H^{2,p}} \leq Ct^{-(1-r/2)}|\mathbf{u}|_{H^{r,p}}.$$

(The complex interpolation spaces $H^{r,p}$ are described in [17]; when r is an integer, $H^{r,p}$ is the usual Sobolev space of functions all of whose derivatives up to order r belong to L^p.)

In the following, we denote by $E(t)$ the Euler solution operator, so that $\mathbf{u}(t) = E(t)\mathbf{u}_0$ is the solution of the Euler equations, with initial data $\mathbf{u}_0$, at time t. Let $k = T/m$, m an integer, denote the size of a time step. We set $t_n = nk$. The Euler/Stokes splitting approximation to the Navier-Stokes solution, $\mathbf{v}(t_n)$, in the notation just introduced, is

$$\left(e^{-kA}E(k)\right)^n \mathbf{v}_0.$$

In the case of three space dimensions, even the existence of this operator splitting needs to be demonstrated, for the Euler equations are known to have solutions only for time intervals which are of decreasing duration for increasing values of the initial data. Thus, it needs to be shown that the solutions do not grow unboundedly as $k \to 0$.

Convergence of Euler/Stokes splitting has only recently been established. Convergence of an Euler/Stokes-like splitting was first proved in [1]. The scheme studied in that paper differs from Euler/Stokes splitting in the replacement of an Euler step by a polynomial (in time) approximation to Euler evolution. Alessandrini et al. use the smoothing property of the Stokes operator to prove that their product formula has solutions which exist and are bounded, independently of k, on some interval of time. The authors go on to establish convergence to the Navier-Stokes solution on such a time interval.

In [19], a different variant of Euler/Stokes splitting, in which Euler steps alternate with solutions of modified Stokes equations, was studied. Ying introduces a projection operator, Q,

$$Q : \{\mathbf{u} \in (H^1(\Omega))^2 : \nabla\cdot\mathbf{u} = 0, \int_{\partial\Omega} \mathbf{v}\cdot\mathbf{n} = 0\} \to \{\mathbf{u} \in (H_0^1(\Omega))^2 : \nabla\cdot\mathbf{u} = 0\},$$

which is bounded in H^s. The modified Stokes equations used in Ying's product formula are given by

$$\begin{aligned} \frac{\partial \mathbf{u}}{\partial t} + \nabla\pi &= \Delta\mathbf{u} + \frac{1}{k}(I - Q)\mathbf{u}^{n+1/2} \\ \nabla\cdot\mathbf{u} &= 0 \\ \mathbf{u} &= 0 \qquad \text{on } \partial\Omega \\ \mathbf{u}(nk) &= Q\mathbf{u}^{n+1/2} \end{aligned}$$

Here $\mathbf{u}^{n+1/2}$ denotes the velocity field at the conclusion of the Euler part of step $n+1$. The projection of the initial velocity field before the Stokes step is related to vorticity generation in the vortex method; $Q\mathbf{u}$ can be taken to differ from $\mathbf{u}$ in a region near the boundary, where $Q\mathbf{u}$ brings the velocity field to zero. Using the smoothing property of Stokes, Ying establishes existence of solutions of his product formula, and the convergence of these solutions to $\mathbf{v}$, on a time interval which depends on the initial data. The approximate solutions converge in L^2 at a rate $O(k)$.

In more recent work [21] and [3] (hereafter referred to as BG), convergence of the Euler/Stokes splitting, in the form described earlier, is proved. Ying shows convergence in H^s, $s < 1/2$, and BG shows convergence in L^p, for all p, both to $O(k)$. We indicate here an outline of the argument, with an example of how the Stokes smoothing property is used in the proof.

The analysis in BG makes use of an intermediate product formula, in which the Euler part of Euler/Stokes splitting is replaced by a solution of a linear Euler equation,

$$\frac{\partial \mathbf{u}}{\partial t} + \mathbf{v}\cdot\nabla\mathbf{u} + \nabla\pi = 0$$

$$\nabla\cdot\mathbf{u} = 0,$$

$$\mathbf{u}\cdot N = 0 \qquad \text{on } \partial\Omega,$$

where $\mathbf{v}$ is the given Navier-Stokes solution. Denote by $E_{\mathbf{v}}(t+\tau,t)\mathbf{u}_0$ the solution of this problem, with initial condition $\mathbf{u}_0$ at time t, after time τ has elapsed. We denote by $\tilde{\mathbf{u}}^n$ the solution given by the intermediate scheme after n time steps, so that

$$\tilde{\mathbf{u}}^{n+1} = e^{-kA}E_v(t_n+k,t_n)\tilde{\mathbf{u}}^n.$$

(Observe that this does not describe a practical scheme, since the desired limit solution, $\mathbf{v}$, is used in its construction; this scheme is introduced merely to facilitate the analysis.)

In the convergence proof of BG, it is proved first that $\tilde{\mathbf{u}}$ gives on $O(k)$ approximation to $\mathbf{v}$ and then that $\mathbf{u}$ is a first-order approximation to $\tilde{\mathbf{u}}$. Expressing $\tilde{\mathbf{u}}_n$ as the integral of its time derivatives over earlier time intervals, we find that

$$\tilde{\mathbf{u}}_n = e^{-t_nA}\mathbf{v}_0 + \sum_{j=0}^{n-1} e^{-(n-j)kA}\mathbf{f}_j,$$

where

$$\mathbf{f}_j = -\int_{t_j}^{t_{j+1}} P(\mathbf{v}(t)\cdot\nabla)E_v(t_j+\tau,t_j)\tilde{\mathbf{u}}^j\,d\tau.$$

Estimates on E_v then yield

$$|\mathbf{f}_j|_{H^{r,p}} \le Ck|\tilde{\mathbf{u}}_j|_{H^{1+r,p}}.$$

Now the smoothing property of the Stokes equations is used to yield, for r sufficiently small,

$$\begin{aligned}|\tilde{\mathbf{u}}_n|_{H^{2,p}} &\le C_1 + C\sum_{j=0}^{n-1}((n-j)k)^{-(2-r)/2}|\mathbf{f}_j|_{H^{r,p}}\\ &\le C_1 + C\sup_{j\le n-1}|\tilde{\mathbf{u}}_j|_{H^{1+r,p}}.\end{aligned}$$

An interpolation estimate between L^2 and $H^{2,p}$ then gives, for some $b<1$,

$$|\tilde{\mathbf{u}}_n|_{H^{2,p}} \le C_1 + C_2 \sup_{0\le j\le n-1}|\tilde{\mathbf{u}}_j|_{L^2}^{1-b}|\tilde{\mathbf{u}}_j|_{H^{2,p}}^{b},$$

and we conclude that

$$|\tilde{\mathbf{u}}|_{H^{2,p}} \le C.$$

This bound on $\tilde{\mathbf{u}}$ is then used to prove that

$$|\tilde{\mathbf{u}}-\mathbf{v}|_{L^2} \le Ck.$$

In this proof $\tilde{\mathbf{u}}$ is regarded as an approximate solution of the exact problem, with error terms. The difference $\tilde{\mathbf{u}}-\mathbf{v}$ satisfies an evolution equation in which these inhomogeneous error terms can be estimated, again using the smoothing property of the diffusion.

Next, the error, $\mathbf{u}-\tilde{\mathbf{u}}$, is expressed as the solution of an inhomogeneous equation with a right-hand side which depends on $\mathbf{u}$ and $\tilde{\mathbf{u}}$. It is easily shown that this error is of order k in L^2 as long as $\mathbf{u}$ is defined. In the two-dimensional case, this completes the convergence proof in L^2. For the three-dimensional case, the argument continues with a demonstration that if $|\mathbf{u}|_{H^{2,p}}$ is appropriately bounded by $|\tilde{\mathbf{u}}|_{H^{2,p}}$, then

$$|\mathbf{u}-\tilde{\mathbf{u}}|_{H^{2,p}} \leq Ck^{\theta},$$

for some $\theta < 1$. Use of these two results in an inductive argument leads to the result that $|\mathbf{u}_j|_{H^{2,p}}$ is uniformly bounded and

$$|\mathbf{u}-\tilde{\mathbf{u}}|_{L^2} \leq Ck.$$

A final argument in BG uses the earlier results and estimates directly the difference, $\mathbf{u}-\mathbf{v}$, expressing it again as the solution of an inhomogeneous linear equation. Estimates based on smoothing by a perturbed Stokes evolution then yield the main result of the paper,

$$|\mathbf{u}-\mathbf{v}|_{L^p} \leq Ck.$$

The interval of time on which convergence is proved is all of $[0,T]$, the interval of existence of the smooth Navier-Stokes solution.

Acknowledgements

The first author was partially supported by N.S.F. Grant No. DMS-9102782. He and the third author are grateful to the IBM T. J. Watson Research Center for visits during which this work was undertaken.

References

[1] Alessandrini, G., Douglis, A., and Fabes, E., *An approximate layering method for the Navier-Stokes equations in bounded cylinders,* Ann. Mat. Pura Appl., 135, 1983, pp. 329–347.

[2] Anderson, C.R., *Observations on vorticity creation boundary conditions,* in "Mathematical Aspects of Vortex Dynamics," Caflisch, R.E., ed., pp. 144–159, SIAM, 1989.

[3] Beale, J.T. and Greengard, C., *Convergence of Euler-Stokes splitting of the Navier-Stokes equations,* Comm. Pure Appl. Math., 46, 1993, to appear.

[4] Benfatto, G. and Pulvirenti, M., *Generation of vorticity near the boundary in planar Navier-Stokes flows,* Commun. Math. Phys., 96, 1984, pp. 59–95.

[5] Benfatto, G. and Pulvirenti, M., *Convergence of Chorin-Marsden product formula in the half-plane,* Commun. Math. Phys., 106, 1986, pp. 427–458.

[6] Chorin, A. J., *Numerical study of slightly viscous flow,* J. Fluid Mech., 57, 1973, pp. 785–796.

[7] Chorin, A. J., *Vortex sheet approximation of boundary layers,* J. Comput. Phys., 27, 1978, pp. 428–442.

[8] Chorin, A.J., Hughes, T.J.R., McCracken, M.F., and Marsden, J.E., *Product formulas and numerical algorithms,* Comm. Pure Appl. Math., 31, 1978, pp. 205–256.

[9] Cottet, G.-H., *Boundary conditions and deterministic vortex methods for the Navier-Stokes equations,* in "Mathematical Aspects of Vortex Dynamics," Caflisch, R.E., ed., pp. 128–143, SIAM, 1989.

[10] Friedman, A., *Partial Differential Equations,* Holt, Rinehart and Winston, New York, 1969.

[11] Fujiwara, D. and Morimoto, H., *An L_r-theorem of the Helmholtz decomposition of vector fields,* J. Fac. Sci. Tokyo U., Sec 1A-Math., 24, 1977, pp. 685–700.

[12] Giga, Y., *Analyticity of the semigroup generated by the Stokes operator in L^r spaces,* Math. Z., 178, 1981, pp. 297–329.

[13] Hou, T. Y., and Wetton, B., *Convergence of a finite difference scheme for the Navier-Stokes equations using vorticity boundary conditions,* to appear in SIAM J. Numer. Anal.

[14] Kato, T., *Remarks on zero viscosity limit for nonstationary Navier-Stokes flows with boundary,* in "Seminar on Nonlinear Partial Differential Equations", Chern, S.S., ed., pp. 85–98, Springer, 1984.

[15] Puckett, E. G., *The random vortex method with vorticity creation: introduction and guide to parameter selection,* in "Vortex Dynamics and Vortex Methods," Anderson, C. and Greengard, C., eds., pp. 567–84, Amer. Math. Soc., 1991.

[16] Temam, R., *The Navier-Stokes Equations,* North-Holland, 1979.

[17] Triebel, H., *Interpolation Theory, Function Spaces, Differential Operators,* North-Holland, Amsterdam, 1978.

[18] Von Wahl, W., *The Equations of Navier-Stokes and Abstract Parabolic Equations,* Vieweg & Sohn, Braunschweig, 1985.

[19] Ying, L.-A., *Viscous splitting method for the unbounded problem of the Navier-Stokes equations,* Math. Comput., 55, 1990, pp. 89–113.

[20] Ying, L.-A., *Viscosity-splitting scheme for the Navier-Stokes equations,* Num. Methods P.D.E., 7, 1991, pp. 317–338.

[21] Ying, L.-A., *Optimal error estimates for a viscosity splitting formula,* in "Proceedings of the Second Conference on Numerical Methods for Partial Differential Equations," Ying, L.-A., and Guo, B., eds., pp. 139-47, World Scientific, 1992.

VELICITY METHODS: LAGRANGIAN NUMERICAL METHODS WHICH PRESERVE THE HAMILTONIAN STRUCTURE OF INCOMPRESSIBLE FLUID FLOW

TOMAS F. BUTTKE
Courant Institute of Mathematical Sciences
New York University
New York NY 10012 USA

ABSTRACT. We present a Lagrangian numerical method valid in any space dimension for approximating solutions to the Incompressible Euler Equation. The method is based on the canonical Hamiltonian formulation of incompressible flow. The method preserves at least three invariants of the flow: the kinetic energy, the impulse, and the angular momentum. We present numerical results which validate the method and elucidate the structure of the Hamiltonian variables in two and three dimensions.

1. Introduction

The description of the evolution of a system in terms of a Hamiltonian is the basis for the study of many physical systems. Once a system is described in terms of a Hamiltonian many properties of the system can be determined by the symmetries of the Hamiltonian. Oseledets [1] introduced a canonical Hamiltonian for the incompressible Euler equations in any number of space dimensions for the case where the fluid has a constant density. The work of Oseledets was based on the discrete Hamiltonian system introduced by Roberts [2] which asymptotically describes the evolution of vortex dipoles. In two dimensions the Hamiltonian structure introduced by Oseledets [1] is different from the point-vortex Hamiltonian structure introduced by Onsager [3].

In this paper we introduce two dimensional and three dimensional Lagrangian numerical methods based on the Hamiltonian formulation introduced by Oseledets. The numerical methods have several important properties which result from the fact that the methods are derived from an equation which has a Hamiltonian structure. The methods preserve the invariants: kinetic energy, impulse and angular momentum. Although we explicitly introduce the methods in two and three space dimensions, the numerical methods are valid in any number of dimensions and have the same basic properties.

The Hamiltonian structure of incompressible fluid flow is described in terms of a variable which we call the *velicity*. The velicity has units of velocity but satisfies an evolution equation which is similar to the evolution equation for the vorticity.

J.T. Beale et al. (eds.), Vortex Flows and Related Numerical Methods, 39–57.

If we make the analogy between the magnetic field in magnetostatics [4] and the incompressible velocity field in fluid flow, the current density is analogous to the vorticity and the dipole density (magnetization) is analogous to the velicity; for this reason we denote the velicity by M.

2. Definition of the Velicity

Consider the incompressible Euler's equation

$$\frac{\partial u}{\partial t} + u \cdot \nabla u = -\frac{\nabla p}{\rho} \quad ,$$

where u is the velocity, $\nabla \cdot u = 0$, p is the pressure and ρ is the density; we shall assume the density $\rho = 1$ for most of the paper. In two and three dimensions we define the vorticity ω as

$$\omega \equiv \nabla \times u \quad ;$$

the equivalent extension to higher dimensions is to interpret u as a 1-form and $\omega \equiv du$ is the 2-form obtained by taking the differential of u [5]. The velicity M is then defined as any 1-form (vector field) such that $\omega = dM$ or in three dimensions such that $\omega = \nabla \times M$. Since $du = dM$ ($\nabla \times u = \nabla \times M$) we have that u and M are equivalent up to a gradient; that is

$$M = u + \nabla\phi \quad ,$$

where ϕ is a scalar function. This is commonly known as the Helmholtz or Hodge decomposition. The velocity u is uniquely determined from the velicity M by solving Poisson's equation

$$\Delta\phi = \nabla \cdot M$$

for ϕ to obtain $u = M - \nabla\Delta^{-1}\{\nabla \cdot M\}$; that is we solve Poisson's equation for ϕ and use $u = M - \nabla\phi$.

The evolution equation for M is given by

$$\frac{\partial M_i}{\partial t} + u \cdot \nabla M_i = -M_j \frac{\partial u_j}{\partial x_i} \quad , \tag{1}$$

where we have used the implied summation convention and the subscripts indicate the Cartesian components of the vectors; this equation can be motivated by considering the evolution of the impulse in a fluid and the velicity can be interpreted as an impulse density. If we substitute $M_i = u_i + \partial_i\phi$ into eq (1) we obtain that

$$\frac{\partial u}{\partial t} + u \cdot \nabla u = -\nabla\{\frac{1}{2}u^2 + \frac{D\phi}{Dt}\} \quad ,$$

where $D/Dt \equiv \partial/\partial t + u \cdot \nabla$ is the substantial or material derivative. With the identification that $p = \{\frac{1}{2}u^2 + D\phi/Dt\}$, we see that eq (1) is equivalent to the incompressible Euler's equation.

The velicity is not unique; however, this proves to be useful since it is convenient to choose M differently for different circumstances. A possible choice for M initially is to take $M(x,0) = u(x,0)$; however, unless the flow is steady $M(x,t) \neq u(x,t)$ at later times.

For flows in unbounded domains it is convenient to choose the velicity so that it has compact support; if the vorticity is compact this can always be done (in two dimensions we must make the additional assumption that the flow has finite kinetic energy). We can show this result via a simple construction. Assume that the support of the vorticity is inside of a ball of radius R centered at the origin. Outside of the ball the velocity can be written as $u = -\nabla\psi$ where $\Delta\psi = 0$, for some ψ. With this in mind we define a velicity M_c as

$$M_c \equiv u - \nabla \int_{\infty}^{r} (\hat{r} \cdot u) dr \quad , \tag{2}$$

where $\hat{r}$ is the unit vector in the radial direction. M_c is compact since outside of the ball of radius R, $M_c = 0$; we can show this by writing ψ as a sum of spherical harmonics (in whatever dimension we are considering). By taking the divergence of eq (2) we find that

$$\phi = -\int_{\infty}^{r} (\hat{r} \cdot u) dr \quad ,$$

we see that ψ, the potential part of the velocity field, is the same as the ϕ in the relation $M = u + \nabla\phi$. Thus we have shown that another candidate for the velicity M is obtained by subtracting off the potential part of the flow field.

If the velicity field is chosen initially to be compact it will remain compact for all later times; since from eq (1) velicity cannot be generated in the fluid. As is the case with the vorticity, the velicity can only be created at the boundaries of a fluid. In fact, generically, one can choose the velicity so that the support of the velicity coincides with the support of the vorticity. There are special cases, however, in which the support of the velicity must differ from the support of the vorticity; a symmetric jet or smoke ring for which the vorticity vanishes in the central region must have a nonzero velicity everywhere inside of the jet or smoke ring.

A simple argument shows that the velicity must be nonzero at the center of a jet or smoke ring. Consider the circulation $\Gamma \equiv \int_C u \cdot d\mathbf{l}$ in the fluid, where the integral is taken about a closed curve C. Now consider a curve C_0 which encircles a smoke ring but lies outside of the support of the vorticity, such that the circulation will have some nonzero value Γ_0. One can show that for an arbitrary closed curve C,

$$\int_C M \cdot d\mathbf{l} = \int_C u \cdot d\mathbf{l} \quad .$$

Consider calculating the circulation about C_0 using the velicity M instead of the velocity u; one would obtain that the circulation of the smoke ring is zero if we could pick the support of the velicity to coincide with the support of the vorticity in this case.

Having proven that the support of the velicity cannot coincide exactly with the support of the vorticity for all flows, we wish to argue that physically this is essentially what happens. Consider a sphere in an infinite fluid at rest initially, we can pick the velicity to be identically zero in this case. Let the fluid have a viscosity so that there will be vorticity generated at the surface of the sphere, but let the viscosity be small enough so that once the vorticity is generated there is essentially no diffusion of the vorticity. Now start accelerating the sphere through the fluid. The velicity and vorticity will be generated at the surface of the sphere; the vorticity and velicity both flow with the fluid velocity so that for all later times the support of the vorticity will coincide with the support of the velicity. This is the generic situation for unbounded flows and thus the support of the velicity will be as small as the support of the vorticity in real flow situations.

In the previous paragraph we considered the velicity in a viscous fluid, but eq (1) is valid only for Euler's equation; we now generalize eq (1) for the case of a viscous fluid. Let the evolution of M be given by the equation:

$$\frac{\partial M_i}{\partial t} + u \cdot \nabla M_i = -M_j \frac{\partial u_j}{\partial x_i} + \nu \Delta M_i \quad . \tag{3}$$

By substituting $M_i = u_i + \partial_i \phi$, we can show that eq (3) is equivalent to the Navier-Stokes equation:

$$\frac{\partial u}{\partial t} + u \cdot \nabla u = -\nabla\{\frac{1}{2}u^2 + \frac{D\phi}{Dt} - \nu\Delta\phi\} + \nu\Delta u \quad .$$

We see that the pressure becomes $p = \{\frac{1}{2}u^2 + D\phi/Dt - \nu\Delta\phi\}$.

We now consider the inviscid equations again. Constant density incompressible fluid flows have many invariants. We wish to present some of those invariants in terms of the velicity.

The first invariant we consider is the impulse of the fluid [6]. The impulse I is defined in three dimensions as

$$I \equiv \frac{1}{2}\int_{R^3} r \times \omega \, dx \quad .$$

We substitute $\omega = \nabla \times M$ and integrate by parts, assuming M is compact, which follows if ω is, to find that

$$I = \frac{1}{2}\int_{R^3} r \times \omega \, dx = \int M dx \quad ;$$

this result also motivates describing the velicity as an impulse density. The generalization to any number of dimensions is to define the impulse as

$$I \equiv \int M dx \quad ; \tag{4}$$

one can show directly then from eq (1) that the impulse is an invariant the flow. This result is the motivation for describing the velicity as an *impulse density*.

The next invariant we consider is the kinetic energy. From the fact that $M = u + \nabla\phi$ and $\nabla \cdot u = 0$ and after an integration by parts we see that

$$\frac{1}{2}\int u^2 \, dx = \frac{1}{2}\int M \cdot u \, dx \quad . \tag{5}$$

We can show that this is an invariant directly from eq (1). We have that

$$\begin{aligned}\frac{d}{dt}\int \frac{1}{2}u^2 dx &= \int u \cdot \frac{\partial u}{\partial t} dx = \int u \cdot \frac{\partial M}{\partial t} dx \\ &= -\int_\Omega \partial_i (u_i u_j M_j) dx = -\int_{\partial\Omega} (\hat{n} \cdot u)(u \cdot M) da = 0 \quad ,\end{aligned}$$

where we have assumed the usual boundary conditions $u \cdot \hat{n} = 0$ for an Eulerian flow and $\hat{n}$ is the outward normal to the domain under consideration.

The third invariant we consider is the angular momentum of the fluid [6]. The angular momentum Ω in three dimensions is defined as $\Omega \equiv \frac{1}{3}\int r \times (r \times \omega) d^3x$. Substitution of $\omega = \nabla \times M$ and integration by parts yields immediately

$$\Omega = \frac{1}{3}\int r \times (r \times \omega) d^3x = \int r \times M d^3x \quad . \tag{6}$$

The generalized invariant is defined as

$$\Omega \equiv \int (r \wedge M)\, dx \quad ,$$

which can be shown to be an invariant of the motion in any dimension directly from eq (1).

The final invariant we consider is the helicity. The helicity η is defined as $\eta \equiv \int (\omega \cdot u) dx$. An integration by parts yields immediately that $\eta = \int (\omega \cdot u) dx = \int (\omega \cdot M) dx$. Oseledets [1] noted that in three dimensions the *helicity density* is a scalar invariant when it is written in terms of the velicity. Using the vorticity equation $\partial \omega / \partial t + u \cdot \nabla \omega = \omega \cdot \nabla u$, and eq (1) we have immediately that

$$\frac{D(\omega \cdot M)}{Dt} = 0 \tag{7}$$

We can generalize eq (7) to any number of dimensions by noting that in three dimensions vortex lines evolve as material curves in the fluid [7]. Thus in order to generalize eq (7) we consider the evolution of a material curve. Let $x(\alpha, t)$ denote the usual flow map which gives the position x at time t of the fluid particle which was located at position α at time $t = 0$. The Jacobian J of the flow map is given by

$$J_{ij} = \frac{\partial x_j}{\partial \alpha_i} \quad .$$

Since $\partial x / \partial t = u$ we have immediately that

$$\frac{\partial J}{\partial t} = J \cdot \nabla u \quad ,$$

where $(J \cdot \nabla u)_{ij} = J_{ik} \partial_k u_j$ (using the implied summation convention). Thus we obtain the important generalization of eq (7) that

$$\frac{D(J \cdot M)}{Dt} = 0 \quad , \tag{8}$$

where $(J \cdot M)_i = M_j \partial x_j / \partial \alpha_i$. Thus we immediately obtain a solution for M in Lagrangian coordinates in terms of the Jacobian of the flow map:

$$M(\alpha,t) = J^{-1}(\alpha,t)J(\alpha,0)M(\alpha,0) \quad , \tag{9}$$

where we have not assumed that the Jacobian at $t = 0$ is the identity. The inverse in eq (9) exists for all times as long as it does initially since the flow is incompressible and therefore $\partial \mid J \mid / \partial t = 0$ [7], where $\mid J \mid$ denotes the determinant of J. In three dimensions the evolution of the velicity eq (9), is in a sense the inverse of the evolution of the vorticity, where

$$\omega(\alpha,t) = J(\alpha,t)J^{-1}(\alpha,0)\omega(\alpha,0) \quad .$$

Equation (8) also has an interesting physical interpretation. If we consider a parametrization such that initially the velicity is perpendicular to the surfaces $\alpha_1 =$ constant, eq (8) shows that the surfaces evolve so as to remain perpendicular to the velicity. We use eq (8) in presenting our numerical results in two dimensions by showing the evolution of the surfaces $dx \cdot M = 0$; see figs (1-10).

The numerical work presented here deals with constant density flows, however, before we conclude this section on background material for the velicity variable we wish to extend eq (1) to variable density flows. Variable density flows do not have the Hamiltonian structure introduced by Oseledets, however, the formulation of incompressible fluid mechanics in terms of the velicity is extremely useful and so we include the variable density equations here for completeness.

In order to derive an equation to describe the evolution of the velicity for variable density incompressible flows we follow the derivation of energy conservation following eq (5). Since kinetic energy is also conserved in variable density flows, the correct equation describing the evolution of the velicity should be identical to eq (1) with the addition of a term to account for the fact that gradients in the density generate vorticity in an incompressible flow. Also the Helmholtz decomposition must be modified appropriately. For variable density flows we relate the velicity M to the velocity u linearly as follows:

$$M = u + \frac{\nabla \phi}{\rho} \quad . \tag{10}$$

We substitute eq (10) into eq (1) and require that the resultant equation have the form of the variable density Euler's equation. As expected we find we must add a term to eq (1) so that the resultant equation has the desired form. We find that M must satisfy

$$\frac{\partial M_i}{\partial t} + u \cdot \nabla M_i = -M_j \frac{\partial u_j}{\partial x_i} - \frac{1}{2} u^2 \frac{\partial \ln \rho}{\partial x_i} \quad . \tag{11}$$

Substitution of eq (10) into eq (11) yields

$$\frac{\partial u}{\partial t} + u \cdot \nabla u = -\frac{\nabla \{\frac{1}{2}\rho u^2 + D\phi/Dt\}}{\rho} \quad , \tag{12}$$

where we have used the continuity equation $D\rho/Dt = 0$. Thus we see for the general variable density case the pressure of the fluid is given by

$$p = \{\frac{1}{2}\rho u^2 + \frac{D\phi}{Dt}\} \quad ,$$

where now ϕ is defined by eq (10). Energy conservation now follows from eqs (10,11) and the continuity equation:

$$\begin{aligned}
\frac{d}{dt}\int \frac{1}{2}\rho u^2 dx &= \int \frac{1}{2}u^2 \frac{\partial \rho}{\partial t}\, dx + \int \rho u \cdot \frac{\partial u}{\partial t} dx \\
&= -\int \frac{1}{2}u^2 (u \cdot \nabla\rho)\, dx + \int \rho u \cdot \frac{\partial M}{\partial t}\, dx - \int \rho u \cdot \frac{\partial(\nabla\phi/\rho)}{\partial t}\, dx \\
&= -\int u^2(u\cdot\nabla\rho)\, dx - \int \rho u \cdot \nabla(u\cdot M)\, dx - \int \{u\cdot(M-u)\}(u\cdot\nabla\rho)\, dx \\
&= -\int_\Omega \partial_i(\rho u_i u_j M_j)\, dx = -\int_{\partial\Omega} \rho(\hat{n}\cdot u)(u\cdot M)\, da = 0 \quad ,
\end{aligned}$$

where once again we assume $u\cdot\hat{n} = 0$ as the boundary condition for Euler's equation.

3. The Lagrangian Numerical Method

In this section we present the Lagrangian numerical methods which are valid in any number of dimensions for constant density incompressible fluid flows. We shall try to make the presentation independent of the dimension of the space. The methods are based upon a discretization of eq (1). The methods preserve three invariants of incompressible fluid flow: kinetic energy, impulse, and angular momentum. In this paper we assume the fluid is in a unbounded domain. We shall deal with bounded domains in a future paper. The discretization is based on the discretization introduced and refined in [8; 9; 10; 11; 12].

We wish to approximate the velicity field $M(x,t)$ which evolves according to eq (1) with initial conditions $M(x,0)$; we denote our approximation to M as $\tilde{M}$. Given $\tilde{M}(x,t)$ we associate the incompressible velocity field $\tilde{u}(x,t)$ which is our approximation to $u(x,t)$ and is related to $\tilde{M}$ via the Helmholtz decomposition $\tilde{M} = \tilde{u} + \nabla\phi$. We define an approximation to $M(x,0)$ and then evolve the approximation according to eq (1).

We first define a mollification (blob) function f_δ as

$$f_\delta(x) = \frac{1}{\delta^d} f_0(x/\delta) \quad ,$$

where f_0 is a smooth function which has unit mass $\int f_0 dx = 1$, and whose first $p-1$ moments vanish

$$\int x^\alpha f_0 dx = 0, \quad \text{for} \quad 0 < |\alpha| \leq p-1 \quad ,$$

where α is the multi-index $x^\alpha \equiv x_{i_1}^{\alpha_1} x_{i_2}^{\alpha_2} \cdots x_{i_d}^{\alpha_d}$ and d is the space dimension. We define our initial approximation to M as follows:

$$\tilde{M}(x,0) = \sum_{i=1}^{N} f_\delta(x - x_i) m_i \quad , \tag{13}$$

where the x_i are equally spaced points on a regular mesh of spacing h, and $m_i \equiv h^d M(x_i, 0)$; points on an irregular mesh can be discretized appropriately, but for clarity we assume a regular mesh. The sum in eq (13) is over all points such that $M(x_i, 0) \neq 0$. The convergence properties are given in [12] and the approximation in eq (13) converges with order p as $h \to 0$ provided $h = \delta^\beta$, where $\beta < 1$. The approximation to M at later times is defined by defining evolution equations for the x_i and m_i and defining $\tilde{M}(x,t)$ as:

$$\tilde{M}(x,t) = \sum_{i=1}^{N} f_\delta(x - x_i(t)) m_i(t) \quad , \tag{14}$$

where the x_i satisfy

$$\frac{dx_i}{dt} = \tilde{u}(x_i) \quad , \tag{15}$$

and the m_i satisfy

$$\frac{d(m_i)_k}{dt} = -(m_i)_j \frac{\partial \tilde{u}_j(x_i)}{\partial (x_i)_k} \quad , \tag{16}$$

where the notation $(m_i)_k$ denotes the k-th component of the vector m_i and likewise for x_i; the notation is somewhat complicated but the right hand side of eq (16) is the same as appears in eq (1). The velocity $\tilde{u}$ is defined by

$$\tilde{u} = \tilde{M} - \nabla \phi \quad , \tag{17}$$

and the fact that $\nabla \cdot \tilde{u} = 0$; we note that $\tilde{u}$ is spatially as smooth as is f_δ. We expect that $\tilde{M} \to M$ as $h \to 0$ and therefore also $\tilde{u} \to u$, as is proven for vortex methods in [10; 11; 12]; our numerical results indicate that this is indeed the case.

We now construct $\tilde{u}$ explicitly and prove the invariance of the kinetic energy, impulse and angular momentum of the numerical scheme. One can show that

$$\tilde{u}(x) = \sum_{j=1}^{N} \{ m_j f_\delta(x - x_j) - (m_j \cdot \nabla) \nabla \psi(x - x_j) \} \quad , \tag{18}$$

where ψ satisfies $\Delta \psi = f_\delta$; for simplicity we assume that f_δ is radially symmetric. Outside of the support of f_δ eq (18) gives a dipole field for the velocity induced by a single blob of velicity. We define the discrete kinetic energy H of the method as an approximation to the kinetic energy $1/2 \int (M \cdot u)\, dx$;

$$\begin{aligned} H &\equiv \frac{1}{2} \sum_{k=1}^{N} m_k \cdot \tilde{u}(x_k) \\ &= \frac{1}{2} \sum_{j=1}^{N} \sum_{k=1}^{N} \{ (m_j \cdot m_k) f_\delta(x_j - x_k) + (m_j \cdot \nabla_j)(m_k \cdot \nabla_k) \psi(x_j - x_k) \} \quad , \end{aligned} \tag{19}$$

where ∇_k indicates the gradient with respect to the x_k. In order to obtain the second equality in eq (19) we substituted eq (18) and used the fact that $\nabla_k\psi(x_j - x_k) = -\nabla_j\psi(x_j - x_k)$. We note that the discrete kinetic energy is finite because f_δ is a smooth function; this is in contrast to the singular form one obtains from Roberts' original work [2]. One can obtain the self-energy of a blob by evaluating the velocity at the origin for a single blob which is also located at the origin. One finds that the self-velocity u_0 of the blob located at x_k is

$$u_0 = \frac{d-1}{d} m_k f_\delta(0) = \frac{h^d}{\delta^d}\frac{d-1}{d} M_k f_0(0) \quad , \tag{20}$$

where d is the space dimension. Thus we see that in the limit $h \to 0$ with $h = \delta^\beta$ the self-velocity of a blob vanishes; for nonzero h and δ, however, $u_0 \neq 0$; the self-energy then is

$$\rho\frac{h^{2d}}{\delta^d}\frac{d-1}{2d} M_k^2 f_0(0) \quad ,$$

where we have included ρ the density of the fluid to make the self-energy dimensionally correct.

The discrete system can be written as a Hamiltonian system if we observe that

$$\frac{dx_k}{dt} = \nabla_{m_k} H \quad , \tag{21}$$

and

$$\frac{dm_k}{dt} = -\nabla_{x_k} H \quad , \tag{22}$$

where $\nabla_{m_k} H$ and $\nabla_{x_k} H$ indicate taking gradients with respect to the variables m_k and x_k. By a direct computation eq (15) is identical to eq (21) and eq (16) is identical to eq (22).

The preservation of the discrete invariants is an immediate consequence of the Hamiltonian structure of the system where H is identified as the Hamiltonian. The discrete kinetic energy is an invariant since $dH/dt = 0$ if the Hamiltonian does not explicitly depend on time [13]. Since the Hamiltonian is translation invariant we have that

$$I \equiv \sum_{i=1}^{N} m_i \quad , \tag{23}$$

is a constant of motion; we see that eq (23) is an approximation to the impulse $\int M\,dx$. The third invariant Ω is an approximation to the angular momentum $\int x \wedge M\,dx$,

$$\Omega \equiv \sum_{i=1}^{N} x_i \wedge m_i \quad , \tag{24}$$

and it is an invariant because the Hamiltonian is rotationally invariant. The fact that I and Ω are discrete invariants of motion can also be verified directly by calculating $[I, H] = 0$ and $[\Omega, H] = 0$ where I and Ω are defined by eqs (23,24) and $[\cdot, \cdot]$ denotes the Poisson bracket [13].

It is interesting to note that this finite particle system describing incompressible fluid flow has the same invariants as N bodies interacting via a central force such as gravity. In particular the asymptotic motion of two velicity particles colliding is completely determined by the three invariants H, I and Ω. This observation and the Hamiltonian structure make an alternative description of incompressible superfluid Helium possible; we will address this in future work.

4. Numerical Results

In this section we present some numerical results obtained using the approximation described in the previous section. We present results in both two and three dimensions. The results presented here are only meant to demonstrate the basic properties of the method and to validate it. In particular we do not advocate that the two dimensional method is better or worse than other two dimensional Lagrangian methods; however, we wish to emphasize the fact that the methods are formally the same independent of dimension. Therefore once the method is developed in two dimensions it is simple to extend the method to any number of dimensions. In the work presented here we developed the three dimensional method first and modified it for two dimensions. In future work, however, which will include boundaries and viscosity, it will be much simpler to develop the method in two dimensions where visualization is easier and then extend the method to three dimensions. The fact that the structure of the vorticity equations changes in going from two to three dimensions has made the development of three dimensional Lagrangian vorticity methods difficult.

The numerical method has several unique features as a Lagrangian method besides the obvious feature it shares with other methods; the fact that it produces a smooth velocity field $\tilde{u}$ which is exactly divergence-free at every time. The first important feature is that the vorticity $\nabla \times \tilde{u}$ is compact. Although this can be done for many methods in two dimensions, it is not possible to produce a compact vorticity field in three dimensions when the vorticity is used as the computational variable. Another important feature of the method is the fact that discrete energy, impulse and angular momentum invariants are preserved. We observe that the preservation of these invariants causes the numerical method to preserve many of the important physical symmetries which are present in Eulerian flows; flows do not become unphysical because of numerical discretization artifacts.

For the three dimensional calculations presented here we used the following blob function:

$$f_0(x) = \frac{21}{31\pi} \begin{cases} 1, & \text{if } r \leq 1/2; \\ 1 - 10(r-1/2)^3 + 15(r-1/2)^4 - 6(r-1/2)^5, & \text{if } 1/2 < r \leq 3/2; \\ 0, & \text{if } r > 3/2, \end{cases}$$

where $r = |x|$. For the two dimensional calculations we used

$$f_0(x) = \frac{7}{2\pi} \begin{cases} 1 - 10r^3 + 15r^4 - 6r^5, & \text{if } r \leq 1; \\ 0, & \text{if } r > 1, \end{cases}$$

where again $r = |x|$. Both of the blob functions have continuous second derivatives. There is no particular reason for choosing the more complicated form for the three dimensional blob function; choosing a form similar to the two dimensional case should work just as well and the formulas would be slightly shorter. The spatial derivative which must be calculated in eq (16) was evaluated exactly via eq (18). The velocity was also evaluated exactly via eq (18) using a direct summation technique. The equations of motion eqs (15,16) were solved exactly using fourth order Runge-Kutta. The error in solving the equations of motion was monitored by calculating the invariants H, I and Ω. Since the kinetic energy H is positive definite, it proves to be the most sensitive measure of error in integrating the system and we choose our time step so as to preserve the kinetic energy invariant to an accuracy of 1×10^{-7}. The impulse invariant and angular momentum invariant were constant to machine accuracy (1×10^{-15}) due to the exact algebraic cancellation of those invariants.

There are many things which can be done to improve the computational efficiency of the method, such as using a fast summation method [14; 15] to calculate the velocity, or introducing an approximation to calculate the spatial derivatives; however, in this paper we deliberately avoid any such approximations so that the basic method can be validated without the added complication of other numerical approximations. These improvements to the method will be addressed in future work.

In the two dimensional calculations, see figs (1-10), we present the results by showing the evolution of the surfaces which are perpendicular to M initially and remain perpendicular to M for all later times, as given by eq (8). We choose for the initial conditions a velicity field which points in the y-direction and thus the surfaces in two dimensions are lines of constant y value; see figs (1,5). In figs (1-10) these surfaces are drawn by simply connecting the locations of the velicity particles which are originally on the same surface. The circles in figs (1,5) are additional Lagrangian markers chosen to help show the evolution of the fluid and to emphasize the fact that the initial velicity magnitudes are chosen radially symmetric, as is discussed below.

The initial conditions for the two dimensional calculations, $M(x, 0)$, is chosen to be a smooth dipole given in velicity variables as

$$M_2(x,0) = \begin{cases} 1 - 10r^3 + 15r^4 - 6r^5, & \text{if } r \le 1; \\ 0, & \text{if } r > 1, \end{cases} \tag{25}$$

where $r = |x|$, and $M_1(x, 0) = 0$. This produces an initial velocity which is given by eq (18) for a single particle. From eq (20) we see that the velocity of the point at the origin should be 1/2, at least initially; from fig (1) and fig (2) this is seen to be the case, although the velocity of the curve bounding the support of the velicity is considerably slower.

In figs (1-4) we show the results of approximating eq (25) with an initial grid with a spacing $h = 0.05$. This results in approximating eq (25) with 1245 points evenly spaced initially. We choose the smoothing parameter $\delta = 0.4$. In fig (3) we plot the direction of the velicity vectors m_i at their locations. We see that indeed the approximate velicity field evolves so as to remain perpendicular to the surface on which it was perpendicular initially.

In figs (5-10) we show the results of approximating the same initial conditions, eq (25), with a grid spacing $h = 0.025$ and leaving $\delta = 0.4$. This results in 5013 particles approximating the flow. In figs (5-9) we plot half of the surfaces $J \cdot M = 0$ and in fig (10) we plot a third of the surfaces. In comparing figs (1-4) with figs (5-10) we see that the results have converged to the limiting value as $h \to 0$. Because of the large number of points necessary to perform the calculations it is not possible to show the convergence as $\delta \to 0$ as accurately except for extremely short times.

In figs (11-16) we show the results of some generic three dimensional calculations. The lines shown in figs (11-16) are actual vortex lines of the continuous velocity field $\tilde{u}$. For instance in fig (11) a Lagrangian curve C, initially a circle, is evolved with the fluid velocity $\tilde{u}$. At any desired time the vortex lines through C are calculated by numerically integrating the vorticity field $\nabla \times \tilde{u}$. Although there is no a priori numerical constraint which guarantees that the vortex lines which close initially will close at later times, we see that the approximation preserves this property in figs (11,12). No swirl is numerically introduced.

The initial conditions for the calculations shown in figs (11-12) are those which produce a vortex ring of large radius $R = 1$, with constant vorticity inside the torus of small radius $a = 0.4$. We choose the velicity so that it initially has only a z component, which is given by

$$M_3(x,0) = \begin{cases} (1 + \sqrt{a^2 - z^2} - r)/2a, & \text{if } (r-1)^2 + z^2 \leq a^2; \\ \sqrt{a^2 - z^2}/a, & \text{if } z^2 \leq a^2 \text{ and } r \leq 1 - \sqrt{a^2 - z^2}; \\ 0, & \text{elsewhere}; \end{cases} \qquad (26)$$

where $r^2 = x^2 + y^2$. The velicity is normalized so that $\mid M \mid = 1$ at the origin. We can verify that eq (26) produces the correct vorticity by taking the curl of the velicity; note that the velicity is continuous whereas the vorticity is discontinuous.

The initial velicity eq (26) is approximated by 5696 points on a grid which is uniform in cylindrical coordinates r,θ,z; the average spacing is $h = 0.1$ and $\delta = 0.4$. The vortex lines which are shown in figs (11,12) are those which intersect a circular curve of radius 0.2 initially. Since the chosen initial conditions are not a steady state solution the vortex ring must evolve. Conservation of impulse and angular momentum prevent the ring from stretching in the radial direction. The ring evolves by stretching in the z direction while maintaining its cylindrical symmetry (see fig (12)) and eventually forms a thin urn-like structure. There is no cylindrical symmetry assumed in the calculation once the initial conditions are chosen.

In figs (13,14) we show the evolution of a flow in which the velocity has compact support inside of a torus. We choose a smooth velocity such that the support lies inside of a torus of large radius $R = 1$ and small radius $a = 0.2$. We define a divergence-free velocity u in terms of toroidal coordinates ρ, θ, ϕ as

$$\begin{aligned} u_\theta &= r\chi^2 f(\rho\chi/a) \\ u_\rho &= \frac{-n\epsilon\chi\rho}{r}\cos(n\theta) f(\rho\chi/a) \\ u_\phi &= \frac{-n\epsilon\chi\rho^2}{r}\cos(n\theta)\sin\phi\left(2f(\rho\chi/a) + \frac{\chi\rho}{a}f'(\rho\chi/a)\right) + \Omega_0\rho^2 f(\rho\chi/a) \quad , \end{aligned} \qquad (27)$$

where f is given by the right hand side of eq (25), $\chi = 1 + \epsilon \sin(n\theta)$, $r^2 = x^2 + y^2$, $\rho^2 = (r-1)^2 + z^2$, n is an integer, and a, ϵ and Ω_0 are parameters. For the flow in figs (13,14) we choose $a = 0.2$, $\epsilon = 0.0$, and $\Omega_0 = 25.0$. We approximated the flow with 5807 points, $h = 0.05$ and $\delta = 0.4$. In this example we choose $M(x, 0) = u(x, 0)$ initially. Although the velocity does not have a compact support for later times the velicity does. In fig (13) we show a single vortex line; note that now the vortex lines do not close initially and therefore should not close at later times either. The vortex lines wrap around torii; a single vortex line eventually covering a particular torus. In fig (14) we show the evolution of the vortex line we showed initially in fig (13). Because of the complicated structure of the vorticity in this calculation it would be difficult to do a similar calculation with a vorticity based method.

In the calculation shown in figs (15,16) we again choose an initial velocity given by eq (27) with compact support and pick $M = u$ initially. The choice of parameters is $a = 0.2$, $\epsilon = 0.25$, $n = 1$ and $\Omega_0 = 25.0$. The difference between the initial conditions in fig (15) and those in fig (13) is that in fig (15) we vary the small radius of the torus as a function of θ, the angle around the torus. Thus as we see in fig (13) the vortex lines now change their pitch and radius as they wrap around the toroidal surface. In fig (16) we show the evolution of the vortex line shown originally in fig (15). Once again we see that performing a similar calculation in vorticity variables would be difficult. We wish to emphasize that the three invariants: kinetic energy, impulse and angular momentum are discrete invariants of the calculations shown here.

5. Conclusions

We have presented a Lagrangian numerical method which approximates the incompressible Euler's equation in any number of space dimensions and preserves the three important invariants: kinetic energy, impulse and angular momentum. We have presented calculations which validate and show the convergence of the numerical method. Although the results shown here are only meant to validate the method, some of the advantages of basing the numerical method on the Hamiltonian formulation of incompressible flow is apparent from the complex structures we have been able to calculate in three dimensions. Work is under way to improve the Lagrangian algorithms so as to handle flows with boundaries and viscosity. We are adding other numerical approximations to increase the computational efficiency of the method. We have also developed a finite difference code based on the velicity formulation which has similar properties to the Lagrangian methods [16].

6. Acknowledgements

We acknowledge support from the Department of Energy under contract #DEFG02-88ER25053 and from the National Science Foundation under grant DMS-9100383.

References

[1] V. I. Oseledets, *On a new way of writing the Navier-Stokes equation: The Hamiltonian formalism*, Commun. Moscow Math. Society, (1988); translated in Russ. Math. Surveys **44**,

210-211 (1989).
[2] P. H. Roberts, *A Hamiltonian theory for weakly interacting vortices*, Matematica **19** 169-179 (1972).
[3] L. Onsager, *Statistical hydrodynamics*, Nuovo Cimento, suppl. to vol. 6, 279-287 (1949).
[4] J. D. Jackson, *Classical Electrodynamics*, John Wiley & Sons, New York, 1975.
[5] M. Spivak, *Calculus on Manifolds*, W. A. Benjamin Inc., Menlo Park, California, 1965.
[6] G. K. Batchelor, *Fluid Dynamics*, Cambridge University Press, New York, 1973.
[7] A. J. Chorin, J. E. Marsden, *A Mathematical Introduction to Fluid Mechanics*, Springer-Verlag, New York, 1979.
[8] A. J. Chorin, *Numerical study of slightly viscous flow*, J. Fluid Mech., **57** 785-796 (1973).
[9] A. J. Chorin, *The evolution of a turbulent vortex*, Comm. Math. Phys., **83** 517-535 (1982).
[10] O. H. Hald, *Convergence of vortex methods for Euler's equations, II*, SIAM J. Numer. Anal., **16** 726-755 (1979).
[11] J. T. Beale, A. Majda, *Vortex methods II: higher order accuracy in two and three dimensions*, Math. Comp., **39** 29-52 (1982b).
[12] C. Anderson, C. Greengard, *On vortex methods*, SIAM J. Numer. Anal., **22** 413-440 (1985).
[13] H. Goldstein, *Classical Mechanics*, Addison-Wesley, Menlo Park, California, 1950.
[14] L. Greengard, V. Rokhlin, *A fast algorithm for particle simulations*, J. Comp. Phys., **73** 325-348 (1987).
[15] T. F. Buttke, *Fast Vortex Methods in Three Dimensions*, Lectures in Applied Mathematics: volume 28. Editors: C. Anderson and C. Greengard, AMS 1991.
[16] T. F. Buttke, *A Finite Difference method based on the velicity formulation of incompressible flows*, in preparation.

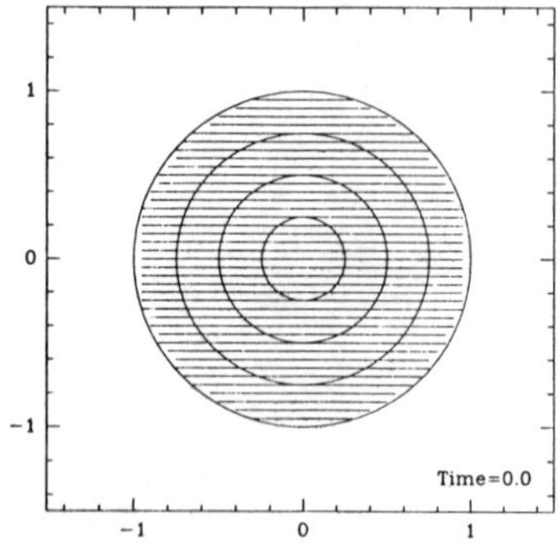

Figure 1. The Initial Material Surfaces. The material surfaces are shown which are orthogonal to the velicity field for a grid spacing of $h = 0.05$. Each surface is at the position of the initial discretization. The circles are additional Lagrangian markers inserted to aid in the visualization of the flow.

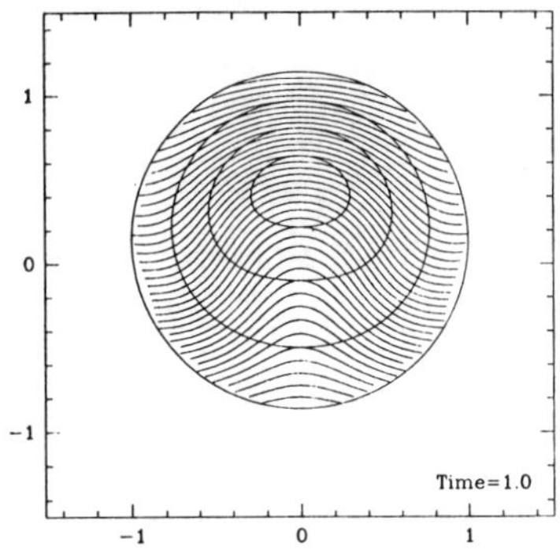

Figure 2. The Evolution of the Material Surfaces. The evolution of the material surfaces are shown, $h = 0.05$.

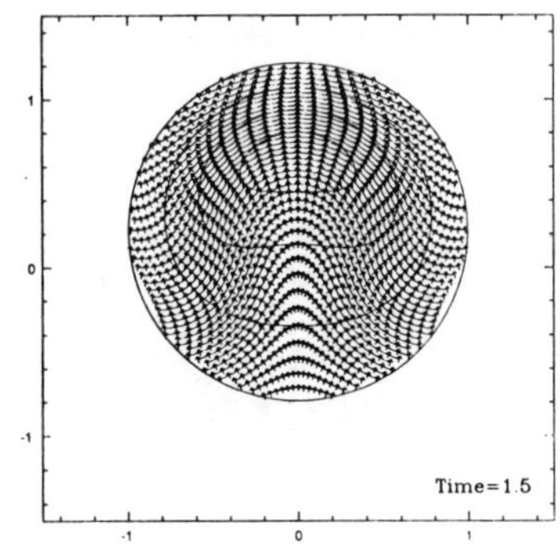

Figure 3. The Direction of the Velicity Field. The direction of the velicity field is shown, demonstrating that it evolves so as to stay perpendicular to the material surfaces.

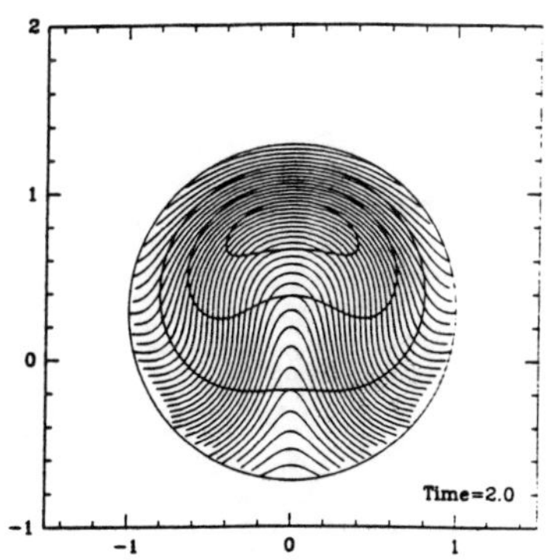

Figure 4. The Evolution of the Material Surfaces. The further evolution of the material surfaces are shown for $h = 0.05$.

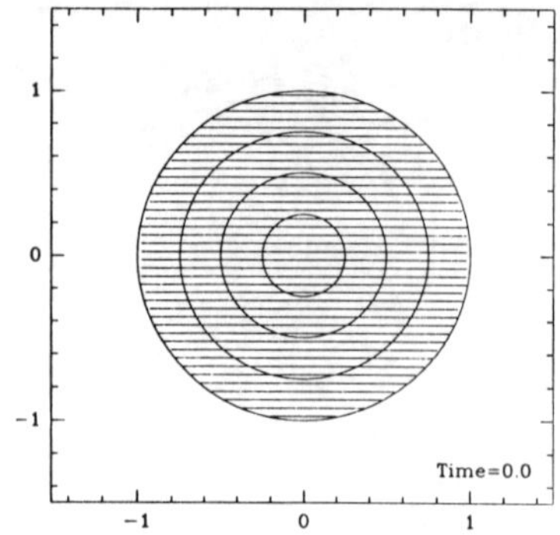

Figure 5. The Evolution of a Dipole with Higher Resolution. The initial surfaces $dx \cdot M = 0$ are shown for the initial velicity field with $h = 0.025$. Only half of the surfaces are shown.

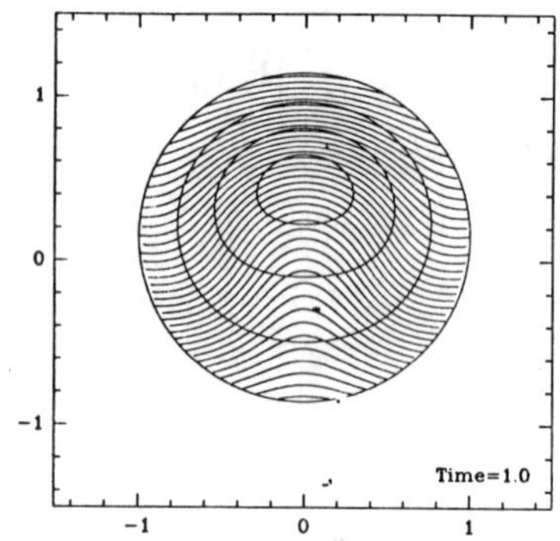

Figure 6. The Evolution of the Material Surfaces. The further evolution of the material surfaces is shown for the higher resolution calculation $h = 0.025$. This figure should be compared with fig (2).

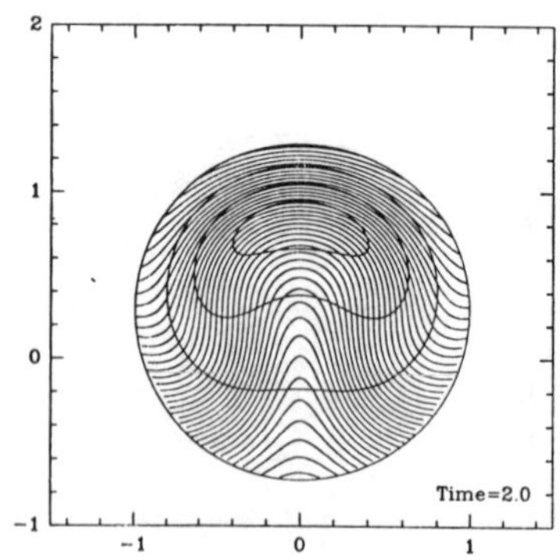

Figure 7. The Evolution of the Material Surfaces. The further evolution of the material surfaces is shown for the higher resolution calculation $h = 0.025$. This figure should be compared with fig (4).

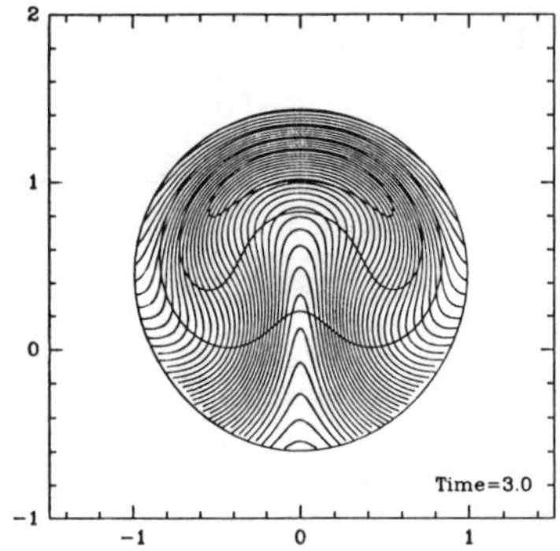

Figure 8. The Evolution of the Material Surfaces. The further evolution of the material surfaces is shown for the higher resolution calculation $h = 0.025$. The magnitude of the velicity is larger in regions where the surfaces are closer together and smaller in regions where the surfaces separate as can be seen from eq (9).

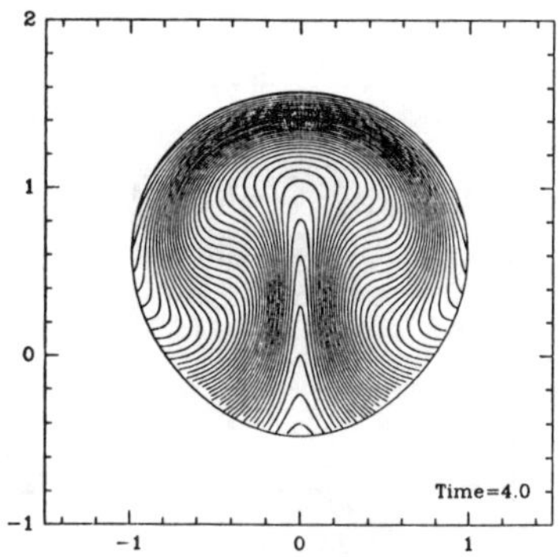

Figure 9. The Evolution of the Material Surfaces. The further evolution of the material surfaces is shown for the higher resolution calculation $h = 0.025$. We have removed the inner Lagrangian markers which were originally circularly shaped.

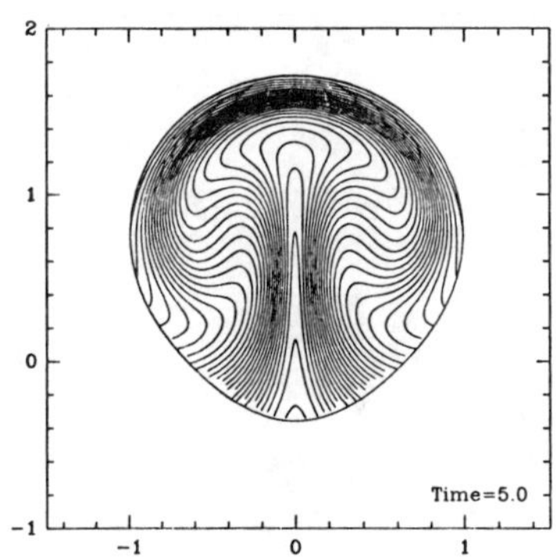

Figure 10. The Evolution of the Material Surfaces. The further evolution of the material surfaces is shown for the higher resolution calculation $h = 0.025$. We show only a third of the surfaces corresponding to the original discretization whereas in figs (5-9) we show half of the surfaces.

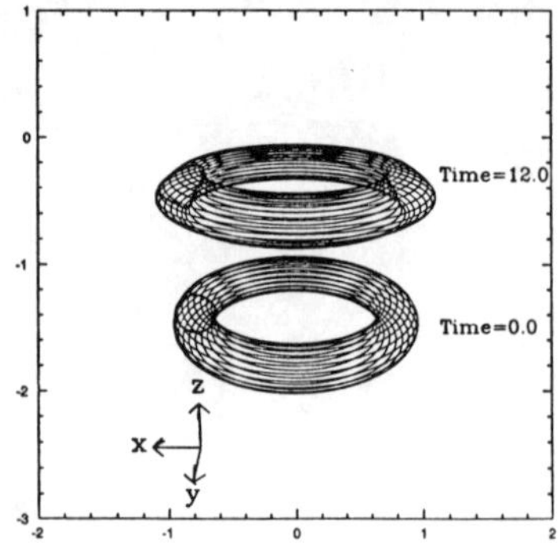

Figure 11. The Evolution of a Vortex Ring of Constant Vorticity. The vortex tube corresponding to a small radius of 0.2 is shown in a vortex ring of radius 0.4.

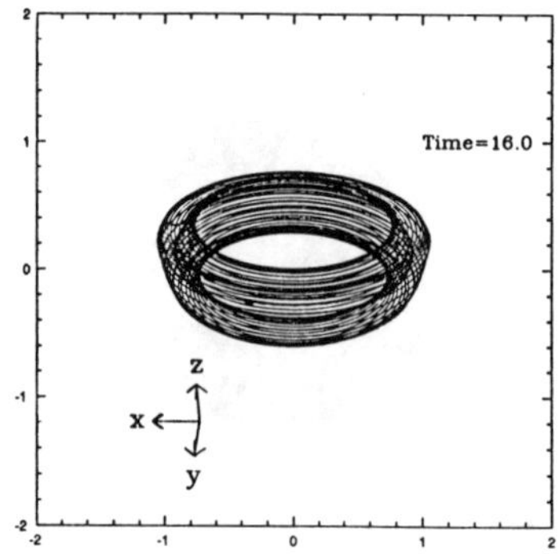

Figure 12. Deformation of a Vortex Ring. The highly deformed stage of a vortex tube is shown after it has evolved from an initially toroidal shape.

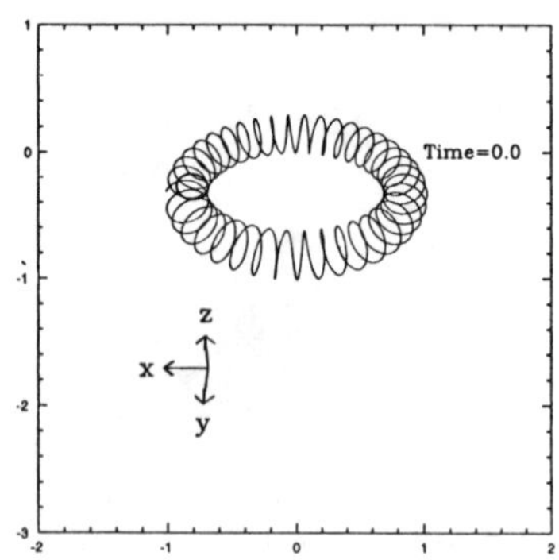

Figure 13. An Initially Swirling Flow. The initial shape of a vortex line is shown in a swirling flow.

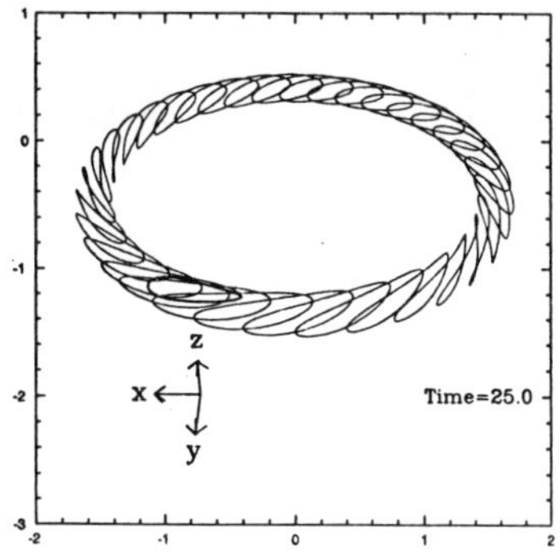

Figure 14. The Evolution of a Swirling Flow. The evolved state of the vortex line shown in fig (13) is shown at a later time.

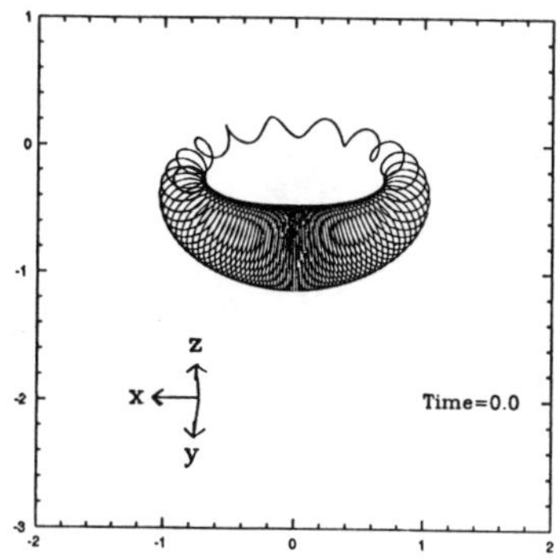

Figure 15. A Swirling Flow of Variable Radius. The initial shape of a vortex line in a swirling flow of variable radius is shown.

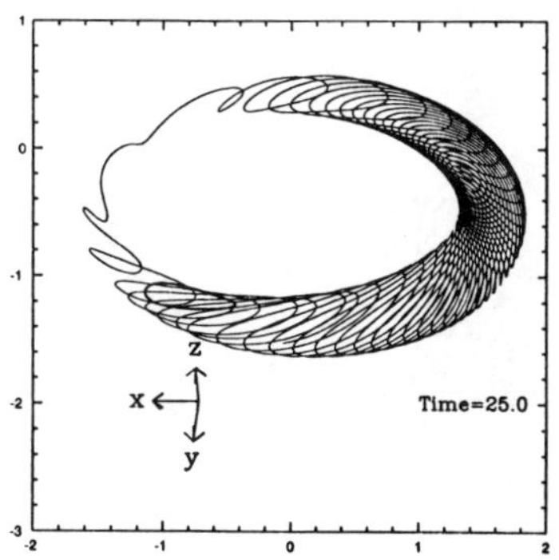

Figure 16. The Evolution of Swirling Vortex Line. The vortex line is shown which has evolved from the one shown in fig (15).

STATISTICAL MECHANICS FOR THE VORTEX MODEL

E. CAGLIOTI
Dipartimento di Matematica
Universita La Sapienza
Piazzale A. Moro 2
Roma ITALIA

1. Introduction

In this note we wish to summarize some rigourous results on statistical mechanics for the vortex model and discuss the possible connection between these results and the long time behaviour of 2-D Euler and Navier-Stokes equations.

The idea of obtaining information on the long time behavior of a two dimensional fluid using statistical mechanics is due to Onsager [15]. The vortex model is a finite dimensional hamiltonian system that in some sense describes the Euler 2-D equations when the vorticity field ω is a linear superposition of δ-functions:

$$\omega(dx) = \sum_{i=1}^{N} \delta_{x_i}(dx) \tag{1}$$

where $\delta_{x_i}(dx)$ is the Dirac measure concentrated at the point x_i. In particular it is possible to show that

$$\omega(dx,t) = \sum_{i=1}^{N} \delta_{x_i(t)}(dx) \tag{2}$$

is a weak solution of the Euler equations in $\Lambda \subset R^2$,[9; 10], where

$$\frac{dx_i}{dt} = \nabla^{\perp} H \ ; \ H(x_i) = N^{-2} \sum_{i \neq j} V(x_i - x_j) + N^{-2} \sum_{i=1}^{N} \gamma_{\Lambda}(x_i) \tag{3}$$

where

$$V(x,y) = -\frac{1}{2\pi} \log|x-y| + \gamma_{\Lambda}(x,y). \tag{4}$$

In his pioneering paper, Onsager considers the microcanonical ensemble for the vortex model, i.e. the microcanonical measure

$$\mu_{N,E}(dx_1, \cdots, dx_N) = \frac{\delta(H-E)}{\int_{\Lambda} \delta(H-E)}, \tag{5}$$

J.T. Beale et al. (eds.), Vortex Flows and Related Numerical Methods, 59–66.

where H is the hamiltonian of the vortex model and E is a certain value of the energy.

By simple considerations,[15], it is possible to show that for this system a range of negative temperature is allowed. This corresponds to the fact that at high energy it is expected that vortices with the same sign tend to clusterize.

It is natural to study the microcanonical ensemble for the vortex model in the so-called Mean Field Limit (MFL in the sequel). The idea is that if we have an initial data for the Euler equations ω_0, continuous for a sake of simplicity, we can think of approximating this scalar field by a distribution obtained as a sum of N δ-functions. So we find

$$\sum_{i=1}^{N} \sigma_i \delta(x - x_i) \to \omega(x)\, dx \Longrightarrow \sum_{i=1}^{N} \sigma_i \to \int_\Lambda \omega(x)\, dx. \tag{6}$$

This idea, in particular in the case $\sigma_i = 1/N$, has been first proposed by Montgomery [12] and Montgomery and Joyce [13], and it has been studied by many authors ([16; 17; 18; 19] and others). This is the case we will discuss in this paper.

It is important to point out that the right ensemble to be used in this case is the microcanonical ensemble. Indeed the justification of the canonical ensemble is based on the equivalence of the two ensembles in the usual thermodynamic limit. In this case, that is in the MFL, it can very well be not the case. In particular for the vortex model it is possible to prove that at high vaues of the energy the microcanonical ensemble and the canonical ensemble can be non-equivalent and the equivalence can depend upon the shape of the domain Λ [4]. This fact may be relevant for the long time behavior of the solution of the Euler and Navier-Stokes equations [5].

A rigourous proof of the convergence to the mean field limit for the canonical ensemble is reported in [3; 8], while in [3] is studied the corresponding mean field equation.

The mean field equation for the microcanonical ensemble is studied in [4] and in [7]. In particular in [4] is given a description of the thermodynamical properties of the system both in the case of the canonical and microcanonical ensemble. The rigourous derivation of the MFL for the microcanonical ensemble has been obtained in [4] in the particular case of the equivalence of the 2 ensembles (for the vortex-model), in [7] for any bounded regular pair potential by means of a large deviation technique, and in [6] also for the case of the vortex model in which the potential is not bounded.

2. Statistical mechanics of the vortex model in the mean field limit

2.1. MFL FOR THE MICROCANONICAL ENSEMBLE

Let us consider the microcanonical ensemble for the vortex model

$$\mu_{N,I}(dx_1, \cdots, dx_N) = \frac{\chi(H \in I)}{\int_{H\in I} dx_1 \cdots dx_N} \tag{7}$$

where I is an energy interval $(E, E + \Delta E)$ and χ_I is the characteristic function of the set I; the denominator is thus only a normalization. The fact that we must take an interval of energy is motivated by the singular feature of the definition (6). however it is possible to take the limit as $\Delta E \to 0$ after the MFL as we will see later. We are interested in the limit of the measure when $N \to \infty$.

2.2. RIGOUROUS RESULTS FOR THE MICROCANONICAL ENSEMBLE

i) It is possible to prove that in this limit the microcanonical measure is concentrated on the solution of the following entropy-energy variational principle

$$S(I) = \sup_{\varrho \in R_I} - \int_\Lambda \varrho \log \varrho; \tag{8}$$

where

$$R(I) = \left\{ \varrho | \int_\Lambda \varrho \, dx = 1, E(\varrho) \equiv \frac{1}{2} \int_{\Lambda^2} \varrho(x)\varrho(y)V(x,y) \, dx \, dy \in I \right\}$$

ii) Then, it is possible to show that if $S(E, E + \Delta E)$ is a continous function of E, $E + \Delta E$ then we can take the limit

$$S(E) = \lim_{\Delta E \to 0} S(E, E + \Delta E), \tag{9}$$

and $S(E)$ satisfies the variational principle

$$S(E) = \sup_{\varrho \in R_E} - \int_\Lambda \varrho \log \varrho \tag{10}$$

where $R(E) = \left\{ \varrho | \int_\Lambda \varrho \, dx = 1, E(\varrho) = E \right\}$. Imposing the stationarity of the functional $S - \beta E$, where β is the lagrange multiplier associated to the constraint of constant energy, we obtain the mean field equation

$$\varrho = \frac{\exp -\beta \int_\Lambda V(x,y)\varrho(y) \, dy}{Z}, \tag{11}$$

where Z is a normalization constant. Taking the laplacian of both sides of (11) we obtain an equation for the stream function ψ

$$-\Delta \psi = \frac{e^{-\beta\psi}}{\int_\Lambda e^{-\beta\psi}} \text{ in } \Lambda \tag{12}$$

$$\psi = 0 \text{ on } \partial\Lambda. \tag{13}$$

This means that, in some sense, in the MFL a vortex interacts with the other vortices only through their mean density.
iii) We have seen that the MFL exists for any value of the energy $E > 0$. We are now interested in the behavior for $E \to \infty$. It is possible to prove [4; 5] that in this limit the solution of the mean field variational principle concentrates on a delta function. This is true for any domain Λ bounded with smooth boundary. The support of the delta function is the points where the function γ_Λ is maximum. For instance in the case of a disk, this delta function is located at the center of the disk. This makes sense if we think that the solution of the mean field equation must be a steady-state solution of the Euler equation (for this solution indeed we have that ω is a function

of ψ). Thus if this solution tends to a delta function, that is a vortex, this vortex must be a steady-state solution of the Euler equations and so must stay in one of the extremal points of the function γ_Λ.
Remark. We have just discussed the case when the vorticity has a ditinguished sign. In the general case, including the case when the vorticity has mean value 0, it is possible to develop the same statistical mechanics formalism. The mean field equation can be put in the form [13] (see also [6] for some rigourous results in this case),

$$-\Delta\psi = \sinh(-\beta\psi)$$

which is generally called the Sinh-Poisson equation.

3. Long time behavior of the Navier-Stokes equations

We now want to show how the theory developped above gives a good description of the Navier-Stokes equations with small viscosity. We will try to motivate this fact *euristically* by direct dynamical arguments and not as above by statistical mechanics arguments. The word euristically is emphasized because we want to discuss here some non rigourous ideas, in contrast with the previous exactly defined mathematical models.

3.1. NUMERICAL RESULTS

In a recent paper [14] it has been shown how by integrating the Navier-Stokes equations over a long time with small viscosity, the vorticity field is well described by a solution of the mean field equation. In particular, in this numerical simulation of the Navier-Stokes equation, it has been shown that there exists a certain set of times for which the solution of the Navier-Stokes equations agrees well with a particular solution of the mean field equation. Let us briefly discuss the results of [14]. These simulations concern the Navier-Stokes equations at high Reynolds number (viscosity $\nu = 10^{-4}$) on the torus T^2 on a time scale of $3\,10^2$. The initial vorticity is ditributed in a random way. In a first stage, the vorticity forms a certain number of blobs of vorticity of both signs with various sizes. These blobs remain stable except when 2 blobs with the same sign collide, in which case they give rise to a greater blob. In a next stage, for $40 < t < 220$, this collision process keeps going on, until only 2 blobs remain. This configuration is a quasi steady-state solution of the Euler equations and fits very well with a particular equation of the Sinh-Poisson equation. It has been proved that this configuration remains stable for the remainder of the numerical simulation and it has been conjectured that it is so as long as the viscous process does not become predominant (up to $t \approx 10^4$). Therefore it seems that the statistical mechanics of the vortex model describes very well the long time behavior of the Navier-Stokes equations when the initial data are not smooth.

3.2. TENTATIVE EXPLANATION

If we go back to the definition of the statistical mechanics solution we notice that this solution has the property of maximizing the entropy functional S

$$S = -\int \omega \log \omega \, dx$$

at a fixed energy E,

$$E = \frac{1}{2} \int \omega(x) V(x, y) \omega(y) \, dx \, dy$$

This means that we must look for a dynamical process for which the entropy S increases very fast while the energy varies slowly. The entropy at time t is $S_t = -\int \omega_t \log \omega_t \, dx$. We want now to stress the following important points
i) It is reasonable that the energy, except for an initial period of time for which the highest modes loose energy faster, varies on a time scale of the order of ν^{-1} (corresponding to the eigenvalues of order 1 of the laplacian)
ii) The entropy cannot increase under the action of the inertial terms, since when $\nu = 0$, $dS/dt = 0$ (the entropy is the integral of a function of the vorticity and thus is conserved by the Euler equations).
iii) The entropy increases under the action of the viscosity, but if the inertial terms were not present then the Navier-Stokes equations would reduce to the heat equation, with an increase of the entropy on a time scale of order ν^{-1}. Let us now show how the inertial process and the viscous process can cooperate to produce a drastic growth of the entripy on a time sscale much smaller than the viscous time-scale.

If we consider the lagrangian point of view, we observe that the vorticity field is convected by an hamiltonian dynamical system, with hamiltonian H, and is diffused by the viscosity. So if we are very far from a stationary solution of the Euler equation we can assume that this motion is chaotic with a positive Liapounov exponent. This implies the possibility of stretching in the flow. So if we consider for example that the vorticity is 1 inside a square and 0 outside, it may happen that the support of the vorticity is stretched by the inertial motion, with an increase of the perimeter, while the area is conserved. We can now see how the viscosity, as small as it is, can drastically increase the entropy. On Figure 1 we can define euristically an effective area A^* where the diffusion acts, as the set on which $\omega > 1/2$. This effective area is of the order of $\sqrt{\nu t} \exp \lambda t$ and the entropy, which can be thought of as the logarithm of this area, would increase linearly like λt, up to some corrections of lower order involving the viscosity. This means that the entropy varies on an inertial time scale, the inverse of a Liapounov number, which is an inertial characteristic of the evolution. So in the limit of a vanishing viscosity the entropy would increase while the energy would remain constant, as we wished to show. It is important to notice that this process can happen only if we start from an initial condition well apart from an equilibrium. Actually, if we were starting near a steady-state solution of the Euler equation, we would most likely remain close to this solution.

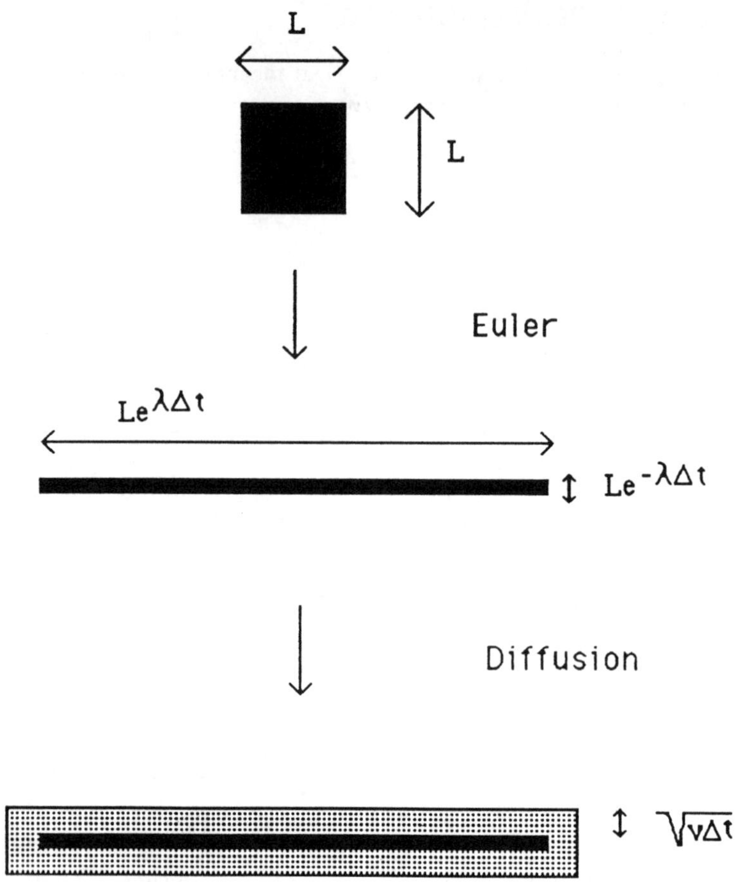

Fig. 1. Increase of entropy by combined actions of Euler flow and Diffusion

3.3. CORRECTION TO THE DYNAMICAL PROCESS

The process we just described is simplistic for the following reason: if we go back to the experiment of [14] we must notice that we did not account for the period of time when the motion is well described by the motion of a certain number of blobs of vorticity. This behavior, corresponding to the early stage of the simulations in [14] has been confirmed numerically by other authors (see for instance the papers of Benzi [1; 2]). This phenomenon is easily explained if we define a local Liapounov exponent as the largest eigenvalue (if positive) of the matrix of the second derivatives of the hamiltonian ψ

$$\lambda = \sqrt{(\psi_{xy})^2 - \psi_{xx}\psi_{yy}}$$

where x, y are the plane coordinates. Let us assume that ψ is the superposition of an external hyperbolic field ψ_{ext} modelling the effect of other vortices, for example we can take $\psi_{\text{ext}} = \lambda_0 xy$, and of the field due to the local vorticity $\psi_{\text{loc}} = -\Delta^{-1}\omega$

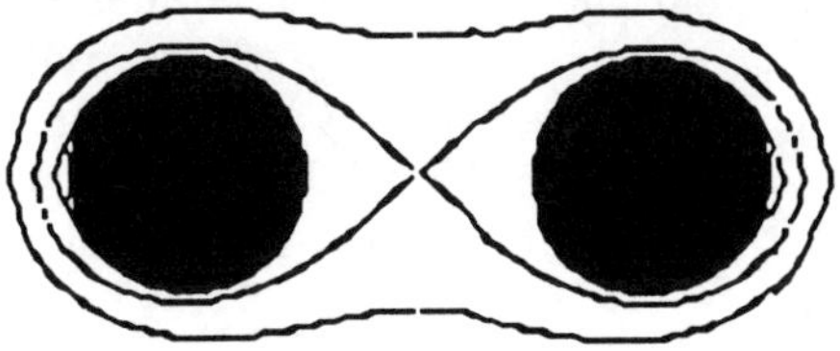

Fig. 2. Streamlines corresponding to 2 circular vortex patches

which we choose for a sake of simplicity, of the form

$$\psi_{\text{loc}} = -\frac{\omega}{4}(x^2 + y^2).$$

If we substitute this field in the expression of λ we find

$$\lambda = \sqrt{\lambda_0^2 - \omega^2}.$$

So there is a competition between the external field and the intensity of the blob and it is reasonable that, in a first stage of the motion, the sets in which the vorticity is not sufficiently high are stretched by the motion giving rise to an increase of the entropy. These blobs remain stable until they collide with one another.

It is important to point out that the collision process also increases the entropy. If we look, on Figure 2, where are plotted two equal vortex blobs and the level curves of ψ, at these level curves, we observe that if the blobs are close enough, there is an overlap between the blobs and the hyperbolic zone of the stream function.

3.4. SUMMARY

We therefore think that the dynamical process can be summarized in 3 steps.
i) At $t = 0$ the vorticity is *random*, that is, it is distributed among many blobs with different shapes and strength. The sets where the vorticity is not big enough are stretched, inducing either more stable blobs or cancellations of blobs of opposite signs through diffusion. In this process there is a growth of entropy.
ii) After this process the vorticity is distributed in blobs which move like vortex points (a fact which has been numerically demonstrated by Benzi and al [1]). This process also produces an increase of the entropy.
iii) At the end of this collision process the vorticity is well described by a solution of the Sinh-Poisson equation which consists in 2 large blobs of vorticity of opposite sign.

These arguments, as we mentioned, are vague and of an euristic nature, but we think that is is possible to test them by simulating the collision of 2 similar blobs and checking some important thermodynamic parameters. For example it would be interesting to plot the entropy and energy of the resulting blobs as a function of the parameters of the colliding blobs. This kind of numerical experiments could allow in our opininon to investigate one of the fundamental processes which give rise to an increase of the entropy. Moreover it should not be very expensive from a numerical point of view.

References

[1] Benzi R., Paternello,S. and Santangelo, P.: 1987, Europhys. Lett., **3**, p. 729
[2] Benzi R., Paternello,S. and Santangelo, P.: 1988, J. Phys. A, **21**, p. 1221-1237.
[3] Caglioti, E., Lions, P.L., Marchioro, C., Pulvirenti, M.: to appear in Comm. Math. Phys.
[4] Caglioti, E., Lions, P.L., Marchioro, C., Pulvirenti, M.: to appear.
[5] Caglioti, E., PhD Thesis.
[6] Caglioti, E., to appear.
[7] Eynik, G.L. and Spohn, H, preprint.
[8] Kraichnan, R.H., Montgomery, D.: 1980, Rep. Prog. Phys., **43**, pp. 547-619.
[9] Marchioro, C. and Pulvirenti, M.: 1984, Lecture Notes in Phys., **203**.
[10] Marchioro, C. and Pulvirenti, M.: 1984, Mathematical theory for an incompressible non-viscous fluid, Springer Verlag, to appear.
[11] Messer, J. and Spohn, H.: 1982, J. Stat. Phys., **29**, pp. 561-578.
[12] Montgomery, D.: 1972, Phys. Lett, **39A**, pp. 7-8.
[13] Montgomery, D. and Joyce, G.: 1974, Phys. Fluids, **17**, pp. 1139-1145.
[14] Mattaeus, W.H. and al.: 1991, Phys. Rev. Lett.
[15] Onsager, L., Suppl. Nuovo Cimento, p. 279.
[16] Poyntin, Y.B., and Lundgren, T.S.: 1976, Phys. Rev., **13A**, pp. 1274-1275.
[17] Poyntin, Y.B., and Lundgren, T.S.: 1976, Phys. Fluids., **19**, pp. 1459-1470.
[18] Lundgren, T.S. and Poyntin, Y.B.: 1977, Phys. Fluids, **20**, pp. 356-363.
[19] Lundgren, T.S. and Poyntin, Y.B.: 1977, J. Stat. Phys., **17**, pp. 323-355.
[20] Taylor, J.B.: 1972, Phys. Lett. A, **40**, pp 1-2.

ON SINGULAR SOLUTIONS OF THE VLASOV-POISSON EQUATIONS

GEORGE MAJDA
Department of Mathematics
Ohio State University
Columbus Ohio 43210–1174
USA

A vortex sheet is, loosely speaking, a surface in a fluid such that the normal velocity of the fluid is continuous along the surface but the tangential velocity of the fluid is discontinuous across the surface. Many theoretical, numerical and analytical investigations have been done to try to understand the properties of solutions to the incompressible Euler Equations with non-smooth (vortex sheet) initial data, see [1], [3], [4], [5], [6], [8], [9], [10], [13], [14], [15], [16], [17] and [18] for example. Despite this research, many fundamental open mathematical problems about the nature of these solutions still exist. In order to get a handle on some of these open problems, Andrew Majda proposed that one should study a simpler problem, namely, the one-component Vlasov Poisson Equations (VPE) from plasma physics. Hopefully, insight gained by studying this model problem will provide new insights into the original problems about incompressible fluids. In this paper I will present this system of equations and some connections between the VPE and the vorticity equation for a 2-D incompressible fluid with vortex sheet initial data.

The work which I will present constitutes one small part of an on-going collaboration with Andrew Majda and Yuxi Zheng. See [11], [12] and [19] for a complete description of the results.

In this paper we consider the single component 1–D Vlasov Poisson Equations (VPE) for a collisionless plasma of electrons in a uniform background of ions. Assume that the problem is periodic in x and that the initial electron density is a Dirac delta function supported over a curve in $x-v$ space (phase space). Let $f(x,v,t)$ denote the density of electrons, $E(x,t)$ the electric field, $\rho(x,t)$ the charge, $\mathbf{x} = (x,v)$ and $\mathbf{c}$ a curve in $x-v$ space. In dimensionless variables, the problem is defined by the system of equations

$$\frac{\partial f}{\partial t} + v\frac{\partial f}{\partial x} - E(x,t)\frac{\partial f}{\partial v} = 0 \text{ for } t > 0 \text{ and } x \in [0,1], \tag{1}$$

J.T. Beale et al. (eds.), Vortex Flows and Related Numerical Methods, 67–75.

$$-\phi_{xx} = \rho(x,t) = 1 - \int_{-\infty}^{\infty} f(x,v,t)\,dv, \quad E = -\phi_x, \tag{2}$$

with the initial condition

$$f(x,v,0) = g(x,v)\ \delta_{\mathbf{C}}, \tag{3}$$

the periodic boundary conditions

$$E(0,t) = E(1,t), \ \ f(0,v,t) = f(1,v,t), \tag{4}$$

and the zero-mean electric field condition

$$\int_0^1 E(x,t)\,dx = 0. \tag{5}$$

Initial data with the form (3) is called **electron sheet initial data**, and the solution of problem (1)–(5) is called an **electron sheet**, to emphasize the close connection with vortex sheets which we now explain.

I briefly highlight some of the features which problem (1)–(5) has in common with the two dimensional vorticity form of the Euler equations with vortex sheet initial data. The reader should consult [4] or [5] for a precise definition of the latter problem.

Let $\mathbf{v} = (v, -E(x,t))^t$ denote a "velocity" field, $\mathrm{div}_{\mathbf{X}} = (\frac{\partial}{\partial x}, \frac{\partial}{\partial v})$, and introduce the particle trajectory equations

$$\frac{d}{dt}\begin{pmatrix} X \\ V \end{pmatrix} = \mathbf{v}(X,V) = \begin{pmatrix} V \\ -E(X,t) \end{pmatrix}.$$

Now just make a correspondence between the electron density f and the vorticity for the 2-D Euler equations. Equation (1) can be interpreted as a transport equation for f by a divergence free velocity field $\mathbf{v}$ since $\mathrm{div}_{\mathbf{X}}(v, -E(x,t))^t = 0$. Furthermore, f is constant along the particle trajectory equations. The 2-D vorticity equation has a similar interpretation with the electron density replaced by the vorticity. Using this correspondence, it easily follows that the electron density f is the analogue of the vorticity, and the initial condition (3) is the direct analogue of vortex sheet initial data.

To demonstrate another analogy, we recall that for an incompressible fluid the velocity field is recovered from the vorticity field through the Biot-Savart Law. This law establishes a global relationship between the velocity field and vorticity. The elliptic equation (2) implies that there is a global relationship between the electric field and the electron density. This, in turn, establishes a global relationship between the velocity field for the VPE and the electron density. Other similarities than the ones presented here are mentioned later in this paper and in the papers [11], [12] and [19].

We believe that given the close relationship between problem (1)–(5) and the vorticity form of the Euler Equations with vortex sheet initial data, one will gain important insights about the open problems for the latter problem by thoroughly understanding the properties of the solutions of problem (1)–(5). Furthermore, the electron sheet problem for the VPE has several important advantages over the corresponding vortex sheet problem, and we list two of them.

1. One can construct explicit solutions for problem (1)–(5) for a perturbed uniform electron sheet, the analogue of a perturbed vortex sheet. (We describe one solution below.) These solutions can form an algebraic singularity at a finite time, just like the predicted behavior of perturbed vortex sheets. The perturbed vortex sheet problem has no similar explicit solutions.
2. For the VPE, the global relationship between the electric field and the electron density involves an integral with a discontinuous kernel. The Biot-Savart Law for an incompressible fluid contains a kernel with a stronger singularity. Consequently, one believes that a mathematical analysis of problem (1)–(5) should be easier than the corresponding vortex sheet problem.

I now present an explicit solution, derived by Dziurzynski [7], corresponding to the case of a sinusoidally perturbed uniform electron sheet. A **uniform electron sheet** is a time-independent, exact weak solution of problem (1)–(5) given by

$$\begin{aligned} Z(\alpha) &= (x(\alpha), v(\alpha)) = (\alpha, 0) = \mathbf{c}(0) \quad \text{for} \quad 0 \le \alpha \le 1, \\ f(Z(\alpha)) &= \delta_{\mathbf{c}(0)}, \\ E(x(\alpha)) &\equiv 0. \end{aligned}$$

Now consider problem (1)–(5) with initial data corresponding to a sinusoidally perturbed uniform electron sheet. Let $\varepsilon > 0$ denote a parameter and $0 \le \alpha \le 1$. This initial data is given by

$$Z(\alpha, 0) = (x(\alpha, 0), v(\alpha, 0)) = (\alpha, \varepsilon \sin(2\pi j\alpha)) = \mathbf{c}(0) \tag{6}$$

$$f(Z(\alpha, 0)) = \left(1 + (\varepsilon 2\pi j)^2 \cos^2(2\pi j\alpha)\right)^{-\frac{1}{2}} \delta_{\mathbf{c}(0)}. \tag{7}$$

The exact weak solution of (1)–(5) with this initial data is

$$\left\{ \begin{aligned} Z(\alpha, t) &= (x(\alpha, t),\ v(\alpha, t)) \\ &= (\alpha + \varepsilon \sin(2\pi j\alpha) \sin t,\ \varepsilon \sin(2\pi j\alpha) \cos t) = \mathbf{c}(t) \\ f(Z(\alpha, t)) &= \left| \tfrac{dZ(\alpha,t)}{d\alpha} \right|^{-1} \delta_{\mathbf{c}(t)} \\ E(x(\alpha, t)) &= \varepsilon \sin(2\pi j\alpha) \sin t. \end{aligned} \right. \tag{8}$$

Dziurzynski [7] shows that if $|2\pi j\varepsilon| < 1$, then solution (8) is defined for all $t \ge 0$. If $|2\pi j\varepsilon| > 1$, then there exists a time $t^* < \infty$ such that solution (8) is defined for the finite time interval $[0, t^*]$ and the charge $\rho(x(\alpha, t^*))$ forms an algebraic singularity.

This exact solution illustrates another striking similarity between the vortex sheet problem and the electron sheet problem. The algebraic singularity in the charge for $|2\pi j\delta| > 1$ is the analogue of the singularity observed in the vortex sheet case before rollup, see [9], [10], [13] and [17].

Just as in the vortex sheet problem, several questions are natural to ask including the following:

1. Does the analytic formula continue to describe the electron sheet after the singularity time?
2. What are the qualitative properties of the solution past the singularity time?
3. What are the mathematical properties of the solution past the singularity time?

In order to answer these questions and many others, I developed a particle method to solve (1)–(5) based on the method analyzed by Cottet and Raviart in [2]. I needed to modify this method to make it efficient and to be able to discretize singular data like electron sheets. These details are described in [11].

To answer questions (1)–(3), I computed the solution of problem (1)–(5) with initial data of the form (6)-(7) for $\varepsilon = 2.0$ and $j = 1$. The singularity time for the solution is $t^* = .079$ to 3 decimal places. I computed the solution with a time step $\Delta t = .0158$, and used enough particles to converge to a solution with 3 decimal place accuracy both before and after the singularity time.

In Figure 1 and Figure 2 I plotted the electron sheet $(Z(\alpha, t))$ and the electric field $E(x(\alpha, t))$, respectively, for 0, 3 and 5 time steps. We note that the singularity time occurs at the fifth time step, so Figures 1 and 2 show graphs of the solution before and at the time of singularity formation. In Figure 3 and Figure 4 I plotted the electron sheet and electric field for 6, 8 and 10 time steps, so these figures show graphs of the solution past the singularity time. These numerical results indicate the following:

(i) The analytic formula (8) does not describe the solution past the singularity time.

(ii) The singularity time is precisely the time when the electron sheet ceases to be the graph of a single-valued function.

(iii) The electron sheet and electric field become multivalued functions past the singularity time.

(iv) The electric field loses smoothness past the singularity time.

Again, we point out the analogy between electron sheets governed by the VPE and vortex sheets. For vortex sheets, the singularity occurs just before the sheet begins to roll up. The folded, multivalued electron sheet is the analogue of a rolled-up vortex sheet. Furthermore, the electron sheet forms a singularity in the charge just before it begins to fold.

In this short paper I have outlined a number of analogies between the Euler equations with vortex sheet initial data and the VPE with electron sheet initial data. These analogies indicate that problem (1)–(5) should be a very useful model for trying to understand the open questions about vortex sheets. The discussion presented here is just the starting point for a much more detailed study of problem (1)–(5). Complete details can be found in the papers [11], [12] and [19].

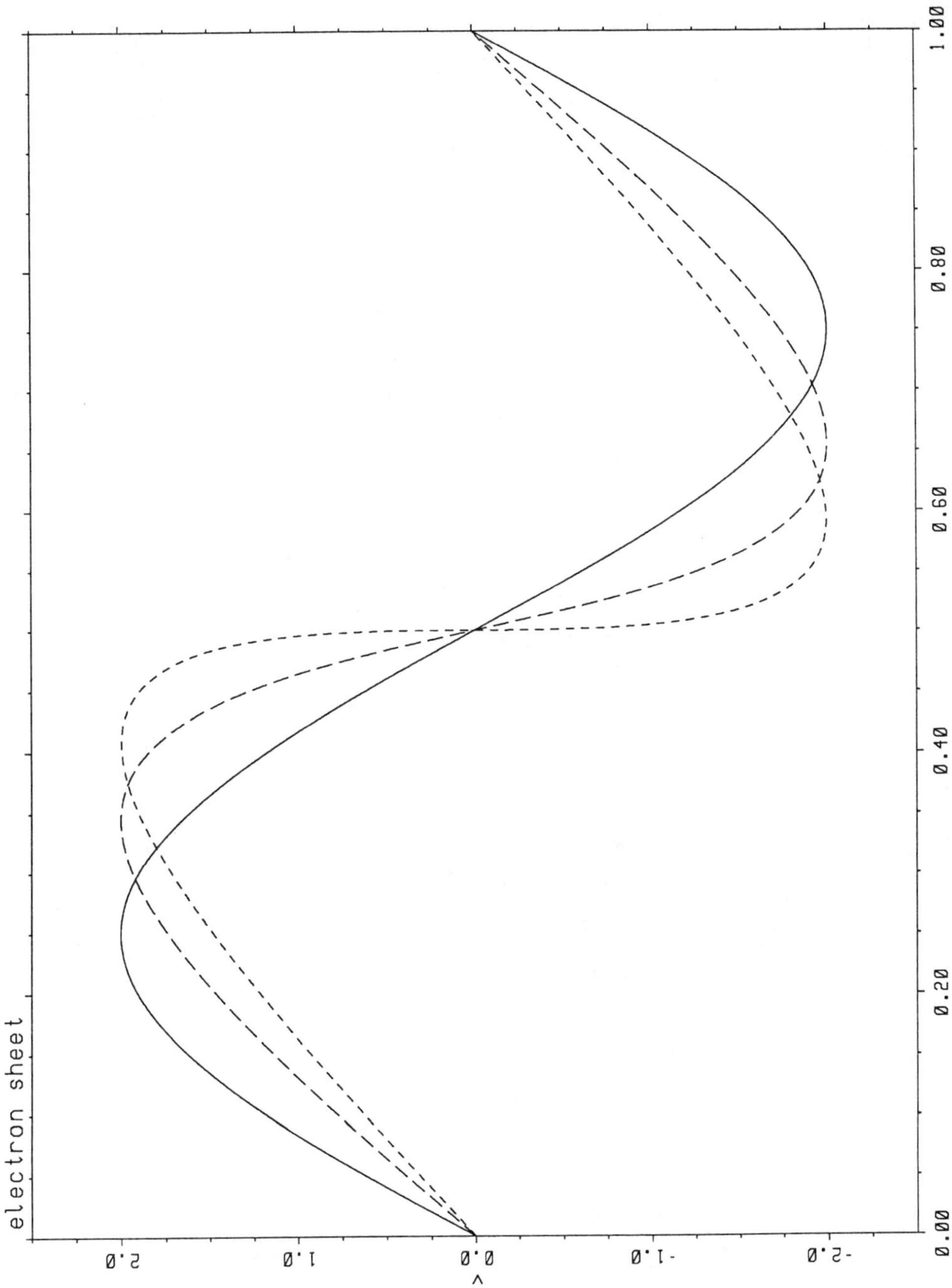

Fig. 1. Electron sheet at 0(solid), 3 (long dashes) and 5 (short dashes) time steps

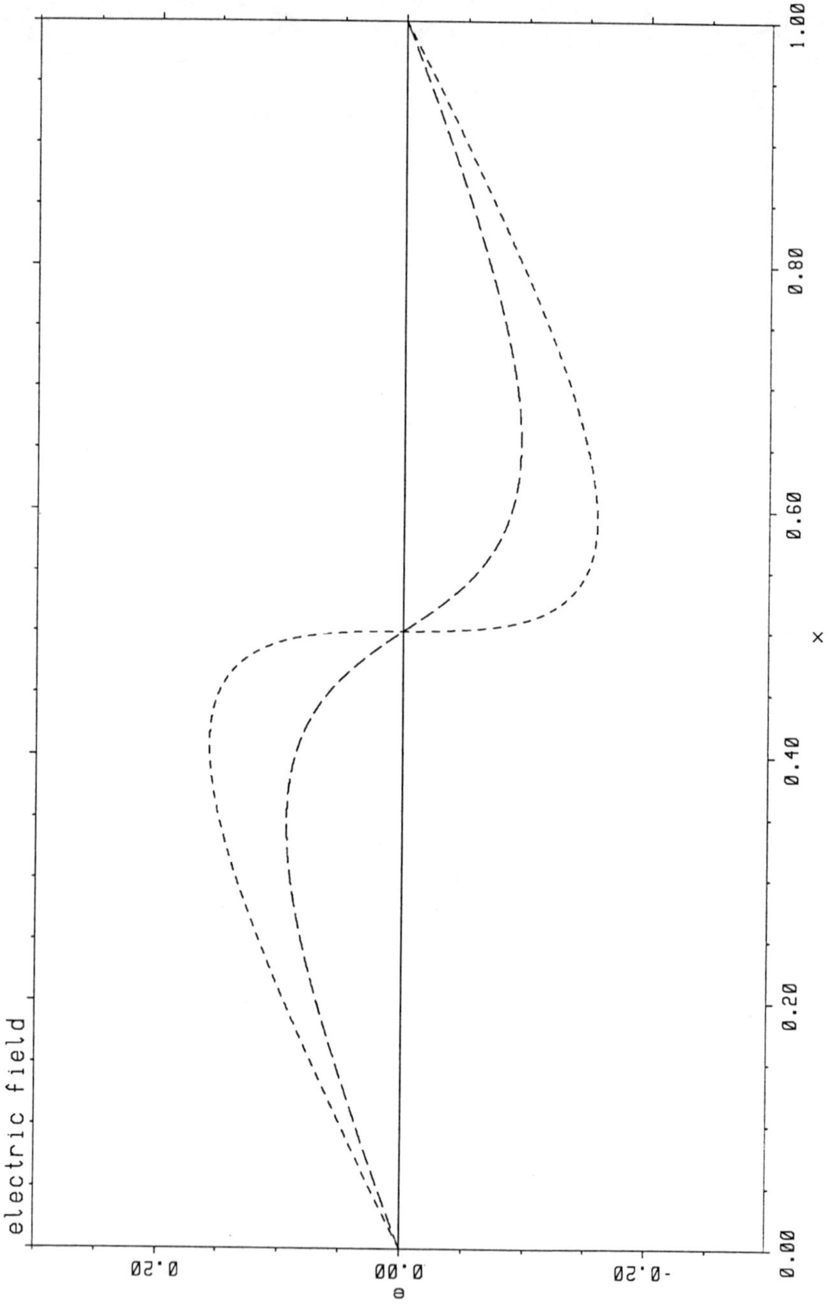

Fig. 2. Electric field at 0 (solid), 3 (long dashes) and 5 (short dashes) time steps

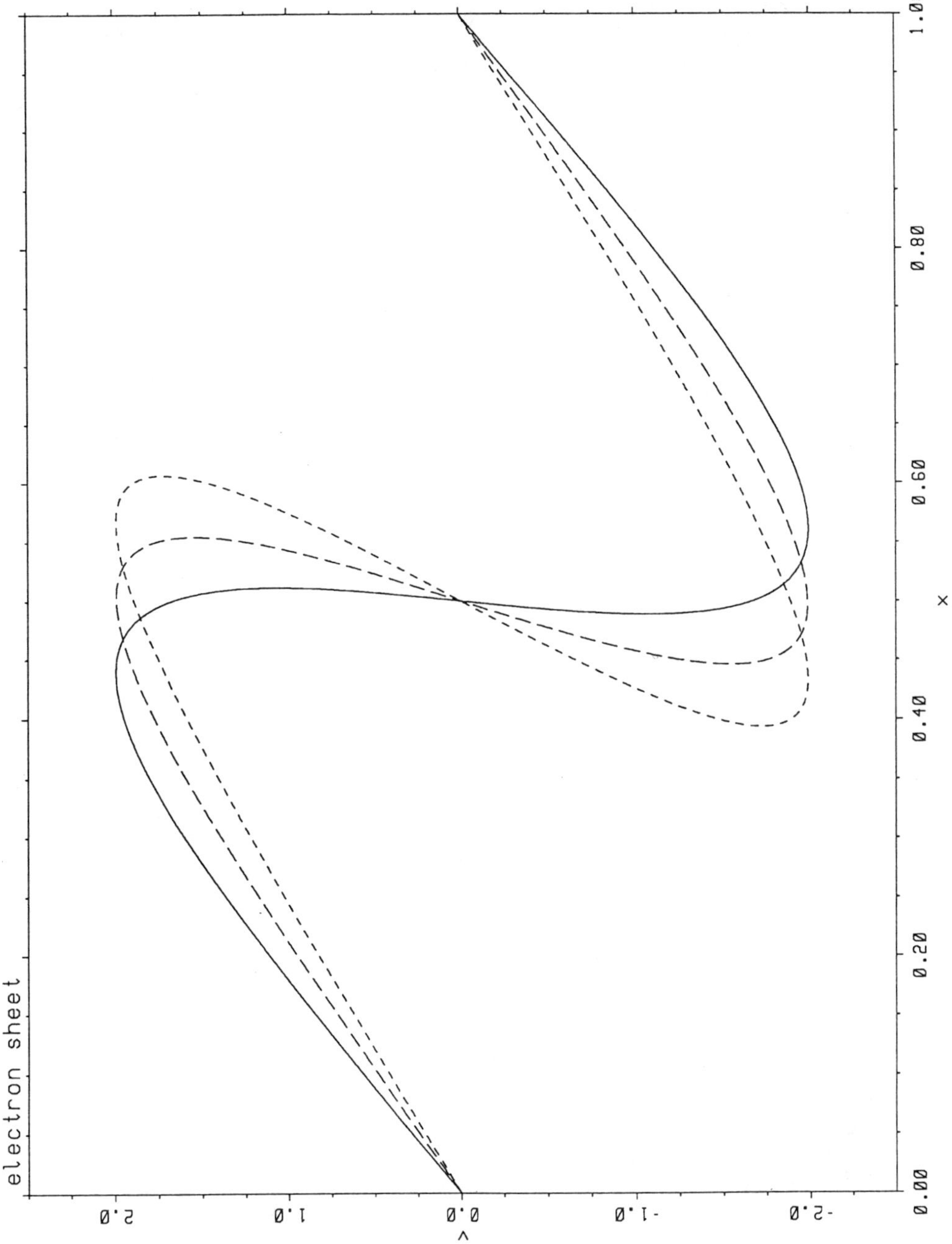

Fig. 3. Electron sheet at 6 (solid), 8 (long dashes) and 10 (short dashes) time steps

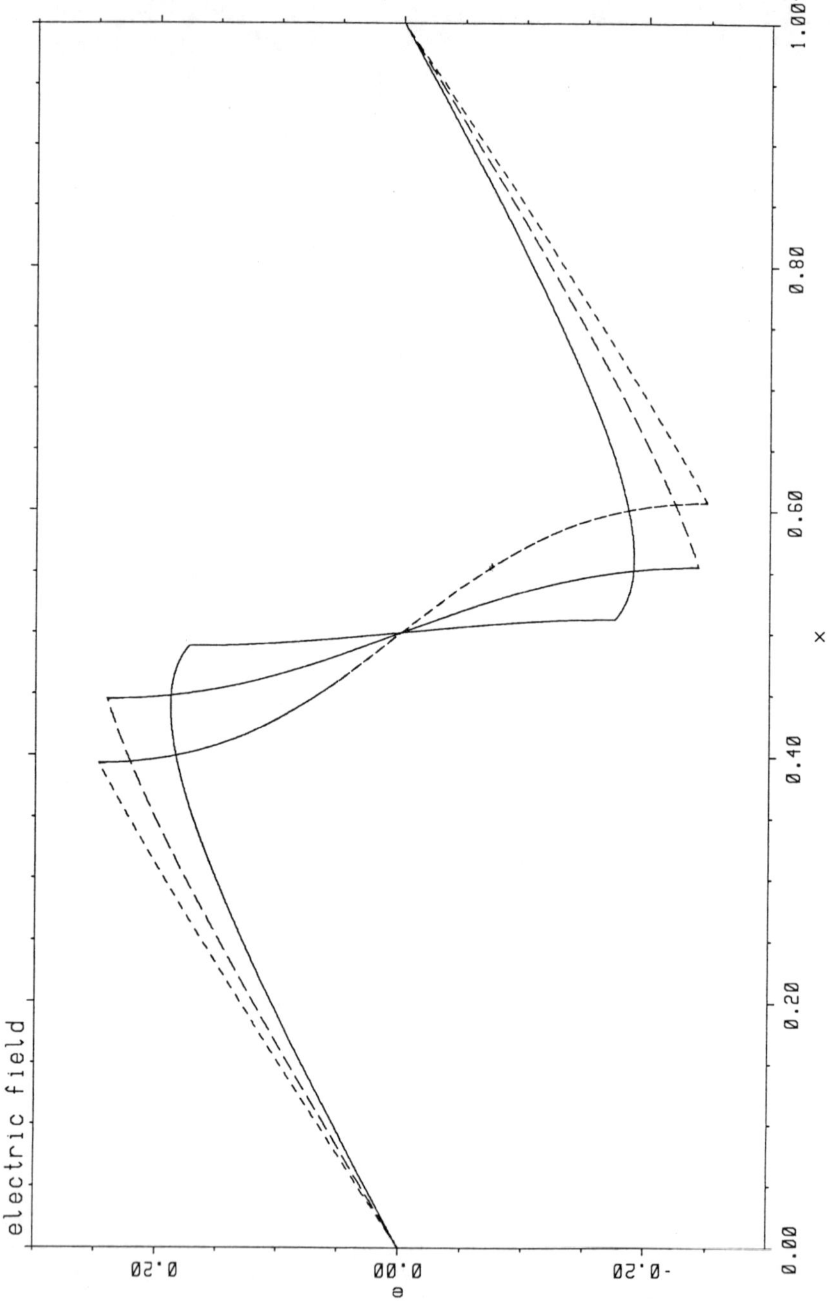

Fig. 4. Electric field at 6 (solid), 8 (long dashes) and 10 (short dashes) time steps

Acknowledgements

I would like to thank Andrew Majda, Yuxi Zheng and Ed Overman for many helpful discussions about this work.

My research was initially supported by the National Science Foundation under grant DMS-8903484, and is currently supported by the Air Force Office of Scientific Research under grant AFOSR–91–0309. The numerical computations are supported by the Ohio State Supercomputer Center under grant PAS330.

I am a visiting member of the Institute for Advanced Study in Princton, New Jersey, and a visitor in the Program in Computational and Applied Mathematics at Princeton University, for the 1992–93 academic year. I would like to thank the people at the I.A.S. for their hospitality. I would also like to thank the members of the PCAM for their hospitality and financial support under grants NSFDMS9001-805.A02 and ONRN00014–89–J–1044.P00005.

References

[1] R. Caflisch, *"Analysis for the evolution of vortex sheets"*. Lectures in Applied Mathematics, **28** (1991), p. 67.

[2] G.- H. Cottet and P.-A. Raviart, *"Particle methods for the 1-D Vlasov-Poisson equations"*. SIAM J. Num. Anal., **21** (1984), p. 52.

[3] J.-M. Delort, *"Existence de nappes de tourbillon en dimension deux"*. Preprint.

[4] R.J. DiPerna and A.J. Majda, *"Oscillations and concentrations in weak solutions of the incompressible fluid equations"*. Comm. Math. Phys., **108** (1987), p. 667.

[5] R.J. DiPerna and A.J. Majda, *"Concentrations in regularizations for 2-D incompressible fluid equations"*. Comm. Pure Appl. Math., **60** (1987), p. 301.

[6] R.J. DiPerna and A.J. Majda, *"Reduced Hausdorff dimension and concentration – cancellation for two-dimensional incompressible flow"*. J. Am. Math. Soc., 1, **1** (1988), p. 59.

[7] R.S. Dziurzynski, *"Patches of electrons and electron sheets for the 1-D Vlasov-Poisson equation"*. Ph.D. dissertation, Dept. of Math., University of California, Berkeley, CA (1987).

[8] C. Greengard and E. Thomann, *"On DiPerna-Majda concentration sets for two dimensional incompressible flow"*. Comm. Pure Appl. Math., **41** (1988), p. 295.

[9] R. Krasny, *"A study of singularity formation in a vortex sheet by the point vortex approximation"*. J. Fluid Mech., **167** (1986), p. 65.

[10] R. Krasny, *"Desingularization of periodic vortex sheet roll-up"*. J. Comp. Phys., **65** (1986), p. 292.

[11] G. Majda, *"Particle approximation of weak solutions to the Vlasov Poisson and Fokker Planc equations"*. In preparation.

[12] A. Majda, G. Majda and Y. Zheng, *"1-D Vlasov-Poisson equations"*. In preparation.

[13] D.I. Meiron, G.R. Baker and S. Orszag, *"Analytic structure of vortex sheet dynamics Part I, Kelvin-Helmholtz instability"*. J. Fluid Mech., **114** (1982), p. 283.

[14] D.W. Moore, *"The equation of motion of a vortex layer of small thickness"*. Stud. Appl. Math., **58** (1978), p. 119.

[15] D.W. Moore, *"The spontaneous appearance of a singularity in the shape of an evolving vortex sheet"*. Proc. R. Soc. London .A, **365** (1979), p. 105.

[16] D.W. Moore, *"Numerical and analytical aspects of Helmholtz instability"*. Theoretical and Applied Mechanics, Proc. XVIIUTAM, Niodson and Olhoff, eds., p. 263.

[17] M.J. Shelley, *"A study of singularity formation in vortex sheet motion by a spectrally accurate method"*. Appeared in J. Fluid Mech.

[18] C. Sulem, P.L. Sulem, C. Bardos and U. Frisch, *"Finite time analyticity for two and three dimensional Kelvin-Helmholtz instability"*. Comm. Math. Phys., **80** (1981), p. 485.

[19] Y. Zheng and A. Majda, *"Existence of global weak solutions to the one-component VlasovPoisson and Fokker-Planck-Poisson systems in one space dimension with initial data of measures"*. Preprint.

POINT VORTICES AND LOCALIZATION IN EULER FLOWS

CARLO MARCHIORO

Dipartimento di Matematica

Università La Sapienza

Piazzale A. Moro 2

00185 Roma ITALY

ABSTRACT. We study the time evolution of an inviscid incompressible two-dimensional fluid when the initial vorticity is sharply concentrated in N small regions of diameter ε. We give a rigourous proof of the validity of the point vortex model as ε vanishes. Then we discuss the three-dimensional case and other possible generalizations.

1. Localization and point vortex system

It is a very old and natural challenge to study incompressible non-viscous fluid flows when the vorticity is concentrated in a subset of a physical space of low dimensionality. For planar situation the first study of a fluid when the vorticity is sharply concentrated in very small regions (namely points) has been done in the last century, leading to the point vortex system. A formal connection between this model and the Euler equations is discussed in many textbooks. However only recently a rigourous connection has been proved (for the first definition of the point vortex system, see [1]; for a review on the topic, see [2; 3]). In particular it has been proved a sort of "localization" of the motion in the sense that the evolved vorticity is concentrated in N regions of diameter d which vanishes when ε (the initial diameter) vanishes for any fixed time. For smooth problems this is a direct consequence of the continuity with respect to the initial data, but our problem is singular because the initial data is sharply concentrated and the time evolution (via the Euler equation) has a logarithmic divergent kernel. Actually, when ε is small, the field in each blob becomes very large and it is difficult to exclude that thin filaments of vorticity are pushed away. We can prove that this fact does not happen. As a main consequence, we can prove a rigourous connection between the Euler equation and the point vortex theory.

Some results in this direction have been obtained in the last ten years [4; 5; 6; 7] but the problem has been completely solved only recently [8].

J.T. Beale et al. (eds.), Vortex Flows and Related Numerical Methods, 77–82.

More precisely, consider an initial vorticity of the form

$$\omega_\varepsilon(x,0) = \sum_{i=1}^{N} \omega_{\varepsilon;i}(x,0) \tag{1}$$

where $\omega_{\varepsilon;i}(\cdot,0)$ is a function with a definite sign supported in a region $\Lambda_{\varepsilon;i} \subset \boldsymbol{R}^2$ such that

$$\Lambda_{\varepsilon;i} \equiv \text{supp } \omega_{\varepsilon;i} \subset \Sigma(z_i,\varepsilon) \ ; \ \ \Sigma(z_i,\varepsilon) \cap \Sigma(z_j,\varepsilon) = \emptyset \text{ if } i \neq j \tag{2}$$

for ε small enough ($\Sigma(z,r)$ denotes the disk with center z and radius r). Moreover

$$\int \omega_{\varepsilon;i}(x,0)\,dx = a_i \in \boldsymbol{R} \text{ (vortex intensity)} \tag{3}$$

and

$$|\omega_{\varepsilon;i}(x,0)| \leq \text{const } \varepsilon^{-\eta} \ ; \ \eta < \frac{8}{3} \tag{4}$$

A simple example of $\omega_{\varepsilon;i}(\cdot,0)$ satisfying (3)-(4) is given by: $\omega_{\varepsilon;i}(\cdot,0) = \varepsilon^{-2} a_i \chi(\Lambda_{\varepsilon;i})$ where meas $\Lambda_{\varepsilon;i} = \varepsilon^2$ and $\chi(\Lambda)$ is the characteristic function of Λ. We can prove the following result

THEOREM 1. *Denote by $\omega_\varepsilon(\cdot,t)$ the time evolution of $\omega_\varepsilon(\cdot,0)$ via the Euler equations. Then*
i) for all $d > 0$ there exists $\varepsilon_0(d,T)$ such that, if $\varepsilon < \varepsilon_0(d,T)$

$$\textit{supp } \omega_{\varepsilon;i}(\cdot,t) \subset \Sigma(z_i(t),d) \ \ \textit{for any } t \in [0,T]$$

where $z_i(t)$ denotes the solution of the ordinary differential system (called point vortex system)

$$\frac{d}{dt} z_i(t) = -\nabla_i^{\perp} \frac{1}{2\pi} \sum_{j=1; j\neq i}^{N} a_j \log|z_i(t) - z_j(t)| \ ; \ z_i(0) = z_i. \tag{5}$$

Moreover: $d \to 0$ as $\varepsilon \to 0$ (the so-called localization property).
ii) for any bounded function f

$$\lim_{\varepsilon\to 0} \int \omega_\varepsilon(x,t) f(x)\,dx = \sum_{i=1}^{N} a_i f(z_i(t))$$

Let us shortly comment on this theorem. Assertion i) states that the blobs of vorticity remain localized until time T for any d and T provided we choose ε small enough. Assertion ii) states that

$$\omega_\varepsilon(\cdot,t) \rightharpoonup \sum_{i=1}^{N} a_i \delta_{z_i(t)}$$

weakly as $\varepsilon \to 0$, where δ denotes the Dirac measure. This last statement gives a rigourous justification of the point vortex model.

Remark 1. We observe that the singular nature of the right hand side of (5), which blows up when two vortices come close to each other, does not gaurantee the existence of the solutions to (5) for all time. In many cases (for instance when $a_i > 0$) collapses are forbidden by the first integrals of the motion, but there are cases for which singularities do happen. However, it can be proved that the collapses are exceptional [2]. In the general case we can say that theorem 1 holds up to any time T for which the solutions to the system (5) exist.

Remark 2. The above theorem does not assert that the motion of the fluid particles supporting the vorticity ω_ε converges, in the limit $\varepsilon \to 0$, to the trajectories $z_i(t)$. In general this is false. Due to the singularity of the kernel, the motion of the fluid particles is very irregular and is not converging at all. However the motion of the center of vorticity of the blobs converges to the motion of the point vortices.

The proof of the result is rather intricated and can be found in [8]. It is based on the well known properties of invariance with respect to translation and rotation of the Euler evolution on the one hand and on a non trvial iterative method on the other hand.

2. Generalizations

In the previous section we have shown the the localization property is a sufficient condition for a rigourous justification of the point vortex model. When all the vorticies have strength with the same sign such a derivation is still possible without localization. In this case we can prove the following result

THEOREM 2. *Consider an initial vorticity of the form*

$$\omega_\varepsilon(x,0) = \sum_{i=1}^{N} \omega_{\varepsilon;i}(x,0)$$

where $\omega_{\varepsilon;i}(\cdot,0) \in L^1(\mathbf{R}^2) \cap L^\infty(\mathbf{R}^2)$, $\omega_{\varepsilon;i}(\cdot,0) \geq 0$. *Moreover assume that for any bounded function* f

$$\lim_{\varepsilon \to 0} \int \omega_{\varepsilon;i}(x,0) f(x)\, dx = \sum_{i=1}^{N} a_i f(z_i)$$

then

$$\lim_{\varepsilon \to 0} \int \omega_{\varepsilon;i}(x,t) f(x)\, dx = \sum_{i=1}^{N} a_i f(z_i(t))$$

where $\omega_{\varepsilon;i}(\cdot,t)$ *is the solution of the Euler equation with initial datum* $\omega_{\varepsilon;i}(\cdot,0)$ *and* $z_i(t)$ *are the solutions of the system (5).*

The proof is essentially as in [5] (see also the techniques of [9]).

When a_i have different signs we do not have, at present, a result like the previous one. However some weaker generalization is possible, as discussed in [8].

The techniques at work in the proof of theorem 2 also yield a rigourous justification of the so called vortex-wave system. This is a model in which point vortices and a smooth flow (wave) coexist. More precisely, let us consider the initial value problem in $\boldsymbol{R}^2$

$$\frac{d}{dt}\phi_t(x) = u(\phi_t(x),t) + \sum_{i=1}^{N} a_i K(\phi_t(x) - z_i(t)) \; ; \; x \neq z_i(0) \tag{6}$$

$$\frac{d}{dt}z_i(t) = u(z_i(t),t) + \sum_{j=1, j\neq i}^{N} a_j K(z_j(t) - z_i(t)) \tag{7}$$

$$u(\cdot,t) = K \star \omega(\cdot,t) \tag{8}$$

$$\omega(\phi_t(\cdot),t) = \omega_0 \in L^1(\boldsymbol{R}^2) \cap L^\infty(\boldsymbol{R}^2) \; ; \; \phi_0(x) = x \; ; \; z_i(0) = z_i \tag{9}$$

where $K(x) = -\frac{1}{2\pi}\nabla^\perp \log|x|$. This describes a system composed by N points z_i, each of them moving along the velocity field produced by the other vortices and by an incompressible fluid with a bounded vorticity. At the same time this fluid moves along the whole velocity field. When the point vortices are absent, this system reduces to the usual Euler equations. The system (6)-(9) has been studied in [10] where a global existence result has been proved (still for a_i with a definite sign). Furthermore, under the additional assumption that the $z_i(0)$ lie outside the support of ω_0, the uniqueness and regularity of the solutions has also been proved.

We can justify this model by a result analogous to theorem 1:

THEOREM 3. *Consider an initial vorticity of the form*

$$\omega_\varepsilon(x,0) = \omega_0(x) + \sum_{i=1}^{N} \omega_{\varepsilon;i}(x,0)$$

where $\omega_{\varepsilon;i}(x,0)$ is a function with a definite sign defined as in section 1, satisfying (2),(3),(4). Assume that for all i $d(z_i, \mathrm{supp}\,\omega_0) > r_0 > 0$. Then for all $d > 0$, $d \ll r_0$, there exists $\varepsilon_0(d,T)$ such that, if $\varepsilon < \varepsilon_0(d,T)$

$$\mathrm{supp}\; \omega_{\varepsilon;i}(\cdot,t) \subset \Sigma(z_i(t),d) \;\; \text{for any } t \in [0,T]$$

and $d \to 0$ as $\varepsilon \to 0$.
Moreover for any continuous bounded function f

$$\lim_{\varepsilon\to 0} \int \omega_\varepsilon(x,t) f(x)\, dx = \int \omega(x,t) f(x)\, dx + \sum_{i=1}^{N} a_i f(z_i(t))$$

where ω_ε is the solution of the Euler equation with initial vorticity $\omega_\varepsilon(\cdot,0)$ and ω, z_i are the solution of the vortex-wave system.

In three dimensions the situation is less satisfactory. The naive generalization of point vortices would be filaments of vorticity. However a simple calculation show that they move in general with an infinite speed so that theorems like the previous ones are hopeless. Some point systems, called vortons, have been studied, but they do not conserve some important quantities and so are less fundamental than the point vortices in two dimensions. However they can be useful for numerical purposes.

Another case in which the vorticity is concentrated in some region of low dimensionality is given by the so-called vortex sheet. In this case also there are successful efforts to justify the model in terms of the Euler equation [11]. More precisely, in the vortex sheet model, the vorticity is concentrated on a line parametrized by $y = \phi(x), x \in \boldsymbol{R}$:

$$\omega(x, y) = \eta(x)\delta(y - \phi(x)) \ ; \ (x, y) \in \boldsymbol{R}^2$$

where η (the strength) and ϕ evolve according to the Birkhoff-Rott equation (see [3; 12] for instance). Consider now an initial profile of the form

$$\omega_\varepsilon(x, y) = \frac{1}{\varepsilon}\chi(\Lambda_\varepsilon) \tag{10}$$

where $\chi(\Lambda_\varepsilon)$ is the characteristic function of the set $\Lambda_\varepsilon = \{(x, y) | \phi_\varepsilon^-(x) < y < \phi_\varepsilon^+(x)\}$, the functions $\phi_\varepsilon^\pm$ being choosen such that

$$\lim_{\varepsilon \to 0} \frac{\phi_\varepsilon^+ - \phi_\varepsilon^-}{2} = \eta$$

$$\lim_{\varepsilon \to 0} \phi_\varepsilon^\pm = \phi.$$

Assuming analyticity for (η, ϕ), the time evolution of the initial data (10) via the Euler equation can be proved to converge to the time evolution of (η, ϕ) via the Birkhoof-Rott equations as long as this solution remains analytic.

References

[1] Helmholtz, H.: 1867, 'On the integrals of the hydrodynamical equations which express vortex motion', *Phil. Mag.*, **33**, 485

[2] Marchioro, C. and Pulvirenti, M.: 1984, 'Vortex methods in two-dimensional fluid dynamics', Lecture Notes in Physiscs, **203**, Springer Verlag.

[3] Marchioro, C. and Pulvirenti, M.: 'Mathematical theory of incompressible non-viscous fluids', *to appear.*

[4] Marchioro, C. and Pulvirenti, M.: 1983, 'Euler evolution for singular data and vortex theory', *Comm. Math. Phys.*, **91**, 563.

[5] Marchioro, C.: 1988, 'Euler evolution for singular data and vortex theory: a global solution', *Comm. Math. Phys.*, **116**, 45.

[6] Turkington, B.: 1987, 'On the evolution of a concentrated vortex in an ideal fluid', *Arch. Rat. Mech. An.*, **97**, 75.

[7] Marchioro, C. and Pagani, E.: 1986, 'Evolution of two concentrated vortices in a two-dimensional bounded domain', *Math. Meth. Appl. Sci.*, **8**, 328.

[8] Marchioro, C. and Pulvirenti, M.: 1992, 'Vortices and localization in Euler flows', *to appear in Comm. Math. Phys.*

[9] Marchioro, C.: 1990, 'On the vanishing viscosity limit for two-dimensional Navier-Stokes equations', *Math. Meth. Appl. Sci.*, **12**, 463.

[10] Marchioro, C. and Pulvirenti, M.: 1991, 'On vortex-wave system', in Mechanics, Analysis and Geometry: 200 years after Lagrange, M. Francaviglia ed., Elsevier, Science Publishers, 79.

[11] Benedetto, D. and Pulvirenti, M.: 1992, 'From vortex layers to vortex sheets', *SIAM J. Appl. Math.*, **52**, 1041-1051.

[12] Sulem, P., Sulem, C., Bardos, C., and Frisch, U.: 1981 'Finite time analyticity for the two and three dimensional Kelvin-Helmholtz instability', *Comm. Math. Phys.*, **80**, 485.

TURBULENCE MODELING FOR INCOMPRESSIBLE VORTEX FLOW

PHILIPPE PASCAL
8 Rue de la Haye
57000 Metz FRANCE

ABSTRACT. Flows resulting from the development of fluid sheet, generated by a separation, give birth to rolled up vortex sheets. In aerodynamics, these vortices can have some important consequences on the performances of aircraft flying at high angles of attack and missiles equipped with highly swept control surfaces. Indeed, under the effects of pressure fields associated to the airfoil, these vortices are likely to undergo strong dislocation, a phenomenon usually known under the name of vortex breakdown. This phenomenon produces an ultimate modification of the flow upon the wing, the prediction of which remains very dubious. The purpose of this work is to propose a computational method able to describe this kind of flow. A theoretical study has been realized in order to compare the different methods in their attempt to predict the vortex evolution during its breakdown process. The proposed methods are based on the assumption that the flow can be described with an axisymmetric model. A classical statistical treatment leading to a one point closure model is also used. Three different approaches are presented:
- the first one combines the statistical equations of motion written with the boundary layer assumptions and an algebraic turbulence model,
- the second one is based on the same equations, but uses the algebraic stress model (A.S.M.),
- the third one uses an algebraic model to solve directly the statistical equations without approximations.
A confrontation of computations with experiments allows to validate several turbulence models.

1. Introduction

Vortices are the key point for the understanding of a lot of phenomena. In aerodynamics, they can have large effects on the flying performances of aircraft and missiles. Indeed, a strong dislocation better known as *vortex breakdown*, can happen under the effects of pressure fields which are developped on the airfoil, leading to effects which can be desirable (acceleration of wing tips vortices in the case of large aircraft) or dangerous (like onset of unstabilizing and unsteady stresses on the aircraft equipped with highly swept control surfaces and flying at high angle of attack).

It seems necessary to be able to modelize the vortex phenomena in their unity. The theoretical and experimental works already made, particularly at O.N.E.R.A., have brought a good knowledge of the mean velocity field and turbulence field in a vortex subject to a breakdown [1]. On the other hand, the numerical simulations [2] give very promising results for simple shapes. The parametric study, derived

J.T. Beale et al. (eds.), Vortex Flows and Related Numerical Methods, 83–104.

from numerical simulations, should allow to improve the prediction of the vortex breakdown. These results show clearly that the breakdown phenomenon itself is a process essentially related to the vortex dynamical properties (velocity field) and the intensity of disturbance (external pressure field). For Reynolds numbers corresponding to ususal problems in aeronautics, the pressure gradients induced by viscosity effects are far from being sufficient to trigger the breakdown.

The actual dependence of breakdown upon viscosity and turbulence is through a memory effect. This means that the fields for the tangential and axial velocity follow from the former evolution of a vortex. During the transition towards breakdown, the velocity profiles, in the vortex core, are progressively changed until they reach a critical shape which suddenly leads to the breakdown.

The purpose of this study is to compare different models applied to the evolution of a turbulent incompressible vortex just before its breakdown. The proposed models are based on the two following assumptions:
- the mean flow can be approximated by an axisymmetric one,
- the turbulence can be modeled by classical one point closure models.

Three different approaches have been studied:
the first on uses the statistical equations of motion, written with the boundary layer assumptions and derived from a turbulence model based on the mixing length [3].
-the second one is based on the same equations, but uses the algebric stress model (A.S.M.) to take into account the turbulence.
- the third one couples the statistical equations without boundary layer assumption, with a turbulence model using a mixing length [4].

2. Solution methods

In the two first approaches, the studied flow is considered like quasi-cylindrical (revolution flow + boundary layer assumptions), turbulent, steady and devoided of mass forces. In these conditions, the statistical equations of motion written in cylindrical coordinates, lead to the so-called *quasi-cylindrical approximation* of Navier-Stokes equations, which can be written as follows:

$$\frac{\partial V_z}{\partial z} + \frac{1}{r}\frac{\partial r V_r}{\partial r} = 0 \tag{1}$$

$$V_z \frac{\partial v_z}{\partial z} + V_r \frac{\partial V_z}{\partial r} = -\frac{\partial p/\rho}{\partial z} + \frac{1}{r}\frac{\partial}{\partial r}\left(r\nu \frac{\partial V_z}{\partial r} - r\overline{v_z' v_r'} \right) \tag{2}$$

$$\frac{V_\theta^2}{r} = \frac{\partial p/\rho}{\partial r} \tag{3}$$

$$v_z \frac{\partial V_\theta}{\partial z} + V_r \frac{\partial V_\theta}{\partial r} + \frac{V_r V_\theta}{r} = \frac{1}{r^2}\frac{\partial}{\partial r}\left(\nu r^3 \frac{\partial}{\partial r}\left(\frac{V_\theta}{r}\right) - r^2\overline{v_r' v_\theta'} \right) \tag{4}$$

This system of equations involve the Reynolds stresses $< \mathbf{v'} \otimes \mathbf{v'} >= \overline{v'v'}$ which is the time averaged tensor product of the velocity fluctuations **v'**. It is very important to note that the Reynolds terms are obtained through explicit calculations from the non-linearity of the Navier-stokes equations. This system, completed by the initial and boundary conditions is underdetermined. It is therefore necessary to insert a closure equation.

2.1. TURBULENCE MODELS

To close the system (1)-(4) two turbulence models have been considered:
- an algebraic model, with anisotropic mixing lentgh
- a model using two transport equations, the Algebraic Stress Model (A.S.M.).

2.1.1. Anisotropic mixing length model

The anisotropic mixing length model is based upon the introduction of an anisotropic turbulent viscosity $\nu_{t_{ij}}$ such that:

$$\overline{v_i'v_j'} = -2\nu_{t_{ij}} S_{ij}$$

In the context of the quasi-cylindrical approximation, we obtain the two correlations arising in the momentum equations:

$$\overline{v_z'v_r'} = -\nu_{t_{zr}} \frac{\partial V_z}{\partial r}$$

$$\overline{v_r'v_\theta'} = -\nu_{t_{r\theta}} \frac{r\partial}{\partial r}\left(\frac{V_\theta}{r}\right)$$

The expression of the turbulent viscosity ν_t relies on a length scale which must be properly chosen . In the case of the model under consideration, two characteristic lengths are introduced: δ_ϵ and r_c. For the axial-radial correlation $\overline{v_z'v_r'}$, the scale δ_ϵ which represents the value of the radius r such that:

$$V_z(r = \delta_\epsilon) = K V_{\text{ext}}$$

where$K$ is a constant very close to one, for example $K = 0.99$, and $V_{\text{ext}} = V_z(r \gg \delta_\epsilon)$. For the radial-tangential correlation $\overline{v_r'v_\theta'}$, the scale r_c which represents the value of the radius where the value of V_θ is maximum.

By analogy with a classical mixing length model, we will write

$$\nu_{t_{zr}} = 2(K_{zr}\delta_\epsilon)^2 \left|S_{zr}\right|$$

$$\nu_{t_{r\theta}} = 2(K_{r\theta}r_c)^2 \left|S_{r\theta}\right|$$

where K_{zr} and $K_{r\theta}$ are two constants of the model.

2.1.2. Algebraic Stress Model

The algebraic stress model is an intermediate closure model, between first order $[k, \epsilon]$ models and the second order model in which a transport equation has to be solved for all components of the Reynolds stress tensor. It relies on the evolution equations of second order moment.

The Reynolds stress components can be determined in a simple way proposed by Launder (1971) then Rodi (1972), without solving an elaborate algebraic system.

This classical approach consists in writing the transport equation of the correlation $\overline{v_i'v_j'}$ as follows [5]

$$\frac{D\overline{v_i'v_j'}}{Dt} = P_{ij} + \Phi_{ij} - \epsilon_{ij} + D_{ij}$$

where P_{ij} is the production term, Φ_{ij} the redistribution by the pressure field, ϵ_{ij} the dissipation, D_{ij} the diffusion, and $\frac{D.}{Dt} = \frac{\partial.}{\partial t} + V_k \frac{\partial.}{\partial x_k}$ denotes the total differentiation along the mean flow.

The left term of the above equation, shows again the variation rate of the turbulent stress during the mean motion. It does not introduce new unknowns; so it can be computed accurately. It is the same for P_{ij} which yields the production of $\overline{v'_i v'_j}$, by the mean kinematic gradients.

Finally, the terms to model are: the pressure-deformation correlations, the dissipation and the diffusion. The expressions of these different terms are given here without proofs; we refer to [6] for details.

1- The production term need not be approximated; it has the expression:

$$-P_{ij} = \overline{v'_i v'_k}\frac{\partial V_j}{\partial x_k} + \overline{v'_j v'_k}\frac{\partial V_i}{\partial x_k}$$

2- The term of pressure-deformation correlation is classically splitted in two terms $\Phi_{ij,1}$ and $\Phi_{ij,2}$, related respectively to the turbulence itself, and to the linear interaction of the mean kinematic gradients with velocity fluctuations:

$\Phi_{ij,1}$ is approximated assuming a linear comeback to isotropy

$$\Phi_{ij,1} = -C_1\frac{\epsilon}{k}\left(\overline{v'_i v'_j} - \frac{2}{3}\delta_{ij}k\right)$$

$\Phi_{ij,2}$ is approximated assuming that the production is isotropic

$$\Phi_{ij,2} = -C_2\left(P_{ij} - \frac{1}{3}\delta_{ij}P_{kk}\right)$$

3- The dissipation term is given by $\frac{2}{3}\epsilon\delta_{ij}$.

4- the triple correlation terms $\overline{pv'_i}$ are obtained through a diffusion model

$$D_{ij} = \mathrm{Diff}\left(\overline{v'_i v'_j}\right) = C_s\frac{\partial}{\partial x_k}\left(\frac{k}{\epsilon}\overline{v'_k v'_l}\frac{\partial\overline{v'_i v'_j}}{\partial x_l}\right)$$

where Diff indicates the diffusion operator.

Finally, the algebraic stress model proposes to connect the kinetic energy of turbulence to Reynolds stresses transport (convection-diffusion), by using the two following relations:

$$\frac{D\overline{v'_i v'_j}}{Dt} = \frac{\overline{v'_i v'_j}}{k}\frac{Dk}{Dt}.$$

$$\mathrm{Diff}\left(\overline{v'_i v'_j}\right) = \frac{\overline{v'_i v'_j}}{k}\mathrm{Diff}(k)$$

From these two relations, we obtain the following equation:

$$\frac{D\overline{v'_i v'_j}}{Dt} - \mathrm{Diff}\left(\overline{v'_i v'_j}\right) = \left(\frac{Dk}{Dt} - \mathrm{Diff}(k)\right)\frac{\overline{v'_i v'_j}}{k}$$

This simplification is based on the assumption of *small variation of rates* $\frac{\overline{v_i'v_j'}}{k}$ in all the flow.

This assumption, according to which the turbulence varies only in total level and keeps its anisotropy in all the field, seems quite simplistic. However it is very convenient because it allows to cancel the *redistribution terms*, by infering , from the turbulent kinetic energy transport equation, the relation:

$$\frac{Dk}{Dt} - \mathrm{Diff}(k) = \frac{P_{kk}}{2} - \epsilon$$

whence

$$\frac{D\overline{v_i'v_j'}}{Dt} - \mathrm{Diff}\left(\overline{v_i'v_j'}\right) = \left(\frac{P_{kk}}{2} - \epsilon\right)\frac{\overline{v_i'v_j'}}{k}$$

The transport equations become algebraic.

Altogether, these assumptions lead to the algebric stress model formulation, which is obtained easily with the help of the above relations:

$$\frac{\overline{v_i'v_j'}}{k} - \frac{2}{3}\delta_{ij} = \frac{\Lambda}{\epsilon}(P_{ij} - \frac{P_{kk}}{3}\delta_{ij} \tag{5}$$

where

$$\Lambda = \frac{1 - C_2}{C_1 - 1 + \frac{P_{kk}}{2\epsilon}}$$

This system must be supplemented with the transport equations for k and ϵ.

The evolution equation for k is deduced from the Reynolds stress equations. through the summation of the diagonal terms from the Reynolds stress equations and we obtain (see [7])

$$\frac{Dk}{Dt} = \frac{P_{kk}}{2} - \epsilon + C_s\frac{\partial}{\partial x_l}\left(\frac{k^2}{\epsilon}\frac{\partial k}{\partial x_l}\right)$$

$$\frac{D\epsilon}{Dt} = \frac{\partial}{\partial x_j}\left(C_\epsilon\frac{k^2}{\epsilon}\frac{\partial \epsilon}{\partial x_j}\right) + C_{\epsilon 1}\frac{P_{kk}}{2}\frac{\epsilon}{k} - C_{\epsilon 2}\frac{\epsilon^2}{k}$$

The values chosen for the different constants are:

$$C_1 = 1.8; C_2 = 0.6; C_S = 0.2; C_\epsilon = 0.06; C_{\epsilon 1} = 1.44; C_{\epsilon 2} = 1.9$$

The third approach is a method based on the unsteady Navier-Stokes equations without quasi-cylindrical approximation.

Our goal has been to develop a computation code solving the Navier-Stokes equations in an incompressible axisymmetrical turbulent flow, derived from an existing code which solves the same equations in laminar flow. Initially, this code has been developped in order to treat the confined vortex case in a cylindrical pipe [4], then extended in the case when the vortices are located in an unbounded flow and submitted to an adverse pressure gradient susceptible to burst them [2]. It introduces new dependent variables: ω the mean curl component normal to the meridian plane, the pseudo mean stream function ψ in meridian motion, the mean circulation Γ as well as $\Omega = \omega/r$ and $\phi = \psi/r^2$.

It has been decided to develop a turbulent version of this code using the same variables. The argument used to work out the equations, as shown below, stems from a choice to write the stream function-vorticity formulation of the time averaged planar Navier-Stokes equations in cartesian coordinates [8]. In this case, we use an algebraic turbulence model involving a single length scale. In order to make the model isotropic, a magnitude has been searched, linking the transversal and longitudinal motions. This led us to the following choice of turbulent viscosity in terms of the vorticity

$$\nu_t = K_t r_c^2 |\mathbf{curl\,V}|$$

where K_t is a constant and r_c represents the radius where the tangential velocity is maximum. This model enables to write the Navier-Stokes equations as an unsteady system of three equations with three unknowns (Ω,ψ,Γ) [12]:

$$\frac{D\widetilde{\Gamma}}{D\widetilde{t}} = \frac{1}{R_0}\left(\widetilde{\nu}\Delta^{(1)}\widetilde{\Gamma} + \widetilde{S}_\Gamma\right) \tag{6}$$

$$\frac{\partial\widetilde{\Omega}}{\partial\widetilde{t}} + \frac{1}{\widetilde{r}}\frac{\partial\widetilde{\psi}}{\partial\widetilde{r}}\frac{\partial\widetilde{\Omega}}{\partial\widetilde{z}} - \frac{1}{\widetilde{r}}\frac{\partial\widetilde{\psi}}{\partial\widetilde{z}}\frac{\partial\widetilde{\Omega}}{\partial\widetilde{r}} = \frac{\partial}{\partial\widetilde{z}}\left(\frac{\widetilde{\Gamma}^2}{\widetilde{r}^4}\right) + \frac{1}{R_0}\left(\frac{\Delta\left(\widetilde{\nu}\widetilde{r}\widetilde{\Omega}\right)}{\widetilde{r}} - \widetilde{\nu}\frac{\widetilde{\Omega}}{\widetilde{r}^2} + \frac{\widetilde{S}_\Omega}{\widetilde{r}}\right) \tag{7}$$

$$-\widetilde{\Omega} = \frac{1}{\widetilde{r}}\frac{\partial}{\partial\widetilde{r}}\left(\widetilde{r}^3\frac{\partial\widetilde{\varphi}}{\partial\widetilde{r}}\right) + \frac{\partial^2\widetilde{\varphi}}{\partial\widetilde{z}^2} \tag{8}$$

The notation $\widetilde{\ }$ denotes dimensionless variables, $\Delta^{(1)}$ is the Laplace operator in the $(\widetilde{z},\widetilde{r})$ plane, R_0 is the Reynolds number based on $U-0$ and r_0 and

$$\widetilde{S}_\Gamma = \frac{\partial\widetilde{\Gamma}}{\partial\widetilde{z}}\frac{\partial\widetilde{\nu}}{\partial\widetilde{z}} + \left(\frac{\partial\widetilde{\Gamma}}{\partial\widetilde{r}} - \frac{2\widetilde{\Gamma}}{\widetilde{r}}\right)\frac{\partial\widetilde{\nu}}{\partial\widetilde{r}}$$

and

$$\widetilde{S}_\Omega = \left(-\frac{2}{\widetilde{r}^2}\frac{\partial\widetilde{\psi}}{\partial\widetilde{r}} + \frac{2}{\widetilde{r}}\frac{\partial^2\widetilde{\psi}}{\partial\widetilde{r}^2}\right)\frac{\partial^2\widetilde{\nu}}{\partial\widetilde{z}^2} + \left(\frac{2}{\widetilde{r}}\frac{\partial^2\widetilde{\psi}}{\partial\widetilde{z}^2}\frac{\partial^2\widetilde{\nu}}{\partial\widetilde{r}^2}\right) + 2\frac{\partial^2\widetilde{\nu}}{\partial\widetilde{z}\partial\widetilde{r}}\left(\frac{1}{\widetilde{r}^2}\frac{\partial\widetilde{\psi}}{\partial\widetilde{z}} - \frac{2}{\widetilde{r}}\frac{\partial^2\widetilde{\psi}}{\partial\widetilde{z}\partial\widetilde{r}}\right).$$

2.2. SOLUTION MODES

The system (1)-(4) will be solved by a finite difference method, using an implicit iterative space marching procedure along the longitudinal direction [9]. This scheme is accurate to second order for a uniform mesh. Now, as Benay has shown [3] (1979, 1980), there are two possible solution modes according to the type of boundary conditions which are imposed:

2.2.1. Direct mode

The first solution, which is called direct mode, consists in writing the external axial component of the velocity on the boundary of the computational domain

($V_z = We(z)$), where the flow is assumed to be irrotational and non viscous. This method does not require informations on the viscous flow internal structure. However, it seems that in the case where the vortex moves towards the breakdown, the numerical computation does not converge when approaching the point where the axial component of longitudinal velocity is going to cancel out.

At this location the breakdown can be explained by a Goldstein type singularity [10], observed in the boundary layer equations at the separation point.

A.S.M. case

As mentioned in section 2.1.2, the A.S.M. is constitued by the system (1)-(4), supplemented by two transport equations for k, ϵ, and the six closure non linear algebraic relations resulting from the projection of the equation (5) onto the axial, radial and tangential directions.

The linearization and the discretization of momentum equations, lead to a tridiagonal linear system, which is solved by Thomas algorithm. Then, we deduce k, ϵ and the Reynolds stresses by a Newton method.

The initial conditions for both the fluctuations and the mean flow velocity are derived from an experimental work of Phillips and Graham [11]. These data were obtained for weak vorticity field by means of wire anemometry techniques.

For the dissipation rate ϵ, we have made the assumption that the transport by convection of the kinetic energy k is equal to its dissipation through diffusion:

$$\frac{Dk}{Dt} = \mathrm{Diff}(k).$$

This assumption involves the isotropy of the turbulence in the inlet zone, and the fact that the internal turbulence production is balanced by the dissipation of this turbulence ($\frac{P_{kk}}{2} = \epsilon$).

The following boundary conditions are imposed:

<u>on the axis</u> ($r = 0$): $V_\theta = V_r = 0$

<u>on the external boundary</u> ($r = R$): $V_\theta = \frac{\Gamma_0}{R}, V_z = We(z)$

where Γ_0 is the potential flow circulation, which is assumed to be constant, and $We(z)$ is a velocity law deduced from the experiment.

In the external zone, in principle, k and ϵ tend to zero. These conditions lead to numerical difficulties, since some quantities tend to infinity. To avoid this problem, we replace $k = \epsilon = 0$ by the weaker conditions $\frac{\partial k}{\partial r} = \frac{\partial \epsilon}{\partial r} = 0$.

2.2.2. Inverse mode

The second method is similar to the classical inverse mode of boundary layer theory. It is based on an integral quantity representative of the vorticity evolution. This quantity can be either a displacement thickness or the longitudinal velocity component on the vortex axis. If in first approximation we suppose that the data of the axial velocity on the axis summarizes the influence of the downstream on the upstream, the lack of informations resulting from the parabolization and which leads to the singular behaviour of direct mode will be compensated by the adoption of W on the z-axis, as boundary conditions for the axial velocity. This approach is similar to the one used to compute the separated boundary layers.

In contrast with the diresct mode, the inverse mode must allow to approach better the breakdown, and go beyond the singularity at the breakdown point.

Mixing length type algebraic turbulence model

The approach which consists in writing down the longitudinal velocity component on the axis has been chosen. To solve the system (1)-(4) we use algorithm.

A.S.M. case

A method working in inverse mode has been improved [12]. Some important modifications have to be made in the code derived for the direct mode.

Indeed, because of the inverse type boundary conditions which consist in imposing the axial velocity on the axis, $V_z = Waxe(z)$, it is not possible, in contrast with the direct mode, to compute the external pressure and to obtain the whole pressure field.

So, it is necessary to write down a relation allowing the pressure computation on the revolution axis. For this, the axial and radial momentum equations have been solved together, but separately from the continuity and tangential momentum equations. After linearization and discretization of the axial and radial momentum transport equations, the two remaining unknowns are the axial velocity V_z and the pressure p. Thomas algorithm leads to a block tridiagonal linear system.
Next, an approach similar to the one followed in direct mode allows to determine the radial velocity V_r, the dissipation ϵ, the specific turbulent kinetic energy k and the six components of the Reynolds stress tensor.

With the exception of the axial velocity data on the axis, the boundary and initial conditions (Phillips-Graham experiments) are similar to the direct mode case.

2.2.3. Solution mode of the method based on the unsteady Navier-Stokes equations

The complete Navier-Stokes equations represented by the system (6)-(8) has been solved by a conventional "time-marching" method , begining with a more or less arbitrary initial state and proceeding until a steady state (if one exists) is reached [4].

The numerical scheme is an implicit second order finite difference scheme. A better accuracy can be obtained by using an adapted mesh. The terms $\widetilde{S}_\Gamma$ and $\widetilde{S}_\Omega$ which result from the introduction of the turbulence are dealt with explicitely, and the operator $\Delta^{(1)}$ relative to the Γ equation is discretized implicitely by centered finite differences. The non linear terms of the convection operator $\frac{D}{D\tilde{t}}$ are treated as follows: the velocity is computed explicitely; the first derivatives are dealt with implicitely and discretized by an upstream scheme.

At time $\tilde{t} = n\delta\tilde{t}$ the following system of three equations is obtained

$$[I + \delta\tilde{t}A]\,\widetilde{\Gamma}^{(n)} = \widetilde{\Gamma}^{(n-1)} + \delta\tilde{t}\left[\widetilde{S}_\Gamma\right]^{(n-1)}$$

$$[I + \delta\tilde{t}B]\,\widetilde{\Omega}^{(n)} = \widetilde{\Omega}^{(n-1)} + \delta\tilde{t}\left[\frac{\partial}{\partial \tilde{z}}\left(\frac{\widetilde{\Gamma}^2}{\tilde{r}^4}\right)\right]^{(n)} + \delta\tilde{t}\left[\frac{\widetilde{S_\Omega}}{\tilde{r}}\right]^{(n-1)}$$

$$[C]\,\tilde{\varphi}^{(n)} = -\widetilde{\Omega}^{(n-1)}$$

where A, B, and C are $M \times N$ matrices (M and N are respectively the number of grid points in the axial and radial direction) depending on the axial and radial velocity at preceding time step, and on the radius.

The computation is made in a rectangular domain, bounded on the one hand by an inlet and an exit cross-sections, and on another hand by an external boundary and the axis of symmetry. In the inlet section, the profiles of mean velocity from Phillips-Graham are imposed. In the exit section, the flow is assumed to be axially symmetric.

On the external boundary, we must impose a particular distribution of axial velocity. But it is easier to take a particular distribution of pressure and to deduce the velocity by the Bernoulli relation. The chosen distribution is given by [2]:

$$K_p(r_0, z) = \frac{\Delta K_p}{2}\left(1 + \tanh\left\{\frac{2}{\Delta K_p}\frac{dK_p}{dz}|_{\max}(z - z_0)\right\}\right)$$

where ΔK_p is the total jump of pressure between the upstream infinity and the downstream infinity and $\frac{dK_p}{dz}|_{\max}$ denotes the maximum value of the pressure gradient, obtained at $z = z_0$.

On the other hand, we also impose that the circulation remains radially constant. On the axis of symmetry, the conditions are written as follows:

$$V_r(0, z) = V_\theta(0, z) = \frac{\partial V_z}{\partial r}(0, z) = 0$$

which leads to:

$$\frac{\partial \varphi}{\partial r}(0, z) = \frac{\partial \Omega}{\partial r}(0, z) = \Gamma(0, z) = 0.$$

3. Applications

3.1. INITIALIZATION OF THE CODES

Figure 1 shows an analytical representation of the initial profiles for the azimuthal and longitudinal components of the velocity. These profiles are taken from the Phillips-Graham experiments [11] and they allow to initialize the various computational codes.

In these experiments, the vortex is produced by two identical blades set at equal and opposite (as viewed from the axis of symmetry) angles of incidence and connected to a cylindrical central body. The chord of the blades is $c_0 = 0.0508m$, and the upstream velocity is equal to $U_0 = 20.6m/s$. For a good adjustment of computation, it is better to take smoothed experimental profiles since it is not always easy to use the original experimental datas. For this reason, we have chosen to impose, in the inlet zone, distributions of mean velocity profiles similar to those described by Batchelor [13] as a solution of Navier-Stokes equations in the case of a vortical wake, which agree with the experimental profiles. So, we have taken:

$$\frac{V_\theta(r)}{U_0} = \frac{\Gamma_0/U_0}{r}\left(1 - \exp\left[-1.12\left[\frac{r}{r_c}\right]^2\right]\right)$$

$$\frac{V_z(r)}{U_0} = 1 + (V_{z_0} - 1)\exp\left[-1.12\left[\frac{r}{r_c}\right]^2\right]$$

where r_c denotes the value of the radius where the velocity is maximum, V_{z_0} is the axial velocity on the axis, in the inlet zone, divided by U_0. A similar approach has led to take analytical profiles for the six components of the Reynolds stress tensor; this is shown in figure 2.

These profiles are written as follows:

- if $i = j$

$$\left(\overline{v_i' v_j'}\right)^{1/2} = U_1 \exp\left(-\left[\frac{r - r_0}{r_1}\right]^2\right) + U_2 \exp\left(-\left[\frac{r - r_0}{r_2}\right]^2\right)$$

- if $i \neq j$

$$\overline{v_i' v_j'} = U_1 \exp\left(-\left[\frac{r - r_0}{r_1}\right]^2\right) + U_2 r \exp\left(-\left[\frac{r - r_0}{r_2}\right]^2\right)$$

The various coefficients U_1, U_2, r_0, r_1, r_2 are calculated by a least square method.

3.2. DIRECT METHOD- A.S.M CASE - VORTEX UNBURST

In this paragraph a comparison is shown between the computation in direct mode using the A.S.M., and the experimental results for the cross-sections $z = 1.68m$ and $z = 3.25m$.

In the cross-section $z = 1.68m$, the computed and experimental profiles of tangential velocity are very similar as seen on figure 3A. The agreement is not as good for the profiles of axial velocity; indeed there is a slight velocity defecit on the axis, and in the region where the shear is maximum the curve does not agree so well with the experiment; this results from a poor prediction of the A.S.M. method for the axial shear. This discrepancy is explained by the scale chosen on the graph (see figure 3B).

Figure 4 shows the profiles of tangential and axial velocity for the cross-section $z = 3.25m$. The computed tangential velocity agrees with the experiment, with however an over-valuation of tangential velocity maximum and of viscous radius which designates the region where the rotation is of solid body type. Then, there is a slight discrepancy which disappears very quickly when the potential region is reached; this is shown in figure 4A.

Figure 4B shows a good agreement between the numerical simulations and the experiment. Moreover, the recession of the axial velocity on the axis, in comparison with this same velocity on the preceding cross-section $z = 1.68$ m, is in good agreement with experiment.

Figure 5 shows the six components of the Reynolds stress tensor in the cross-section $z = 1.68m$. The computed location of the maximum of $\overline{v_r' v_\theta'}$ is fairly well predicted, despite the lack of accuracy resulting from the smallness of this correlation. The computation-experiment agreement is also good for the correlation $\overline{v_r' v_z'}$.

Concerning the correlation $\overline{v'_\theta v'_z}$, the computation shows a small over-valuation from $r = 0m$ up to $r = 0.10m$; further on, the computation gives results identical to the experiment.

As for to the three variances connected to the velocity fluctuations, the profiles of $Var(v'_r)$ and $Var(v'_\theta)$ are nearly identical for the computation and for the experiment. But, in the radial region expanding form $r = 0m$, to $r = 0.10m$, there is an over-evaluation of computation for the three components of the Reynolds stress tensor. However, we can point out that this model represents qualitatively very well the anisotropy of the turbulence production.

Figure 6 still shows the six components of the Reynolds stress tensor, but in the cross-section $z = 3.25m$. The radial-tangential correlation is nearly zero, as we can see on the experiment; the experimental points are few here and near zero. Concerning the radial-axial and tangential-axial correlations, the computation predicts very well their behaviour in all the analytical domain. The variances for the radial and tangential velocity fluctuations (normal stresses) have still the same behaviour; we observe a decrease of the level of these variances, exactly like for the axial velocity fluctuation, in comparison with the results obtained for the cross-section $z = 1.68m$. The computation of these three components still give an over estimation of the experimental values.

3.3. INVERSE MODE - A.S.M. CASE - VORTEX UNBURST

Here we compare the experimental datas to the results of a code based on inverse mode with A.S.M. still for the cross-sections $z = 1.68m$ and $z = 3.25m$.

In figure 7, there is a good agreement between computations and experiments , for the tangential velocity in the cross-section $z = 1.68m$. On the other hand, the error in the velocity profile due to a poor evaluation of the external velocity (figure 7B), which in turn results from the fact that the axial velocity is given in inverse mode. Concerning the cross-section $z = 3.25m$, the tangential velocity prediction in inverse mode is the same as in direct mode (figure 8A).

In figure 8B, the calculated profile for the axial velocity coincides much better with the experiment than the one previously computed, in particular in the external region where the axial velocity is well connected to the external axial velocity. The results given by the computation in inverse mode are in agreement with the computations in direct mode. Similarily, concerning the Reynolds stress tensor, the computation gives the same results in inverse and direct mode.

3.4. INVERSE MODE - TWO MIXING LENGTHS MODEL - VORTEX UNBURST

The following applications concern a comparison between a code working in inverse mode and using an anisotropic algebraic turbulence model, and the previous experimental results. Figure 9 shows the tangential and axial velocities in the cross-section $z = 1.68m$. In the viscous region, we note an over estimate of the tangential velocity which disappears when the viscous radius is exceeded, as we can see in Figure 9A. Figure 9B shows a good agreement between the computed and experimental axial velocities. In the cross-section $z = 3.25m$, the accuracy is better than in the previous cut, even in the viscous region. One can also observe a recession of tangential velocity maximum (Figure 10A). Figure 10B shows a good profile of the axial velocity calculated with a slight decrease of this velocity on the axis, compared to the one observed in the cut $z = 1.68m$, which is in agreement with the experiment.

The use of an anisotropic turbulence model using two mixing lengths allows to calculate only two correlations, the radial-tangential and the radial-axial correlations. Figure 11 shows the latter for the cross-section $z = 1.68m$. About the correlation $\overline{v'_r.v'_\theta}$, we can note a very important over estimate of the computation. Concerning the correlation $\overline{v'_r.v'_z}$, the computation is in agreement with the experiment. In the cross-section $z = 3.25m$ the computation still overestimates the correlation $\overline{v'_r.v'_\theta}$ whereas it is satisfactory for $\overline{v'_r.v'_z}$ (see Figure 12).

3.5. COMPARISON OF RESULTS OF A TURBULENT UNSTEADY NAVIER-STOKES CODE WITH THE EXPERIMENTAL RESULTS OF PHILLIPS-GRAHAM

In the first place the turbulence model is tested against the experimental velocity profiles, in the reference case of Phillips-Graham corresponding to a uniform external pressure ($K_p\,(r_0, z) = 0$).

Figure 13 shows the calculated and experimental profiles for the tangential and axial component of the velocity in the cut $z = 1.68m$. The agreement is very good between the computation and the experiment in particular for the tangential velocity field, with however a slight under estimation in the region where the shear is maximum. The axial velocity field tends correctly to the experimental external axial velocity. The the turbulence model constant K_t has been adjusted to the studied flow; it has been taken equal to 10^{-2}. In the cross-section $z = 3.25m$ (see figure 14), the code produces a slight over estimation of the tangential velocity in the transition region located between the viscous core and the potential flow zone. We also note a decrease of tangential velocity maximum compared to the previous cut and an increase of the viscous radius r_c. Concerning the axial velocity in the cross-section $z = 3.25\ m$, there is a good agreement between the computation and the experiment, with in the meantime a slight decrease of the axial velocity on the axis as compared to the cross-section $z = 1.68m$. This is in agreement with the experiment (see figure 14).

The determination of fluctuating quantities calculated with the help of velocity gradients proves to be very difficult owing to the stream function-vorticity formulation used for the computation. Indeed, the double derivatives of the stream-function required in this calculation introduce a strong error on the estimate of gradients Therefore the efficiency of turbulence model has only been tested on the good agreement which it yields for the velocity profiles.

3.5.1. Evolution of the axial velocity on the axis for two values of V_{z0}

We now turn to a different flow, with a Reynolds number of 170. The boundary conditions are the same as previously with the exception of tangential and axial velocity profiles at the inlet boundary of the computational domain which are still given by

the analytical form, but with the following parameter values: $\Gamma_0 = 0.3\ r_0 U_0$, $r_c = 0.289\ r_0$, and $V_{z_0} = 1.2$. Concerning the parameters governing the evolution of the external pressure, they can be written as follows: $\Delta K_p = 0.4$, $\frac{dK_p}{dz}|_{\max} = 0.2$ and $z_0 = 4r_0$.
The constant for the mixing length model K_t has been chosen equal to 10^{-4}. Figure 15 shows the evolution for the axial component of the velocity, $V_{z_{axe}}$, when it is taken equal to 1.20 ($V_{z_0} = 1.20$) in the inlet section. This component, under the effect of adverse pressure gradient, becomes negative, a fact which characterizes the so-called vortex breakdown point. From this point, there is a recirculation zone, close to the domain located between $\frac{z}{r_0} = 2.75$ and $\frac{z}{r_0} = 6$ where the pressure gradient is imposed; this zone has been visualized in Figure 16 as iso-ψ plots in the meridian plane. The flow shows an upstream zone which is roughly cylindrical; then the streamlines of the meridian plane progressively diverge, and outline a recirculation bubble characterized by negative values of ψ. Downstream this recirculation zone, the axial velocity on the axis becomes again positive, but remains smaller than that observed upstream of the breakdown, as seen on figure 15. Now, we are in a region which is again becomes cylindrical; the recirculation bubble is closed again, in $\frac{z}{r_0} = 6$ as shown on figure 16. In the next case, V_{z_0} has been fixed to 1.15, while keeping constant all the other parameters. The breakdown point is pushed upstream, and the recirculation zone is stretched in the upstream direction (Figures 17 and 18). So, when the turbulence is strongly under-estimated, this indicates that the vortex bursts under the effect of pressure gradient, with an important influence of the upstream axial velocity profile on the bubble stretching.

3.5.2. Flow stability under the effects of the increase of the turbulent viscosity

Next we investigate the influence of a realistic turbulence on the vortex breakdown. For this, we take again the first case above studied, now with $K_t = 10^{-2}$. We immediately note that the imposed pressure gradient does not produce the formation of a recirculation bubble, as shown on figure 19. So, it appears clearly from our numerical experiments that a fully developped turbulence has a stabilizing effect on vortical flows.

4. Conclusion

Two kinds of computation methods for turbulent axisymetrical vortex flows have been developped and tested. For methods using the boundary layer approximations, the Algebraic Stress Model (A.S.M.) has been introduced but its prediction quality does not appear to be better than that using an anisotropic algebraic model, with mixing length.

The direct solution of statistical equations without approximations and with the addition of an isotropic algebraic model using a mixing length is the most satisfactory. This shows that the approximations of the boundary layer, introduced in the first methods, are too drastic in order to describe correctly a vortex and to give some acceptable results.

On the other hand, the Navier-Stokes code has allowed a theoretical evaluation of the effect of developped turbulence on the vortex stability when it is submitted to a pressure gradient. A comparison of calculation results with some experimental datas allows to show the agreement between the computation and the experiment.

These remarks must be tempered by the fact that the only available experimental datas correspond to a case with a weak rate of rotation. More complete experimental datas would therefore be highly desirable in order to carry further these validations.

Acknowledgements

This work was done while the author was working at ONERA, Chatillon (France).

References

[1] Pagan D. & Solignac J.L. (1986) Etude expérimentale de l'éclatement d'un tourbillon engendré par une aile delta. La Recherche Aérospatiale 1986-3, 197-219.

[2] Pagan D. & Benay, R. (1988) Etude numérique de l'éclatement de tourbillons soumis à des gradients de pression. La Recherche Aérospatiale, 1988-1, 251-263.

[3] Solignac, J.L. & Benay, R. (1980) Etude expérimentale et théorique de la turbulence dans les écoulements décollés. Application aux sillages tourbillonnaires incompressibles. Recherches effectuées en 1979-1980. R.T. O.N.E.R.A. 18/1147 AY.

[4] Benay, R. (1984) Modélisation numérique de l'éclatement d'un tourbillon en écoulement laminaire de révolution. La Recherche Aérospatiale 1984-4,15-27.

[5] Launder B.E. (1979) Reynolds-stress closures. Status and prospects. AGARD CP 271/13 (September 1979).

[6] Escande B. (1986) Modélisation de l'interaction onde de choc/ couche limite turbulente. Comparaison calcul-expérience. Thèse de Doctorat, Ecole Centrale de Lyon.

[7] Launder B.E. (1980) Turbulence transport for computation of complex turbulent flows, V.K.I. Lecture Series 1980-3.

[8] Lazareff, M. (1980) Expériences numériques sur un écoulement turbulent décollé - Analyse des conditions de paroi. Thèse de Docteur Ingénieur, Université Pierre et Marie CURIE (Paris 6).

[9] Mirande, J., Solignac, J.L. & Benay, R. (1980-1981) Etude expérimentale de la turbulence dans les écoulements tourbillonnaires - R.T. O.N.E.R.A. 22/1147 AY.

[10] Goldstein, S. (1948) On Laminar boundary-layer flow near a position of separation- The quartely Journal of Mech. and Applied Math., vol. 1, part. 1, p43-69.

[11] Phillips, W.R.C. & Graham, J.A.H. (1984) Reynolds-stress measurements in a turbulent trailing vortex. J. Fluid Mech. vol.147, p353-371.

[12] Pascal P. (1992) Contribution à l'étude d'un écoulement tourbillonnaire en fluide incompressible. Thèse de Doctorat de l'Université Louis Pasteur de Strasbourg.

[13] Batchelor G.K. (1964) - Axial flow in trailing line vortices. J. Fluid. Mech., vol.20, p.645-658.

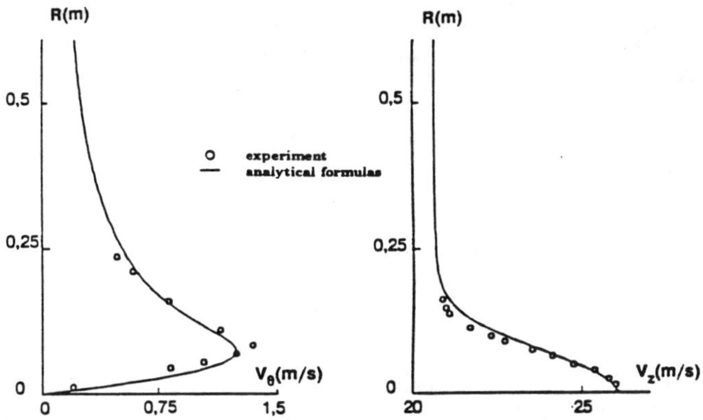

Fig.1 : Initialisation of mean flow velocity profiles, by using analytical formulas

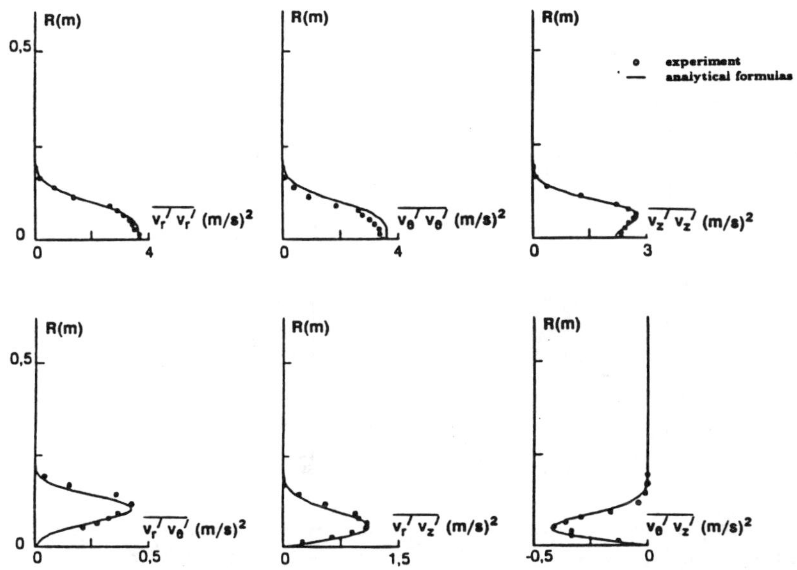

Fig.2 : Initialisation of the Reynolds stress tensor, by using analytical formulas

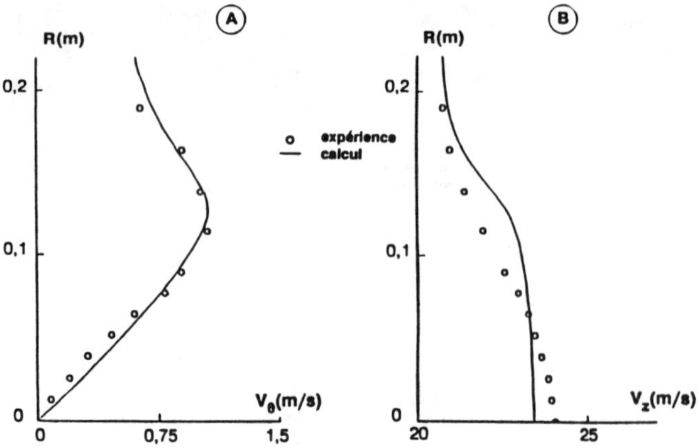

fig.3 : Comparison computation-experiment for the mean velocity field, using the direct mode, with the A.S.M., in the section z = 1.68 m.

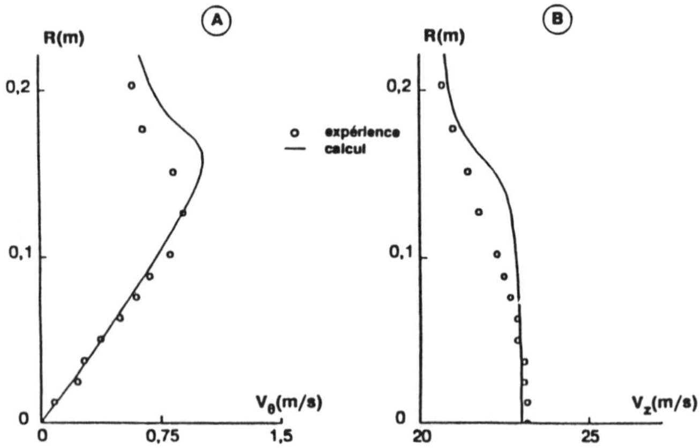

fig.4 : Comparison computation-experiment for the mean velocity field, using the direct mode, with the A.S.M., in the section z = 3.25 m.

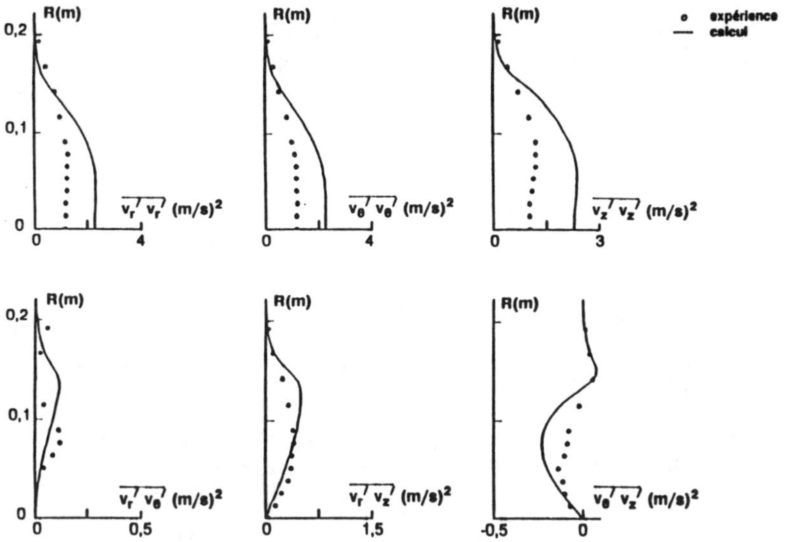

fig.5 : Comparison computation-experiment for the Reynolds stress tensor, using the direct mode, with the A.S.M., in the section z = 1.68 m.

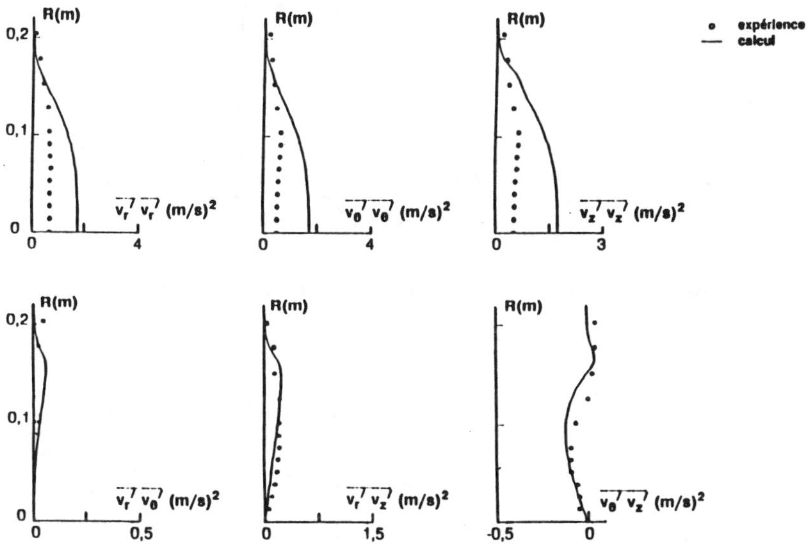

fig.6 :Comparison computation-experiment for the Reynolds stress tensor, using the direct mode, with the A.S.M., in the section z = 3.25 m.

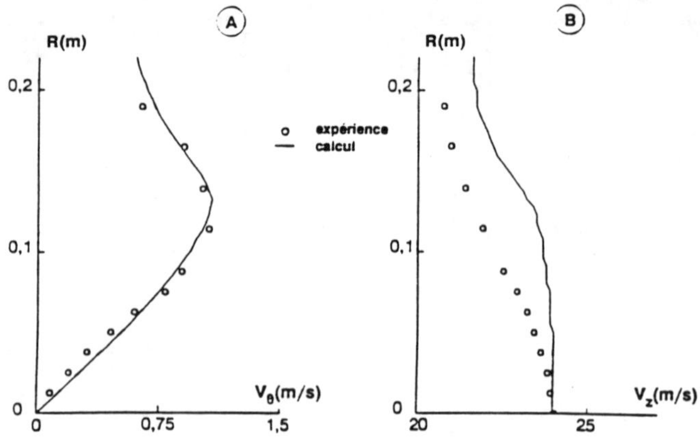

fig.7 : Comparison computation-experiment for the mean velocity field, using the inverse mode, with the A.S.M., in the section $z = 1.68$ m.

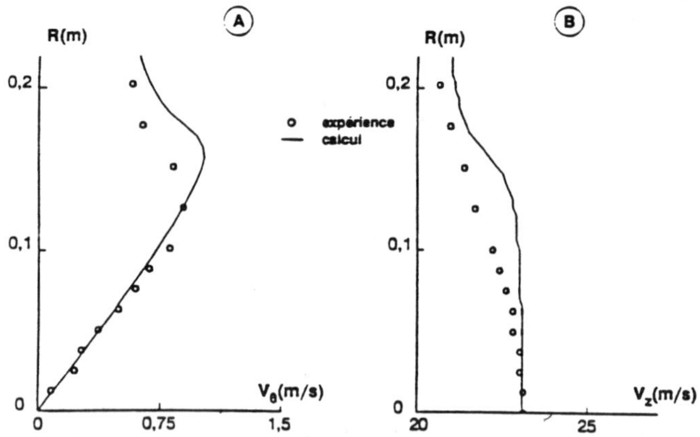

fig.8 : Comparison computation-experiment for the mean velocity field, using the inverse mode, with the A.S.M., in the section $z = 3.25$ m

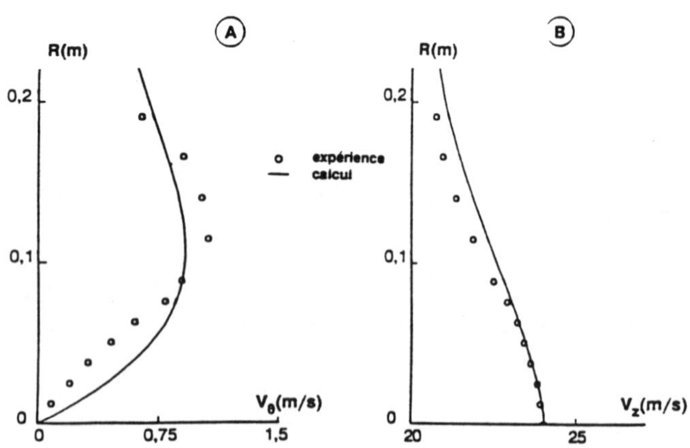

fig.9 : Comparison computation-experiment for the mean velocity field, using the inverse mode, with an algebrical model, in the section $z = 1.68$ m.

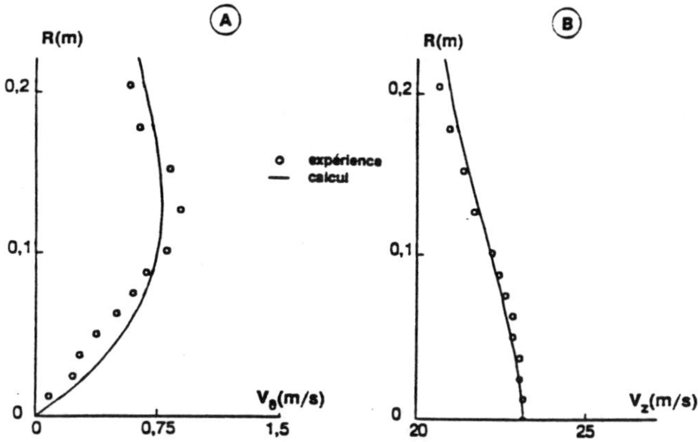

fig.10 : Comparison computation-experiment for the mean velocity field, using the inverse mode, with an algebrical model, in the section z = 3.25 m.

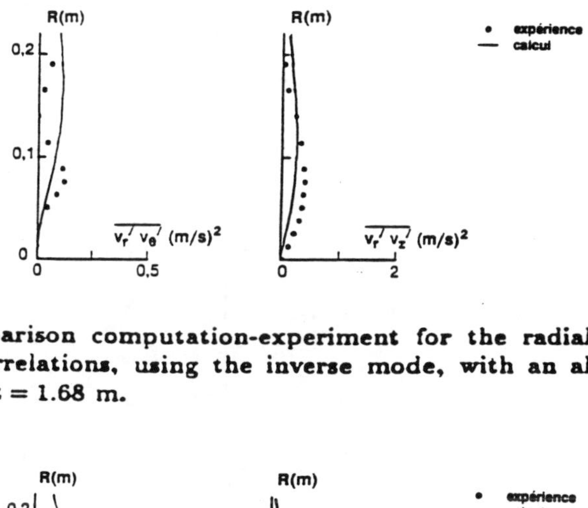

fig.11 : Comparison computation-experiment for the radial-tangential and radial-axial correlations, using the inverse mode, with an algebrical model, in the section z = 1.68 m.

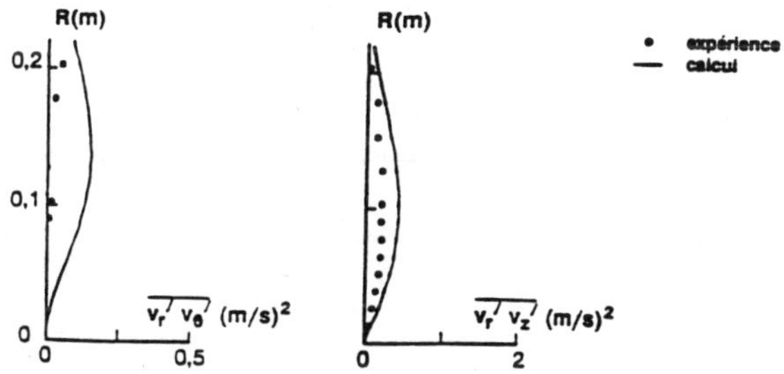

fig.12 :Comparison computation-experiment for the radial-tangential and radial-axial correlations, using the inverse mode, with an algebrical model, in the section z = 3.25 m.

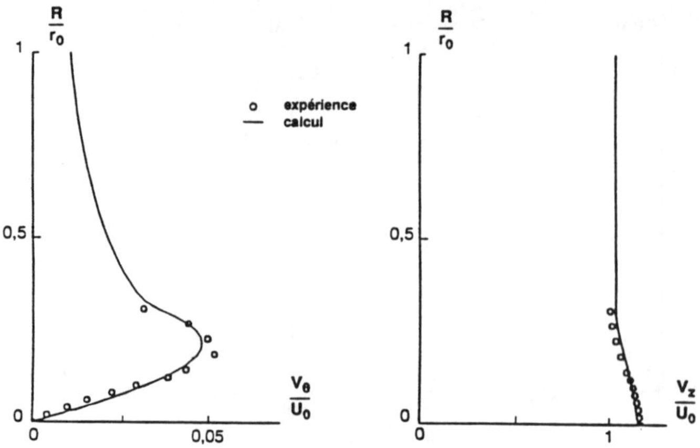

fig.13 : Comparison computation-experiment for the mean velocity field, using an unsteady Navier-Stokes code, and an algebrical model, in the section z = 1.68 m.

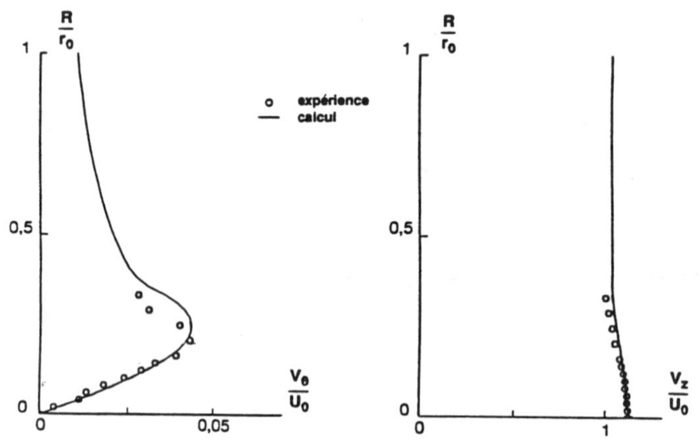

fig.14 : Comparison computation-experiment for the mean velocity field, using an unsteady Navier-Stokes code, and an algebrical model, in the section z = 3.25 m.

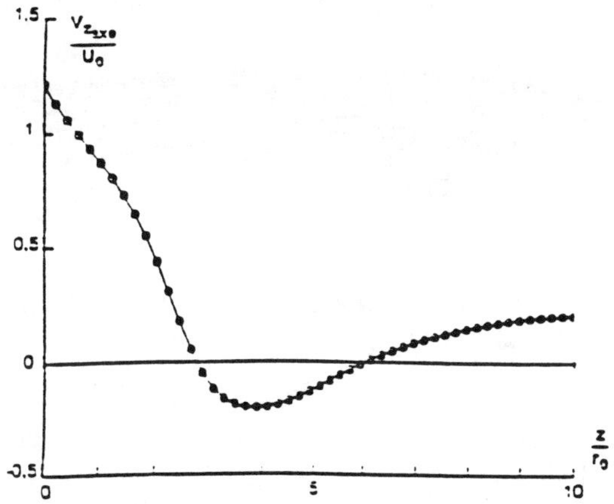

Fig.15 : Unsteady Navier-Stokes compuatation. Evolution of $V_{z_{axe}}$ in a weakly turbulent form. $K_t = 10^{-4}$ and $V_{z_0} = 1.2$

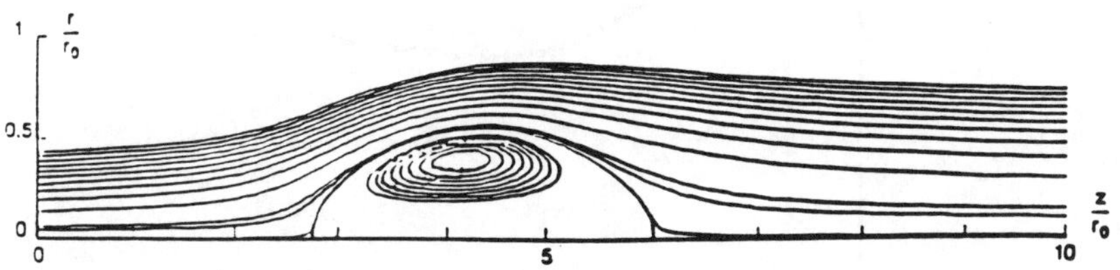

Fig.16 : Unsteady Navier-Stokes computation - Iso-ω lines plot in the plane (r,z) for $K_t = 10^{-4}$ and $V_{z_0} = 1.2$

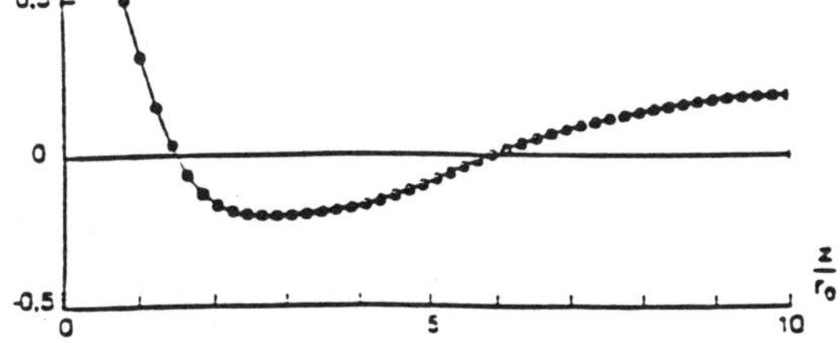

Fig.17 : Unsteady Navier-Stokes compuatation. Evolution of $V_{z_{axe}}$ in a weakly turbulent form, $K_t = 10^{-4}$ and $V_{z_0} = 1.15$

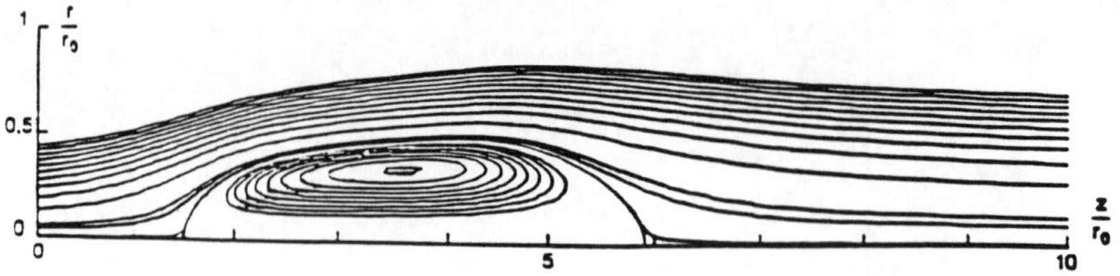

Fig.18 : Unsteady Navier-Stokes computation - Iso-ψ lines plot in the plane r,z; for $K_t = 10^{-4}$ and $V_{z_0} = 1,15$

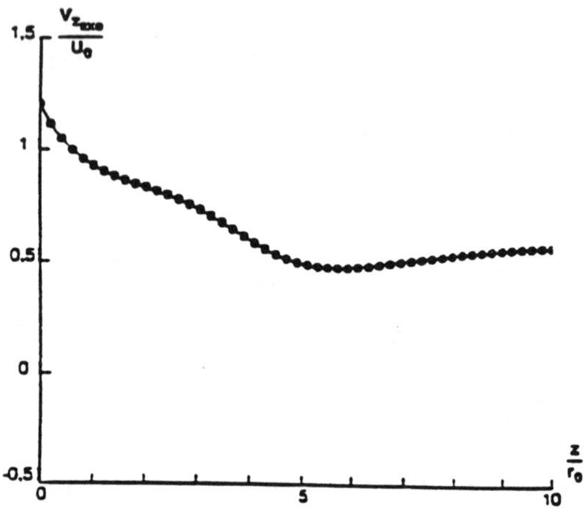

Fig.19 : Unsteady Navier-Stokes compuatation. Evolution of $V_{z_{axe}}$ in a weakly turbulent form, $K_t = 10^{-2}$ and $V_{z_0} = 1.2$

INVESTIGATION OF THE USE OF PRANDTL/NAVIER-STOKES EQUATION PROCEDURES FOR TWO-DIMENSIONAL INCOMPRESSIBLE FLOWS

CHRISTOPHER R. ANDERSON, and MARC REIDER
Department of Mathematics
UCLA
Los Angeles California 90024 USA

ABSTRACT. We investigate the technique of combining solutions of the Prandtl equations with solutions of the Navier-Stokes equations to compute incompressible flow around two dimensional bodies. We present computational evidence which shows that if the "obvious" coupling is used to combine the solutions, then the resulting solution is not accurate. We describe an alternate coupling procedure which greatly improves the accuracy of the solutions obtained with the combined equation approach.

1. Introduction

The purpose of this paper is to discuss the accuracy of a procedure used in the computation of the motion of a viscous incompressible fluid past objects in two dimensions. In particular, we will focus on those methods where the Prandtl boundary layer equations are used to describe the flow near an object and the Navier-Stokes equations are used to describe the flow away from an object. A schematic diagram of the situation is presented in Figure 1. In this figure, corresponding to the domain for the computation of flow about a cylinder, the Prandtl equations are solved in an inner region from $r_a \leq r \leq r_a + \delta = r_\delta$ while the full Navier-Stokes are solved in the external region $r > r_\delta$.

One may wonder why a dual equation approach is considered. One reason is that the Prandtl boundary layer equations have a simpler structure than the full Navier-Stokes equations, and so it may be easier to obtain accurate solutions. An example where this aspect has been exploited is in the vortex sheet/vortex blob method [3],[4], [5]. This method is a Lagrangian scheme in which small segments of vorticity (or sheets) are used to approximate the vorticity distribution in the boundary layer region and vortex blobs are used to approximate the vorticity distribution in the external region. The sheet discretization of the Prandtl equations was introduced in order to remove inaccuracies near the boundaries present in vortex blob discretizations of the Navier-Stokes equations.

Another example where a combined equation approach has been used are vortex sheet/potential flow methods. In such methods the flow field is represented as an

J.T. Beale et al. (eds.), Vortex Flows and Related Numerical Methods, 105–115.

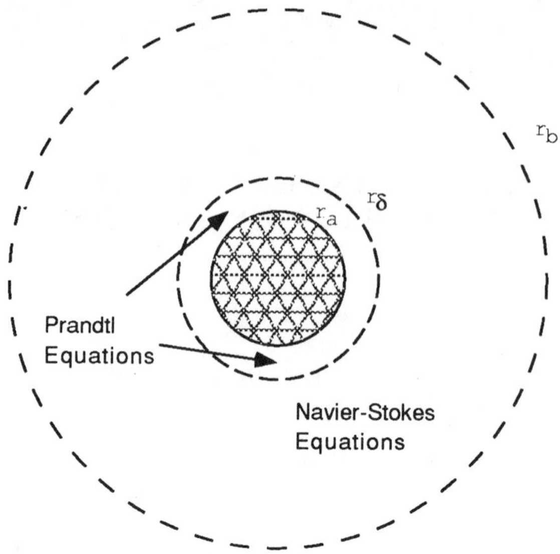

Fig. 1. The decomposition of the domain for the computation of flow about a circular cylinder using a Prandtl/Navier-Stokes equation procedure.

irrotational flow with embedded vortex sheets. A component of such methods is the necessity to determine the point at which the vortex sheet attaches to the body (or alternately where the boundary layer separates). Techniques for determining the separation point often utilize solutions of the Prandtl boundary layer equations.

Aside from questions about specific methods, there is the general question about the validity of a procedure which utilizes a combination of the Prandtl and Navier-Stokes equations. This is an interesting question because in many problems of two-dimensional fluid flow the boundary layer separates, and thus the solutions violate the assumptions which are used to derive the Prandtl equations. The fact that the flow separates does not necessarily invalidate the combined equation approach (the Prandtl equations may be applicable even if their method of derivation is inappropriate) but it does suggest that the accuracy is questionable.

To investigate the combined equation approach we first constructed a finite difference method for solving the complete Navier-Stokes equations for flow about a circular cylinder. The solutions obtained with this code served as benchmarks. We then constructed a finite difference method which utilized the combined equation approach. Specifically a finite difference method which solved the Prandtl equations in the inner region was coupled to a finite difference method which solved the Navier-Stokes equations in the external region. There was the issue of how one couples the solutions to these two sets of equations. Our initial coupling procedure was that used in the vortex sheet/vortex blob method. (This coupling is the "obvious" one.)

We found that the solutions obtained with the combined equation approach differed from those obtained with the full Navier-Stokes equations. (The solutions in both cases were fully converged.) We also found that the solutions obtained with the combined equation approach depended significantly upon the size of the domain where the Prandtl equations were solved. Since the size of the inner region is somewhat arbitrary this behavior was undesirable. For both of these reasons we could only conclude that as implemented, the combined equation approach is not a good idea. After reaching this conclusion, we began to examine procedures which could possibly remove this inaccuracy. We discovered that if one changes the way in which the coupling is performed, then the dual equation approach yields solutions which are close to the solutions of the Navier-Stokes equations. We now provide some of the details of this investigation and indicate how we changed the coupling to yield more accurate solutions. For more information on this investigation one should see [6].

2. The Test Problem and the Results

Our test problem was that of the fluid motion induced by an impulsively started circular. (At $t = 0$ a cylinder at rest is accelerated to unit velocity.) The Navier-Stokes equations were expressed the vorticity form and polar coordinates were used. The equations were thus

$$\frac{\partial \omega}{\partial t} + (u, \frac{v}{r}) \cdot (\frac{\partial \omega}{\partial r}, \frac{\partial \omega}{\partial \theta}) = \frac{1}{\mathbf{Re}} \left(\frac{1}{r}\frac{\partial}{\partial r}(r\frac{\partial \omega}{\partial r}) + \frac{1}{r^2}\frac{\partial^2 \omega}{\partial \theta^2} \right) \tag{1}$$

$$u = \frac{1}{r}\frac{\partial \Psi}{\partial \theta} \qquad v = -\frac{\partial \Psi}{\partial r} \tag{2}$$

$$\frac{1}{r}\frac{\partial}{\partial r}(r\frac{\partial \Psi}{\partial r}) + \frac{1}{r^2}\frac{\partial^2 \Psi}{\partial \theta^2} = -\omega \tag{3}$$

for $r \geq r_a$ and $0 \leq \theta \leq 2\pi$. Here (u, v) are the radial and tangential components of the velocity respectively, ω is the vorticity, and Ψ is the stream function. The non-dimensionalization is based on the cylinder diameter $2r_a$ and the velocity at infinity; $\mathbf{Re} = \frac{2r_a U_\infty}{\nu}$. Initial and boundary conditions for Ψ and ω which corresponded with the problem of an impulsively started cylinder motion were used.

We computed solutions with a range of Reynolds numbers between 500 and 9,600. Since the conclusions we reached did not depend upon Reynolds number, we choose to report those at Reynolds number 1000. (At this Reynolds number much of the complicated boundary layer features seen in higher Reynolds number flows are present. However it is still possible to resolve Re 1000 flow using a reasonable number of grid points and computing time.)

The numerical method for the full Navier-Stokes equations was a fourth order finite difference scheme. The convective derivative terms and the Laplacian were approximated with fourth order centered differences. (Sufficiently many grid points were used so that the problem of spatial oscillations of the computed solutions was avoided.) The grid was a uniform polar grid from $r = r_a$ to $r = r_b$. r_b was chosen so that the vorticity, except for an exponentially small amount, was contained

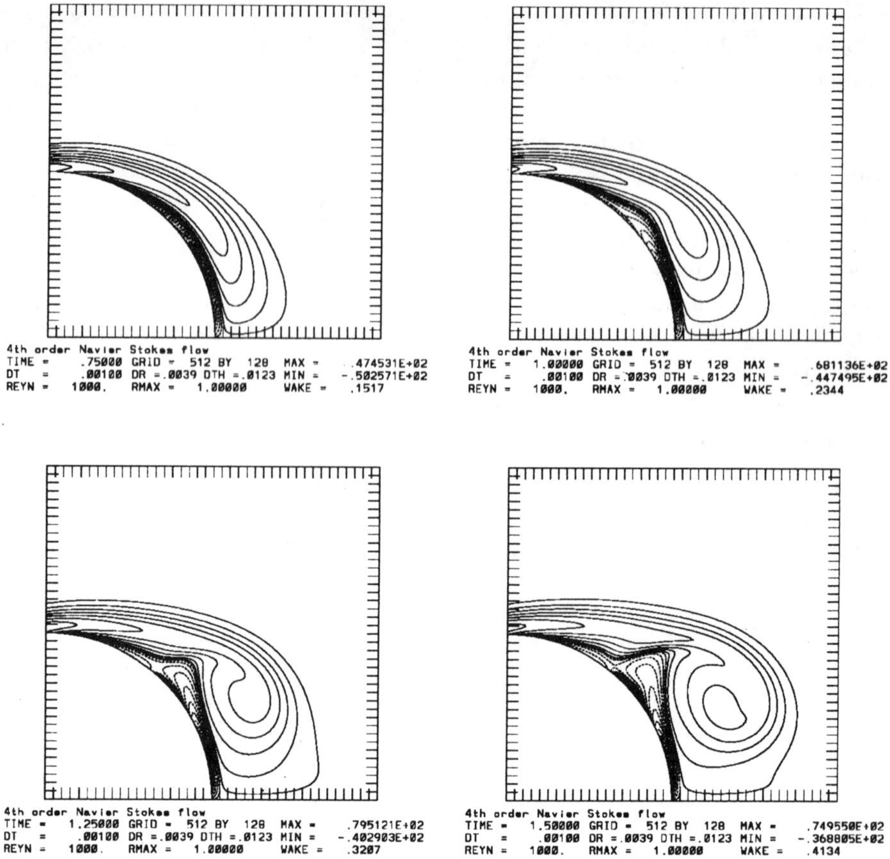

Fig. 2. Solutions of the Navier-Stokes equations at **Re** = 1000. Solid contours indicate negative vorticity, dashed contours positive vorticity.

within the region $r_a \leq r \leq r_b$ over the times which we computed. We implemented "infinite domain" boundary conditions for the determination of the stream function (these are described in [2]) as well as high order vorticity boundary conditions at the surface of the cylinder (extending the procedure of [1]). The time-stepping method was explicit fourth order Runge-Kutta. The cylinder radius was 0.5. For a full description of the method and verification of its accuracy, see [6].

The development of the solution for this problem can most easily be seen by considering the dynamics of the vorticity. In Figure 2 we show contours of the vorticity at times t = .75, 1.0, 1.25 and 1.5. Since the flow is symmetric and the most interesting part is near the rear of the cylinder, we just show the upper rear corner. In this figure, and in all the remaining figures, the solid lines correspond to negative vorticity and the dashed lines correspond to positive vorticity. At time

$t = .75$ the vortex sheet which forms along the cylinder is just beginning to separate. This process continues as time evolves, and by $t = 1.5$ the boundary layer has clearly separated and acts as a source of vorticity for the interior flow. At this later time we also see the roll-up of the vortex sheet occurring behind the cylinder.

The method utilizing both the Prandtl and Navier-Stokes equations was based on a decomposition of the domain as indicated in Figure 1. In the region external to the boundary layer region the Navier-Stokes equations were approximated in the same way as the full Navier-Stokes equations. In the boundary layer region, that region with $r_a \leq r \leq r_\delta$ the equations solved were the vorticity form of the Prandtl boundary layer equations,

$$\frac{\partial \omega}{\partial t} + (u, \frac{v}{r}) \cdot (\frac{\partial \omega}{\partial r}, \frac{\partial \omega}{\partial \theta}) = \frac{1}{\mathbf{Re}} \frac{\partial^2 \omega}{\partial r^2} \tag{4}$$

$$\frac{\partial v}{\partial r} = \omega \tag{5}$$

$$\frac{\partial (ru)}{\partial r} + \frac{\partial v}{\partial \theta} = 0 \tag{6}$$

Here u is the normal velocity and v the tangential velocity. These velocity fields are determined from the vorticity by calculating the integrals

$$v(r,\theta) = v_\infty(\theta) + \int_r^\infty \omega(r,\theta)\, dr \tag{7}$$

$$u(r,\theta) = -\frac{1}{r} \int_0^r \frac{\partial v(r,\theta)}{\partial \theta}\, dr \tag{8}$$

The discretization of these equations was similar to that employed for the full Navier-Stokes - i.e. a fourth order centered difference approach. The boundary conditions for the vorticity at the cylinder surface were also discretized in a similar fashion (again see [1]).

An important aspect of the combined equation approach is the determination of the coupling between the two solutions at the interface $r = r_\delta$. When solving the Navier-Stokes equations in the external region one must specify the vorticity at the interface $r = r_\delta$ as well as decide where to specify the stream function boundary condition. When one solves the Prandtl boundary layer equations one treats $r = r_\delta$ as infinity, and thus one needs to specify v_∞, the tangential velocity there. One also needs to specify the boundary conditions for the transport and diffusion of vorticity.

Our initial approach to the coupling was that implicitly used in the vortex sheet/vortex blob method. In such an approach one assumes that the vorticity is continuous across the interface so that any data necessary to close the equations for the transport and diffusion of vorticity for either equation is obtained by using the vorticity on "the other side" of the interface. This external velocity was obtained by

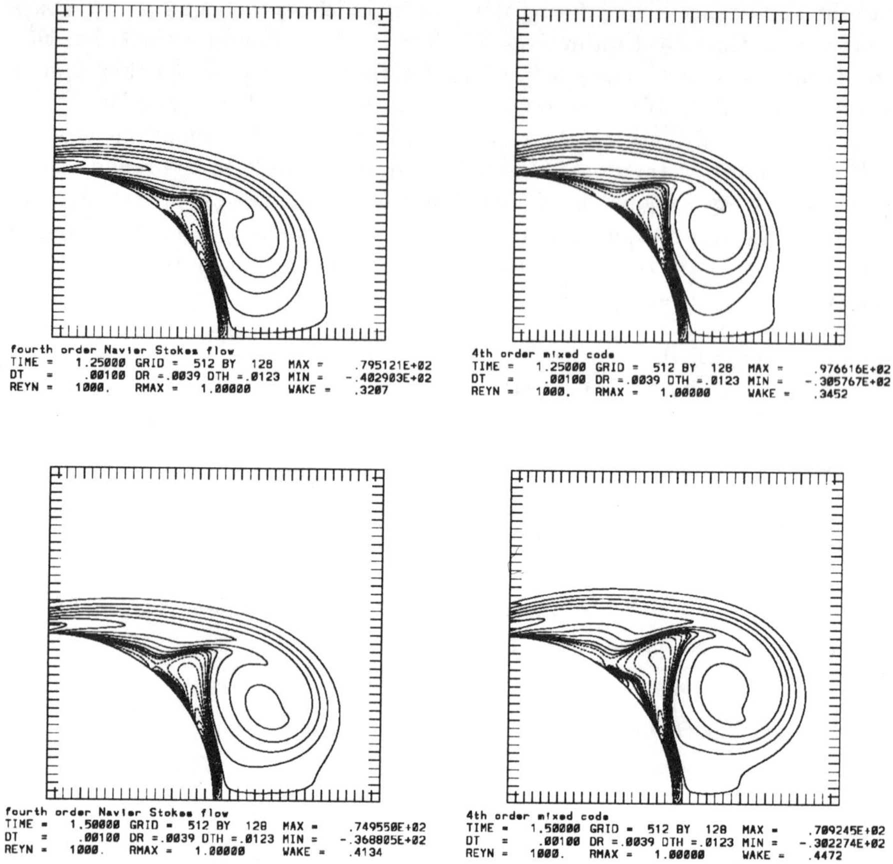

Fig. 3. A comparison of the solutions of the Navier-Stokes equations (left) with the solutions of Prandtl/Navier-Stokes equations (right) at $\mathbf{Re} = 1000$ with $\delta = \frac{3}{32}$.

differentiating the solution to (3) with the boundary condition $\Psi = 0$ at the cylinder surface $r = r_a$. The tangential velocity v_∞ needed for the Prandtl equations was just taken to be the external velocity evaluated at the point $r = r_\delta$. The net result of this process was a velocity field which satisfies the differential equations (5)-(6) in the inner region and (2)-(3) in the external region. At the interface, the tangential velocity is continuous.

One free parameter in this Prandtl/Navier-Stokes approach is δ the width of the inner region. In our experiments we used values on the order of $\frac{1}{\sqrt{\mathbf{Re}}}$, the order of the size of the boundary layer [7]. This choice for δ is that which has been previously used in calculations with a vortex sheet/vortex blob method.

At early times the solutions for the Navier-Stokes equations and those obtained with the combined Prandtl/Navier-Stokes equations did not differ appreciably. This

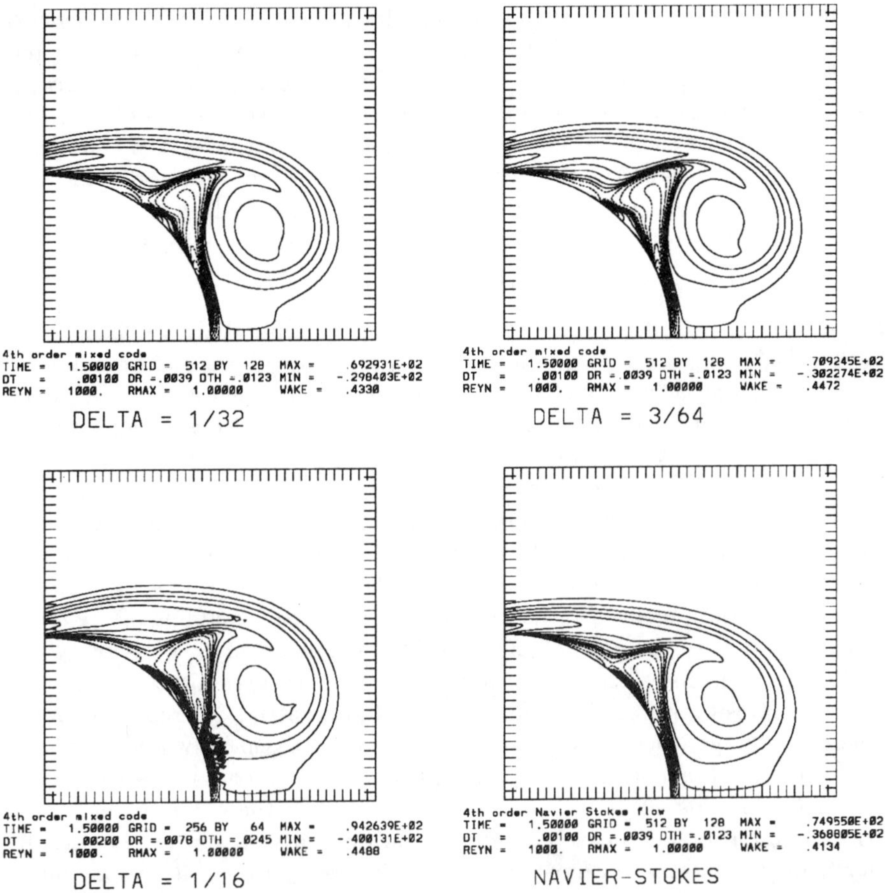

Fig. 4. Solutions of the Prandtl/Navier-Stokes equations for values of the inner region width $\delta = \frac{2}{64}, \frac{3}{64}$, and $\frac{4}{64}$.

is to be expected since the boundary layer had not separated, and it is reasonable to expect that approximations made in deriving the Prandtl boundary layer equations are appropriate. At later times the situation is rather different. In Figure 3 we show a comparison of the vorticity distributions obtained with the Prandtl/Navier-Stokes equations verses the solution obtained with the full Navier-Stokes equations. These solutions were obtained with a value of $\delta = \frac{3}{64} \approx 1.5 \frac{1}{\sqrt{\mathbf{Re}}}$. While each solution exhibits a vortex sheet which forms and leaves the surface there are significant quantitative differences. Most noticeable is that the vortex sheet associated with the Prandtl/Navier-Stokes equations separates from the body earlier in time, separates at a different angle, and induces a much larger amount of opposite signed vorticity beneath it. They are clearly not the same solution.

In Figure 4 we show a comparison of the solutions to the Prandtl/Navier-Stokes equations at a fixed time $t = 1.5$, but with values for the width of the inner region of $\delta = \frac{2}{64}, \frac{3}{64}$, and $\frac{4}{64}$. The large scale features do not change much for this range of δ, but the small scale features do - in particular the structure of the vorticity which forms beneath the separated sheet. For reference purposes, the cylinder is of radius 0.5, and so each of the tick marks on the figures represent a distance of $\frac{1}{40}$. At $\delta = \frac{4}{64}$ the width is equal to 2.5 tick marks. With larger values of δ the solutions differed even more (in fact with $\delta = \frac{6}{64} \approx \frac{3}{\sqrt{\mathbf{Re}}}$ the solution "blew up" and the computation could not be continued).

These figures lead one to conclude that if one has separated flow, using a combined Prandtl/Navier-Stokes equation is not appropriate. A question remains however, "Can this approach can be fixed ?". After some experimentation, we concluded that the error was in large part due to the manner in which the solutions were coupled at $r = r_\delta$. To obtain better solutions we therefore changed the way in which the solutions were coupled. The coupling described above enforced a continuity of the vorticity and a continuity of tangential velocity. While this coupling is simple and easy to implement, it gives velocity fields whose normal component is not necessarily continuous across the line $r = r_\delta$. It is clear that a continuous velocity field is a reasonable requirement, and we therefore determined a coupling of the two calculations so that both the normal and tangential components of the velocity would be continuous at the interface. The technique for doing this is similar to that used in domain decomposition procedures for solving elliptic partial differential equations. (See the proceedings where [2] appears for references). Essentially, given a vorticity distribution and a tangential velocity along $r = r_\delta$ one can compute a velocity field in the inner region using (7)-(8) and a velocity field in the external region using (2)-(3) with a normal derivative boundary condition on Ψ at $r = r_\delta$. The resulting velocity field will have a continuous tangential velocity by construction, but not necessarily a continuous normal velocity. However, we can ask "Is there a tangential velocity at $r = r_\delta$ so that, after the velocity field is constructed we also get a continuous normal velocity?". It turns out that the answer to this question is "yes" and in fact one can construct a linear system of equations to find it. Our procedure was therefore to construct and solve the set of linear equations which determine this special tangential velocity at $r = r_\delta$. With this tangential velocity one uses (7)-(8) and (2)-(3) to compute the velocity field in the inner and external regions. The net result of this process is the determination of a velocity field which satisfies the differential equations (5)-(6) in the inner region and (2)-(3) in the external region. At the interface, both the tangential and normal velocities are continuous.

The results of this approach are seen in Figure 5. It is clear that the improved coupling removes much of the error seen previously. While there are still slight differences, there is good general agreement. Moreover, as can be deduced from Figure 6 we see that the improved method has only a weak dependence on the thickness of the inner region. With large values of δ the solutions were still significantly different from the Navier-Stokes equations, but this is to be expected, because, away from the body, the terms neglected in the Navier-Stokes equations in order to derive the Prandtl equations become important.

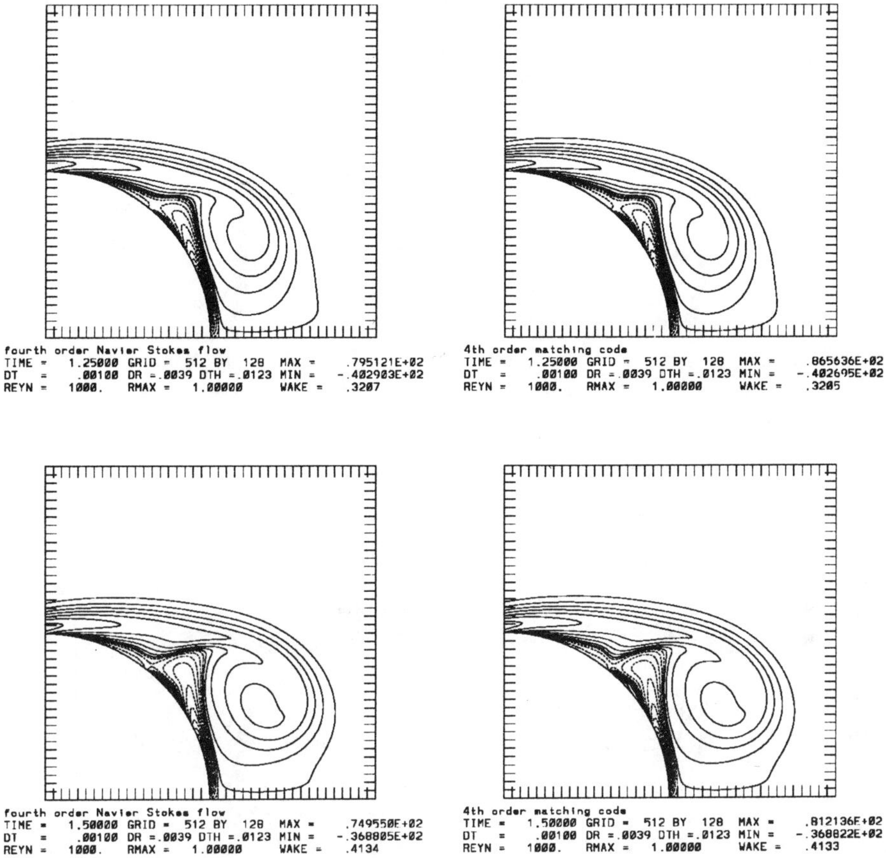

Fig. 5. A comparison of the solutions of the Navier-Stokes equations (left) with the solutions of Prandtl/Navier-Stokes equations (right) at **Re** = 1000 with $\delta = \frac{3}{32}$. The Prandtl/Navier-Stokes equations were coupled so that both components of the velocity field were continuous across $r = r_\delta$.

3. Conclusion

We have described an investigation in which we sought to determine the accuracy of using a Prandtl/Navier-Stokes equation approach to computing flows about bodies in two dimensions. From our initial experiments in which a simple coupling was used, it was clear that when separation is present, the combined equations yield solutions which are significantly different from those obtained with the Navier-Stokes equations. We have also shown that if one improves the coupling, then the solutions to the Prandtl/Navier-Stokes equations appear to be acceptable, even in cases in which there is separation. The computational experiments we presented were carried out at Re=1000, but experiments at higher Reynolds numbers indi-

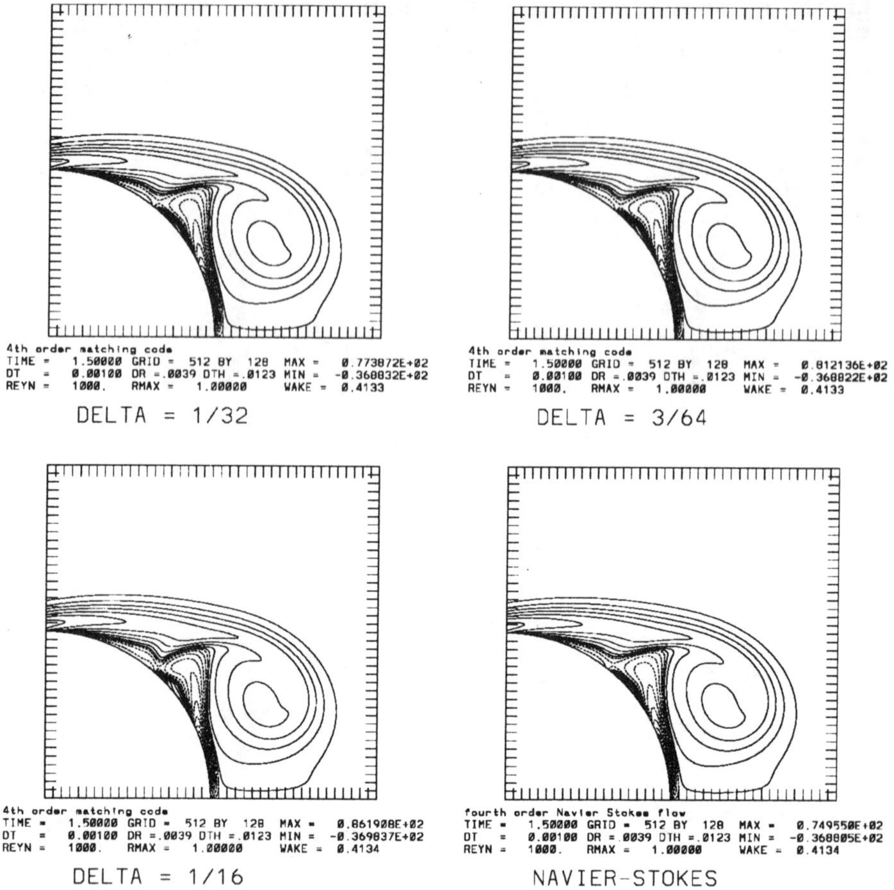

Fig. 6. Solutions of the Prandtl/Navier-Stokes equations for values of the inner region width $\delta = \frac{2}{64}, \frac{3}{64}$, and $\frac{4}{64}$. The Prandtl/Navier-Stokes equations were coupled so that both components of the velocity field were continuous across $r = r_\delta$.

cate that our conclusions are not sensitive to the Reynolds number. Another aspect of this work is that it provides evidence that the inaccuracies incurred with the use of a Prandtl/Navier-Stokes approach are problem specific. (When the flow wasn't separated the simple coupling approach was o.k..) Since this is the case, we cannot immediately generalize this work to Prandtl/Navier-Stokes approaches to flows where separation occurs at a sharp edge. It is our expectation that simple matching is sufficient, but further numerical work (which we are in the process of carrying out) should provide information regarding this question.

Acknowledgements

The research of both authors was supported in part by Army Research Office Grant #DAAL03-91-G-0162. The second author was also supported by a graduate fellowship from the National Science Foundation.

References

[1] C. R. Anderson, "Vorticity Boundary Conditions and Boundary Vorticity Generation for Two Dimensional Viscous Incompressible Flows", J. Comput. Phys., **80**, 1989, pg. 72-97.

[2] C. R. Anderson, " Domain Decomposition Techniques and the Solution of Poisson's Equation in Infinite Domains", in *Domain Decomposition Methods*, SIAM publications, 1989, pg. 129-139.

[3] A.Y. Cheer, "Unsteady Separated Wake Behind an Impulsively Started Cylinder in Slightly Viscous Fluid", J. Fluid Mech., **201**, 1989, pg. 485-505.

[4] A.J. Chorin, "Vortex Models and Boundary Layer Instability", SIAM J. Sci. Statist. Comput., **1**, 1980, pg. 1-21.

[5] A.J. Chorin, "Vortex Sheet Approximation of Boundary Layers", J. Comput. Phys., **27**, 1978, pg. 428-442.

[6] M. B. Reider, "Development of Higher Order Numerical Methods for Two Dimensional Incompressible Flows with Applications to Flows Around Circular Cylinders and Airfoils", Ph.D. Thesis, UCLA, 1992.

[7] H. Schlichting, *Boundary Layer Theory*, McGraw-Hill, New York, 1960.

VORTICITY BOUNDARY CONDITIONS FOR THE NAVIER–STOKES EQUATION IN VELOCITY–VORTICITY FORMULATION

Olivier DAUBE
LIMSI–CNRS ; BP 133 ; 91403 ORSAY CEDEX ; FRANCE

ABSTRACT.
In this paper, vortex flows are computed by means of a finite difference resolution of the 2D unsteady incompressible Navier–Stokes equations cast in the vorticity function ω – velocity vector $\mathbf{v}$ formulation. The boundary conditions on the function ω (i.e the definition of ω as the curl of the velocity) are enforced by an influence matrix technique *at each time step*, a fact that enables us to enforce everywhere in the domain the definition of the function ω as the curl of the velocity and consequently the zero divergence of the velocity field. The efficiency of the proposed method is supported by the resolution of the driven cavity problem and by the resolution of the flow around a circular cylinder. In both cases, the definition of the vorticity and the continuity equation are shown to be satisfied within machine accuracy and round–off errors of library routines. In the later case the computational domain is not simply connected, a fact that implies that an additionnal condition has to be imposed.

1. Introduction

In the last decade, the velocity–vorticity formulation of the Navier–Stokes equations appeared to an increasing number of people an attractive alternative to the usual primitive variables formulation. In constrast with the well known stream function – vorticity function approach that cannot be easily extended to 3D flows, this $(\mathbf{v}, \omega)$ formulation is valid both in 2D and 3D. Moreover this formulation deals directly with the vorticity that is the quantity of primary interest in many flows. At last, it is expected that for external flows, "open boundary conditions" would be easier to implement for the vorticity than for the pressure.

2. The governing equations

For the sake of simplicity, the computational domain Ω is in this section assumed to be simply connected and bounded with boundary Γ. The additional specific difficulties that arise in the case of multiply connected domains will be outlined in the last section. The governing equations read then :

$$\frac{\partial \omega}{\partial t} + \nabla.(\omega \mathbf{v}) = \frac{1}{R}\nabla^2 \omega \qquad (1)$$

$$\nabla.\mathbf{v} = 0 \qquad (2)$$

J.T. Beale et al. (eds.), Vortex Flows and Related Numerical Methods, 117–127.

$$\nabla \times \mathbf{v} = \omega \mathbf{k} \tag{3}$$

In addition a set of boundary and of initial conditions have also to be specified.

$$\left.\begin{aligned} \mathbf{v} &= \mathbf{v}_\Gamma && \text{on } \Gamma \\ \mathbf{v}(M,0) &= \mathbf{v}^0(M) && \forall M \in \Omega \\ \nabla \times \mathbf{v}^0 &= \omega^0 \mathbf{k} && \text{in } \Omega \\ \nabla . \mathbf{v}^0 &= 0 && \text{in } \Omega \end{aligned}\right\} \tag{4}$$

where $\mathbf{v}_\Gamma$ is a known velocity distribution on Γ and $\mathbf{k}$ an unit vector normal to the plane of the flow.

It has been shown in [11] that this set of equations is equivalent to the original 2D Navier–Stokes equations written in primitive variables. Some people (Gatski *et al* [1], Osswald *et al* [2]) have chosen to solve simultaneously the three equations (1,2,3) but most of the published works deal with a vector Poisson equation

$$\nabla^2 \mathbf{v} = -\nabla \times (\omega \mathbf{k}) \tag{5}$$

to compute the velocity. This is also the case in this paper and therefore its main goal is, within the context of a Poisson equation for the velocity, to provide a method to ensure that the relations (2,3) are satisfied *at each time step* and not only at steady state, everywhere in the domain, included at the boundaries.

For the details of the proposed method the reader is referred to [3]. Only the main results which allow for the design of the method are described hereafter.

3. The Poisson Equation for the Velocity

The main difficulties which occur when using a vector Poisson equation to compute the velocity arise in the specification of the correct boundary conditions in order to ensure that it is equivalent to solve the Poisson equation (5) or the equations (2,3). This problem was adressed for instance by Anderson [4] in the context of finite difference methods and by Cottet [5] in the context of vortex methods. In this section, the results that constitute the basis of the influence matrix technique to be described later and that specify the correct boundary conditions for the vorticity are reminded.

PROPOSITION 1. *Let ω and $\mathbf{v}$ be a function and a vector field defined on Ω which satisfy :*

$$\left.\begin{aligned} \nabla^2 \mathbf{v} &= -\nabla \times (\omega \mathbf{k}) && \textit{in } \Omega \\ \mathbf{v} &= \mathbf{v}_\Gamma && \textit{on } \Gamma = \partial\Omega \\ \int_\Gamma \mathbf{v}.\mathbf{n}\, dl &= 0 && \end{aligned}\right\} \tag{6}$$

Then,

$$\omega \mathbf{k} = \nabla \times \mathbf{v} \textit{ in } \Omega \Longrightarrow \nabla . \mathbf{v} = 0 \textit{ in } \Omega$$

PROPOSITION 2. *Let us consider ω and $\mathbf{v}$ that satisfy the relations (6). Then we have:*

$$\omega \mathbf{k} = \nabla \times \mathbf{v} \ in\ \Omega \Longleftrightarrow \omega \mathbf{k} = \nabla \times \mathbf{v} \ on\ \Gamma = \partial\Omega$$

As a consequence of these propositions, the Navier–Stokes equations in vorticity velocity formulation can be written :

$$\left.\begin{aligned} \frac{\partial \omega}{\partial t} + \nabla.(\omega \mathbf{v}) &= \frac{1}{R}\nabla^2\omega && \text{in } \Omega \\ \nabla^2 \mathbf{v} &= \mathbf{k} \times \nabla\omega && \text{in } \Omega \\ \nabla \times \mathbf{v} &= \omega \mathbf{k} && \text{on } \Gamma \\ \mathbf{v} &= \mathbf{v}_\Gamma && \text{on } \Gamma \end{aligned}\right\} \tag{7}$$

4. The Influence Matrix Method

Since the pionner work by Kleiser and Schummann [6], influence matrices have been widely used, especially by people using spectral methods [7; 8] to solve the Navier–Stokes equations cast in the primitive variables formulation but also in the context of the vorticity stream function formulation [9]. This technique makes use of the superposition principle for linear operators and therefore a time linearization has first ot be performed (see next section). This technique permits to compute the boundary values of the vorticity through the resolution of an auxiliary "small" linear system so that the definition of the vorticity is enforced on the boundary.

4.1. THE TIME DISCRETIZATION

At each discrete time $t_n = n\Delta t$, the time derivative is approximated by a backward finite difference. The non linear terms are explicitely evaluated by means of a linear extrapolation from their known values at the previous instants whereas the diffusion terms are implicitely evaluated. Thus, taking into account propositions 1 and 2 we have at each time step to solve a Helmholtz type equation for the vorticity and two Poisson equations for the velocity components :

$$[P]\left\{\begin{aligned} (\sigma \mathrm{I} - \nabla^2)\omega &= S && \text{in } \Omega \\ \nabla^2 \mathbf{v} &= \mathbf{k} \times \nabla\omega && \text{in } \Omega \\ \nabla \times \mathbf{v} &= \omega \mathbf{k} && \text{on } \Gamma \\ \mathbf{v} &= \mathbf{v}_\Gamma && \text{on } \Gamma \end{aligned}\right. \tag{8}$$

where I is the identity operator and σ a real positive constant. We will assume that this problem has an unique solution.

4.2. THE SPATIAL DISCRETIZATION

The underlying numerical scheme is based upon classical second order centered differences on a regular grid. An important point to be noticed is that it is necessary to use a MAC staggered grid. The reason for doing so is that the discrete difference operators are thus defined in such a way that the discrete analog of propositions (1) and (2) hold (see [3]).

4.3. THE INFLUENCE MATRIX

The difficulties to solve the problem [P] arise from the fact that the values of the vorticity are not known on the boundary. Let us consider the auxiliary problem $[\tilde{\mathrm{P}}]$ defined as follows :

$$[\tilde{\mathrm{P}}]\begin{cases}(\sigma \mathrm{I}-\nabla^2)\tilde{\omega} = S & \text{in } \Omega\\ \tilde{\omega} = \tilde{\omega}_\Gamma & \text{on } \Gamma\\ \nabla^2\tilde{\mathbf{v}} = \mathbf{k}\times\nabla\tilde{\omega} & \text{in } \Omega\\ \tilde{\mathbf{v}} = \mathbf{v}_\Gamma & \text{on } \Gamma\end{cases}$$

where $\tilde{\omega}_\Gamma$ is an arbitrary given distribution on Γ. This problem $[\tilde{\mathrm{P}}]$ has clearly an unique solution $(\tilde{\omega},\tilde{\mathbf{v}})$ which generally does not satisfy the continuity equation (2) and the definition of the vorticity (3).

Let $(\omega,\mathbf{v})$ be the unique solution of the problem [P]. The difference $(\overline{\omega},\overline{\mathbf{v}}) = (\omega-\tilde{\omega},\mathbf{v}-\tilde{\mathbf{v}})$ is then a solution of the homogeneous problem :

$$[\overline{\mathrm{P}}]\begin{cases}(\sigma \mathrm{I}-\nabla^2)\overline{\omega} = 0 & \text{in } \Omega\\ \nabla^2\overline{\mathbf{v}} = \mathbf{k}\times\nabla\overline{\omega} & \text{in } \omega\\ \overline{\mathbf{v}} = 0 & \text{on } \Gamma\end{cases}$$

From proposition (2), the determination of the solution $(\omega,\mathbf{v})$ of the problem [P], i.e. the solution of the Navier–Stokes problem, is thus equivalent to finding *the* couple $(\overline{\omega},\overline{\mathbf{v}})$ which satisfies the additional condition :

$$\overline{\omega}\mathbf{k}-\nabla\times\overline{\mathbf{v}} = -(\tilde{\omega}\mathbf{k}-\nabla\times\tilde{\mathbf{v}}) \quad \text{on } \Gamma \tag{9}$$

For the continuous case, the solutions of $[\overline{\mathrm{P}}]$ are generated by the elementary Green functions of the problem. From a discrete point of view, let $\{\gamma_k\}_{k=1,\dots,N_\Gamma}$ be the grid points located on the boundary Γ. Then, the solutions of the problem $[\overline{\mathrm{P}}]$ belong to the space spanned by the solutions $(\overline{\omega}_k,\overline{\mathbf{v}}_k)$ of the elementary problems $[\overline{\mathrm{P}}_k]$:

$$[\overline{\mathrm{P}}_k]\begin{cases}(\sigma \mathrm{I}-\nabla^2)\overline{\omega}_k = 0 & \text{in } \Omega\\ \overline{\omega}_k(\gamma_l) = \delta_{kl} & \forall\gamma_l\in\Gamma\\ \nabla^2\overline{\mathbf{v}}_k = \mathbf{k}\times\nabla\overline{\omega}_k & \text{in } \omega\\ \overline{\mathbf{v}}_k = 0 & \text{on } \Gamma\end{cases}$$

where δ_{kl} is the Kronecker symbol. Note that these problems do not depend on the actual time level and can therefore be solved in a preprocessing stage.

The solutions $(\overline{\omega},\overline{\mathbf{v}})$ of problem $[\overline{\mathrm{P}}]$ read then :

$$\begin{pmatrix}\overline{\omega}\\ \overline{\mathbf{v}}\end{pmatrix} = \sum_{\gamma_k}\lambda_k\begin{pmatrix}\overline{\omega}_k\\ \overline{\mathbf{v}}_k\end{pmatrix} \tag{10}$$

The determination of the set of parameters $\{\lambda_k\}$ in order to satisfy the conditions (9) is then achieved by solving a linear system :

$$\mathbf{M}.\boldsymbol{\lambda} = \begin{bmatrix} & \vdots & \\ \dots & \overline{\zeta}_k(\gamma_l) & \dots \\ & \vdots & \end{bmatrix} \begin{bmatrix} \vdots \\ \lambda_l \\ \vdots \end{bmatrix} = - \begin{bmatrix} \vdots \\ \widetilde{\zeta}(\gamma_l) \\ \vdots \end{bmatrix} = -\widetilde{\zeta} \tag{11}$$

The terms $\overline{\zeta}_k$ (resp $\widetilde{\zeta}$) are defined as in the previous sections by :

$$\overline{\zeta}_k \mathbf{k} = \overline{\omega}_k \mathbf{k} - \nabla \times \overline{\mathbf{v}}_k$$

Because of the use of a staggered grid and of the uniqueness of the solution to (8), the system (11) can be inverted. From a practical point of view, the inverse of the matrix $\mathbf{M}$ is computed in the preprocessing stage.

5. The Driven Cavity Problem

The computations which are reported here were performed for R=400 and R=1000. The time evolution of $||\nabla.\mathbf{v}||_\infty$ and of $||\omega - (\nabla \times \mathbf{v}).\mathbf{k}||_\infty$ are reported on figure (1). It is clearly seen that the desired conditions are satisfied at each time step within machine accuracy and within the accuracy of the subroutine that computes the inverse of the matrix $\mathbf{M}$. This last point is evidenced by comparing the results on grids with different sizes which clearly show the degradation of the quality of the matrix inversion routine with an increasing order of the matrix.

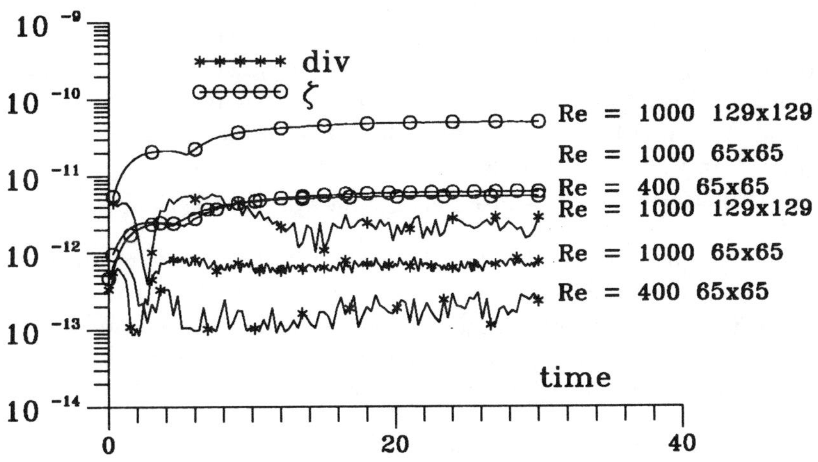

Fig. 1. Time evolution of $||\nabla.\mathbf{v}||_\infty$ and of $||\zeta||_\infty$ for R=400 and R=1000

Concerning the physical features of the flow, the u-velocity along the centerline $x = 0.5$ computed by the present method is reported on figure (2) together with the results by Ghia *et al* [10]. The comparison is quite satisfactory.

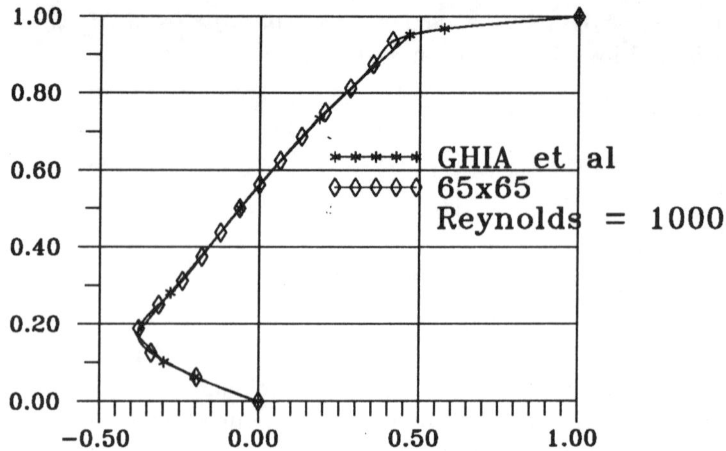

Fig. 2. R = 1000 ; u–component of the velocity along the centerline

6. The Flow Around a Circular Cylinder

This case was chosen to illustrate the additional conditions that have to be imposed when using the vorticity velocity formulation to deal with flows in a non simply connected domain Ω which is defined here by :

$$\Omega = [0, \infty[\times[0, 2\pi]$$

For further details, the reader is referred to [11]. Roughly speaking, the difficulties arise when the momentum equation in primitive variables is to be recovered starting from the transport equation for the vorticity. The gradient of a function ϕ has then to be introduced in the *simply connected* domain Ω' that is obtained from the whole domain Ω by removing a number of crosscuts equal to the connectivity degree of Ω. In the considered case, this domain Ω' is defined by :

$$\Omega' = [0, \infty[\times]0, 2\pi[$$

Obviously, this function ϕ is the pressure. In order to get meaningful physical solutions, the function ϕ has to be chosen so that it can be expanded in an *single valued* function in the whole domain Ω. This is possible if ϕ does not undergo any jump across the different crosscuts. Therefore, an additional condition has to be imposed. In the case of the flow around a circular cylinder, this condition reads :

$$\oint_C \nabla\phi . dl = 0$$

along an arbitrary closed path C surrounding the cylinder. The obvious choice for C is the boundary Γ of the cylinder, and thus the previous condition reads in polar coordinates:

$$\oint_\Gamma \frac{\partial\omega}{\partial n}\, dl = \int_0^{2\pi} \frac{\partial\omega}{\partial r}(0, \theta)\, d\theta = 0 \tag{12}$$

This condition was already used by Stella and Guj [12].

6.1. THE NUMERICAL METHOD

The cylindrical coordinates are used but the r–coordinate is defined as $r = \exp(\xi)$ which is the classical exponential conformal mapping that leads to a more tractable form of the Navier-Stokes equations [14]. Because of the natural periodicity of the problem, the dependent variables at the discrete points (r_i, θ_j) are expanded in truncated Fourier series:

$$\omega(r_i,\theta_j) = \sum_{k=-N/2}^{k=+N/2} \widehat{\omega}_k(r_i)\exp(-ik\theta_j)$$

$$\mathbf{v}(r_i,\theta_j) = \sum_{k=-N/2}^{k=+N/2} \widehat{\mathbf{v}}_k(r_i)\exp(-ik\theta_j)$$

In addition to accuracy considerations, another advantage of such a scheme is that the resolution of each Fourier mode is decoupled from the others and involves the resolution of a scalar tridiagonal system when the radial derivatives are discretized by means of classical centered finite differences. Therefore the influence matrix reduces to a 2×2 matrix per mode. Actually, it even reduces to a single scalar quantity per mode since the condition at "infinity" that is used for the vorticity is of Dirichlet type. In fact, since our goal was not here to design a "best" open boundary condition, the irrotational flow condition $\omega = 0$ was used at a distance $r = r_\infty$ which was assumed to be sufficiently large. It must be noticed that we would still have a Dirichlet condition in the case where a value of the vorticity deduced from an open boundary condition $D\omega/Dt = 0$ (see Daube *et al* [13]) would be used.

Let u and v be the radial and tangential components of the velocity $\mathbf{v}$. The scalar influence matrix is obtained by imposing the conditions :

$$\widehat{\omega}_k\big|_{r=1} = \frac{1}{r}\left[\frac{\partial}{\partial r}(r\widehat{v}_k) - ik\widehat{u}_k\right]_{r=1} \qquad \text{for each mode } k \geq 1$$

$$\left.\frac{\partial\widehat{\omega}_0}{\partial r}\right|_{r=1} = 0 \qquad \text{for the } 0^{th} \text{ mode}$$

The last condition is the condition (12) for the pressure to be single valued.

6.2. COMPUTATION OF THE VELOCITY

At first sight, an important drawback of the velocity–vorticity formulation is the number of equations to be solved: three for 2D problems and six for 3D problems. However, this diffulty may be partly overcome by using for instance the zero divergence relation (2) to compute the v–velocity once the u–component is calculated by the resolution of a Poisson type equation. More precisely, the kth mode of the u

and v components of the velocity are solution of:

$$\left.\begin{aligned}\left(\sigma\,\mathrm{I}-\frac{\partial^2}{\partial\xi^2}\right)\widehat{u}_k &= -ik\widehat{\omega}_k\\ \widetilde{v}_k &= -\frac{i}{k}\frac{\partial\widehat{u}}{\partial\xi}\end{aligned}\right\}\qquad \text{for } k\neq 0$$

The second relation is the zero divergence applied to the kth mode. It can easily be shown that this process is equivalent to the original one.

7. Results

Once again the zero divergence and the definition of the vorticity are satisfied within machine accuracy as it can be seen on figure (3) for a value of the Reynolds number equal to 10^3. It must be noticed that the results are by almost two decades better than in the driven cavity case, especially for the difference ζ. This due to the fact that the influence matrix reduces to scalar quantities and that therefore no matrix inversion is to be performed.

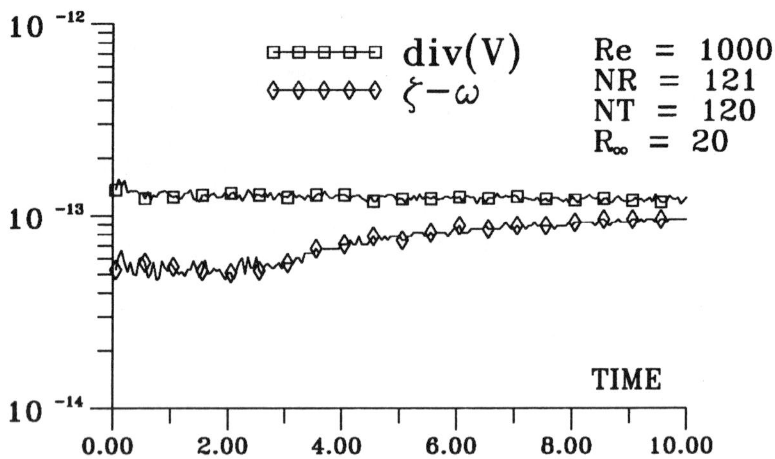

Fig. 3. Time evolution of $||\nabla.\mathbf{v}||_\infty$ and of $||\zeta||_\infty$; R=1000

Concerning the physical features of the flow, the comparisons with previous numerical results by Ta Phuoc Loc [14] are quite satisfactory. An example of such results is given on figure (4) for the vorticity distribution along the cylinder for R=1000.

8. On the Use of Fully Implicit Schemes

The influence matrix technique which is presented in this paper requires the use of a time discretization with an explicit treatement of the advection terms in order that the matrix could be computed and inverted in a preprocessing step. This semi-implicit discrtization can be an imortant drawback due to the restrictions on the stability criteria that it may involve.

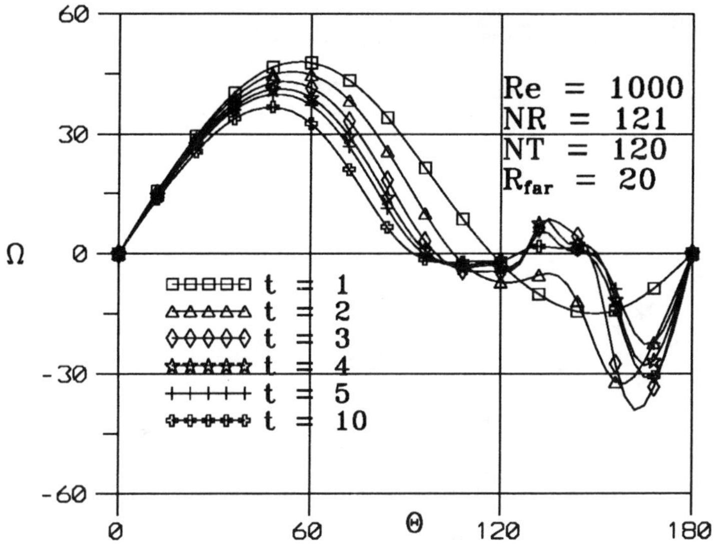

Fig. 4. Vorticity on the cylinder ; R=1000 ; 121×120 grid

On the other hand, let us recall that the computation of this matrix requires the resolution of N_Γ Navier–Stokes like problems. Therefore, even if formally, this technique could be used in the context of an implicit treatement of the non linear terms (with a time linearization of the velocity components), it would lead to a much too prohibitive computation time because the improvememt on the stability criteria would not compensate the necessity to solve a large number of Navier–Stokes problems at each instant $t_n = n\Delta t$ to compute and invert the associated influence matrix $\mathbf{A}^{(n)}$. If doing so the correct boundary values of the vorticity at time $(n+1)\Delta t$ would be given by:

$$\omega_\Gamma^{(n+1)} = \omega_\Gamma^{(n)} - \left(\mathbf{A}^{(n)}\right)^{-1} \dot{\zeta}_\Gamma^{(n)} \tag{13}$$

where ζ as the same definition as in the previous sections.

Instead of using equation (13), we have used an iterative process to get an approximation of the correct boundary values of ω, a process which is actually a preconditionning of equation (13) by the influence matrix $\mathbf{M}$ of the semi-implicit scheme which was used in the previous section. Therefore at each iteration(p) within a time step, the values ω_Γ are updated by means of:

$$\omega_\Gamma^{(n,p+1)} - \omega_\Gamma^{(n,p)} = -\mathbf{M}^{-1}\dot{\zeta}_\Gamma^{(n,p)} \qquad p = 0, \ldots, P$$

Each iteration of this process requires the solution of a Navier–Stokes problem and therefore only two or three such iterations are performed in order not to loose the improvement in computational time due to better stability criteria owing to the implicit time discretization. The time evolution of the divergence and of ζ are reported on figure (5) in the case of the driven cavity at a Reynolds number equal to 400 and show that despite the fact that relations (2) and (3) are "exactly" satisfied only at steady state, it is possible to get at each time step a significant improvement with respect to the classical approach (1 iteration) in which, the boundary conditions for the vorticity would be:

$$\omega_\Gamma^{(n+1)} = \nabla \times \mathbf{V}_\Gamma^{(n)}$$

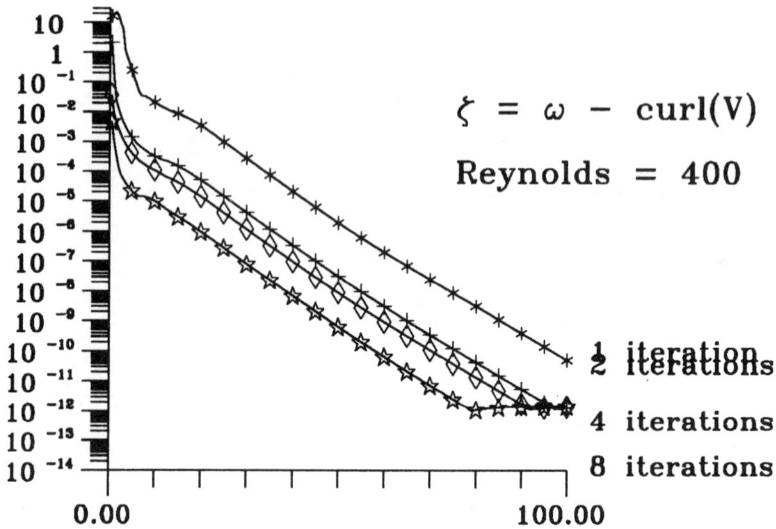

Fig. 5. The driven cavity ; time evolution of $||\zeta||_\infty$

Conclusions

In this paper we have shown that it is possible to solve the 2D Navier–Stokes equations written in velocity–vorticity formulation by means of an influence matrix technique. At each time step, a strong coupling between the vorticity and the velocity field is enforced and the continuity equation is also satisfied within machine accuracy.

Beyond this fact, we want to emphasize that meaningful physical results can still be obtained even if an "exact" definition of ω as the vorticity of the velocity field $\mathbf{V}$ on the solid boundaries is not exactly ensured at each time step. This depends on the level at which it is believed that the mass conservation equation must be satisfied.

The main question that arises now is the extension of this technique to 3D flows for which it must be noticed that the boundary conditions that have to be imposed are not so "obvious" as in the 2D case. In particular, special care has to be taken in order to ensure that the vorticity vector field $\boldsymbol{\omega}$ is solenoïdal. By virtue of remaks similar to those that were done in section 6, the number of equations whose resolution is time consuming is not 6 but 3 or 4, depending on the coupling between the different components.

Acknowledgements

The CPU time on the Fujitsu VP200 at CIRCE was provided by the "Direction Scientifique du Département SPI du CNRS".

References

[1] T.B. Gatski, C.E. Grosch, and M.E. Rose: A Numerical Study of the Two–Dimensional Navier–Stokes equations in Vorticity–Velocity Variables, *J. Comp. Phys.* **82**, 1989

[2] G.A. Osswald, K.N. Ghia, and U. Ghia: Direct Method for Solution of Three Dimensional Unsteady Incompressible Navier–Stokes Equations, in Proceeding of the 11th Int. Conf. on Num. Meth. in Fluid Dynamics, (Williamsburg, USA), ed(s) D.L. Dwoyer *et al.* Springer–Verlag, Berlin, 1989

[3] O. Daube: Resolution of the 2D Navier-Stokes Equations in Velocity–Vorticity Form by Mens of an Influence Matrix Technique, *J. Comp. Phys.* **103**, 1992

[4] C. Anderson: Vorticity Boundary Conditions and Boundary Vorticity Generation for Two Dimensional Viscous Incompressible Flows, *J. Comp. Phys.* **80**, 1989

[5] G.H. Cottet: Boundary Conditions and the Deterministic Vortex Method for the Navier Stokes equations, in Mathematical Aspects of Vortex Dynamics, ed R.E. Caflish, SIAM, Philadelphia, 1988

[6] L. Kleiser and U. Schumann: Treatment of Incompressibility and Boundary Conditions in 3–D Numerical Spectral Simulations of Channel Flows, in Proceedings of the 3rd GAMM–Conf. on Num. Meth. in Fluid Mechanics, Köln, West Germany, edited by E.H. Hirschel, (Vieweg, Wiesbaden), 1980

[7] P. Le Quéré and T. Alziary de Roquefort: Computation of Natural Convection in Two Dimensional Cavities with Chebyshev Polynomials, *J. Comp. Phys.* **57**, 1985

[8] L. Tuckerman: Divergence Free Velocity Field in Non Periodic Geometries, *J. Comp. Phys.* **80**, 1989

[9] J.M. Vanel, R. Peyret, and P. Bontoux: A Pseudo–Spectral Solution of Vorticity–Stream Function Equations Using the Influence Matrix Technique, in Proceedings of the Int. Conf. on Num. Meth. for Fluids Dynamics, Reading, United Kingdom, ed(s). K.W. Morton and M. J. Baines, (Clarendon Press, Oxford), 1986

[10] U. Ghia, K.N. Ghia and C.T. Shin: High–Re Solutions for Incompressible Flow Using the Navier–Stokes Equations and a Multi Grid Method, *J. Comp. Phys.* **48**, 1982

[11] O. Daube, J.L. Guermond and A. Sellier: Sur la formulation vitess–tourbillon des équations de Navier–Stokes en écoulement incompressible, *C.R. Acad. Sci. Paris* **313** Série II, 1991

[12] F. Stella and G. Guj: Vorticity–Velocity Formulation in the Computation of Flows in Multi-connected Domains, *Int. J. Num. Meth. in Fluids* **9**, 1989

[13] O. Daube, L. Ta Phuoc, A. Dulieu, M. Coutanceau, K. Ohmi and A. Texier: Numerical Simulation and Hydrodynamic Visualization of Transient Viscous Flow Around an Oscillating Airfoil, *Int. J. Num. Meth. in Fluids* **9**, 1989

[14] L. Ta Thuoc: Numerical Analysis of Unsteady Secondary Vortices Generated by an Impulsively Started Circular Cylinder, *J. Fluid Mech.* **100**, 1980

A COUPLED POTENTIAL-BOUNDARY LAYER CALCULATION METHOD FOR UNSTEADY FLOWS AROUND AIRFOILS

M. KERMAREC, A. F. DECAIX, P. RENON, D. FAVIER, C. MARESCA
IM2-IMFM
CNRS
163 Av. de Luminy
13009 Marseille FRANCE

1. Introduction

The present work deals with a numerical modelling of subsonic flows around 2D airfoils, submitted to pitching motions. The goal of this modelling is to get accurate prediction of the global airloads on the airfoil (lift, drag and pitching moment), and also to provide an estimate of local quantities such as friction coefficient or velocities inside the boundary-layer. The application fields of such model may be very large, since many mechanical modern devices are concerned by this kind of problems : rudders and keels of boats or submarine crafts, wings of manoeuvring planes, helicopter rotor blades, etc ...

Two kinds of methods are currently used on that subject :

- Full Navier-Stokes equations approximation methods, which are more and more efficient (see for example [Peyret & Taylor, 1983], [Daube and al., 1986], [Deane & Mavriplis, 1991], [Jong Yu and al., 1991]), but which still have difficulties to deal with large time calculations and realistic Reynolds numbers. However, the increasing power of computers is very encouraging for the future of these methods.
- Potential-boundary layer coupling methods, which allow faster calculations with less limitations on the Reynolds number values (see for example [Moses et al., 1978], [Gilmer & Bristow, 1982], [Veldman & Lindhout, 1985], [Le Balleur & al., 1986]). This kind of methods is particularly efficient when the physical basis of the modelling are respected, e.g. when the viscosity effects are located near the airfoil wall, all along it, but it also has difficulties for flows where more complex phenomenon occurs, such as static or dynamic boundary-layer separation, and static or dynamic airfoil stall.

The main feature of the coupled potential-boundary layer method presented here, based on Henderson works for steady flows ([Henderson, 1978]) lies in the way the separation zone, and the extentions of the boundary-layer downstream of the trailing edge, are taken into account. First, boundary-layer calculations are only

J.T. Beale et al. (eds.), Vortex Flows and Related Numerical Methods, 129–143.

performed until prediction of a separation point, and not downstream of that point. Secondly, the domain where the flow is assumed to be incompressible and potential is bounded by a line, the location of which is known for attached boundary-layer (displacement thickness line), but a priori unknown downstream of the separation point and of the trailing edge. For these parts of the bounding line, the location is considered as an unknown of the potential problem, and the following physical conditions are imposed :

- the line is considered as a streamline in the frame linked to the moving airfoil
- a constant pressure value is assumed on the separated boundary-layer part of the line, equal to the value at the trailing edge.
- pressure values across the wake behind the trailing edge must be equal.

The first condition is analogue to the condition imposed along the line corresponding to the attached boundary-layer. The second one comes from experimental data : for steady flows, measurements show a quasi-constant pressure along the wall where the boundary-layer is separated. This is also assumed here for unsteady flows. The last one results in the local equilibrium of the wake downstream of the trailing edge.

The panel method used for potential calculations, the vortex shedding downstream of the coupling line and the finite difference approximations used for the boundary-layer equations are described in section 2. Results obtained using this method are discussed in section 3.

2. Numerical model

2.1. COUPLING PROCEDURE BETWEEN POTENTIAL AND BOUNDARY-LAYER CALCULATIONS

The procedure is initialized by a potential calculation around the airfoil boundary, which gives the tangential velocities at the wall. Boundary-layer calculations are then performed, using these velocities as boundary condition at the outer edge of the boundary-layer. Instantaneous displacement thickness is thus obtained up to the separation point, where the boundary-layer calculation is stopped. The displacement thickness is used to modify the geometry of the airfoil until the separation point, and an initial guess is made for the geometry of the coupling line downstream that point and the trailing edge. The actual instantaneous geometry of this line is found as a result of the potential calculation, which also gives the tangential velocities along it. A new boundary-layer calculation is then performed, and so on until convergence of the results is reached.

2.2. POTENTIAL FLOW

2.2.1. Equations of the problem and approximation at a fixed time step

Outside the boundary-layer, the fluid is assumed to be incompressible and inviscid, and the flow is non rotational. In that domain, the perturbation potential ϕ has to verify the Laplace equation. The boundary conditions that are to be verified on the final geometry of the coupling line between the two calculation domains are the

following (see figure 1) :
- No-flow condition on (Σ), which is the coupling line corresponding to the attached boundary-layer, and on $(\Sigma_d), (\Sigma^+)$ and (Σ^-) which denote respectively the separation zone, and the upper and lower sides of the boundary-layer extensions downstream of the trailing edge (see figure 1). This condition may be written as:

$$\frac{\partial \phi}{\partial n}|_{\text{boundary}} = \mathbf{u}_L \cdot n - \mathbf{V}_\infty \cdot n \tag{1}$$

where $\mathbf{u}_L$ is the airfoil driving velocity for (Σ), and the displacement velocity of boundary $(\Sigma_d), (\Sigma^+)$ and (Σ); $\mathbf{V}_\infty$ is the freestream velocity.
- on (Σ_d), the pressure condition $p_d(x,t) = f(t) = p_I(t)$, which may be written as:

$$\frac{\partial \phi_d}{\partial t} - \mathbf{u}_{L_d} \cdot \mathbf{V}_d + \frac{1}{2}|\mathbf{V}_d|^2 = \frac{\partial \phi_I}{\partial t} - \mathbf{u}_{L_I} \cdot \mathbf{V}_I + \frac{1}{2}|\mathbf{V}_I|^2 \tag{2}$$

where the subscripts d and I refer respectively to quantities defined on the coupling line (Σ_d) and at point I (see figure 1), and $\mathbf{V}$ is the absolute velocity.
- on (Σ^+) and (Σ^-), the condition $p^+(M^+,t) = p^-(M^-,t)$, which gives :

$$\frac{\partial \phi^+}{\partial t} - \mathbf{u}_L^+ \cdot \mathbf{V}^+ + \frac{1}{2}|\mathbf{V}^+|^2 = \frac{\partial \phi^-}{\partial t} - \mathbf{u}_L^- \cdot \mathbf{V}^- + \frac{1}{2}|\mathbf{V}^-|^2 \tag{3}$$

where superscripts + and - refer to opposite points on boundaries (Σ^+) and (Σ^-). The geometry of boundaries $(\Sigma_d), (\Sigma^+)$ and (Σ^-) where the conditions (1), (2), and (3) are fullfilled is unknown. The calculation is started with an initial location of these boundaries, and the numerical procedure leading to the final geometry is the following :
- Until convergence is reached, pressure conditions (2) and (3) are imposed, but there is no reason for the no-flow equation (3) to be verify. This equation is thus replaced by :

$$\frac{\partial \phi}{\partial n}|_{\text{boundary}} = \mathbf{u}_L \cdot n - \mathbf{V}_\infty \cdot n + v_n \tag{4}$$

where v_n is the normal velocity across the boundaries.
- To insure a zero value of flow across (Σ^+) and (Σ^-), we also impose :

$$v_n^+ = -v_n^- \tag{5}$$

on opposite points M^+ and M^- of the boundaries (Σ^+) and (Σ^-).
- For a given geometry of $(\Sigma_d), (\Sigma^+)$ and (Σ^-), the solution of the Laplace equation on the potential domain, with the boundary conditions (2) to (5), is approximated by a panel method, resulting into a discrete set of doublets μ on panels over the whole boundary, and discrete values of v_n on panels of $(\Sigma_d), (\Sigma^+)$ and (Σ^-).
- A new geometry of $(\Sigma_d), (\Sigma^+)$ and (Σ^-) is provided by rotating each panel in the relative velocity direction calculated at its middle, and a new calculation is performed around this new geometry.

The procedure is repeated until each discrete normal velocity reaches a quasi-zero value. Generally, only three or four iterations on the geometry are necessary to converge.

2.2.2. Vortex shedding downstream the coupling boundary

Time approximation of the potential problem is made by a step by step method. Between two time steps, a vortex shedding procedure, similar to the one introduced by Basu and Hancock [Basu & Hancock, 1978] is performed. At a given time t_n, convergence in the geometry iterations gives the locations of $(\Sigma_d), (\Sigma^+)$ and (Σ^-). The orientation of a so-called vortex shedding panel (S) and the locations, downstream the point F, of vortices shed at previous times are also known. The starting geometry at time t_{n+1} is obtained as follows:
- (S), $(\Sigma_d), (\Sigma^+)$ and (Σ^-) are moved with the airfoil driving velocity.
- previous vortices are convected. The convection velocity is taken as the fluid velocity at the centre of each vortex.
- (S) panel is convected with the absolute fluid velocity in its middle, calculated at time t_n. This panel supports a constant value of doublets equivalent to two vortices located at the ends of the panel. The panel is then replaced by these two vortices.
- a new (S) panel is created, and is defined as the segment joining the vortex of the previous panel to the point F (see figure 2). During the iterations for finding the coupling boundary geometry, the new (S) panel is also rotated in the direction of the absolute fluid velocity calculated at its middle.

2.3. APPROXIMATION OF BOUNDARY-LAYER EQUATIONS

2.3.1. Equations and approximations

Eddy viscosity turbulence models give the following non dimensional equations of the boundary-layer, expressed in terms of pressure and relative velocity in the frame linked to the airfoil [Monavon, 1986]:

$$\frac{\partial p}{\partial y} = 0 \tag{6}$$

$$\frac{\partial u}{\partial t} + u\frac{\partial u}{\partial x} + \frac{\partial u}{\partial y}(v - \frac{\partial}{\partial y}(\frac{1}{Re_t})) - (\frac{1}{Re} + \frac{1}{Re_t})\frac{\partial^2 u}{\partial y^2} = \frac{\partial u_p}{\partial t} + u_p\frac{\partial u_p}{\partial x} \tag{7}$$

$$\frac{\partial u}{\partial x} + \frac{\partial v}{\partial y} = 0 \tag{8}$$

where u and v are respectively the streamwise and normal velocities in the boundary-layer, normalized by $\mathbf{V}_\infty$, u_p is the velocity at the outer edge of the boundary-layer, normalized by $|\mathbf{V}_\infty|$, Re and Re_t are the Reynolds numbers, based on the chord c and respectively the kinematic viscosity ν and the eddy viscosity ν_t. The boundary conditions are: $u = u_p$ at the outer boundary-layer limit, and $u = v = 0$ at the wall. A frequency approach is used for the time approximation, that significantly reduces the calculation time. Each unknown is written as:

$$f = f_0 + f_1 \cos \omega t + f_1 \sin \omega t \tag{9}$$

where f stands either for u, v, $1/Re_t$, or u_p so that equations (7) and (8) lead to a non linear system, of 6 equations and 6 unknowns. An approximate solution of this system is obtained by means of a finite difference method, with an upwind scheme in the wall tangential direction. Non linearities are taken into account by a fixed point algorithm.

2.3.2. Eddy viscosity model

For steady flows, the classical mixing length expressions of Cebeci & Smith ([Cebeci & Smith, 1974]) have been used. For unsteady flows, comparisons between calculations and experiments have suggested (see [Favier & al, 1991]) to introduce an empirical time delay for the phases of the eddy viscosity first harmonic. If $(\nu_{t0})_{cbs}$ is the eddy viscosity mean term of the Cebeci and Smith model, and $(\nu_{t1})_{cbs}$ the complex number defined by $(\nu_{t1})_{cbs} = (\nu_{t1}^{*})_{cbs} + i(\nu_{t1}^{**})_{cbs}$, where (ν_{t1}^{*}) and $(\nu_{t1}^{**})_{cbs}$ are the first harmonic terms of Cebeci and Smith, the following expressions have been used:

$$\nu_{t0} = (\nu_{t0})_{cbs} \tag{10}$$

$$\nu_{t1} = (\nu_{t1})_{cbs} \exp i\phi_{\text{delay}} \tag{11}$$

where $\phi_{\text{delay}} = -\beta \cdot k$. β is an empirical constant, which has been adjusted to give a good agreement between calculations and experiments. Satisfactory results were obtained with the value $\beta = 4$.

3. Results

3.1. TESTING THE COUPLING METHOD FOR STEADY FLOWS

Results obtained by the present numerical method have been compared to wind tunnel measurements performed in 2-D flows on NACA0012 and ONERA OA209 airfoils. Experiments have been conducted in the same S2-Luminy wind-tunnel [Favier et al , 1992] on airfoils models with the following dimensions: for the NACA0012 airfoil: span 0.495m, chord 0.3m; for the OA209 airfoil: span 0.495m, chord 0.2m.

3.1.1. Measured and computed lift coefficients.

A main interest of the proposed model is its efficiency to predict lift coefficient for angles of attack near or greater than the static stall incidence. For such angles of attack, lift coefficient is mainly dependent of the location of the upper side separation point, as is illustrated in figure 3. The computational results of this figure correspond to arbitrary separation point locations, imposed in the potential calculation without taking into account the boundary-layer result. Figure 4 shows two different geometries of the coupling line, associated to different separation point locations. A good agreement between calculated and measured lift coefficients may thus be obtained only if the location of the separation point is well predicted. This location depends on :
- boundary-layer separation criterion
- accuracy of the approximate solution of the boundary-layer equations
- transition criterion.

Concerning the first point, different criteria have been checked :
- $H > 2.35$ where H is the shape factor
- Smith criteria ([Smith, 1975])
- Non convergence of the boundary-layer approximation scheme.

All these criteria give approximately the same results, within a variation of 1% of the chord length on the separation point location. The accuracy of the boundary-layer calculations has been checked by comparisons with calculated and measured velocity profiles inside the boundary-layer developing around a flat plate (see [Maresca et al., 1990] and [Favier et al., 1991]). The results of the computations agree well with experiments if the path of the finite differences approximations is sufficiently small. The convergence of the approximation, with good agreement with experiments, has been reached for example at the following values of the non dimensional paths :
- $2.5\,10^{-4}$ for Reynolds numbers of $3.\,10^5$. (e.g. 4000 geometrical steps on the tangential direction on each side of the airfoil)
- 5.10^{-5} for Reynolds numbers of 3.10^6 (20000 geometrical steps on each side of the airfoil).

Concerning the transition criterion, previous studies, [Favier et al., 1991], have shown the interest of empirical transition criterion based on the energy thickness δ_3' of the boundary-layer, as defined by:

$$(Re\delta_3')_T = a_0 + a_1\left(1 + a_2 \exp\left(-\frac{Re_s}{a_3}\right)\right)(Re_s)^{a_4} \tag{12}$$

where $\delta_3' = \int_0^\delta \left(\frac{u}{u_p}\right)^2 (1 - \frac{u}{u_p})\, dy$, and Re_S is the Reynolds number based on local curvilinear coordinates. The coefficients a_i are empirical constants which have been set to the following values :

$$a_0 = 12 \quad a_1 = 0.015 \quad a_2 = 1.34 \quad a_3 = 10^7 \quad a_4 = 0.68$$

Finally, the whole numerical procedure described in section 2 has been checked by comparing measured and computed lift coefficient, for two airfoils shapes (NACA0012 and ONERA OA209 airfoils), and two values of the Reynolds number. Figure 5 shows that even for angles of attack strongly larger than the static stall incidence, there is a good agreement between measurements and computations.

3.1.2. Computations vs measurements for drag coefficients

Drag coefficients directly depend on the transition criterion which is used. In order to check this influence, calculations have been made with the following different criteria :
- Michel criterion ([Michel, 1952])
- Transition imposed when the non-convergence of the laminar boundary-layer approximation scheme is reached
- Criterion defined by equation (12), with the previous values of empirical constants.

For a NACA0012, at $Re = 3.\,10^6$, figure 6 shows the discrepancies between experiments and computations with the first two criteria. Good results are obtained for the two airfoils shapes, at two different Reynolds numbers, with the third one (see figures 7 and 8).

3.2. UNSTEADY FLOWS

3.2.1. Pitching motion below dynamic stall

The airfoil is now submitted to a pitching motion. The angle of attack is a function of time, given by $\alpha = \alpha_0 + \Delta\alpha \cos\omega t$.The values of α_0 and $\Delta\alpha$ are chosen in the first place in order to avoid dynamic stall of the airfoil. For easy analysis, quantities such as the lift coefficient may be written as :

$$Cl = Cl_0 + Cl_1 \cos\omega t + \phi_1$$

Figure 9 shows a comparison between the measured lift coefficient ([Coulomb, 1971]) and the computed ones by two methods, for a NACA0012 airfoil submitted to a pitching motion characterized by $\alpha_0 = 6°, \Delta\alpha = 3°, k = .125$, where k is the reduced frequency ($k = \omega c/2|\mathbf{V}_\infty|$). The "pure potential" curve corresponds to a calculation where there was no boundary-layer or wake. A potential calculation was performed around the exact airfoil geometry, and the vortex shedding was made at the trailing edge in accordance with the Basu & Hancock numerical procedure. The discrepancies between the two calculated results confirm that it is important to take into account the boundary-layer by performing the research procedure of the coupling line location. In terms of the first Fourier coefficients, the effects of coupling are mainly to decrease Cl_0, and to decrease the phase advance (ϕ_1 is positive, and decreases), which is physically coherent : viscosity smoothes down the low pressure peak near the leading edge, which in turn decreases the lift coefficient, and induces a delay in the flow response. Nevertheless, some discrepancies still exist between experiments and calculations with the coupling method, in the right part of the curve. This is not the case in figure 10, where pitching motion is now defined by :$\alpha_0 = 6°, \Delta\alpha = 6°, k = .314$, and where the coupling method results agree well with experiments. The "potential+extensions" curve has been obtained by taking into account a guess for the displacement thickness δ_1 of the boundary-layer along the wall. This calculation is still a potential calculation, but an arbitrary constant value has been given to δ_1, and the geometry of this virtual boundary-layer extension is looked for at each time step by the method described in section 2.2. The discrepancies between the "pure potential" and the "potential+extensions" curves show the influence of the wake, and the discrepancies between this last curve and the results of the complete coupling method the one of viscous effects.

3.2.2. Capability of the method to describe dynamic stall

The capability of the present method to predict lift coefficient value through dynamic stall, has been checked in the following way : for a particular motion case, an empirical law (see figure 11), defining the location of the turbulent separation point of the boundary-layer is introduced in the potential calculation. There was no boundary layer calculation, and an arbitrary constant in time value is given to the displacement thickness. Numerical research of the coupling line geometry downstream the separation point and the trailing edge is performed. Comparison between calculated and measured lift coefficients, for a mean angle-of-attack $\alpha_0 = 9°$, a reduced frequency $k = 0.135$, and an amplitude of variation $\Delta\alpha$ equal to 6°, is illustrated in figure 12. The numerical prediction of lift is quite close to measurements, except for the part of the hysteresis loop corresponding to the boundary-layer reattachment (left part of the curve).

4. Conclusion

Validations and developments of this coupling method are up to now still incomplete, since the present tests did not yet deal with complex flows, such as those where dynamic stall occurs. Nevertheless, tests on simpler flows have revealed the interest of this model, and also the importance of some features of the method. For steady flows case, at quite high Reynolds numbers (from 3.10^5 to 3.10^6), and angles of attack near or larger than the static stall incidence, calculations have shown the importance of a good prediction of the upper side separation point, in order to get good estimates of the lift coefficient. This has been obtained, on the one hand by using an empirical transition criterion coming from velocity measurements inside the boundary-layer around a flat plate model, and on the other hand by an accurate approximation of the boundary-layer equations. Computed lift values are then very close to measured ones. Moreover, drag coefficients are also well predicted by the method, as a function of the incidence. Using the method for unsteady flows around airfoils submitted to pitching motions below dynamic stall, has pointed out the influence of the viscous effects, and of the instantaneous location of the coupling line between the two computational domains. For two different flow cases (e.g. different reduced frequencies and amplitudes of oscillation) the lift coefficients measured and calculated by the present method are in good agreement. Finally, a preliminary calculation has been performed for a pitching motion through dynamic stall, in order to evaluate the capabilities of the model to well reproduce such phenomenon. An empirical location of the separation point during the oscillating period has been introduced in the potential calculation; comparisons between measured and calculated lift coefficients are quite close, which is very encouraging for the interest of this model applied to such flow cases. Next steps will be to continue similar preliminary calculations, but for deeper dynamic stall cases, and to modify the boundary-layer approximation method to account for time dependence, in order to predict dynamic separation of the boundary-layer.

References

[1] Abbott I.H., Von Doenhoff A.E.: 1958, 'Theory of wing sections', Dover Publications, INC., 180 Varick Street, New York, USA.

[2] Basu B.C., Hancock G.J.: 1978, 'The unsteady motion of a two dimensional aerofoil in incompressible and inviscid flow', *Journal of Fluid Mechanics*, **87**, pp.159-178

[3] Belleudy J.: 1991 , 'Influence du couplage de variations de vitesses et d'incidence sur le décrochage dynamique d'un profil: application au roror d'hélicoptère', PhD Thesis, University Aix-Marseilles II.

[4] Bezard H.: 1989 , 'Exploitation des essais complémentaires sur des profils pour roror proincipal', ONERA Final Report OA-41, Mai 1989.

[5] Cebeci T., Smith A.M.O.: 1974, 'Analysis of turbulent boundary-layers', *Appl. Math. and Mech.*, **15**, Academic Press.

[6] Coulomb J.: 1971, 'Mesures instationnaires sur NACA0012 en oscillation de tangage', Internal report CEAT.

[7] Daube O., Loc T. P., Monnet P., Coutanceau M.,:1986, 'Ecoulement instationnaire décollé d'un fluide incompressible autour d'un profil : une comparaison théorie-expérience', AGARD CP 386.

[8] Deane A.E., Mavriplis C.: 1991, 'Low-Dimensional description of the dynamics in separated flow past thick airfoils', AIAA 91-1622, 22nd Fluid Dynamics, Plasma Dynamics and Lasers Conference, June 24-26, Honolulu, Hawai.

[9] Favier D., Agns A., Barbi C., Maresca C.: 1988, 'Combined Translation/Pitch Motion : a New Airfoil Dynamic Stall Simulation', *AIAA Journal of Aircraft*, **25**, pp. 805-814.

[10] Favier D., Maresca C., Renon P.: 1991, 'Etude du comportement de la couche limite sur un modèle oscillant par vélocimétrie laser embarquée', 28ème colloque d'Aérodynamique appliquée de l'Association Aéronautique et Astronautique de France, Saint Louis.

[11] Favier D., Maresca C., Petit L., Vingut G.: 1992, 'Décrochage dynamique de profils oscillants: étude expérimentale et numérique appliquée au rotor d'hélicoptère', STPA contrat, Final Report 90-91004.

[12] Gilmer B.R., Bristow D.R.: 1982, 'Analysis of stalled airfoils by simultaneous perturbations to viscous and inviscid equations', *AIAA Journal*, vol. 20, n9.

[13] Henderson M.L.: 1978, 'Two-dimensional separated wake modeling and its use to predict maximum section lift coefficient', AIAA Paper 78-156.

[14] Jong Yu N., Allmaras S.R., Moschetti K.G.: 1991, 'Navier-Stokes calculations for attached and separated flows using different turbulence models', AIAA 91-1622, 22nd Fluid Dynamics, Plasma Dynamics and Lasers Conference, June 24-26, Honolulu, Hawai.

[15] Le Balleur J. C., Girodroux-Lavigne, P.: 1986, 'A viscous-inviscid onteraction method for computing unsteady transonic separation', Numerical and Physical aspects of aerodynamic flows III, chapter 14, pp. 253-271, Springer Verlag, New York.

[16] Maresca C., Favier D., Belleudy J.: 1990, 'A New LDA System for Unsteady Aerodynamics of Moving Wall', ASME Proccedings of the International Symposium of Nonsteady Fluid Dynamics, Toronto, June 1990, pp. 303-310.

[17] Michel R.: 1952, 'Détermination du point de transition et calcul de la trainée des profils d'aile en incompressible', ONERA Publication 58

[18] Monavon A.: 1986, 'Etude d'une couche limite instationnaire sur une paroi déformable', Thèse de Doctorat d'Etat, Université de Paris.

[19] Moses H.L., Jones R.R., O'Brien W.F. Jr, Peterson R.S.: 1978, 'Simultaneous solution of the boundary-layer and free-stream with separated flow', *AIAA Journal*, vol. 16.

[20] Peyret R., Taylor T.: 1983, 'Computational Methods for Fluid Flows', Springer Series in Computational Physics, Springer Verlag 1983.

[21] Smith A.M.O.: 1975, 'High lift aerodynamics, AIAA Journal of Aircraft', vol. 12, June 1975.

[22] Veldman A.E.P., Lindhout J.P.F.: 1985, 'Quasi-simultaneous calculations of strongly interacting viscous flow', Third Symposium on Numerical and Physical Aspects of Aerodynamic Flows, Long Beach (California)

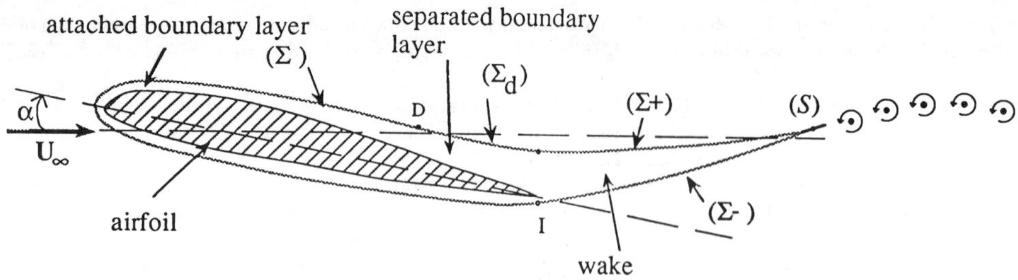

Figure 1: coupling boundary view

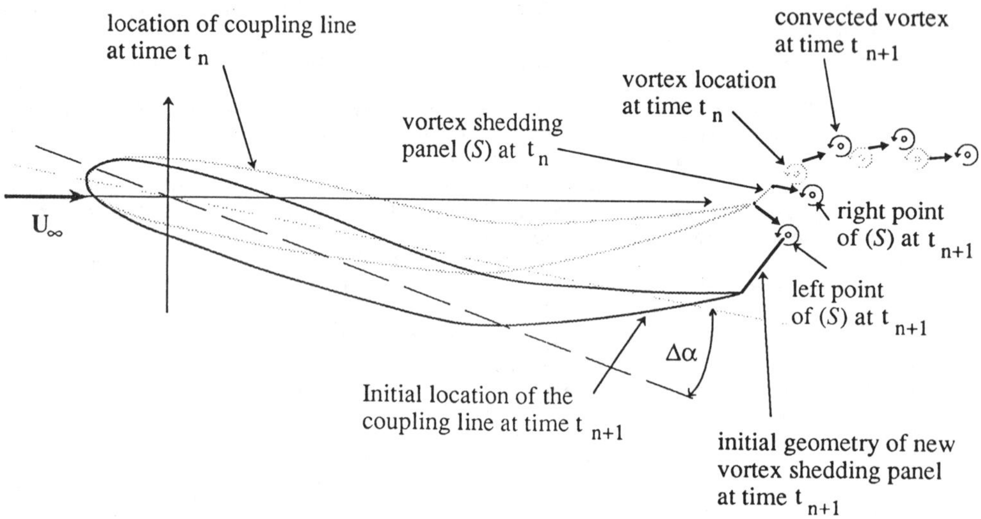

Figure 2: vortex shedding scheme

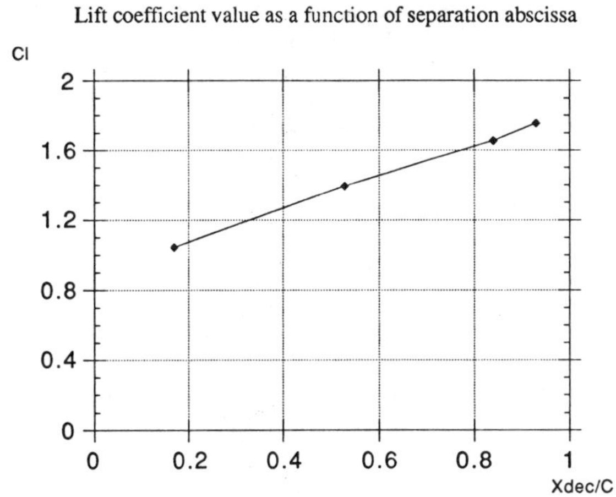

Figure 3: influence of the separation point location on the lift coefficient, for a steady flow around a NACA0012 airfoil, and a large angle of attack ($\alpha = 16°$, $Re = 3.10^5$)

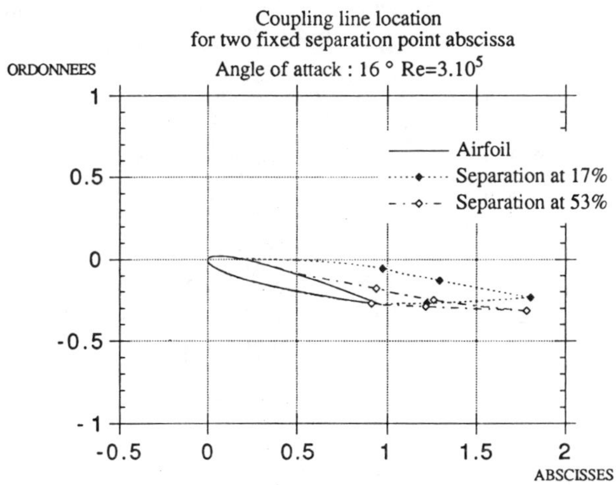

Figure 4: location of the coupling line for two separation point locations (17% and 53% of the chord) on the upper side of a NACA0012 airfoil at $\alpha = 16°$ and $Re = 3.10^5$.

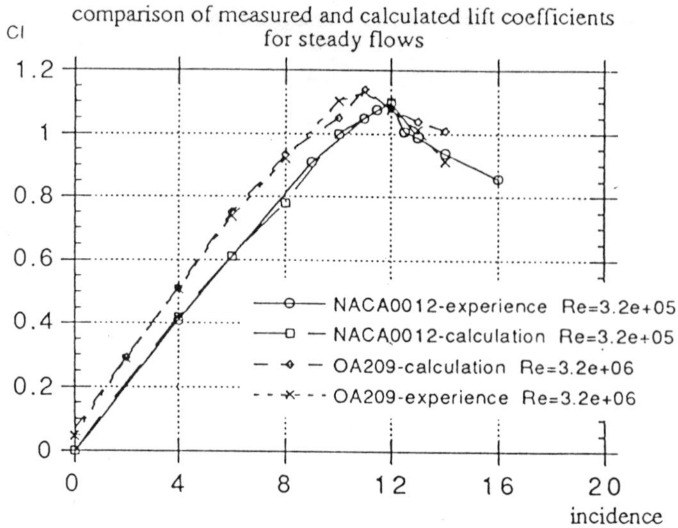

Figure 5: comparison of calculated and measured lift coefficients for a steady flow around NACA0012 and ONERA OA209 airfoils, at moderate values of Reynolds number ($Re = 3.210^5$ and 3.10^6). Experiments [Bezard, 1989] for the OA209 airfoil, and [Favier et al., 1992] for the NACA0012 airfoil.

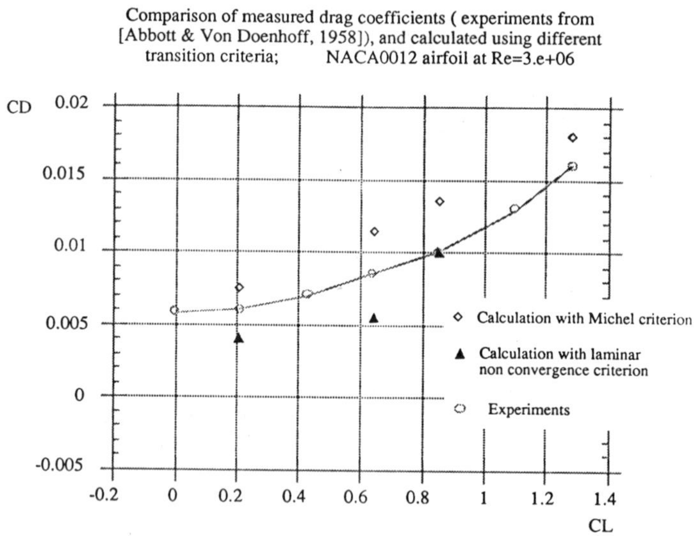

Figure 6: drag coefficients for two different transition criteria ([Michel, 1952] and non convergence of laminar boundary-layer approximation scheme (LNC)). Experiments from [Abbott & Von Doenhoff, 1958].

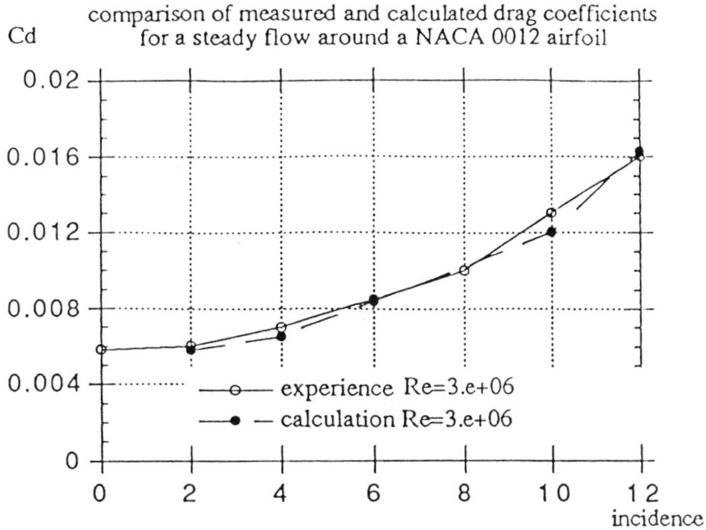

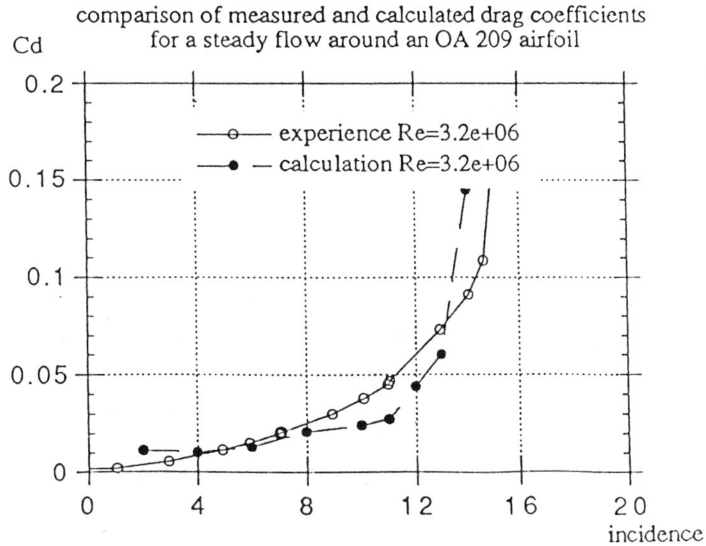

Figure 7-8: drag coefficients of NACA0012 and ONERA OA209 airfoils with the new transition criterion. Experiments [Bezard, 1989] for the OA209 airfoil, and [Abbott & Von Doenhoff, 1958] for the NACA0012 airfoil

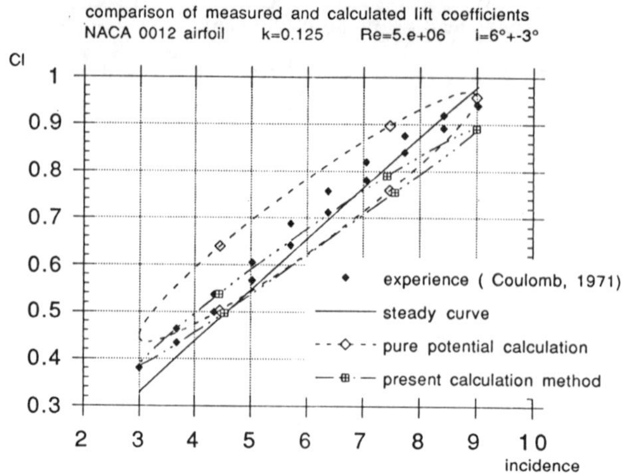

Figure 9: comparisons of measured and calculated lift coefficients for pitching motion of airfoil ($\alpha_0 = 6°, \Delta\alpha = 3°, k = .125$). Measurements from [Coulomb, 1971]

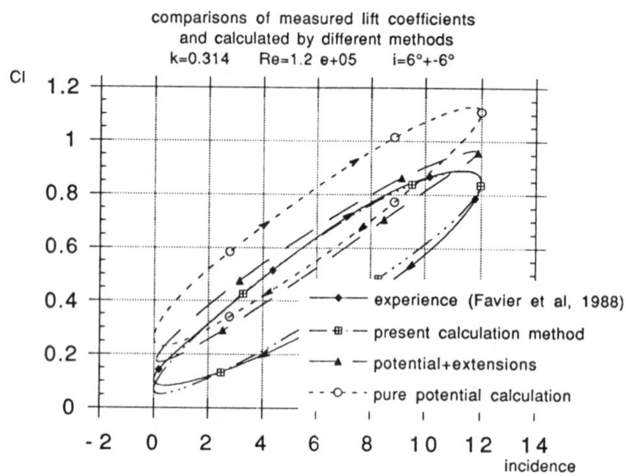

Figure 10: comparisons of measured and calculated lift coefficients for pitching motion of airfoil ($\alpha_0 = 6°, \Delta\alpha = 6°, k = .314$). Measurements from [Favier et al., 1988]

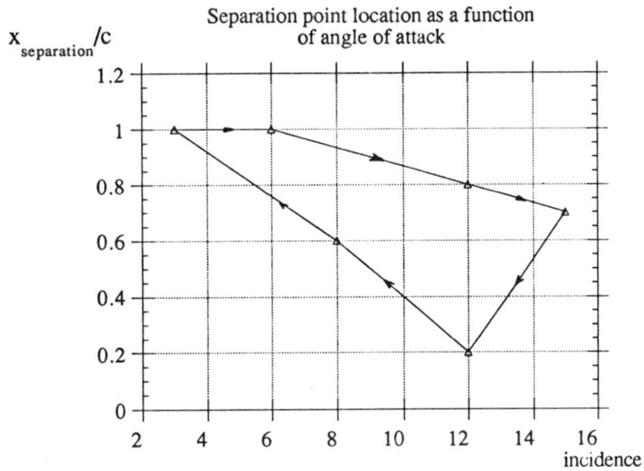

Figure 11: empirical separation point location law

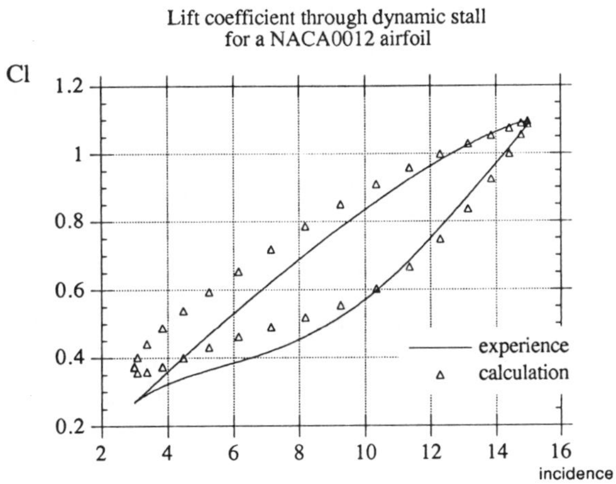

Figure 12: comparisons of measured and calculated lift coefficients for pitching motion of airfoil through dynamic stall. Measurements from [Belleudy, 1991], calculations performed with empirical law of figure 11

VISCOUS SIMULATION OF WAKE PATTERNS

ROBERT KRASNY
Mathematics Department
University of Michigan
Ann Arbor
MI 48109 USA

ABSTRACT. Vortex sheet motion is regularized by smoothing the initial data and solving the Navier-Stokes equations. The underlying flow is induced by a vortex sheet containing positive and negative circulation. Wake patterns are obtained which closely resemble previous experimental results and vortex blob simulations.

1. Introduction

A basic premise in fluid dynamics, due to Prandtl, is that a free shear layer converges to a vortex sheet in the zero viscosity limit. From this point of view, one can gain insight into shear layer dynamics at high Reynolds number by studying vortex sheet motion.

The initial value problem for a vortex sheet is ill-posed in the sense of Hadamard, but an analytic solution exists locally [20]. Asymptotic analysis and numerical work indicate that the sheet develops a singularity in finite time [13,16,17,19]. Several methods of regularization have been proposed to extend the sheet's motion past the critical time, into the roll-up regime. The vortex blob method introduces an artificial smoothing parameter to regularize the Birkhoff-Rott integral [1,5,8,14]. Convergence of the smoothing process prior to the critical time has been established [6]. Numerical results indicate that past the critical time, the regularized solutions converge to a well-defined spiral when the smoothing parameter is reduced [14,15].

An important problem is to determine whether the physical regularization due to viscosity gives the same result as the vortex blob regularization. Numerical evidence indicates that the two regularizations do converge to the same limit for a vortex sheet with circulation of one sign, as in a mixing layer [21]. With vortex sheet initial data of this type, there also exists a global weak solution of the Euler equations [10,11].

Another way to regularize the vortex sheet problem is to replace the sheet by a layer of constant vorticity which evolves according to the Euler equations. The initial value problem for the layer is globally well-posed [4,7]. Before the critical

J.T. Beale et al. (eds.), Vortex Flows and Related Numerical Methods, 145–151.

time, the layer converges to the vortex sheet in the zero thickness limit $H \rightarrow 0$ [2,3]. It seems that past the critical time, the finite-thickness regularization does not converge to the same limit as the vortex blob and the viscous regularizations. However, the motion of a piecewise constant vorticity layer does resemble viscous simulation and experiment [12].

The present article concerns the case in which the underlying vortex sheet has circulation of both signs, as in a wake. This work was motivated by the experiments of Couder and Basdevant on the wake of a solid body in a thin soap-film [9]. They found that vortex couples form in a forced wake, with different patterns developing as the forcing parameters were varied. A previous article showed that the experimental wake patterns can be captured by varying the initial conditions in the vortex blob method [15]. The present article extends the investigation by showing that similar patterns are also obtained by solving the Navier-Stokes equations with the vortex blob initial data.

2. Numerical Method

Let $\omega(x,y,t)$ be the vorticity, $\psi(x,y,t)$the stream function and ν the viscosity. The 2-d Navier-Stokes equations in vorticity/stream form are

$$\omega_t + \mathbf{u} \cdot \nabla\omega = \nu\Delta\omega \quad , \tag{1}$$

$$\Delta\psi = -\omega \quad , \tag{2}$$

$$\mathbf{u} = (\psi_y \ , \ -\psi_x) \ . \tag{3}$$

The initial data is obtained from a vortex sheet $(x_0(a), y_0(a))$ having circulation density $\sigma(a)$. Let δ be the artificial smoothing parameter. The Birkhoff-Rott integral [5] is regularized and differentiated to obtain the initial vorticity

$$w_0(x,y) = \pi\delta^2 \int_0^1 \frac{(\ \cosh 2\pi(y - y_0(a)) + \cos 2\pi(x - x_0(a))\ )\ \sigma(a)\ da}{(\ \cosh 2\pi(y - y_0(a)) - \cos 2\pi(x - x_0(a)) + \delta^2)^2} \ . \tag{4}$$

The viscosity was $\nu=10^{-4}$ and the smoothing parameter was $\delta=10^{-1}$. With $U=||\sigma||_\infty=$ the Reynolds number based on initial layer thickness is

$$Re = \frac{U\delta}{\nu} = 10^3 \ . \tag{5}$$

The domain $0 \leq x \leq 1, \ -1 \leq y \leq 1$ was discretized by a 256 x 512 uniform mesh. Central differences were used in space. Time integration was performed using the 4th order Runge-Kutta method with time step $\Delta t = 10^{-2}$. The Poisson equation (2) was solved using a routine from FISHPACK [18] with periodic boundary conditions at $x = 0, 1$ and free-slip conditions at $y = \pm 1$. The integral in equation (4) was evaluated by the trapezoid rule with increment $\Delta a = 0{\cdot}0025$. The curve $(x_0(a), y_0(a))$ was passively advected for diagnostic purposes. Bilinear interpolation was used to obtain the curve velocity from the fluid velocity on the mesh. To maintain the curve's resolution, additional points were inserted during the computation [15].

case	$x_0(a)$	$y_0(a)$	$\sigma(a)$
1	a	$0.20 \sin 2\pi a$	$\sin 2\pi a$
2	$a + 0.10 \sin 2\pi a$	$0.20 \sin 2\pi a$	$\sin 2\pi a$
3	$a + 0.05 \sin 4\pi a$	$0.05 \sin 4\pi a$	$\sin 4\pi a$; $0.0 \leq a \leq 0.5$
		$-0.05 \sin 4\pi a$	$-\sin 4\pi a$; $0.5 \leq a \leq 1.0$

TABLE I
Initial data: vortex sheet $(x_0(a), y_0(a))$, circulation density $\sigma(a)$.

3. Numerical Results

Table 1 contains (x_0, y_0) and σ for the 3 cases computed. The initial conditions lie within a restricted class of sinusoidal perturbations. The various amplitudes were chosen to obtain close agreement with the experimental results [9]. Computed results on the time interval $0 \leq t \leq 8$ are shown in Figures 1,2,3. The passive curve is plotted on the left and vorticity contours are plotted on the right.

4. Discussion

The passive curves obtained here are in close agreement with experimental wake patterns [9] and vortex blob simulations [15]. This is consistent with the previously observed close agreement between the viscous and vortex blob regularizations for a vortex sheet with circulation of one sign [21]. Such results indicate that the vortex blob method gives the physically correct extension of vortex sheet motion. Close agreement between these two regularizations can only hold for a finite time. The Navier-Stokes equations are dissipative and the circulation decays to zero as $t \to \infty$. On the other hand, the vortex blob method is conservative and the circulation does not decay. A more detailed study of these issues is in preparation.

Acknowledgements

This work was supported by NSF grant DMS-9204271. The computations were performed at the University of Michigan and the San Diego Supercomputer Center.

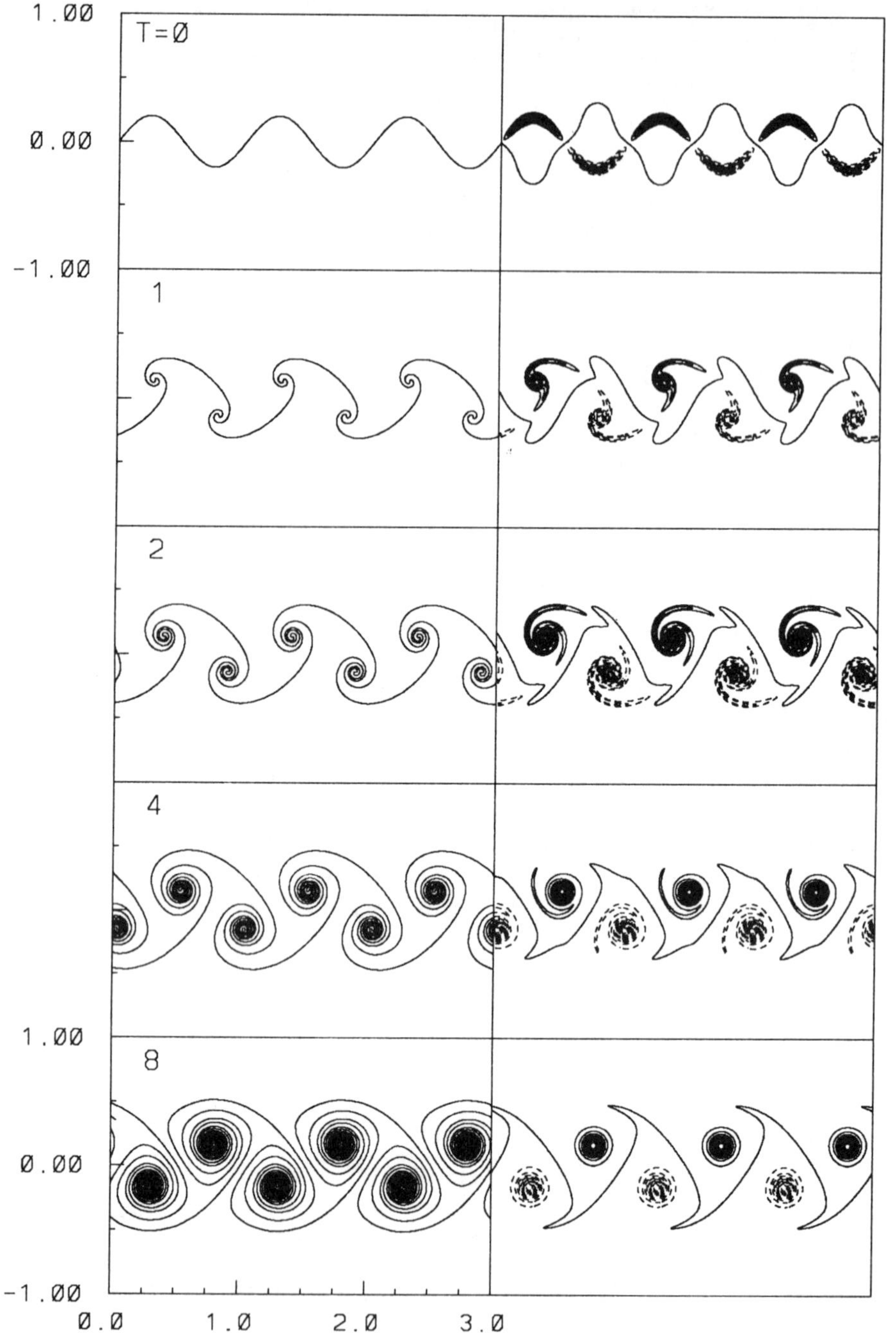

Fig. 1. The initial curve contains a transverse sinusoidal perturbation. The curve rolls up into an array of counter-rotating vortices resembling the unforced experimental wake [9,15]

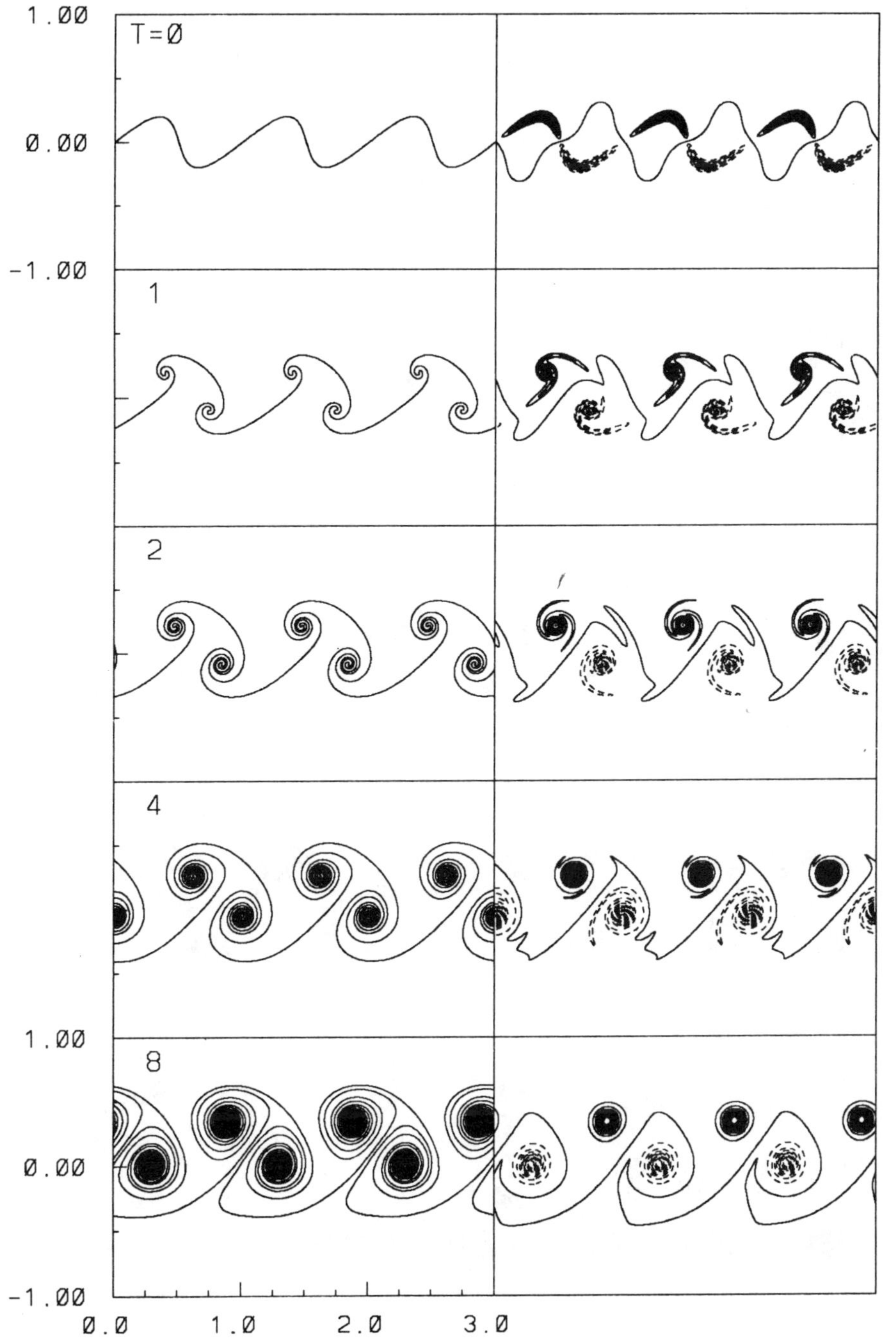

Fig. 2. A longitudinal perturbation has been added to the shape of the initial curve. Vortex couples propagate into the region $y > 0$ as in a forced wake experiment [9,15].

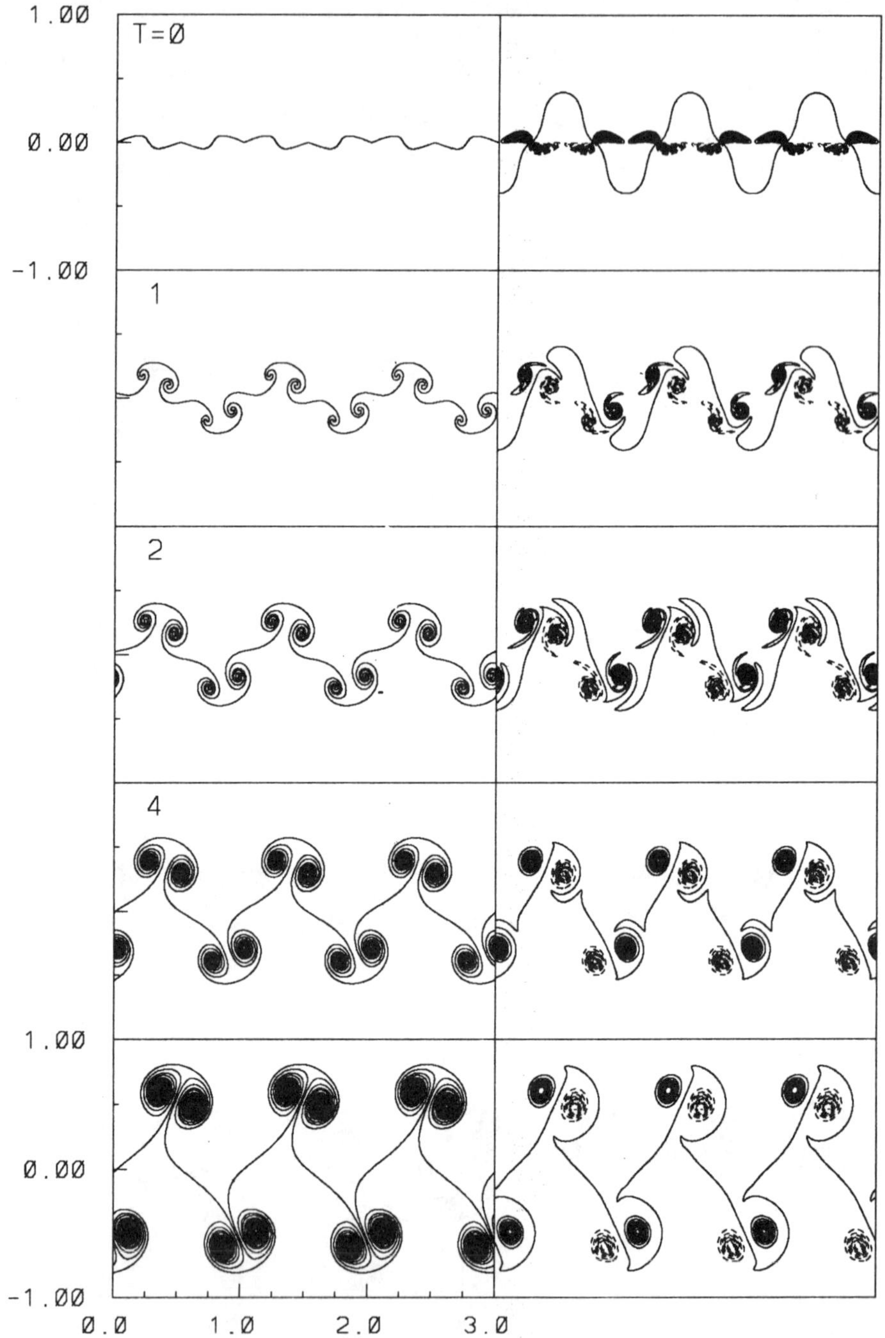

Fig. 3. The initial perturbation is reduced in scale and a reflection about $y = 0$ is added. Vortex couples propagate into the regions $y > 0$ and $y < 0$ as in another forced wake experiment [9,15]).

References

[1] 1. Anderson, C. (1985) A vortex method for flows with slight density variations. *J. Comp. Phys.* **61**, 417.

[2] 2. Baker, G. R. & Shelley, M. J. (1990) On the connection between thin vortex layers and vortex sheets *J. Fluid Mech.* **215**, 161.

[3] 3. Benedetto, D. & Pulvirenti, M. (1992) From vortex layers to vortex sheets. *SIAM J. Appl. Math.* **52**, 1041

[4] 4. Bertozzi, A. L. & Constantin, P. (1992) Global regularity for vortex patches. preprint.

[5] 5. Birkhoff, G. (1962) Helmholtz and Taylor Instability. *Proc. Symp. Appl. Math.* **XIII** , 55.

[6] 6. Caflisch, R. & Lowengrub, J. (1989) Convergence of the vortex method for vortex sheets. *SIAM J. Numer. Anal.* **26**, 1060.

[7] 7. Chemin, J.-Y. (1991) Persistance de structures geometriques dans les fluides incompressibles bidimensionnels. preprint.

[8] 8. Chorin, A. J. & Bernard, P. S. (1973) Discretization of a vortex sheet with an example of roll-up *J. Comp. Phys.* **13**, 423.

[9] 9. Couder, Y. & Basdevant, C. (1986) Experimental and numerical study of vortex couples in two-dimensional flows. *J. Fluid Mech.* **173**, 225.

[10] 10. Delort, J.-M. 1991 Existence de nappes de tourbillon en dimension deux. *J. Amer. Math. Soc.* **4**, 553.

[11] 11. DiPerna, R. J. & Majda, A. (1987) Concentrations in regularizations for 2-d incompressible flow. *Comm. Pure Appl. Math.* **XL**, 301.

[12] 12. Jacobs, P. A. & Pullin, D. I. (1989) Multiple-contour-dynamic simulation of eddy scales in the plane shear layer. *J. Fluid Mech.* **199**, 89.

[13] 13. Krasny, R. (1986) A study of singularity formation in a vortex sheet by the point vortex approximation *J. Fluid Mech.* **167**, 65.

[14] 14. Krasny, R. (1986) Desingularization of periodic vortex sheet roll-up. *J. Comp. Phys.* **65**, 292.

[15] 15. Krasny, R. (1991) Vortex sheet computations: roll-up, wakes, separation. *Lect. 'Appl. Math., AMS* **28**, 385.

[16] 16. Meiron, D. I., Baker, G. R., & Orszag, S. A. 1982 Analytic structure of vortex sheet dynamics. Part I. Kelvin-Helmholtz instability. *J. Fluid Mech.* **114**, 283-298.

[17] 17. Moore, D. W. (1979) The spontaneous appearance of a singularity in the shape of an evolving vortex sheet. *Proc. Roy. Soc. Lond. A* **365**, 105.

[18] 18. Swarztrauber, P. & Sweet, R. (1975) Efficient fortran subprograms for the solution of elliptic equations. *NCAR TN/IA-109.*

[19] 19. Shelley, M. J. (1992) A case of singularity formation in vortex sheet motion studied by a spectrally accurate method. *J. Fluid Mech.* to appear.

[20] 20. Sulem, C., Sulem, P., Bardos, C. & Frisch, U. (1981) Finite time analyticity for the two- and three-dimensional Kelvin-Helmholtz instability. *Comm. Math. Phys.* **80**, 485.

[21] 21. Tryggvason, G., Dahm, W. J. A. & Sbeih, K. (1991) Fine structure of vortex sheet roll-up by viscous and inviscid simulation. *ASME J. Fluids Engin.* **113**, 31.

THE VORTON METHOD

FONS ALKEMADE, and F.T.M. NIEUWSTADT
Laboratory for Aero- and Hydrodynamics
Delft University of Technology
Rotterdamseweg 145
2628 AL Delft
The Netherlands

and

E. VAN GROESEN
Faculty of Applied Mathematics
University of Twente
The Netherlands

ABSTRACT. After a general introduction to the vorton method, which is a 3-D vortex particle method resembling the 2-D point-vortex method, a set of equations describing dynamics of 3-D vortex singularities (vortons) is derived, avoiding the inconsistency in the derivation of former vorton equations by Novikov and Kuwabara. Though inviscid, numerical simulations show reconnection phenomena. Current examples of vorton simulations are shown; they show at least qualitative agreement with results from other methods.

1. Introduction

The background of our research is the investigation of the dynamics of relatively simple vortex structures, like vortex rings, which may be regarded as models of some well-known and much-studied coherent structures in current turbulence research.

Application of vorticity theory to the study of turbulence is not only formally attractive. It can be claimed that the problem of turbulence is a problem of vorticity dynamics. The belief that coherent structures can be regarded as regions of concentrated vorticity embedded in essentially irrotational flow, has led to the view that turbulence should be modelled or described as the creation, evolution, interaction, and decay of these large-scale vortex structures.

It is mathematically correct and often very convenient to consider fluid dynamics in terms of *parcels* of vorticity which induce motion and deformation on each other. This is the basic idea behind the use of **vortex methods** in fluid dynamics.

An additional advantage of describing flow by means of vorticity comes from the fact that specification of the vorticity field may be far more economical than that of the velocity field, since the rotational part of the total flow field usually takes up only a small volume.

J.T. Beale et al. (eds.), Vortex Flows and Related Numerical Methods, 153–158.

From the aim of our research given above, it may be clear that we are not thinking the vorton method is best suited. We are just investigating its value for our purposes. This means that numerical aspects like convergence, stability, computational speed, etc. are not subject of serious study or optimalization.

2. Vorticity Theory

In case the fluid is perfect, the flow is incompressible, and the external forces on the fluid are conservative, vortex elements have material properties (Helmholtz's second theorem).

Vortex deformation (reorientation and stretching) is described by the Helmholtz equation:

$$\frac{Dw_i}{Dt} = \frac{\partial v_i}{\partial x_j} w_j \tag{1}$$

This equation can be shown to be equal to:

$$\frac{Dw_i}{Dt} = \frac{\partial v_j}{\partial x_i} w_j \tag{2}$$

2.1. GENERAL OUTLINE OF VORTEX METHODS

The concept of every vortex method can be summarized by the following questions:

- How do we discretize the vorticity? Which are the vortex *elements*?
- Which parts of the method are based on a local/Lagrangian description, which on a global/Eulerian description?
- Which are the equations for displacement and deformation of the discretized vortices?

The most widely used methods seem to be *molecular*, i.e. the vortex elements have a finite size or core. However, molecular vortex methods require a large amount of detailed information since entire vortex filament structures must be located and tracked, and often local curvatures of the vortex filaments have to be computed as well. Use of coreless or *atomic* vortex elements thus seems to be more attractive.

3. The Vorton Method

Referring to the three questions posed above, the atomic vorton method can shortly be described by:

- The elements are, essentially, 3-D point vortices or singular sources of vorticity.
- The method is completely local/Lagrangian: the deformation of the vortons is followed on their paths through 3-D space. Helmholtz's second theorem allows us to regard vortons as material.
- Vorton displacement is derived from the velocity field caused by the other vortons. Vorton deformation is derived from the Helmholtz equation.

A vorton labelled α has a location vector, $\boldsymbol{r}_\alpha$, and a vorticity intensity vector, $\boldsymbol{\gamma}_\alpha$. For the sake of simplicity, a vorton is represented by an arrow; it starts at $\boldsymbol{r}_\alpha$ and the arrow line itself indicates $\boldsymbol{\gamma}_\alpha$.

For every vorton we need a set of displacement and deformation equations, which can be derived once we have found the vorton velocity and vorticity field.

These fields can be found from the following vorton vector-potential field:

$$\begin{aligned} \boldsymbol{A}(\boldsymbol{x},t) &= \frac{1}{4\pi}\sum_\alpha \frac{\boldsymbol{\gamma}_\alpha}{R_\alpha} \\ &\equiv \sum_\alpha \phi_\alpha(\boldsymbol{x},t)\,\boldsymbol{\gamma}_\alpha(t) \end{aligned} \tag{3}$$

where $R_\alpha \equiv |\, \boldsymbol{x} - \boldsymbol{r}_\alpha \,|$ and summation is over all vortons.

From this vector-potential field we find the following velocity field:

$$\boldsymbol{v}(\boldsymbol{x},t) = \boldsymbol{\nabla}\times\boldsymbol{A} = \sum_\alpha \boldsymbol{\nabla}\phi_\alpha \times \boldsymbol{\gamma}_\alpha \tag{4}$$

Finally, from (4) we get the following vorticity field:

$$\begin{aligned} \boldsymbol{w}(\boldsymbol{x},t) &= \sum_\alpha \{\boldsymbol{\gamma}_\alpha\,\delta(\boldsymbol{R}_\alpha) + (\phi''_\alpha)\circ\boldsymbol{\gamma}_\alpha\} \qquad (5)\\ &= \sum_\alpha \{\boldsymbol{\gamma}_\alpha\,\delta(\boldsymbol{R}_\alpha) - \boldsymbol{\nabla}[\boldsymbol{\nabla}\cdot(\phi_\alpha\boldsymbol{\gamma}_\alpha)]\}\,. \end{aligned}$$

Here (ϕ''_α) is a matrix whose ij-th element is given by $\partial^2\phi_\alpha/\partial x_j\partial x_i$.

Compare this field with the non-divergencefree vorton vorticity field originally introduced by Novikov in [6] (see also [1]):

$$\boldsymbol{w}(\boldsymbol{x},t) = \sum_\alpha \boldsymbol{\gamma}_\alpha(t)\,\delta(\boldsymbol{R}_\alpha(\boldsymbol{x},t)) \tag{6}$$

We see that apparently the second (gradient) part of (5) makes the vorticity field divergencefree. It can be regarded as the nonlocal vorticity field surrounding the original singular vorton represented by the first part.

By applying the above vorton velocity and vorticity fields (4) and (5), we are able to derivé the vorton displacement and deformation equations by inserting them into the Helmholtz equation (1) or its transposed version (2).

Since we reject self-displacement of a vorton, we have to apply the velocity field induced by all other vortons, giving for the displacement of vorton α:

$$\dot{\boldsymbol{r}}_\alpha = -\sum_{\beta\neq\alpha} \boldsymbol{\gamma}_\beta \times (\boldsymbol{\nabla}\phi_\beta)|_{\,\boldsymbol{x}=\boldsymbol{r}_\alpha} \tag{7}$$

As remarked before, the deformation equation for vortons is derived from the Helmholtz equation or its transposed version. Our approach to derive the vorton deformation equation is by integrating the Helmholtz equation about a small sphere with a small radius and centre $\boldsymbol{r}_\alpha$, moving accordingly to the velocity field derived above. It is assumed that the radius is so small that no other vortons are inside the sphere. Remark the resemblance of this approach to the definition of so-called weak solutions.

After elaboration we get:

$$\dot{\boldsymbol{\gamma}}_\alpha = N_\alpha + S_\alpha \tag{8}$$

where

$$N_\alpha \equiv \sum_{\beta \neq \alpha} [(\phi''_{\alpha\beta}) \circ \boldsymbol{\gamma}_\alpha] \times \boldsymbol{\gamma}_\beta$$

and

$$S_\alpha \equiv \frac{1}{2} \sum_{\beta \neq \alpha} \boldsymbol{\gamma}_\alpha \times [(\phi''_{\alpha\beta}) \circ \boldsymbol{\gamma}_\beta] \, .$$

which latter contribution is due to the gradient part of the vorticity field (5).

In much the same way, using the transposed Helmholtz equation (2) one can derive:

$$\dot{\boldsymbol{\gamma}}_\alpha = K_\alpha - S_\alpha \tag{9}$$

where

$$K_\alpha \equiv \sum_{\beta \neq \alpha} (\phi''_{\alpha\beta}) \circ (\boldsymbol{\gamma}_\beta \times \boldsymbol{\gamma}_\alpha) \, .$$

(8) and (9) are completely equivalent. The original equations used by Novikov in [6] (derived by means of vorticity field (6)) are equal to our expression N_α. The equations used by Kuwabara in [5], derived from the transposed Helmholtz equation (2) also by applying the vorticty field (6), are equal to our expression K_α. The inconsistency between Novikov's and Kuwabara's equations, i.e. $N_\alpha \neq K_\alpha$, can now be attributed to their use of the non-divergencefree vorticity field (6).

4. Requirements on the vorton method

Certain requirements should be imposed on vortex methods in general. We will treat them here for the vorton method:

– The conservation of motion-invariants should not be violated.

So far, this requirement has never been investigated seriously for the vorton method. This is partly due to difficulties is expressing invariants by means of vorton quantities.

One should realize that in case of the vorton method complete conservation of energy will never be acquired, since part of the interaction energy will be converted into "self-energy" of the vortons, and *vice versa*. This effect may appear as (positive or negative) supply of energy, which has been named "inviscid dissipation". It may account for reconnection-like phenomenona, which have been observed during vorton simulations (see below).

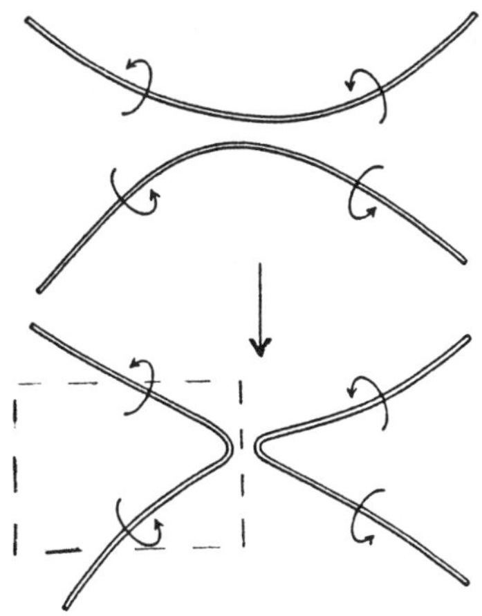

Fig. 1. Two approaching vorton filaments (parts of vorton rings) have reconnected in applying the vorton equations as proposed by Novikov and Kuwabara (left) and the authors (right).

– The discretization by finitely many elements should converge to the exact smooth solutions as the number of elements goes to infinity.

For the relative simple case of the vortex ring, this has been shown analytically by Novikov [6].

A first mathematical proof of convergence of a 3-D vortex particle method (the vortex blob method) was given in [2]. A definitive proof for the case of a 3-D nonsmooth, grid-free method has been given in [3].

– A vortex method has to show correct reconnection phenomena.

Vortex reconnection is an important and much-studied, though still little-understood phenomenon. A remarkable feature of the inviscid vorton method is its ability to simulate this typically viscous phenomenon.

An elementary configuration showing reconnection is shown in fig. 4: two slightly bent vorton filaments (in fact parts of two large vorton rings) approach each other closely. Simulations like these may show the specific characteristics of the several vorton equations which have been proposed. However, strong criteria for "correct" reconnection behaviour do not seem to exist.

Another interesting simulation in this context is that of the behaviour of a elliptical-shaped vorton ring. In contrast to the circular vorton ring, it deforms on its own, dependently of its aspect ratio.

We have performed simulations with pseudo-elliptical vorton rings: two semi-circles connected by two straight filaments. The first results show qualitative resemblance to those of simulations performed by others [4]; see fig. 4.

– The method should not require a disproportionate amount of computational effort and computation time.

So far our numerical simulations could be performed on an ordinary minicomputer (HP9000); computational times were of order of a few hours.

At the moment, we try to study the behaviour of a vorton ring in a shear flow above a flat plate. The first attempt by Novikov [1] showed the formation of a structure resembling a horseshoe vortex.

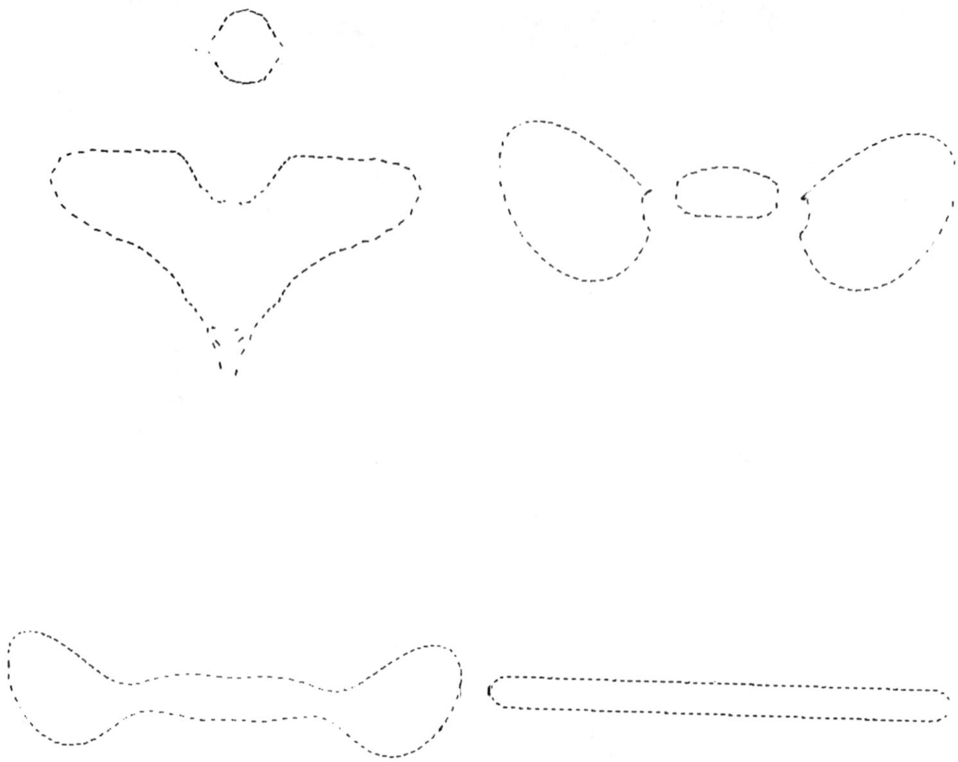

Fig. 2. The deformation of a pseudo-elliptical vorton ring in time (from left to right and from top to bottom). For the aspect ratio used here, splitting of the ring takes place, involving reconnection.

Acknowledgements

The authors thank Gianni Pedrizetti who kindly provided his numerical code for vorton dynamics, on which ours has been based. We also thank Dirkjan van Dam for the simulations of the elliptical ring and Roy DeJong for visualizing our numerical results by means of AVS.

References

[1] M.I. Aksman, E.A. Novikov: 1988, "Reconnections of vortex filaments". *Fluid Dyn. Res.* **3**, 239.

[2] J.T. Beale, A. Majda: 1982, "Vortex methods I: Convergence in three dimensions". *Math. Comp.* **39**, 1.

[3] G.-H. Cottet, J. Goodman, T.Y. Hou: 1991, "Convergence of the grid-free point vortex method for the three-dimensional Euler equations". *SIAM J. Numer. Anal.* **28** no. 2, 291.

[4] M. Kiya, H. Ishii: 1991, "Deformation and splitting of pseudo-elliptical vortex rings". In: *Advances in Turbulence 3* (eds. A.V. Johanson, P.H. Alfredsson). Springer, Heidelberg.

[5] S. Kuwabara: 1988, "Pseudo-canonical formulation of 3-dimensional vortex motion and vorton model analysis". *Fluid Dyn. Res.* **3**, 163.

[6] E.A. Novikov: 1983, "Generalized dynamics of three-dimensional vortical singularities (vortons)". *Sov. Phys. JETP* **57**, 566.

NUMERICAL SIMULATION OF UNSTEADY FLOWS BEHIND CYLINDRICAL STRUCTURES USING A FINITE DIFFERENCE-PARTICLE SUPERPOSITION ALGORITHM

Francis CASSOT
I.N.S.E.R.M. U230
CHU Purpan
31059 Toulouse Cedex

and

Serge HUBERSON
Laboratoire de Mécanique des Fluides
Université du Havre
Le Havre

ABSTRACT. The present work was justified as an essential step in the numerical study of some flow-structure interaction problems such as vortex induced vibrations or flow-resonant sound interaction. In such problems there exists an obvious need for a stable, relatively accurate and easy to set up method to predict the complex unsteady vortex wakes and their effects on the structures. The method used here is based on the solution of the unsteady bidimensional Navier-Stokes equations by the superposition of a finite difference compact method and a particle-grid eulerian-lagrangian scheme. Results are presented in both cases of the flow behind a circular cylinder and around a plate with a semi-circular leading edge in a duct. They are compared with numerical and experimental (flow visualizations) reference data showing a quite good qualitative agreement.

1. Introduction

For a significant class of problems in Numerical Fluid Mechanics, an essential requirement lies in achieving a good quality simulation of the vortex wake. Among them, let us mention some fluid-structure interaction phenomena such as vortex induced vibrations [1] or interaction between a sound wave and a vortex street behind a bluff body [2]. Vorticity formulation and lagrangian vortex scheme make easier the control of the transport of the vortex wake and minimize the numerical diffusion inherent in more classical discretization methods.

On the other hand, "natural" boundary conditions for the vorticity do not exist. Deterministic vortex methods, which have proven their efficiency and accuracy in the absence of boundaries, become otherwise critical to use.

Methods for deriving approximate or exact boundary conditions for the vorticity and implementing them together with vortex-in-cell method have been proposed

J.T. Beale et al. (eds.), Vortex Flows and Related Numerical Methods, 159–170.

recently [3; 4]. They lead to vorticity creation algorithms overcoming the drawbacks of *Chorin*'s algorithm.

Another strategy based on a superposition algorithm, combining finite-difference and grid-particle methods, has been considered for flows with large localized vorticity zones [5]. It is easy, with lesser modifications, to make it suitable for problems in bounded domains. In this paper, such a method is presented and applied to external plane flow problems.

2. The theoretical formulation

We represent the impulsively started flow of a viscous incompressible fluid past a fixed circular cylinder as a two-dimensional flow. There are obvious *Reynolds*-number limitations to these assumptions.

2.1. THE EQUATIONS

The governing equations are the unsteady *Navier-Stokes* equations which in polar co-ordinates, with stream-function ($\tilde{\psi}$) and vorticity ($\tilde{\omega}$) dependent variables are :

$$\begin{cases} \Delta\tilde{\psi} = \tilde{\omega} \\ \frac{D\tilde{\omega}}{D\tilde{t}} = \frac{1}{Re}\Delta\tilde{\omega} \end{cases}$$

where ($\tilde{r}$, θ) are the polar co-ordinates, $\tilde{t}$ the time, ν the kinematic viscosity, and Re, the *Reynolds* number.

The fluid velocities are found from :

$$\tilde{u} = -\frac{1}{\tilde{r}}\frac{\partial\tilde{\psi}}{\partial\theta} \quad \text{and} \quad \tilde{v} = \frac{\partial\tilde{\psi}}{\partial\tilde{r}}$$

Since the major velocity gradients in the flow appear near the cylinder surface, it is necessary to find a co-ordinate transformation which will concentrate the mesh spacing near the surface. This is accomplished by setting :

$$\tilde{r} = ae^{\pi\eta} \quad \text{where} \quad \theta = \pi\eta$$

If we define now the dimensionless variables and parameters :

$$t = U_\infty\frac{\tilde{t}}{a} \;,\; Re = \frac{2U_\infty a}{\nu} \;,\; \psi = \frac{\tilde{\psi}}{U_\infty a}$$

$$\omega = \frac{\tilde{\omega}a}{U_\infty} \;,\quad u = \frac{\tilde{u}}{U_\infty} \;,\; v = \frac{\tilde{v}}{U_\infty}$$

we obtain the unsteady *Navier-Stokes* equations in the new (ξ , η) co-ordinates system in dimensionless (and conservative) form :

$$g(\xi,\eta)\frac{\partial\omega}{\partial t} + \frac{\partial}{\partial\eta}(\frac{\partial\psi}{\partial\xi}\omega) - \frac{\partial}{\partial\xi}(\frac{\partial\psi}{\partial\eta}\omega) = \frac{2}{Re}\Delta_\varsigma\omega \tag{1}$$

$$\Delta_\varsigma\psi = g(\xi,\eta)\omega \tag{2}$$

$$\text{where } \Delta_\varsigma = \frac{\partial^2}{\partial\xi^2} + \frac{\partial^2}{\partial\eta^2} \text{ and } g(\xi,\eta) = \pi^2 e^{2\pi\eta}.$$

2.2. THE BOUNDARY CONDITIONS

To complete this system of equations, boundary and initial conditions given by the physical problem are :

no slip condition at the surface of the cylinder

potential flow at infinity

fluid initially at rest.

These conditions can be written in stream function and vorticity formulation as :

on the cylinder surface ($\xi = 0$)

$$\psi = \frac{\partial \psi}{\partial \xi} = \frac{\partial \psi}{\partial \eta} = 0.$$

These conditions also give $\partial^2\psi/\partial\eta^2 = 0$, and, together with equation (2) :

$$\frac{\partial^2 \psi}{\partial \xi^2} = g(\xi, \eta)\omega \quad \text{for} \quad \xi = 0.$$

This condition allows us to determine a substitute for the boundary condition for ω on the surface of the cylinder. Actually, there is no physical condition for ω. The value of ω may be approximated from ψ using a second order expansion derived from equation (2)(cf. section 3.2.2 below)

The η variation of the dependent variables is taken to be periodic, with the values at $\eta = 0$ and $\eta = 2$ being the same.

The downstream ($\xi = \xi_\infty$) boundary conditions

We assume that the flow is potential, far from the cylinder :

$$\begin{cases} \psi &= 2\text{sh}(\pi\xi)\sin(\pi\eta) \\ \omega &= 0 \end{cases}$$

3. The numerical method

3.1. OUTLINES OF THE METHOD

The method used here is based on a superposition principle. The total vorticity ω at any point of the fluid domain may be represented as the sum of two contributions : the one (ω_1) convected by particles (or discrete vortices), the other ($\omega_2 = \omega - \omega_1$) being sampled on a fixed mesh.

Due to this decomposition of the vorticity, the transport equation (1) is then splitted into two transport equations, respectively for the " grid " and "particle" vorticity components :

$$g(\xi, \eta)\frac{\partial \omega_1}{\partial t} + \frac{\partial}{\partial \eta}(\frac{\partial \psi}{\partial \xi}\omega_1) - \frac{\partial}{\partial \xi}(\frac{\partial \psi}{\partial \eta}\omega_1) = \frac{2}{Re}\Delta_\zeta \omega_1 \tag{3}$$

$$g(\xi,\eta)\frac{\partial\omega_2}{\partial t}+\frac{\partial}{\partial\eta}(\frac{\partial\psi}{\partial\xi}\omega_2)-\frac{\partial}{\partial\xi}(\frac{\partial\psi}{\partial\eta}\omega_2)=\frac{2}{Re}\Delta_\zeta\omega_2. \tag{4}$$

The flow velocities, in the (ξ ,η) co-ordinates system are determined after solving the stream-function *Poisson* equation :

$$\Delta\psi = g\omega = g(\omega_1+\omega_2). \tag{5}$$

The method used to numerically solve this system of three equations is a combination of two Eulerian and one Eulerian-Lagrangian numerical schemes :

a fourth order finite difference compact method for the resolution of the *Poisson* equation

a second order finite difference method is used to numerically solve the "grid " vorticity transport equation

a particle-mesh method for the " particle" vorticity transport equation.

The combination of the first two numerical schemes has been introduced for the numerical study of impulsively started viscous flow past a circular cylinder by *T.P. Loc* and *O. Daube* and called by them combined method of order $\mathcal{O}(h^2)$ and $\mathcal{O}(h^4)$.

The particle-mesh code has been recently proposed by *A. Jolles* and *S. Huberson*.

3.2. THE FINITE-DIFFERENCE SCHEMES

3.2.1. Poisson equation

The compact Hermitian finite-difference scheme enables to achieve a fourth-order accuracy with a three-point approximation by introducing the first and second derivatives of the stream-function (or any variable) as the new unknowns of the problem. Let us recall the well-known Hermitian tridiagonal relationships :

$$f'_{i-1}+4f'_i+f'_{i+1}=\frac{3}{h}(f_{i+1}-f_{i-1})+O(h^4) \tag{6}$$

$$f''_{i-1}+10f''_i+f''_{i+1}=\frac{12}{h^2}(f_{i+1}-2f_i+f_{i-1})+O(h^4) \tag{7}$$

where h is the spatial step and f_i , f'_i and f''_i design the values of the function f and its first and second derivatives at node i.

The finite-difference equations obtained by discretization of the *Poisson* equation and the use of the above tridiagonal relations are resolved by a technique derived from the A.D.I. scheme. This allows, finally, to reduce the problem, at each step of the A.D.I. iterative technique, to a tridiagonal system solved by the factorization algorithm.

3.2.2. "grid" vorticity transport equation

A second-order accurate scheme has been chosen to solve equation (4). As for the treatment of the Poisson equation, the A.D.I. technique is used, each dimensionless time-step Δt being decomposed in two successive half-step.

A second-order approximation is used to determine the value of the vorticity at the boundary. From equation (2) we can write :

$$2g(0,\eta)\omega(0,\eta) = -g(\Delta\xi,\eta)\omega(\Delta\xi,\eta) + \frac{6}{\Delta\xi^2}[\psi(\Delta\xi,\eta) - \psi(0,\eta)] + \mathcal{O}(\Delta\xi^2) \quad (8)$$

3.3. THE PARTICLE-MESH SCHEME

3.3.1. The particle approximation

At a given instant, a particle approximation is used to represent the part of the vorticity distribution above a given treshold, i.e. this part of the vorticity field which absolute value is greater than this threshold is "transferred" to a set of N particles defined by their positions ($\mathbf{X}_k$) and their weight (Ω_k). So, the Lagrangian (particle) method is used to describe the vorticity field in the higher vorticity concentration zones.

In fact, the wall neighbourhood is excluded from the "particle domain". As a consequence, the boundary conditions need not to be taken into account in the computation of the "particle" vorticity ω_1; these conditions must be entirely supported by the "grid" vorticity ω_2.

The initial particle approximation is estimated only after an initial stage during which the starting flow is treated by the finite difference method. This approximation is periodically renewed in order to transfer new vorticity created at the wall to the particle set.

3.3.2. Operator splitting

The transport equation of the "particle" vorticity ω_1 is solved by means of a Lagrangian scheme combined with the splitting of the advection-diffusion operator, so that :

the position ($\mathbf{X}_k$) of each particle, in the lagrangian representation, is solution of the ODE :

$$\begin{aligned} \frac{d\mathbf{X}_k(t)}{dt} &= \mathbf{u}(\mathbf{X}_k(t),t) \\ \mathbf{X}_k(0) &= \mathbf{X}_k^0 \end{aligned} \quad (9)$$

its weight is modified by the viscous diffusion according to :

$$\begin{aligned} \frac{d\Omega_k}{dt} &= \frac{1}{Re}\Delta\Omega_k \\ \Omega_k(0) &= \Omega_k^0. \end{aligned} \quad (10)$$

where $(\mathbf{X}_k^0, \Omega_k^0)$ are the set of values defining the kth particle at an initial instant, $\mathbf{u}(\mathbf{X}_k, t)$ designs the fluid velocity vector at the point of co-ordinates $\mathbf{X}_k$ where is located the particle at the instant t, and $\Delta\Omega_k(t)$ is the Laplacian of the vorticity convected by the particle at the same instant.

3.3.3. Application to the transport of particle vorticity

The particle velocity $\mathbf{u}(\mathbf{X}_k, t)$, as well as the Laplacian of the vorticity is determined as in classical PIC methods by a three-step procedure :

a **projection** step : the "particle" vorticity is distributed on the nine nodes of the Eulerian mesh surrounding the particle (TSC algorithm).

a **resolution** step : values of ψ are obtained from the discretised *Poisson* equation using the fourth-order compact scheme. The same fourth-order accuracy is achieved for the velocity at the mesh points, using the first Hermitian tridiagonal relation (6) . Simultaneously, a second-order finite difference scheme is used to estimate the Laplacian of the "particle" vorticity at the nodes.

an **interpolation** step : the "particle" values of the velocity $\mathbf{u}(\mathbf{X}_k, t)$ and of the Laplacian of the vorticity $\Delta\Omega_k(t)$ are interpolated from the nodal values using the same TSC algorithm.

The ODE (9) and (10) are solved respectively, by :

a second-order *Adams-Bashforth* method for the particle location

a forward *Euler* method for its weight.

3.4. APPLICATIONS TO CYLINDRICAL STRUCTURES

3.4.1 Circular cylinder

Conformal mapping The above co-ordinate transformation (section 2.1) defines the mapping of the domain exterior to the circular cylinder onto the semi-infinite strip $(\xi \geq 0, 0 \leq \eta \leq 2)$. It reduces to the simple mathematical relation, in dimensionless form : $z = e^{\pi\zeta}$, where $z = x + iy$ and $\zeta = \xi + i\eta$.

When a uniform cartesian mesh (of steps $\Delta\xi, \Delta\eta$) is used in the transformed co-ordinate system , this is equivalent to numerically solve on a polar mesh with a non-constant radial step very fine near the cylinder and coarser far from it.

pressure forces on the cylinder The fluid pressure is solution of the *Poisson* equation :

$$\Delta p = 2\rho \left[\frac{\partial^2 \psi}{\partial x^2} \frac{\partial^2 \psi}{\partial y^2} - \left(\frac{\partial^2 \psi}{\partial x \partial y}\right)^2 \right] \tag{11}$$

where ρ is the fluid density.

In the transformed domain of (ξ, η) coordinates the pressure equation reads :

$$\Delta_\zeta p = 2\rho \frac{1}{g} \left[\frac{\partial^2 \psi}{\partial \eta^2} \frac{\partial^2 \psi}{\partial \eta^2} - (\frac{\partial^2 \psi}{\partial \xi \partial \eta})^2 \right]. \tag{12}$$

The relation between the pressure value on the wall of the cylinder is approximated in term of a *Neumann* condition : $\partial p / \partial n = 0$.

The *Green* function $\mathcal{G}$ of the *Neumann* exterior problem of the *Laplace* equation is known as a serie expansion. So, the pressure at any point on the cylinder wall expresses as : $p = \mathcal{G} * \mathcal{S}$, where $\mathcal{S}$, the second member of the above *Poisson* equation, may be easily deduced from the fluid computations since $\partial^2 \psi / \partial \eta^2$ and $\partial^2 \psi / \partial \xi^2$ are computed in the fourth-order compact scheme and $\partial^2 \psi / \partial \eta \partial \xi$ and may be estimated from the first derivatives of ψ by centered finite difference.

Using explicitely the *Green* function expansion one obtains finally :

$$p(\theta) = 2 \sum_{\substack{n \neq 0 \\ n=-\infty}}^{+\infty} \frac{e^{-in\theta}}{\mid n \mid} \int_0^{2\pi} e^{in\theta'} d\theta' \int_1^{\infty} \mathcal{S}(r', \theta') (\frac{1}{r'})^{|n|} r' dr'. \tag{13}$$

3.4.2 Plate in a wind tunnel

Conformal mapping We must define the mapping of the exterior of the plate to the exterior of the unit circle ($\mid Z \mid \geq 1$). This transformation is achieved using :

Karman-Trefftz transformations in order to suppress the sharp corners at the trailing edge of the plate. They are defined by :

$$\left(\frac{z - z_0}{z - z_l} \right) = \left(\frac{z^* - \beta z_0}{z^* - \beta z_l} \right)^{\frac{1}{\beta}}$$

where z_l (z_0) is the complex coordinate of a sharp corner (respectively a point interior to the plate) and where $\beta = \frac{2}{3}$

a conformal mapping defined by :

$$z^* = z + \frac{a_1}{z} + \frac{a_2}{z^2} + \ldots + \frac{a_n}{z^n}$$

The coefficients (a_i) of the above expansion are estimated using an iterative procedure [8] based on the FFT computation of $\log(dz^*/dZ)$.

4. Results

4.1. COMPARISON WITH EXPERIMENTAL AND NUMERICAL DATA ON THE EARLY STAGE OF THE IMPULSIVELY STARTED FLOW BEHIND A CIRCULAR CYLINDER

In order to test the present method we have compared the results obtained for a starting flow behind a circular cylinder at moderate and relatively high *Reynolds* numbers (between 200 and 3000) to experimental data (flow visualizations) [7] and numerical results given by the so-called combined method of *Loc et al.* [6]. Fig.1 shows that our results are in quite good agreement with the experimental and numerical reference data.

4.2. SIMULATION OF THE CYLINDER WAKE

Fig.2 describes the evolution of the vortex wake behind a circular cylinder at high *Reynolds* number (Re = 5000) . The more characteristic features of this wake are the coherence of the vortex structure with a pronounced spatio-temporal periodicity and the distinct characteristic vortex zones, at large distances far from the cylinder included. This indicates both a good representation of the convective effects and a small "numerical diffusion", two properties which were expected from the lagrangian method proposed.

Fig.3 gives a detailed picture of the near wake. We can follow on this velocity fields the growth and interaction of the main alternating *Von Karman* vortices, as well as the formation of small secondary vortices near the points of boundary layer separation. Evolution of the pressure pattern on the cylinder is described on Fig.4. The solid line on the right side on this figure represents the pressure fluctuations at a point on the rear of the cylinder. This emphasizes a rather complex periodic behaviour. Wether this exhibits the real phenomenon at high Re or traduces a numerical artifact would need to be clarified. Fig.5 describes the streamlines behind the plate in a duct at rather high *Reynolds* number (3×10^4). The alternating wake is clearly simulated in good qualitative agreement with the experimental data of *Welsh et al.* .

5. Conclusion

The purpose of this paper was to describe a new approach for achieving an easy to set up and relatively accurate simulation of the vortex wake effects with the perspective of some fluid-structure interaction problems. A superposition algorithm of a particle-mesh code with a fourth-order compact finite difference method has been implemented. Numerical simulations have shown to give results in good qualitative agreement with numerical reference data obtained with more conventional discretization methods of the *Navier-Stokes* equations as well as with the experiments. In the future, it would be useful to implement this method with more efficient outflow boundary conditions.

Acknowledgements

The first author wishes to thank Dr. Paul Filippi and other members of the team " Propagation Acoustique et Protection de l'Environnement" of the "Laboratoire de Mécanique et d'Acoustique" de Marseille, where the main part of this work was done.

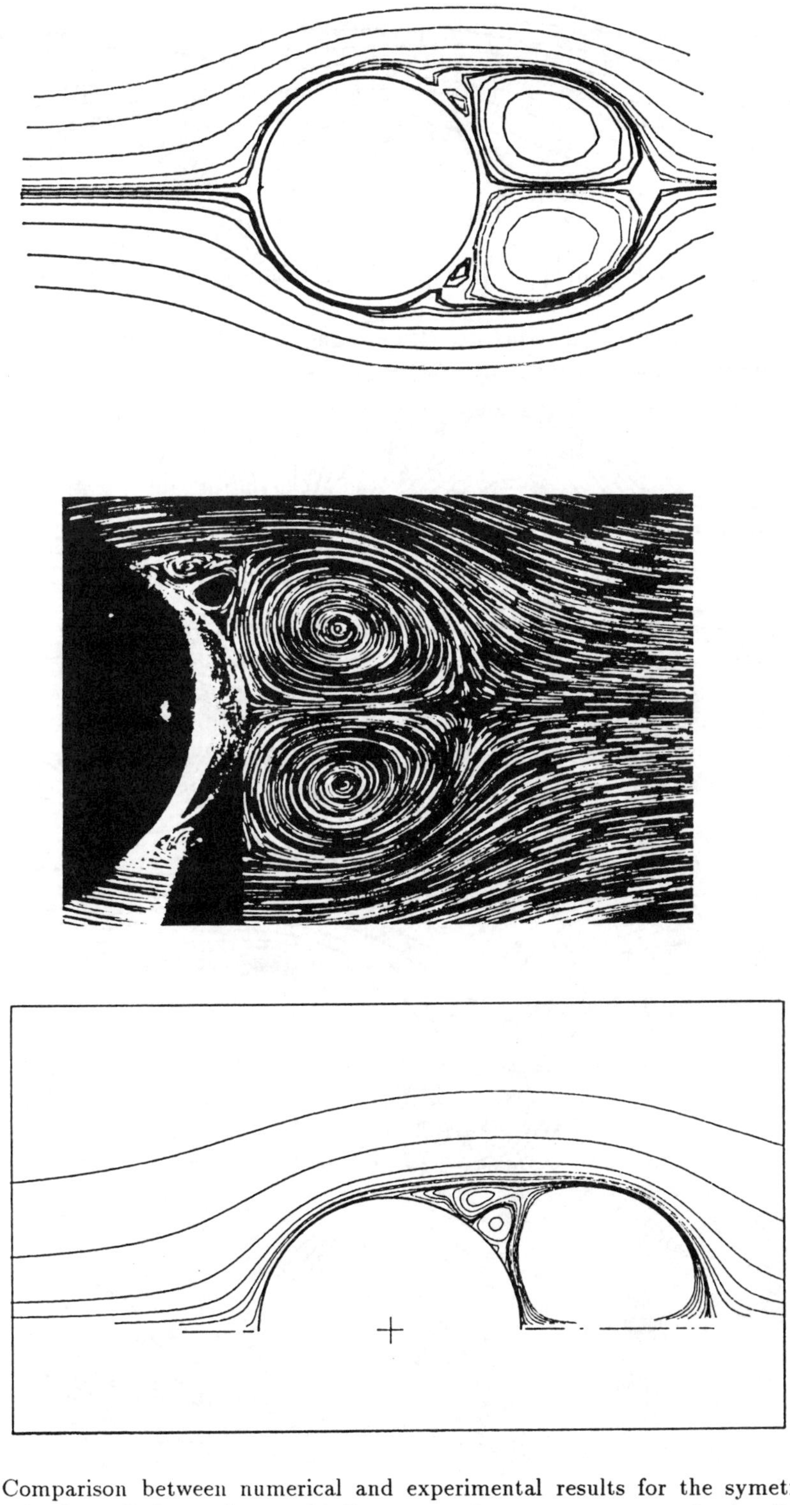

Fig. 1. Comparison between numerical and experimental results for the symetric flow behind a circular cylinder at $Re = 3000$. From top to bottom : present work, visualizations, combined method of *Loc*

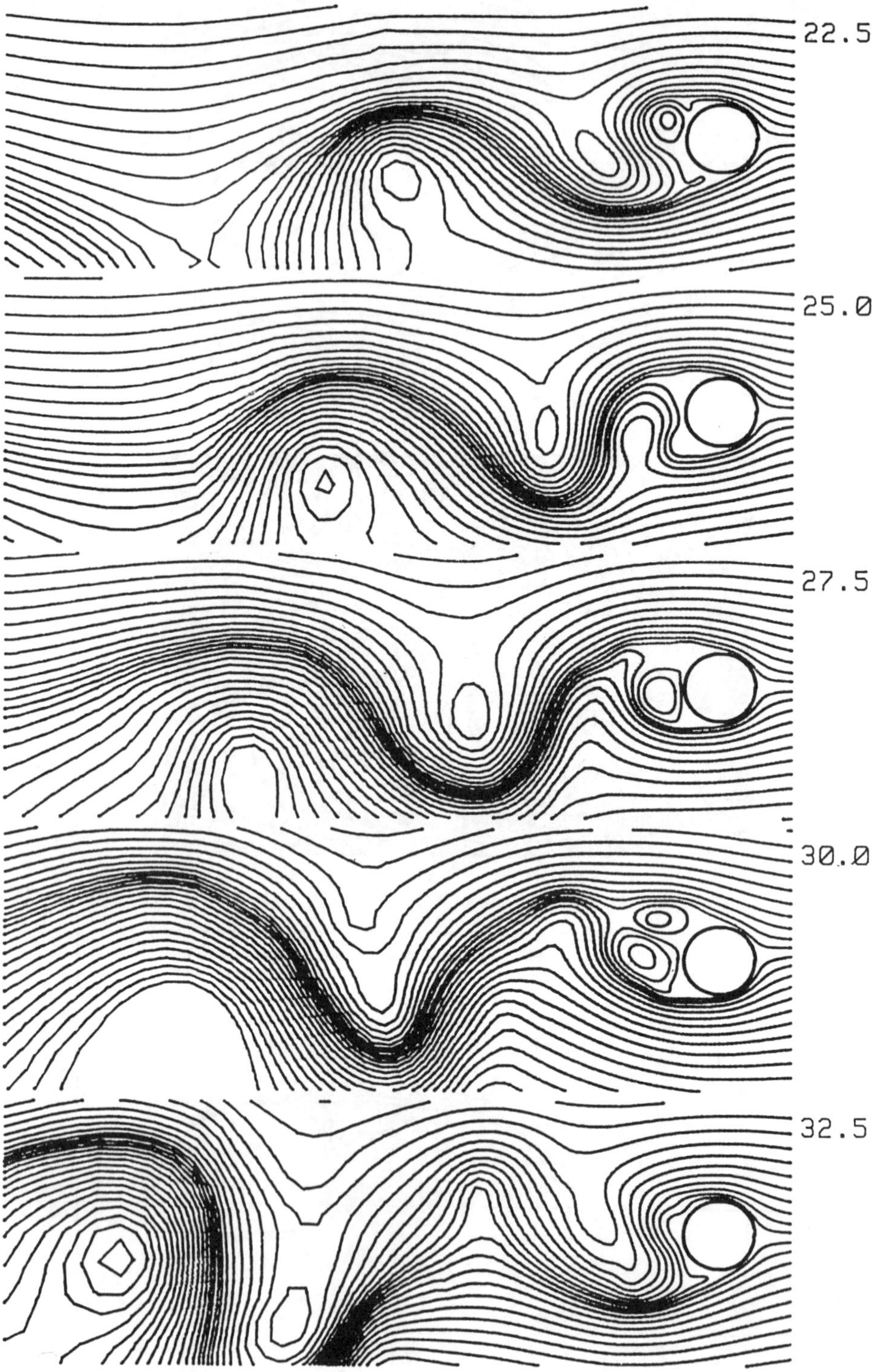

Fig. 2. Temporal evolution of the cylinder asymetric flow pattern at $Re = 5000$

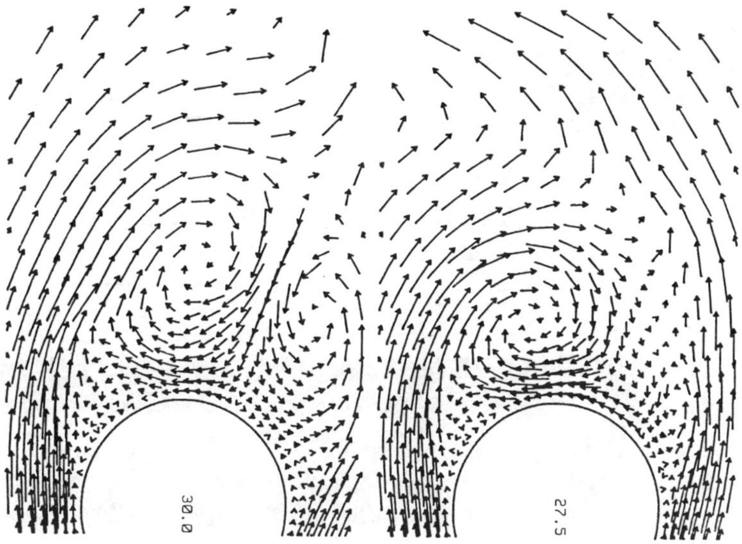

Fig. 3. Instantaneous flow fields in the near wake

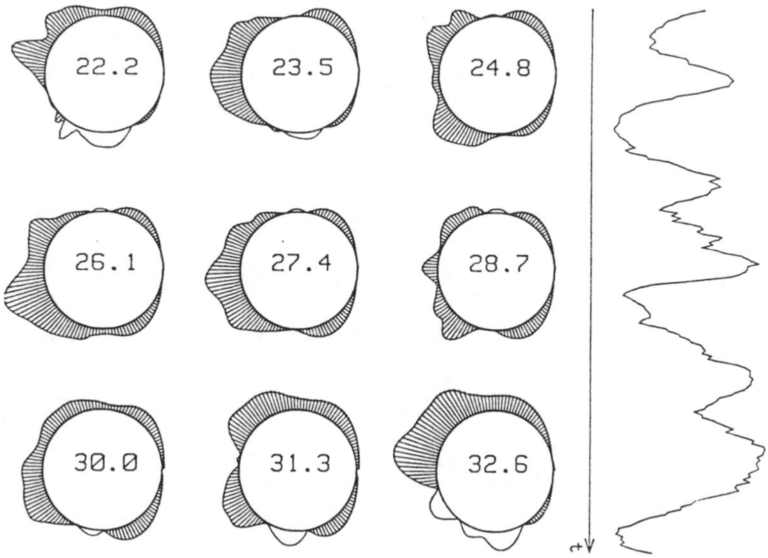

Fig. 4. Flow effects on the wall pressure (hatched zones : negative pressures)

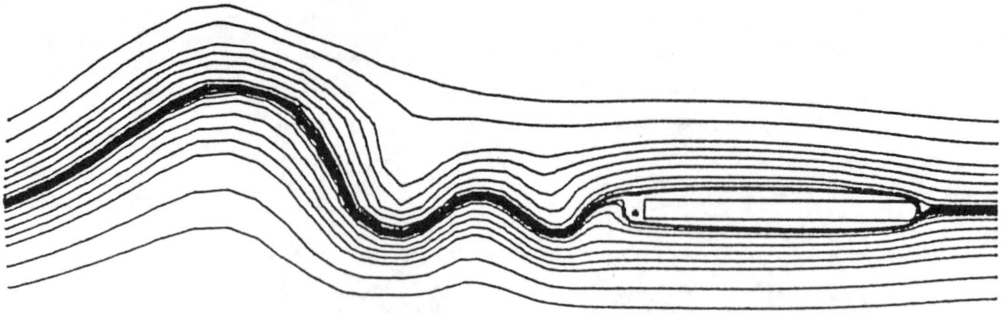

Fig. 5. Vortex shedding from a plate with a semi-circular leading edge and a sharp trailing edge. Numerical (top) vs experimental(bottom) flow patterns.

References

[1] T. Sarpkaya: Vortex-induced Oscillations A selective review. *Journal of Applied Mechanics* **46**, pp. 241-258, 1979

[2] M. C. Welsh, A. N. Stokes and R. Parker: Flow-resonant sound interaction in a duct containing a plate. Part I : semi-circular leading edge*Journal of Sound and Vibration* **93**, pp. 305-323, 1984

[3] C.Anderson: Vorticity boundary conditions and boundary vorticity generation for two dimensional viscous incompressible flows. *Journal of Computational Physics* **80**, pp. 72-97, 1989

[4] G.H. Cottet: Boundary conditions and deterministic vortex methods for the Navier-Stokes equations. in *Proc. Workshop on Mathematical Aspects of Vortex Dynamics*, pp. 128-143, 1988

[5] A. Jolles: Résolution des équations de Navier-Stokes par des méthodes particules maillage, *Thèse*, Université Paris VI, 1989.

[6] O. Daube and Ta Phuoc Loc: Etude numérique d'écoulements instationnaires de fluide visqueux incompressibles autour de corps profilés par une méthode combinée d'ordre $O(h^2)$, $O(h^4)$, *Journal de Mécanique***17**, pp. 651-678, 1978.

[7] Ta Phuoc Loc and R. Bouard: Numerical solution of the early stage of the unsteady viscous flow around a circular cylinder :a comparison with experimental visualization and measurements. *Journal of Fluid Mechanics***160**, pp. 93-117, 1985

[8] J. Halsey: Potential flow analysis of multielement airfoils using conformal mapping. *AIAA Journal* **17**, pp. 1281-1287, 1979

MOMENT ACCELERATED CONTOUR SURGERY

David G. DRITSCHEL
Department of Applied Mathematics and Theoretical Physics
University of Cambridge
Silver Street
Cambridge CB3 9EW
ENGLAND

ABSTRACT.
An accelerated version of the original contour dynamics algorithm is briefly described and demonstrated. This new algorithm, "moment-accelerated contour surgery", makes use of a multipole expansion to reduce the computational cost of evolving a system of many interacting vortices (when the density of vortices is low) to that comparable to evolving a system of the same number of structureless point vortices. This advance has opened up the way to explore essentially inviscid turbulence, and preliminary results show striking differences from those obtained using conventional (pseudo-spectral) methods. The differences can be attributed to an unexpectedly rapid, numerically-driven vortex erosion process operating in conventional methods.

1. Introduction

In this paper, an accelerated contour dynamics method called "moment-accelerated contour surgery" (MACS) is summarized and then applied to an example of 2D turbulence in spherical geometry. Full details of the algorithm and its applications to 2D turbulence can be found in Dritschel (1992 & 1993), hereafter 'D2' and 'D3'. Background information on contour dynamics and its numerical refinement, contour surgery, can be found in Dritschel (1989), hereafter 'D1'

The original reason for developing a faster version of contour dynamics was to examine essentially inviscid 2D turbulence. Already, there existed conventional numerical methods designed for this purpose, the most frequently used being the pseudo-spectral one. The introduction of hyperviscosity was an attempt to confine dissipative effects to the smallest scales; in the pseudo-spectral method, this involved a trivial change to the numerical code: the replacement of the usual k^2 dissipation factor by k^4 or even higher powers of wavenumber. An assumption was made that the small scales dissipated in this way had little effect on the large-scales; hyperviscosity just removed the cascading enstrophy from the system.

When contour dynamics was extended to make contour surgery, a similar means of removing unwanted enstrophy was introduced. The namesake of the algorithm was the introduction of a small-scale cut-off; contours or two distinct parts of the

J.T. Beale et al. (eds.), Vortex Flows and Related Numerical Methods, 171–178.

same contour were automatically fused if they moved to within the cut-off scale. In this way, two vortices could topologically merge; likewise a vortex could split into two pieces. But also, excessively thin (parts of) filaments could be removed altogether. This surgery violated exact conservation of vorticity, and resulted in the non-conservation of exact integral invariants (though only weakly); so, in a sense, the original Euler equations were no longer being strictly modelled. But, they were not being strictly modelled once one decided to represent the vortex boundaries (or contours) by a finite number of points or to use a finite time step in the numerical integration. In fact, surgery was designed to give the same level of error as the basic spatial and temporal discretization, to make a "balanced code". No numerical algorithm exactly solves the original, infinite-degree-of-freedom equations. The true problem is to do it as accurately as the application requires. For this, one has to carefully monitor convergence with increasing numerical resolution (see D2 & D3)

At this time, studies are only beginning to check this convergence for the pseudo-spectral method with hyperviscosity. The original code change to introduce hyperviscosity, though trivial, seems now to have far greater implications than was anticipated. This is the motivation for trying a different approach, one in which the dissipation of vorticity gradients is absent. Contour surgery is such an approach (with dissipation being confined to filamentary structures exclusively), but its original form is not suitable for extended simulations of turbulence. This paper reviews the steps taken to make an accelerated method. It is then illustrated in an application to 2D turbulence on a non-rotating sphere.

2. Summary of the Method

The MACS algorithm is founded on "contour dynamics" (Zabusky, Hughes & Roberts 1979), a Lagrangian description of inviscid, incompressible fluid motion based on piecewise-constant vorticity. The equations of contour dynamics give the velocity of material points (including those on the contours dividing different regions of uniform vorticity) in terms of contour integrals:

$$\frac{d\mathbf{x}}{dt} = \mathbf{u}(\mathbf{x}) = -\frac{\tilde{\omega}'}{4\pi} \oint_{\mathcal{C}'} \log|\mathbf{x}-\mathbf{x}'|^2 d\mathbf{x}' \ . \tag{1}$$

The above equation applies to one patch of vorticity (a closed contour $\mathcal{C}'$ directed in a right-handed sense). $\tilde{\omega}'$ is the jump in vorticity crossing $\mathcal{C}'$ inwards and $\mathbf{x}' \in \mathcal{C}'$. For $\mathbf{x} \in \mathcal{C}'$, (1) describes the motion of $\mathcal{C}'$ itself. Because the functional relationship between velocity and vorticity is linear, (1) can be generalized to any number of contours by simply summing the contributions from each contour.

In practice, (1) has few analytical solutions. It corresponds to an infinite-degree-of-freedom Hamiltonian system and so forces one to discretize it for general practical use. Details can be found in D1, but basically discretization involves using a finite number of "nodes" to approximate each contour and a finite time step. One of the early-seen consequences of (1), even when discretized, is that it characteristically produces small, filamentary scales of motion (Dritschel 1988a, D1). This forced the introduction of a small-scale cut-off, or "surgery" (Dritschel 1988b), to remove filamentary structures and permit the continuation of numerical calculations to moderate times (tens of characteristic vortex rotation periods) while maintaining accuracy on the large scales (see the above references for demonstrations).

The basic contour dynamics algorithm takes approximately $400n^2$ basic operations per time step, where n is the total number of nodes on all contours. The time step is only weakly linked to the spatial resolution, not by a CFL criterion, but by the requirement of balanced accuracy. Typically, a time step equal to a twentieth of a vortex rotation period is used (that period equals 4π divided by the peak vorticity). Because a 4th-order Runge-Kutta time-stepping method is used, the main computational burden arises from the velocity evaluation, and it is here that a new approach, summarized next, has brought substantial savings in computer time.

It is sufficient to consider first the interaction between two well-separated vortices, and look at the one-way interaction between the source vortex, or contour $\mathcal{C}'$, and the recipient vortex, or contour $\mathcal{C}$. The reverse interaction can be obtained by switching primed and unprimed quantities. The basic idea is to expand (1) in a descending series in distance from the center of the source contour, then relate this series to an ascending series in distance from the center of the recipient contour. This latter series is just a sum of all possible non-singular, irrotational solutions to Laplace's equation, with coefficients that depend on linking the descending series of the source contour to the ascending series of the recipient contour.

This sounds complicated, but Laplace's equation in two-dimensions has many nice properties to make this a simple procedure. The first trick is to use the complex form of (1), where $z \equiv x + iy$ replaces the vector $\mathbf{x}$ and so on for the primed quantities. Once this is done, the logarithmic kernel naturally splits as

$$\log(z - z') + \log(z - z')^* .$$

The remaining work is then to expand these terms about the center Z' of the source contour and the center of the recipient contour Z (e.g. use $(z-Z)-(z'-Z')-(Z'-Z)$ in place of $(z-z')$ and expand everything assuming $|[(z-Z)-(z'-Z')]/(Z'-Z)| < 1$). One soon arrives at the final expression for the velocity in the vicinity of the recipient contour induced by the source contour:

$$\frac{dz^*}{dt} = \sum_{n=0}^{\infty} \beta_n (z - Z)^n \ , \tag{2}$$

where

$$\beta_n = \sum_{m=1}^{\infty} \binom{m+n-1}{m-1} \frac{\alpha'_m}{(Z' - Z)^{m+n}} \tag{3}$$

are the local expansion coefficients for the recipient contour and

$$\alpha'_m \equiv (-1)^m \frac{\tilde{\omega}'}{4\pi m} \oint_{\mathcal{C}'} (z' - Z')^m dz'^* \tag{4}$$

are the "moments" of the source contour. These moments depend only on the shape of the source contour. For example, $\alpha'_1 = i\kappa'$ where κ' is the product of the vorticity jump and area of the source contour divided by 2π; $\alpha'_2 = 0$ by definition of the centroid; α_3' is proportional to the squared focal distance times a complex orientation factor of an elliptical patch with the same area and value of α_3' as the source contour. Higher-order moments describe increasingly fine scale shape characteristics of the source contour.

Like the original contour dynamics equations (1), equations (2) through (4) must be approximated in practice. What is done is to truncate the sums in (2) and (3) to the first M terms, and if the error so introduced for a particular interaction exceeds a prescribed tolerance, then the interaction is done using the original expression (1) (discretized of course). The error in this truncation can be estimated rigorously (see D2), leading to the following criterion for performing a particular interaction by the truncated sums:

$$\frac{2|\tilde{\omega}'|}{M\pi\omega_{\max}}\left(\frac{\max(R,R')}{|Z'-Z|-R'-R}\right)^M < \epsilon \tag{5}$$

where R and R' are the maximum extents of the recipient and source contours (i.e., $R \equiv \max_{z\in\mathcal{C}}|z-Z|$), $\omega_{\max}$ is the maximum vorticity magnitude in the flow, and ϵ is the dimensionless "accuracy" of the moment expansion (to be balanced with the errors of discretization, time-stepping and surgery — see D2).

In calculations of 2D turbulence, when many contours are present, most of the interactions can be done with the truncated sums. For these interactions, the effects of all the source contours on a given recipient contour are added up in the local expansion coefficients β_n. This is done for each recipient contour. So, for N contours, this requires $O(N^2)$ operations (for a given value of the truncation M). Notice that the discretization of the contours into nodes does not enter here. Outside of the double loop over contours, the velocity on each contour can be obtained by (1) plus the contributions of nearby contours that require direct contour integration. Since only a small percentage of all interactions need be performed by direct contour integration, and cost of the calculation is proportional to the number of contours squared, and no longer the total number of nodes squared. This represents a substantial reduction in computing cost.

To maximize computational efficiency, the truncation level M is adjusted to balance the estimated cost of all direct contour integrations with the cost of accumulating the local expansion coefficients β_n (see D2 for details). In the limit of a low density of vortices, M tends to unity, implying that the cost of the computation is then only a factor of two greater than a calculation using structureless point vortices. That is, one can carry out a calculation using finite-sized, nontrivially-shaped vortices, capable of merging and other inelastic interactions, for a price comparable to a calculation using the same number of point vortices.

A potentially faster algorithm may result by combining the above method with the nesting procedure of Greengard & Rokhlin (1987) (regarding the contours as the particles in their algorithm). This could reduce the computational cost to $O(N)$.

There is an analogous moment-expansion technique for 2D flows on the surface of a sphere and for generalized 2D flows (general linear operator relationships between streamfunction and vorticity) — see D2 for details.

3. Results

A demonstration of the algorithm is next illustrated for 2D turbulence on the surface of a sphere (see D3 for further results, diagnostics, and comments). The calculation begins with 1000 vortex patches, 500 with $+2\pi$ vorticity and 500 with -2π vorticity, chosen such that the total area occupied by each species is 2.5% of the total surface area. The number density of vortices $n(A)$ is initially inversely proportional to their squared area, with the maximum vortex area being 1000 times the minimum. Initially, the largest positive and negative vortices occupy 0.72% of the surface area.

Figure 1 shows the initial conditions, viewed orthographically from one side of the sphere. Figure 2 shows the state of the flow more than 100 vortex rotation periods later, at $t = 203$ (note: the enstrophy turn-over time equals $\sqrt{10}/\pi \approx 1$). Line-drawn plots are used because much of the structure would be invisible in a shaded rendition. Oddly, the flow has not changed substantially during the evolution, despite the fact that the duration of this simulation is 50% longer (in enstrophy turn-over times) than the calculations reported in D3. That there has been little change is dramatically revealed by the evolution of the vortex number density distribution $n(A)$ — see figure 3.

What is different here from the calculations in D3 is the steeper and broader initial form of $n(A)$; previously $n(A) \propto A^{-1}$ and $A_{\text{max}}/A_{\text{min}} = 10$ initially. It is a remarkable coincidence, at least, that an A^{-2} number density distribution corresponds approximately to the famous k^{-3} energy spectrum, according to Benzi *et al* (1989). Why there is no apparent inverse cascade in this case is an intriguing question.

The calculation just displayed would have been virtually impossible to perform without the extension to moments. The calculation required approximately 17 hours of CPU time on a single processor of a Cray Y-MP (peak performance: 235 Mflops). Had the original algorithm been used, approximately 725 hours would have been required. This represents a speed-up factor of 43.

Acknowledgements

Support for this research has come from a grant by the UK Science and Engineering Research Council.

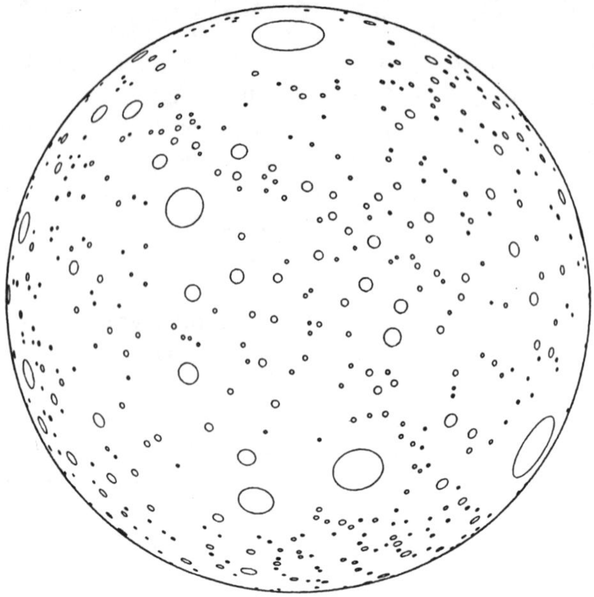

Fig. 1. An orthographic projection of the vorticity distribution at $t = 0$. There are equal numbers of positive and negative vortices. The numerical parameters are the same as used in D2.

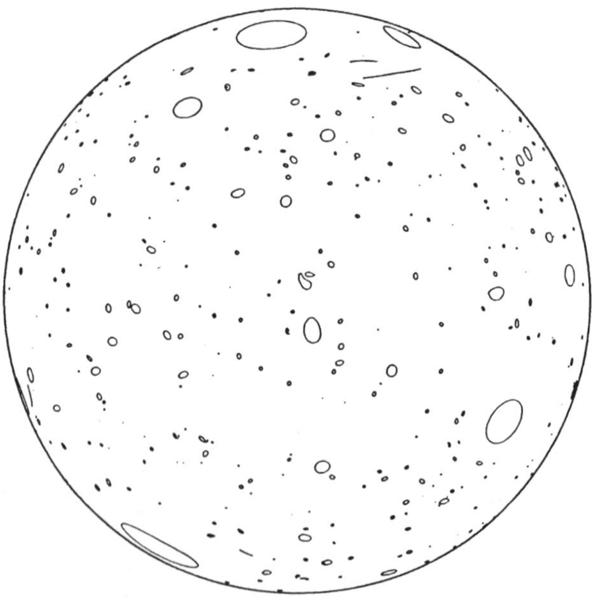

Fig. 2. Same as in figure 1 but at $t = 203$.

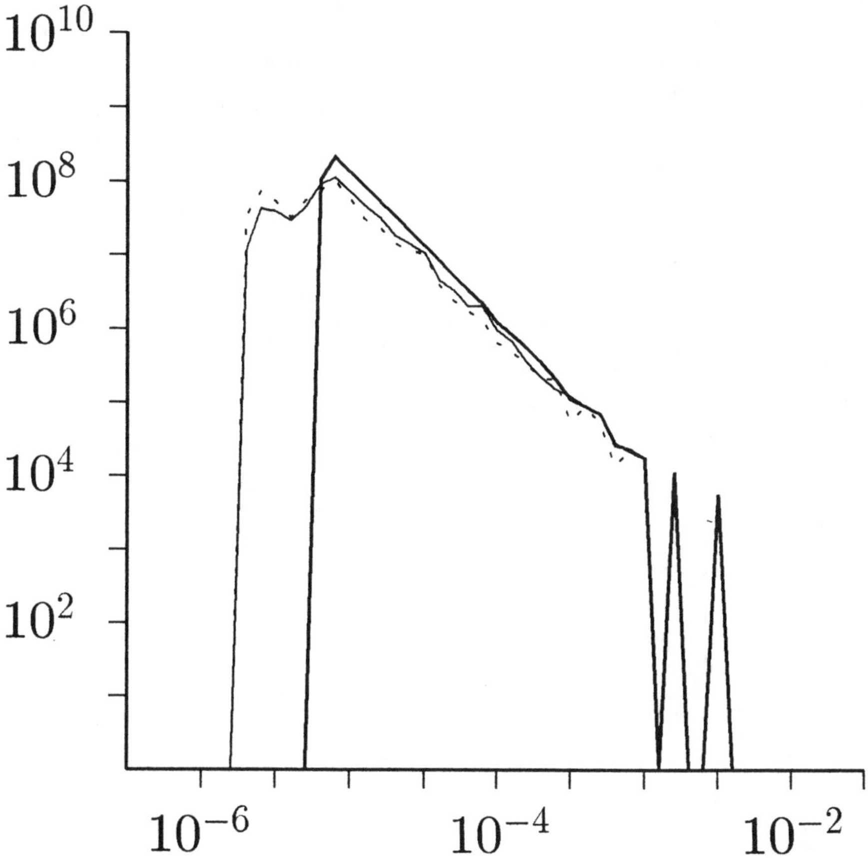

Fig. 3. Number density of vortices as a function of vortex area divided by 4π ($\hat{A}$). The heavy solid curve corresponds to the initial conditions. The thin solid and dashed curves correspond to time averages over the intervals (88,112) and (178,203), respectively. The two humps seen for $\hat{A} > 10^{-3}$ are caused by the low density of large vortices (there are only two positive and two negative vortices of this size, which do not change appreciably over the entire evolution).

References

[1] Benzi, R., Patarnello, S. & Santangelo, P., 1989 Self-similar coherent structures in two-dimensional decaying turbulence. *J. Phys. A: Math. Gen.*, **21**, pp. 1221-1237.

[2] Deem, G.S. & Zabusky, N.J., 1978 Vortex waves: stationary V-states, interactions, recurrence, and breaking. *Phys. Rev. Lett.*, **40**, pp. 859-862. (Also see "Stationary V-states, interactions, recurrence, and breaking," in *Solitons in Action* (ed. K. Lonngren and A. Scott) Academic, pp. 277-293 (1978))

[3] Dritschel, D.G., 1988a The repeated filamentation of two-dimensional vorticity interfaces. *J. Fluid Mech.*, **194**, pp. 511-547.

[4] Dritschel, D.G., 1988b Contour surgery: a topological reconnection scheme for extended integrations using contour dynamics. *J. Comput. Phys.*, **77**, pp. 240-266.

[5] Dritschel, D.G., 1989 Contour dynamics and contour surgery: numerical algorithms for extended, high-resolution modelling of vortex dynamics in two-dimensional, inviscid, incompressible flows. *Comp. Phys. Rep.*, **10**, pp. 77-146.

[6] Dritschel, D.G., 1992 A moment-accelerated contour surgery algorithm for many-vortex, two-dimensional flow simulations. *Phys. Fluids A* (to appear December 1992).

[7] Dritschel, D.G., 1993 Vortex properties of two-dimensional turbulence. *Phys. Fluids A* (to appear March 1993).

[8] Greengard, L. & Rokhlin, V., 1987 A fast algorithm for particle simulations. *J. Comput. Phys.*, **73**, pp. 325-348.

[9] Zabusky, N.J., Hughes, M.H. & Roberts, K.V., 1979 Contour dynamics for the Euler equations in two dimensions. *J. Comp. Phys.*, **30**, pp. 96-106.

DIRECT NUMERICAL SIMULATIONS USING VORTEX METHODS

P. KOUMOUTSAKOS and A. LEONARD
Graduate Aeronautical Laboratories
California Institute of Technology
Pasadena Ca 91125
USA

ABSTRACT. Direct numerical simulations are presented for viscous incompressible flows with and without solid wall boundaries. Our numerical method is based on vortex methods. The classical scheme is enhanced to account for viscous effects (diffusion and no-slip boundary condition) by appropriately modifying the strength of the particles. Computations for extended times are feasible by periodically remeshing the vorticity field. The particles are advanced using the Biot-Savart law for the evaluation of the velocity. Large scale computations using $\mathcal{O}(10^6)$ particles are feasible by efficiently employing fast summation algorithms. We present viscous and inviscid simulations of vortex merging and compare our results with those of spectral methods and contour dynamics. The impulsively started flow around a circular cylinder is also examined for Re = 40 to 9500.

1. Introduction

The adaptive character of vortex methods as well as the straightforward physical interpretation of its results have inspired many studies since the pioneering work of Rosenhead (1931).

However the computational cost and the difficulty of the method to accurately represent viscous effects have limited the method to be used only as an engineering tool in the past to gain physical insight and a rough estimate of unsteady forces for a large class of unsteady separated flows. A review of the foundations of the method and an extensive account of its previous applications may be found in review articles by Leonard (1980) and Sarpkaya (1989). Vortex methods have not had much success in the past in competing with grid based schemes in the arena of direct numerical simulations. Here we discuss some advances in vortex methods that make them much more competitive. A numerical method is developed that while maintaining the attractive features of vortex methods alleviates its restrictions.

The Biot-Savart law is used for the computation of the velocities. We implement fast summation algorithms (Greengard and Rohklin, 1987) to reduce the $\mathcal{O}(N^2)$ computational cost associated with the classical scheme and obtain an efficient $\mathcal{O}(N)$ algorithm. The Lagrangian nature of the method encounters difficulties when dealing with viscous effects. In the past the scheme of random walk has been

J.T. Beale et al. (eds.), Vortex Flows and Related Numerical Methods, 179–190.

used extensively but has the disadvantage of a low convergence rate (proportional to $1/\sqrt{(N)}$). Here we implement the scheme proposed by Degond and Mas-Gallic (1987) that accounts for viscous diffusion by appropriately modifying the strengths of the particles. This method has been shown to enjoy quadratic convergence provided the particles overlap and therefore in the present implementation includes a particle redistribution algorithm that ensures its good convergence properties. The grid free character of the method makes it convenient to simulate flows around complex configurations. However the use of a vorticity formulation complicates the enforcement of the (natural) velocity boundary conditions. We present here a novel technique for the enforcement of the no-slip boundary condition in the context of vortex methods. Unlike other vortex schemes no new particles are generated at the wall but the strength of the existing ones is modified appropriately.

The validity of the present scheme is tested for the direct numerical simulation of incompressible flows. We simulate the merging of two like signed vortices in an axisymmetric strain field and we compare our results with those of a spectral method computation by Buntine and Pullin (1989). As an example of flows around bluff bodies we present results for the initial stages of the flow development around an impulsively started circular cylinder. We compare our results with those of other computational, theoretical and experimental works to assess the validity of the present scheme.

2. Governing Equations

The two dimensional incompressible Navier-Stokes equations may be expressed in terms of the vorticity ($\boldsymbol{\omega} = \omega \mathbf{e}_z = \nabla \times \mathbf{u}$) field as :

$$\frac{\partial \boldsymbol{\omega}}{\partial t} + \mathbf{u} \cdot \nabla \boldsymbol{\omega} = \nu \nabla^2 \boldsymbol{\omega} \tag{1}$$

with boundary conditions at infinity,

$$\mathbf{u} \to \mathbf{U}_\infty \qquad \text{as} \quad |\mathbf{x}| \to \infty \tag{2}$$

and in the presence of a stationary body

$$\mathbf{u}|_{\text{surface}} = 0 \tag{3}$$

In a Lagrangian formulation the vorticity transport equation may be replaced equivalently by the following set of equations for a continuum of vortex particles with locations $\mathbf{x}_a$ and vorticity ω:

$$\begin{aligned} \frac{d\mathbf{x}_a}{dt} &= \mathbf{u}(\mathbf{x}_a, t) \\ \frac{d\omega}{dt} &= \nu \, \nabla^2 \omega \end{aligned} \tag{4}$$

The velocity field $\mathbf{u}(\mathbf{x}, t)$ is given by the solution to

$$\nabla^2 \mathbf{u} = -\nabla \times \boldsymbol{\omega} \tag{5}$$

yielding

$$\mathbf{u} = -\frac{1}{2\pi}\int \frac{(\mathbf{x}-\mathbf{y})\times\boldsymbol{\omega}(\mathbf{y})}{|\mathbf{x}-\mathbf{y}|^2}\,d\mathbf{y} + \mathbf{U}_\infty(t) \tag{6}$$

where we have included the free-stream component of the velocity as a solution to the homogeneous equation 5.

The use of the Biot-Savart law for the computation of the velocity guarantees the enforcement of the far field boundary condition. In the present formulation the no-slip condition is rendered equivalent to a vorticity flux on the surface of the body (see below).

3. Vortex Methods

In vortex methods the vorticity field is represented by a collection of moving computational elements with locally distributed vorticity ($\boldsymbol{\omega} = \omega\mathbf{e}_z$) and variable circulation ($\Gamma(t)$) as follows,

$$\omega(\mathbf{x},t) = \sum_{i=1}^{N}\Gamma_i(t)\,\eta_\sigma(\mathbf{x}-\mathbf{x}_i(t)) \tag{7}$$

where η is the distribution for each particle, and σ is the width of the distribution. The vorticity transport equation is resolved in its Lagrangian form by following the trajectories of the particles.

3.1. PARTICLE VELOCITY - FAST VORTEX METHODS

The Biot-Savart integral when discretized using as quadrature points the locations of the particles results in the following set of differential equations to determine the trajectories ($\mathbf{x}_i$) of the particles ,

$$\frac{d\mathbf{x}_i}{dt} = -\frac{1}{2\pi}\sum_{j=1}^{N}\Gamma_j\,\frac{(\mathbf{x}_i-\mathbf{x}_j)\times\mathbf{e}_z}{|\mathbf{x}_i-\mathbf{x}_j|^2}\,g(|\mathbf{x}_i-\mathbf{x}_j|/\sigma) + \mathbf{U}_\infty(t) \tag{8}$$

where

$$g(y) = 2\pi\int_0^y \eta(z)\,z\,dz$$

The above set of 2N differential equations involves a summation over all particles thus implying an $\mathcal{O}(N^2)$ operations per time step. This has limited in the past the number of vortices used to a few thousands. Hybrid methods (particle-mesh) tried to alleviate this problem by introducing a grid to compute the velocity field and solving a Poisson's equation for the streamfunction. This approach limits the applicability of the method to flows around simple configurations, requires the explicit treatment of the far field boundary condition and introduces otherwise absent numerical diffusion.

However efficient computations may be achieved by using the Biot-Savart law and employing the method of multipole expansions. The velocity field induced , at distances sufficiently far from the center of a cluster of particles, is computed using a certain number of expansions for each cluster. A hierarchical (tree) data structure

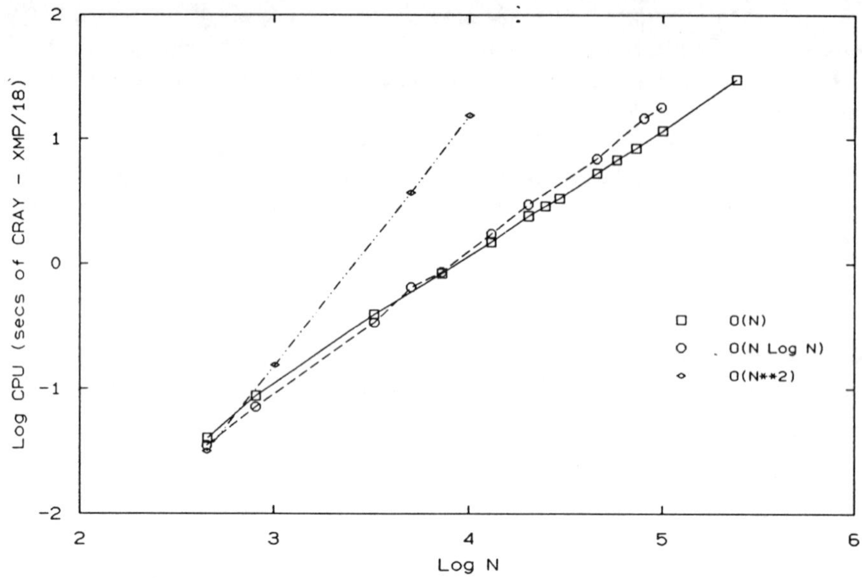

Fig. 1. Computational Cost of the multipole and direct summation algorithms

is implemented in order to determine when those expansions may be used instead of the exact pairwise interactions in order to preserve the accuracy of the solution. This idea was originally devised by Barnes and Hut (1986) for astrophysical simulations and its computational cost scales as $\mathcal{O}(N \log N)$. Greengard and Rohklin (1987) proposed to consider the interactions between clusters of particles as well, by shifting appropriately the centers of expansions, resulting in an $\mathcal{O}(N)$ algorithm.

We have implemented both multipole algorithms as well as the traditional scheme with great care in regards to vectorization. In the present implementation on the CRAY XMP/18 fast methods become advantageous for computations involving only $N \approx 500$ particles with the $\mathcal{O}(N)$ method beating out the $\mathcal{O}(N \log N)$ method for $N \approx 4400$. The time required on a YMP processor is about 2/3 of the time required on an XMP. A timestep requiring one evaluation of all velocities for a million particles requires about one minute on single processor of a CRAY YMP while the N^2 algorithm would require roughly 24 hours.

3.2. VISCOUS DIFFUSION

For viscous diffusion we need to solve the diffusion equation for the vorticity field as it is described by the particles. An explicit solution of the diffusion equation may be expressed in an integral form at time $t + \Delta t$ by

$$\omega(\mathbf{x}, t + \Delta t) = \int \frac{exp(-|\mathbf{x} - \mathbf{y}|^2 / 4\nu\Delta t)}{4\pi\nu\Delta t} \omega(\mathbf{y}, t)\, d\mathbf{y} \tag{9}$$

The method of random walk as described by Chorin (1973) interprets the above integral probabilistically and has been in the past the main tool for the simulation of diffusion effects with particle methods. However it has the disadvantage of being slowly convergent. An alternative scheme has been proposed (Degond and Mas-Gallic (1987)) according to which the Laplacian operator (∇^2) is approximated as an integral operator

$$\nabla^2 \boldsymbol{\omega} \approx G * \boldsymbol{\omega} = \frac{1}{\sigma^2} \int G(|\mathbf{x} - \mathbf{y}|/\sigma) \, [\boldsymbol{\omega}(\mathbf{y}) - \boldsymbol{\omega}(\mathbf{x})] \, d\mathbf{y} \tag{10}$$

where, for example, with gaussian particle distributions,

$$G(z/\sigma) = 4\eta(z/\sigma) = \frac{4}{\pi\sigma^2} \exp(-|z|^2/\sigma^2)$$

When the integral of (10) is discretized on the particle locations this results in a set of differential equations for the strength of the particles:

$$\frac{d\Gamma_i}{dt} = \frac{\nu \, h^2}{\sigma^2} \sum_j G(|\mathbf{x}_i - \mathbf{x}_j|/\sigma)(\Gamma_j - \Gamma_i) \tag{11}$$

Note that when an Euler time step of σ^2/ν is used in (11), and Gaussian particles are implemented the discretized form of (9) is recovered.

3.3. THE NO-SLIP BOUNDARY CONDITION

In the presence of solid bodies the method has to account for the generation of vorticity at the boundaries due to the no-slip boundary condition. Chorin (1973) introduces new particles in the vicinity of the boundary so as to mimic the process of vorticity creation and to enforce the no-slip condition. Mas-Gallic (1990) extended the method of PSE to account for diffusion and boundary conditions in one dimensional domains by modifying the strength of the particles. Lu and Ross (1991) propose an approximate extension of the (one-dimensional) "method of images" to two dimensions and accordingly modify the strength of the particles as well. Here we adopt this idea (of modifying the strength of the particles) to account for the enforcement of the no-slip condition and we present a formulation applicable to two (and three) dimensional viscous flows around solid configurations.

In the present method we model the surface of the body as a vortex sheet with strength $\gamma(s)$ determined so that the flow field follows the prescribed solid body motion. The strength of the vortex sheet is determined by enforcing the kinematic boundary condition of no-through flow $\mathbf{u} \cdot \mathbf{n} = 0$ on the surface of a stationary body. The resulting integral equation is given by :

$$\gamma(s) - \frac{1}{\pi} \oint \frac{\partial}{\partial n} [Log|\mathbf{x}(s) - \mathbf{x}(s')|] \, \gamma(s') \, ds' = -2 \frac{\partial \Psi_f}{\partial n}(\mathbf{x}(s)) \tag{12}$$

where Ψ_f is the streamfunction due to the free stream and the vortices in the wake. The above equation admits a non-unique solution (Prager, 1928) and an extra constraint needs to be imposed on the strength of the vortex sheet. This is provided by Kelvin's theorem :

$$\oint \gamma(s) ds + \sum_j^N \Gamma_j = 0 \tag{13}$$

The above set of equations (12)-(13) is then well posed an admits a unique solution.

The model presented herein relies in the nullification of this spurious vortex sheet at the body surface (Lighthill, 1963) so as to enforce the no-slip boundary condition. Using Kelvin's theorem for the production of circulation at a wall we may determine a vorticity flux at the wall as:

$$\nu \frac{\partial \omega}{\partial n}(s) = -\gamma(s)/\delta t$$

where δt denotes the time step of the present procedure. Subsequently we solve the diffusion equation with homogeneous initial conditions and Neumann type conditions at the wall. The integral solution of the problem is given by (Friedman, 1964)

$$\omega(\mathbf{x}, t) = \int_{t-\delta t}^{t} \oint G(\mathbf{x}, t; \boldsymbol{\xi}, \tau)\, \mu(\boldsymbol{\xi}, \tau)\, ds_{\xi}\, d\tau \tag{14}$$

where $\mu(\mathbf{x}, t)$ is determined by the solution of :

$$-\frac{1}{2}\mu(\mathbf{x}, t) + \int_{t-\delta t}^{t} \oint \frac{\partial G}{\partial n}(\mathbf{x}, t; \boldsymbol{\xi}, \tau)\, \mu(\boldsymbol{\xi}, \tau)\, ds_{\xi}\, d\tau = \nu \frac{\partial \omega}{\partial n}(\mathbf{x}(s), t) \tag{15}$$

with :

$$G(\mathbf{x}, t; \boldsymbol{\xi}, \tau) = \frac{1}{4\,\pi\,\nu\,(t-\tau)} \exp(\frac{|\mathbf{x} - \boldsymbol{\xi}|^2}{4\,\nu\,(t-\tau)})$$

The integrals over the surface of the body are computed using a boundary element method and the vorticity field is updated by discretizing (14) on the particles (see [15] for further discussion).

3.4. REMESHING

The convergence properties of vortex methods with a finitie core dictate that the particles must overlap (i.e. $h < \sigma$) at all times (Beale, 1986). However the local strain field of the flow may distort the particle locations thus producing particle clustering in one direction accompanied by an expansion in another direction , similar to the situation around a hyperbolic stagnation point in a steady flow. When such a situation occurs the particle locations are re-initialized onto a uniform grid while interpolating the old vorticity on the new particle locations [3]. Alternatively, in the context of using the PSE scheme to account for diffusion, Cottet (1990) treated the problem of this inaccuracy by a grid correction procedure.

In our algorithm we use a nine point scheme to perform this remeshing conserving the circulation as well as the linear and angular momentum of the vorticity field. More specifically the ith old vortex with circulation Γ_i and location $(x_i,\ y_i)$ contributes $\Delta\Gamma_{j,i}$ circulation to new mesh point $(\tilde{x}_j, \tilde{y}_j)$ according to

$$\Delta\Gamma_{j,i} = \Gamma_i\, \Lambda(\frac{\tilde{x}_j - x_i}{h})\, \Lambda(\frac{\tilde{y}_j - y_i}{h}) \tag{16}$$

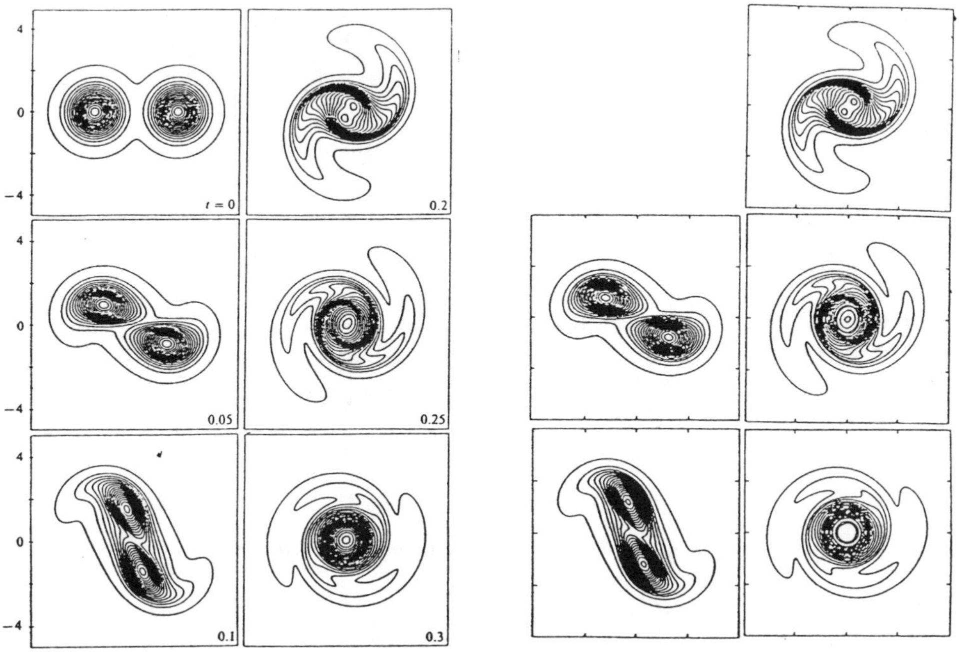

Fig. 2. Comparison between the spectral method of [5] (left columns) and the present vortex method (right columns) for the merging of two like-signed Gaussian vortices in an axisymmetric strain field ($\alpha = 4,\ \beta = 2$) at Re $= 160$. The minimum contour is $|\omega_{\min}| = 3$ and all contour intervals are $|\delta\omega| = 60$.

where the interpolation kernel Λ is given by

$$\Lambda(u) = \begin{cases} 1 - u^2, & 0 \le |u| < 1/2; \\ (1 - u)(2 - u)/2, & 1/2 \le |u| < 3/2; \\ 0, & \text{otherwise.} \end{cases} \tag{17}$$

If the old particle is less then a distance h from a wall the interpolation is modified to maintain the same conservation properties. See [13,15] for further discussion.

4. Results

We present the results of viscous and inviscid computations with and without solid boundaries. We compare our results with those from respective simulations of other

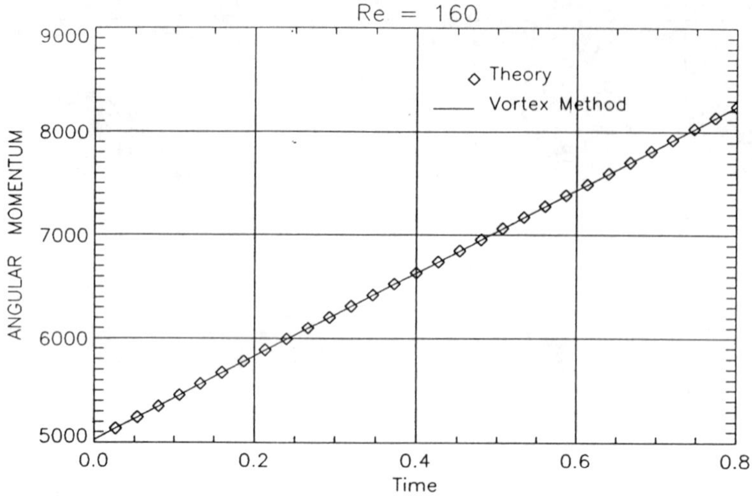

Fig. 3. Angular Momentum for the merging of two like-signed Gaussian vortices in an axisymmetric strain field

numerical methods as well as theoretical and experimental works.

4.1. VORTEX MERGING

First to illustrate the accuracy of the present scheme we simulate the merging of two like-signed vortices in an axisymmetric strain field given by :

$$\mathbf{u}_s = -\beta x\hat{\mathbf{i}} + [\beta - \alpha]y\hat{\mathbf{j}} + \alpha z\hat{\mathbf{k}}$$

The problem is described in detail by Buntine and Pullin (1989) where they present simulations for a variety of cases using a spectral method. We repeat their experiments for $Re = \Gamma/(2\pi\nu) = 160$ and with the strain rates α, β taken to be 4 and 2 respectively.

In Figure 2 we present the contour plots of the vorticity field as computed by the spectral method (left) and by the present vortex method (right). One may observe the excellent quantitative agreement of the results of the two computations. Whereas spectral methods are the yardstick for direct numerical simulations the present results exhibit the accuracy of our scheme. This accuracy of the viscous scheme is further assesed by examining Figure 3 the angular momentum of the vorticity field that is known to increase linearly with time.

We repeated the above simulation using the same initial conditions, in the absence of any viscosity and strain field. In Figure 4 we present the results at $\Gamma t/d^2 = 12.5$ obtained by the present simulation (right) with those obtained by the contour dynamics code of Jacobs (1987) (middle) and a simulation at Re = 1280 in the presence of an axisymmetric strain field by the spectral method (left). Note that the vortex method simulates the inviscid evolution of a smooth vortex

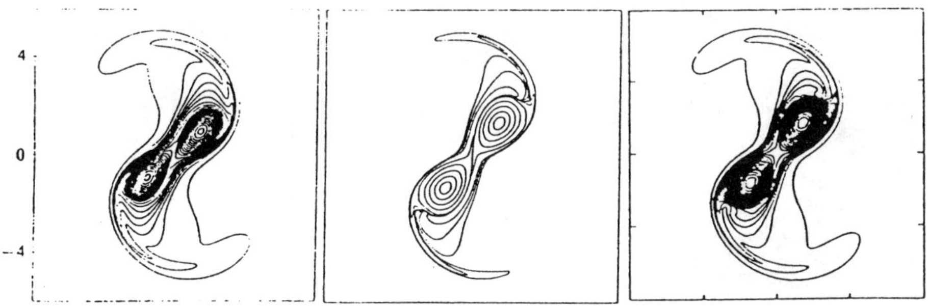

Fig. 4. Comparison of the spectral method [5] (left) at Re = 1280 ($\alpha = 4$, $\beta = 2$) with the contour dynamics [14] (middle) (with $\nu = \alpha = \beta = 0$) and the present vortex method (right) (with $\nu = \alpha = \beta = 0$).

field in the absence of any viscosity whereas the contour dynamics simulate the inviscid evolution of its stepwise approximation and the spectral code approximates a smooth viscous flow.

4.2. THE IMPULSIVELY STARTED CYLINDER

The problem of the flow development around a cylinder started impulsively from rest was studied as a model problem for unsteady flow separation. Extensive theoretical, computational and experimental results exist for this type of flow and we compare our results with those from other investigations. In Figure 5 we present the results (streamlines and vorticity contours) of our computations and we compare with experimental streaklines of Bouard and Coutanceau (1980). However diagnostics such as streamlines are smooth enough to hide defficiencies of a numerical method. As a stronger test we present in Figure 6 the drag coefficient at early times for various Re numbers and we compare with the analytical results of Bar-Lev and Yang (1975) and Collins and Dennis (1973).

The drag coefficient exhbits a $1/\sqrt{\nu T}$ type singularity and the present numerical scheme is able to capture accurately this singular behaviour. In Figure 7 the drag coefficient obtained for $Re = 1000$ by the present scheme is compared with those obtained by various other numerical works ([6] using a vortex in cell scheme, [19] using a random walk method, [20] using a finite difference scheme). Note that there is a significant discrepancy of the results and an agreement is observed with the numerical scheme of [6] for this Re. Finally in Figure 8 we present the drag coefficient for $Re = 9500$. A total of about $.6 \times 10^6$ particles were present at the end of this simulation.

5. Summary

The above results demonstrate that fast vortex methods coupled appropriately to the technique of particle are an efficient tool for direct numerical simulations of

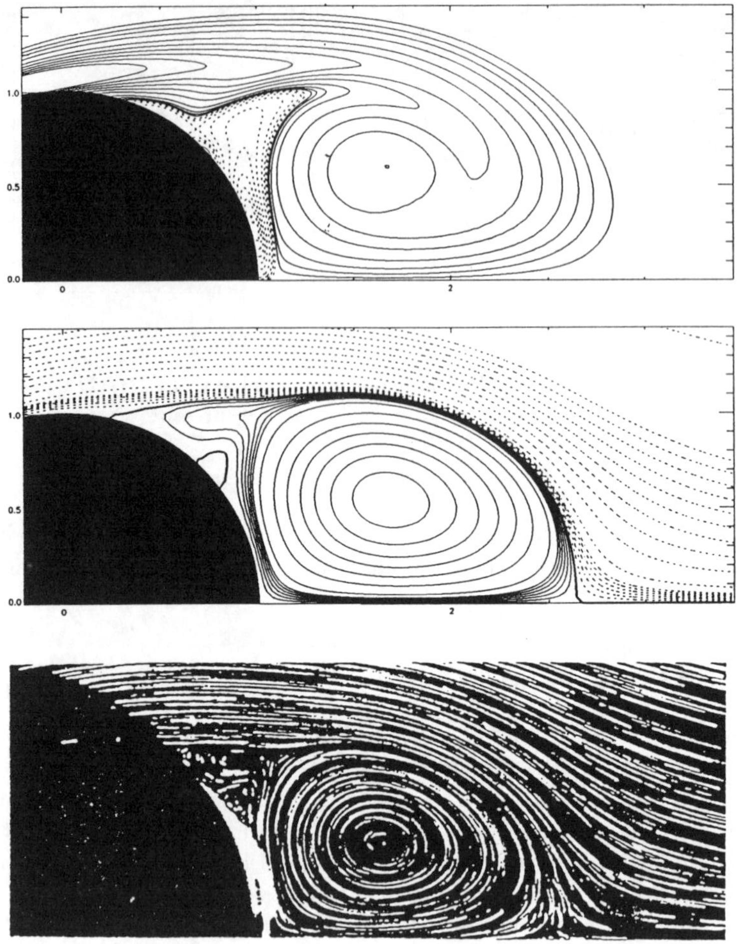

Fig. 5. Vorticity contours (top), Instanteneous Streamlines (middle) and Experimental streaklines [4] for an Impulsively Started Circular cylinder at T = 5.0 for Re = 550.

viscous and inviscid incompressible flows. The adaptivity of the method makes it also a powerful tool for studying unsteady bluff body flows.

We are concentrating our efforts to extend the method to three dimensions and implement fast vortex methods on parallel supercomputers.

Acknowledgements

We are pleased to acknowledge the support of this research by the Office of Naval Research through Grant Nos. N00014-92-J-1072 and N00014-92-J-1189. Computer time was provided by Caltech's Jet Propulsion Laboratory Supercomputing Project.

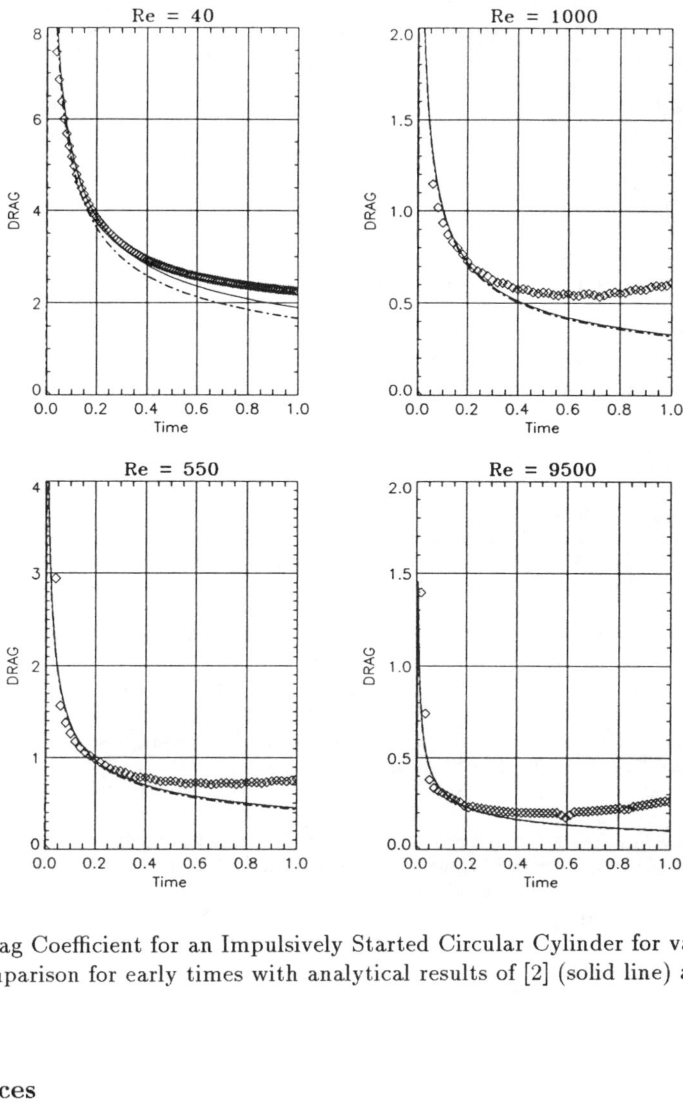

Fig. 6. Drag Coefficient for an Impulsively Started Circular Cylinder for various Re numbers. Comparison for early times with analytical results of [2] (solid line) and [9] (dashed line).

References

[1] Barnes, J.E. and P. Hut, "A hierarchical $\mathcal{O}(N\log N)$ force calculation algorithm", *Nature* Vol. 324, (1986), pp. 446-449.

[2] Bar-Lev, M. and H.T. Yang, "Initial flow field over an impulsively started circular cylinder" , *J. Fluid Mech.* Vol. 72 (1975), pp. 625-647.

[3] Beale, J.T. (1986) , "On the accuracy of Vortex Methods at large times", *Proc. Workshop on Comp. Fluid Dyn. and React. Gas Flows* I.M.A., Univ. of Minnessota.

[4] Bouard, R. and M. Coutanceau, "The early stage of development of the wake behind an impulsively started cylinder for $40 \leq 10^4$", *J. Fluid Mech.* Vol. 101 (1980), pp. 583-608.

[5] Buntine, J.D. and D.I. Pullin, "Merger and Cancellation of Strained Vortices", *J. Fluid Mech.* Vol. 205 (1989), pp. 263-295.

[6] Chang C.C. and Chern R.L. (1991), "A Numerical Study of Flow Around an Impulsively Started Circular Cylinder by a Deterministic Vortex Method", *J. Fluid Mech.* 233, 243-263.

[7] Chorin, A, "Numerical study of slightly viscous flow", *J. Fluid Mech.* Vol. 57, (1973), pp. 380-392.

[8] Collins, W.M. and S.C.R. Dennis, "The Initial flow past impulsively started circular cylinder , *Quart. J. Mech and Appl. Math.* Vol. 26 (1973), pp. 53-75.

[9] Cottet, G.-H., "A Particle-Grid Superposition method for the Navier-Stokes equations" , *J. Comp. Phys.* Vol. 89 (1990), pp. 301-318.

[10] Degond, P. and S. Mas-Gallic, "The weighted particle method for convection-diffusion equations, Part I: the case of isotropic viscosity, Part II: the anisotropic case", *Math. Comp.* Vol. 53, (1989), pp. 485-526.

[11] Friedmann A., *Partial Differential Equations of Parabolic Type*, Prentice-Hall Inc., Englewood Cliffs, N.J. 1964.

[12] Greengard, L. and V. Rohklin, "A Fast Algorithm for particle simulations", *J. Comp. Phy.* Vol. 73 (1987), pp. 325-348.

[13] Hockney, R.W. and J.W. Eastwood , *Computer simulations using particles*, McGraw-Hill, New York, 1981.

[14] Jacobs, P.A. ,1987, *Nonlinear dynamics of piecewise-constant vorticity distributions in an inviscid fluid*, Ph.D. thesis, University of Queensland.

[15] Koumoutsakos, P. ,1992, *Direct Numerical Simulations of Unsteady Separated Flows Using Vortex Methods*, Ph.D. thesis, Caltech.

[16] Leonard, A., "Vortex Methods for flow simulation", *J. Comp. Phys.* Vol. 37, (1980), pp. 289-335.

[17] Lighthill, M.J., *Introduction. Boundary Layer Theory, J. Rosenhead ed.*, Oxford Univ. Press, New-York 1963, pp. 54-61.

[18] LU, Z.Y. and Ross, T.J., "Diffusing-Vortex Numerical Scheme for solving incompressible Navier-Stokes equations" , *J. Comp. Phys* Vol. 95 (1991), pp. 400-434.

[19] Mas-Gallic, S., "Une Méthode Particulaire Déterministe incluant diffusion et conditions aux limites", *C. R. Acad. Sci. Paris* Vol. 310 Série I (1990), pp. 465-468.

[20] Prager W. (1928), Die Druckverteilung an Körpern in ebener Potentialströmung, *Physik. Zeitschr.* 29, 865-869.

[21] Rosenhead, L. 1931, The formation of vortices from a surface of discontinuity, *Proc. Roy. Soc., Ser. A* 134, 170.

[22] Sarpkaya, T., "Computational methods with vortices - The 1988 Freeman scholar lecture", *ASME J. Fluids Eng.* Vol. 111, (1988), pp. 5-52.

[23] Smith P.A. and Stansby P.K. (1988), Impulsively Started Flow Around a Circular Cylinder by the Vortex Method, *J. Fluid Mech.* 194, 45-77.

[24] Ta Phuoc Loc and Bouard R. (1985), Numerical Solution of the Early Stage of the Unsteady Viscous Flow Around a Circular Cylinder: a Comparison with Experimental Visualization and Measurements, *J. Fluid Mech.* 160, 93-117.

NUMERICAL STUDY OF THE MOTION AND DEFORMATION OF TWO-DIMENSIONAL BUBBLES BY A VORTEX METHOD

HENRYK KUDELA
Technical University of Wrocław
50-370 Wrocław
Wybrzeże Wyspiańskigo 27
Poland

ABSTRACT. In this paper are presented numerical results that relate to the motion and deformation of a thermal bubble in a gravity field. The study was done using the vortex blob method. To regularize the initial value problem for the vortex sheet a parameter δ^2 was introduced in the Biot-Savart formula. We point out the viscous character of the δ^2 regularization. The evolution of circular and elliptical bubbles was studied. The numerical experiments showed great sensitivity of the evolved forms to the initial conditions.

1. Introduction

In the field of fluid mechanics it is well-known that a density difference between two fluids causes motion of the fluids. One of the most important problems within the general area of pattern formation and bubble mechanics relates to interface deformations. In this paper our interest concerns the evolution of the two-dimensional bubble in a gravity field when the fluid inside the bubble is slightly less dense than in an ambient fluid. Gravity causes the lighter fluid to accelerate in the direction of the heavier one. The interface of the bubble undergoes the classical Rayleigh-Taylor instability [Sharp, 1984]. The evolution of the bubble progresses to conditions "far from an equilibrium". One can expect that the future shape of the bubble will be very sensitive to the initial condition and will undergo unexpected changes. When the final shape of the bubble does not resemble its initial one, we may speak about morphological instability [Langer, 1980] [Sekerka, 1973]. If we make some simplifying assumptions about the physics (see below), the interface between the bubble and surrounding fluid can be regarded as a vortex sheet. The motion study of the bubble can be reduced to an initial value problem. Throughout the paper, we will assume that density variations are small. This allows us to significantly simplify the evolution equations for the vortex sheet and enables us to avoid some problems related to nonlinearity of these equations (Kudela 1990a,b). Instead of the Fredholm integral equation of the second kind, a simple differential equation has to be solved. This case still has practical interest, .e.g. in the study of the motion of a thermal [Scorer, 1978]. For the solution of the initial value problem we use the

J.T. Beale et al. (eds.), Vortex Flows and Related Numerical Methods, 191–205.

generalized vortex method described by Baker, Merion, Orszag (1980, 1982). It is known that the initial value problem for a vortex sheet is ill-posed in the Hadamard sense [Moore, 1979] [Krasny, 1986a]. At a finite time, the solution has a singularity. This is a consequence of the fact that the linear approximation yields an arbitrarily large growth rate for short-wavelength perturbations. From the practical point of view, the emergence of a singularity is physically unacceptable and shows inadequacy of the mathematical model in describing the problem. So it is reasonable to include certain physical mechanisms like diffusion, interfacial tension or the finite thickness of the interfacial transition region to regularize the problem. Another way to regularize the problem is to slightly change the original equations, as was done by Krasny [Krasny, 1986b] for the Kelvin-Helmholtz instability problem. He introduced into the formulas which describe the velocity induced by the vortex sheet (the Biot-Savart formula) a δ^2 parameter in such a way that the singular integrand is replaced by a smooth one [Krasny, 1986b]. This kind of regularization, by analogy to the Chorin vortex blob method [Chorin & Bernard 1973] is called vortex blob regularization. Krasny showed [Krasny, 1986b] that the δ^2-linearized equations give a bounded rate of growth for short wavelength perturbations. For the Rayleigh-Taylor problem it was shown [Bellman & Pennington, 1954] that viscosity causes the linear growth rate to vanish asymptotically with wavelength, and the same occurs for the δ^2 equations. So, although the introduction of δ^2 is a quite formal procedure and does not correspond precisely to any physical effect, one can interpret the introduction of the δ^2 parameter as a trial to mimic fluid viscosity.

2. Statement of the problem and equations of motion

We assume that the motion of the fluids is potential, two-dimensional, inviscid, and incompressible, and that the pressure across the interface is continuous (no surface tension effects). A rectangular system of coordinates (x,y) is introduced. The constant acceleration of gravity g acts in the negative y-direction. The densities of the fluids are constant and change their value across the interface. The density ρ_1 of the ambient fluid is slightly greater than the density ρ_2 of the fluid inside the bubble. The basic control parameter of the problem is the Atwood number:

$$A = \frac{\rho_1 - \rho_2}{\rho_1 + \rho_2} \tag{1}$$

The interface is described parametrically as $(x(e,t), y(e,t))$, where t is time and e is a real parameter $\in [0,1]$ which is used also as a Lagrangian variable. At the initial instant, the value of e marks a point which follows the motion of the interface. The initial shape of the bubble is described by an ellipse $(x(e,0), y(e,0)) = (a\cos(2\pi\, e), b\sin(2\pi\, e))$, where a and b are the large and small semiaxes of the ellipse. Moreover the ellipse can be rotated by an angle θ, where θ is the angle between the large axis of the ellipse and the horizontal x-axis. So the initial position and shape of the bubble can be characterized by the angle θ and the ratio $\beta = b/a$. Except for the case of a circle ($\beta = 1$), we present results only for the evolution of the ellipse with $\beta = 1/3$, $a = 1$ and for a different angles θ. The interface can be regarded as a vortex sheet with strength equal to the tangential velocity jump across the interface:

$$\Gamma(s,t) = (\mathbf{u}_2 - \mathbf{u}_1)\,\mathbf{s}^{\circ} \tag{2}$$

where $\mathbf{u}_1, \mathbf{u}_2$ are the limiting velocities of the fluids at the interface, $\mathbf{s}^o$ = unit tangential vector, s = arclength. Due to the assumption that $\rho_1 \cong \rho_2$ we apply the Boussinesq approximation $A \mapsto 0$, $g \mapsto \infty$, such that Ag remains finite, $Ag = 1$. Equations describing the evolution of the bubble interface are:

$$\frac{\partial x}{\partial t} = -\frac{1}{2\pi}\int_0^1 \frac{\gamma(e',t)(y(e,t) - y(e',t))}{r^2+\delta^2}\,d\,e' \tag{3}$$

$$\frac{\partial y}{\partial t} = \frac{1}{2\pi}\int_0^1 \frac{\gamma(e',t)(x(e,t) - x(e',t))}{r^2+\delta^2}\,d\,e' \tag{4}$$

$$\frac{\partial \gamma}{\partial t} = 2\frac{\partial y}{\partial e} \tag{5}$$

where $r^2 = (x(e,t)-x(e',t))^2+(y(e,t)-y(e',t))^2)^2$, and $\gamma(e,t) = \Gamma(e,t)(x_e^2+y_e^2)^{1/2}$. Equations (3), (4) describe the velocity induced by the vortex sheet (Biot-Savart law, where we have already introduced δ^2), and equation (5), in the Boussinesq approximation, describes the vortex sheet strength. The units used in (3), (4) are such that length is scaled by the length of the large semiaxis of the ellipse a and time is scaled by $(a/Ag)^{1/2}$. For the rest of the paper we assume $a = 1$.

3. The numerical approximation

The vortex sheet was divided into N segments and the curve $(x(e,t), y(e,t))$ was approximated by $(x_i, y_i) = (x(e_i,t), y(e_i,t))$, $e_i = ih$, $h = 1/N$. The integrals in (3) and (4) were approximated by the trapezoidal rule at the alternate points [Shelley, 1992]. This yields a system of ordinary differential equations for the trajectories of the points and their strengths as follows:

$$\frac{dx_i}{dt} = -\frac{h}{\pi}\sum_{\substack{k=1\\k+i=odd}}^{N} \frac{\gamma_k(y_i - y_k)}{r_{ik}^2+\delta^2} \tag{6}$$

$$\frac{d\,y_i}{dt} = \frac{h}{\pi}\sum_{\substack{k=1\\k+i=odd}}^{N} \frac{\gamma_k(x_i - x_k)}{r_{ik}^2+\delta^2} \tag{7}$$

$$\frac{d\,\gamma_i}{dt} = 2\frac{d\,y_i}{d\,e} \tag{8}$$

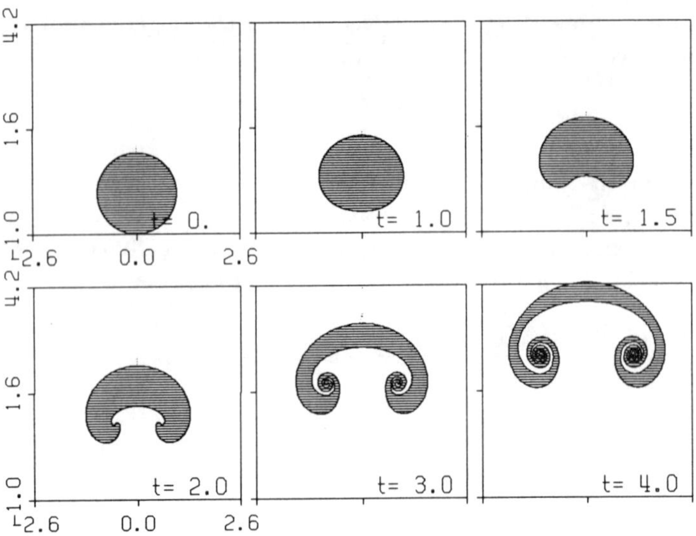

Fig. 1. Successive stages in the evolution of the circular bubble with $\delta^2 = 0.01$, $N = 600$. The time is indicated in the lower right corner of each frame.

The numerical procedure goes as follows [BMO, 1982]: for known $x(e)$, $y(e)$, $\gamma(e)$ the interface is marched forward using (6) and (7); next $\gamma(e)$ is also marched in time by using (8). The time stepping is performed by using the fourth-order Adam-Multon predictor-corrector scheme. To start calculations a Runge-Kutta (RKF45) algorithm was used. The derivative $\mathrm{d}y/de$ was calculated by a cubic spline. To check the accuracy, two invariants of the motion were monitored: the mass S of the bubble (it should be constant) and the mass flux Ω through the interface (it should be zero):

$$S = \int_0^1 y x_e de\,, \qquad \Omega = \int_0^1 \mathbf{u} \cdot \mathbf{n} de \tag{9}$$

When the relative error for S reached the value 0.025, or Ω reached .05, the calculation was stopped.

4. Numerical results

4.1. MOTION OF THE CIRCULAR BUBBLE

A numerical study of this case by the vortex method has already been done by Meng & Thomson (1978) and Anderson (1985) (but by another vortex method). We repeated those calculations in order to gain some understanding of the δ^2 influence. In Fig. 1 successive stages are shown in the evolution of the circular bubble for $\delta^2 = 0.01$. The spiral structure on both sides of the vertical axis of symmetry is clearly visible. In Fig.2 a close-up frame for $t = 4.$ from Fig.1 is presented.

A smaller value of δ^2 causes greater flexibility of the vortex sheet for roll-up. As δ^2 approaches to zero, the number of the turns in the spiral structure grows (see

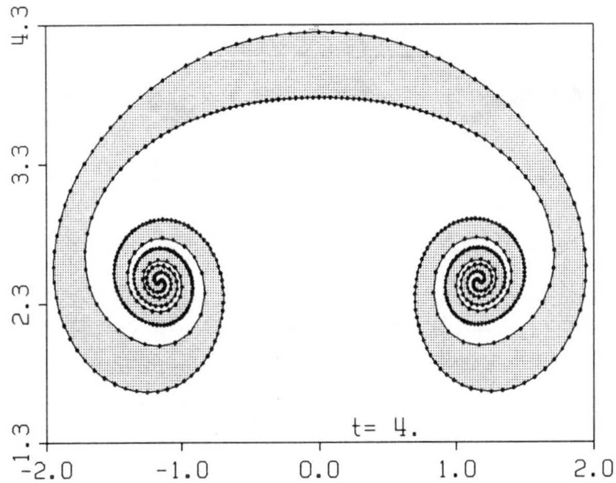

Fig. 2. Close-up of the frame for $t = 4$ from fig.1 . The points on the interface indicate the position of the markers moving with Lagrangian velocity ($N = 600$).

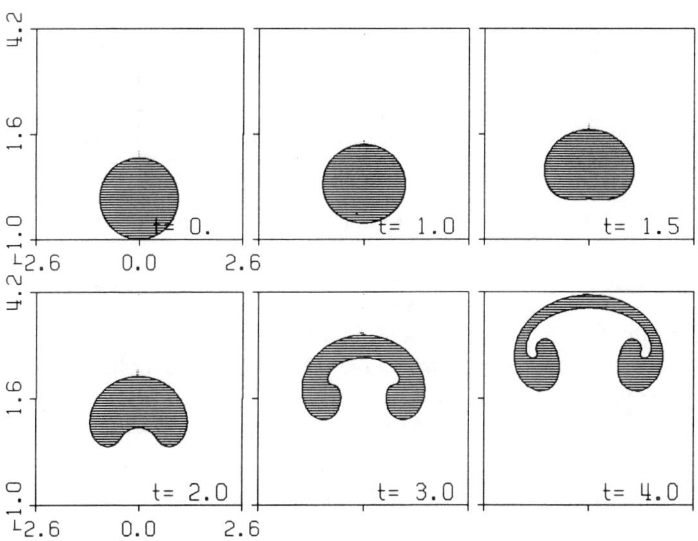

Fig. 3. Successive stages in the evolution of the circular bubble with $\delta^2 = 0.1$.

also [Krasny, 1986b], [Anderson, 1985]). A large value of δ^2 prevents any vortex sheet roll-up at all. In Fig.3 the evolution of the bubble for $\delta^2 = 0.1$ is shown. To illustrate the influence of the value of δ^2, solutions are shown in Fig. 4 which were obtained for a different δ^2 ($= 0.1$, $= 0.04$, $= 0.01$) at the same time $t = 4$.

One can see that while large scales of the bubble suffer only slight changes, the spiral structure becomes more and more complicated when the value of δ^2 is decreased. Intuitively a greater value of δ^2 corresponds to the effect of greater viscosity of the fluid. In Fig.5 we present the velocity of the upper point of the

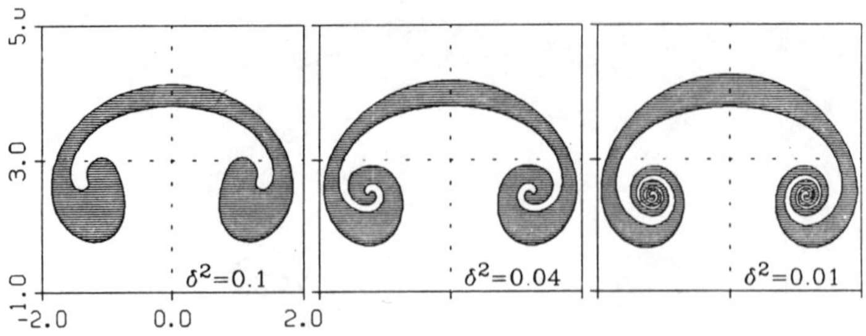

Fig. 4. Effect of δ^2. Interface at $t = 4.$ with δ^2 value as shown.

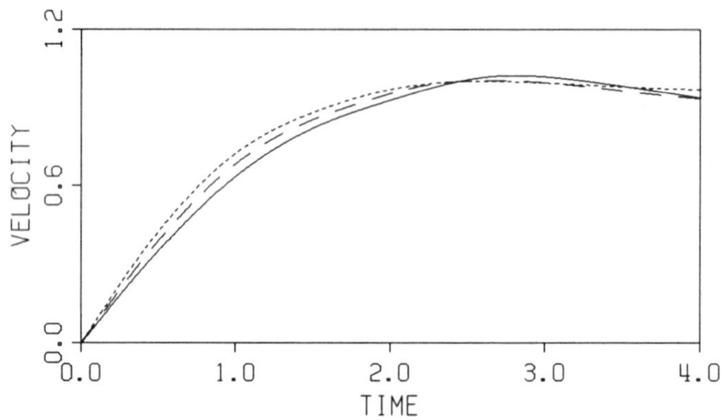

Fig. 5. Velocity of the upper point of the bubble vs. time. — $\delta^2 = 0.1$, – – – $\delta^2 = 0.04$, ··· $\delta^2 = 0.01$.

bubble on the symmetry axis vs. time, and in Fig.6 the thickness of the cap vs. time for different values of the δ^2 parameter. The rate of change of the total vorticity in one half of the curve is measured by the thickness Δy of the cap on the symmetry axis; integration of (5) with respect to e in interval [0, 0.5] gives $\frac{\partial}{\partial t}\int_0^{0.5}\gamma = -2\Delta y$. Fig.5 suggests that a limit for the bubble velocity may exist independently from the δ^2 value. A greater value of δ^2 (more "viscous" fluid) causes a slower raising of velocity (Fig.5), and the total rate of change of vorticity drops ("dissipates") more quickly (Fig.6).

4.2. MOTION OF THE ELLIPTICAL BUBBLE

In actual experiments the initial shape of the bubble would most likely not be circular (of course it would never be two-dimensional as well). For this reason it was interesting to study the case of the evolution of a bubble that was not circular at the initial instant. We assumed that the initial form of the bubble would be an ellipse. Its shape was to be characterized by the ratio of the small axis of the ellipse to the large one $\beta = b/a$ and by the angle θ between the x-axis of the coordinate system and the large axis of the ellipse. In this paper numerical results are presented only

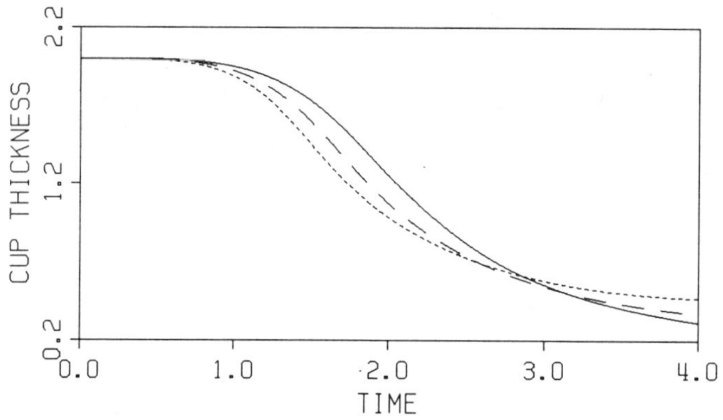

Fig. 6. Cap thickness vs time. — $\delta^2 = 0.1$, - - - $\delta^2 = 0.04$, $\cdots$ $\delta^2 = 0.01$.

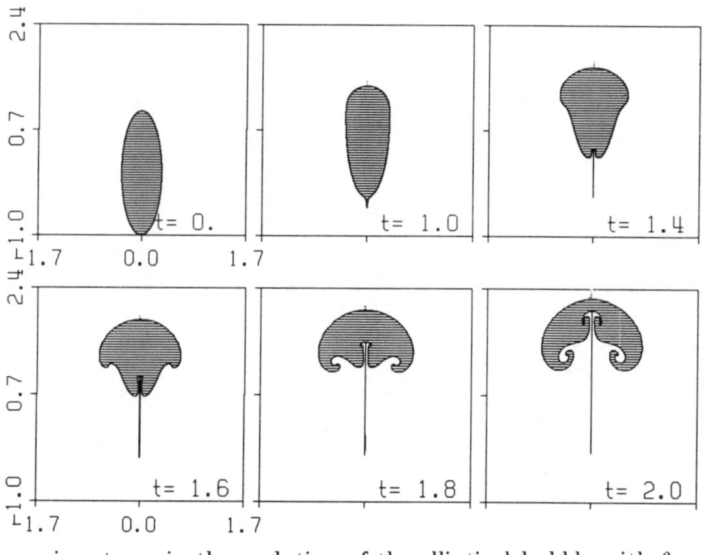

Fig. 7. Successive stages in the evolution of the elliptical bubble with $\theta = 90$, $\beta = 1/3$, $\delta^2 = 0.01$, $N = 600$.

for $\beta = 1/3$ but with different θ. In Fig.7 the successive stages are shown for the evolution of the bubble for $\beta = 1/3$, $\theta = 90^o$ and $\delta^2 = 0.01$. (See close-up portraits for $t = 1.8$ and $t = 2$ in Fig.8.)

At first one may be struck by the development of the tail structure. Such a tail was observed in many experiments [Kojima, Hinch & Acrivos, 1984] [Sparrow, Husar & Goldstein, 1970] [Griffithes, 1986]. The surrounding fluid penetrates the bubble near the base of the tail. The tail (fluid thread) is stretched, and its thickness at some point goes to zero; at this point probably the tail will be separated from the bubble (see[Kojima, Hinch & Acrivos, 1984]). Due to the relatively small value of δ^2, the back side of the bubble undergoes very fine deformation and undulation. The front of the bubble takes the shape of a circle. A greater value of δ^2 prevents

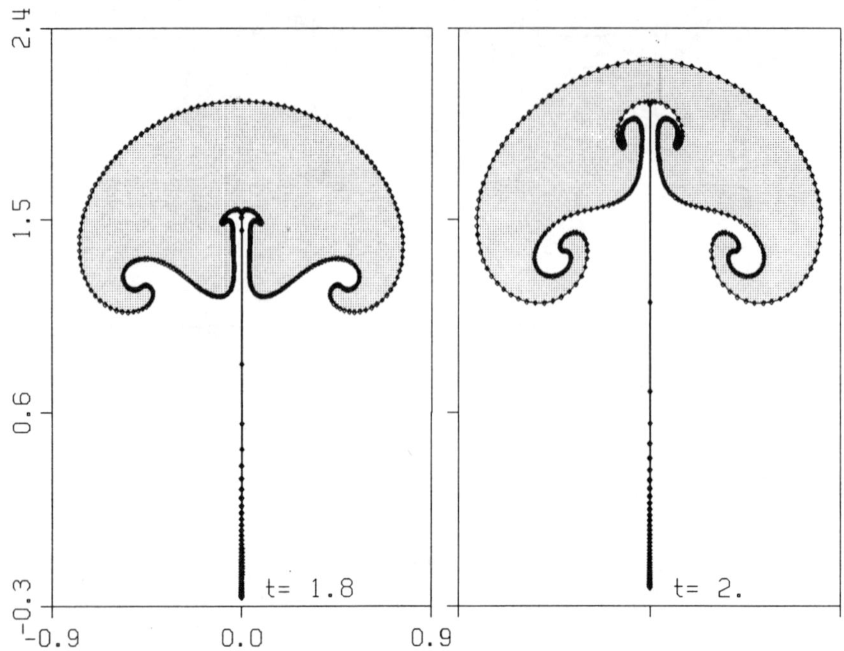

Fig. 8. Close-up of the two frames from Fig.7 for $t = 1.8$ and $t = 2$. ($\delta^2 = 0.01$, $\theta = 60^o$, $N = 600$)

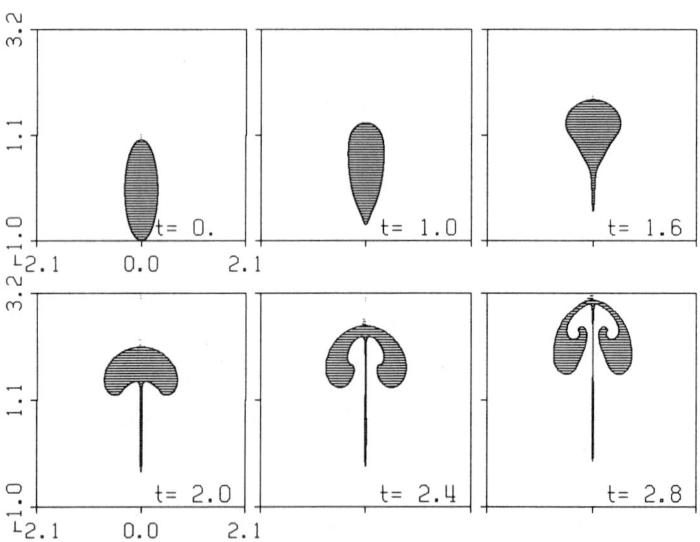

Fig. 9. Successive stages in the evolution of the elliptical bubble with $\theta = 90$, $\beta = 1/3$, $\delta^2 = 0.04$.

such fine undulating deformation of the interface. The tail becomes slightly thicker. Greater values of δ^2 change the time scale of deformation. Evolution progresses more slowly (see Fig.9).

In Fig.10 the evolution of the bubble is presented for $\theta = 60^o$ and $\delta^2 = 0.01$.

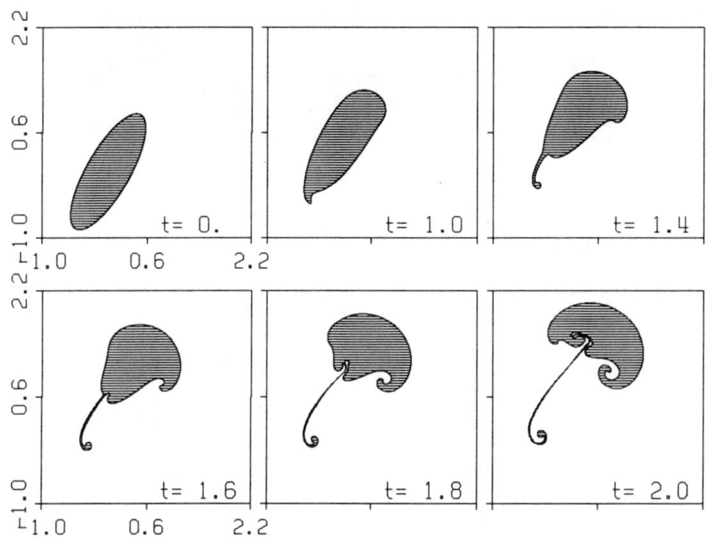

Fig. 10. Successive stages in the evolution of the elliptical bubble with $\theta = 60$, $\beta = 1/3$, $\delta^2 = 0.01$, $N = 800$

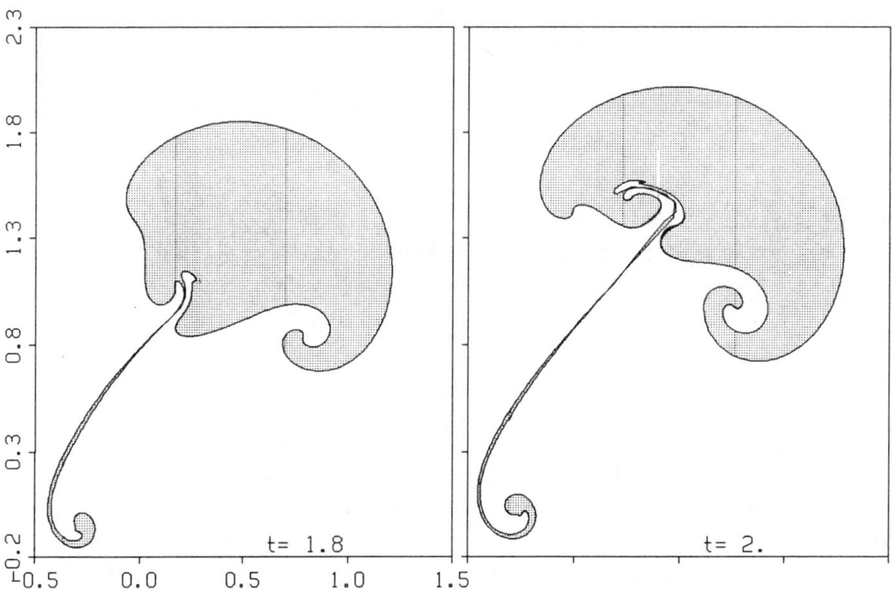

Fig. 11. Close-up of the two frames from Fig.10 for $t = 1.8$ and $t = 2.$($\delta^2 = 0.01$, $\theta = 60^o$, $N = 800$)

Due to the relatively small δ^2 (compare with Fig.7) the rear side of the bubble was strongly undulated (see close-up portraits in Fig.11).

As previously, greater values of δ^2 makes the back side of the bubble smoother

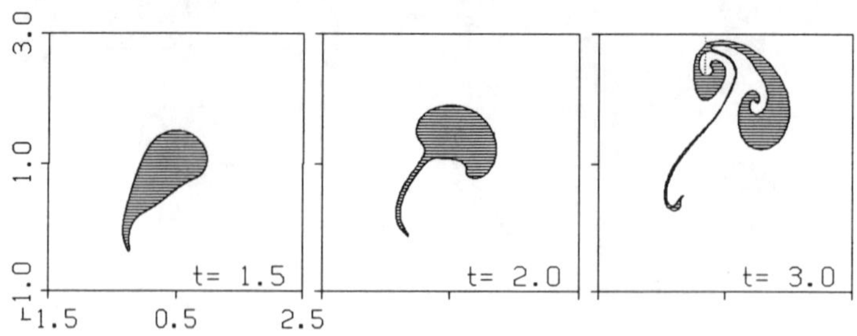

Fig. 12. Final stages of the elliptical bubble evolution with $\theta = 60$, $\beta = 1/3$, $\delta^2 = 0.04$.

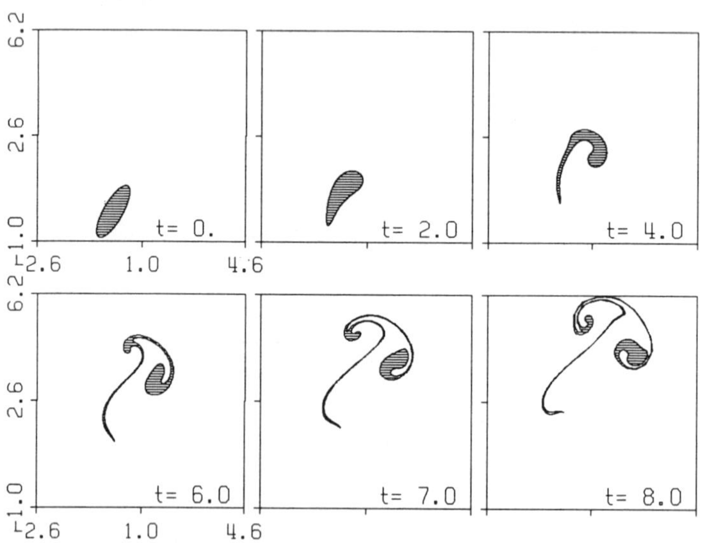

Fig. 13. The successive stages of the bubble in evolution for $\delta^2 = 0.4$, $\theta = 60^o$, $N = 600$.

(Fig.12).

Greater values of δ^2 not only change the time scale of deformation and give a smoothing effect but may also considerably change the shape of the bubble during the evolution. In Fig. 13 we present the evolution of the bubble for $\theta = 60^o$ (as in Fig.10) but with $\delta^2 = 0.4$. Deformation of the bubble progresses very slowly and in the final stages the front of the bubble is composed of a thin thread (see close-up portraits for $t = 7$ and $t = 8$ in Fig.14).

In Fig.15 the evolution of the elliptical bubble for $\theta = 45^o$, and $\delta^2 = 0.04$ is presented.

In the following figures we present results for the bubble evolution for the same initial angle $\theta = 30^o$ and for different δ^2 values, $\delta^2 = 0.01$, .04, and 0.4. Once more we can notice the influence on the evolution of the initial angle θ (compare with Fig.10 and 15) and the modification caused by the different values of δ^2.

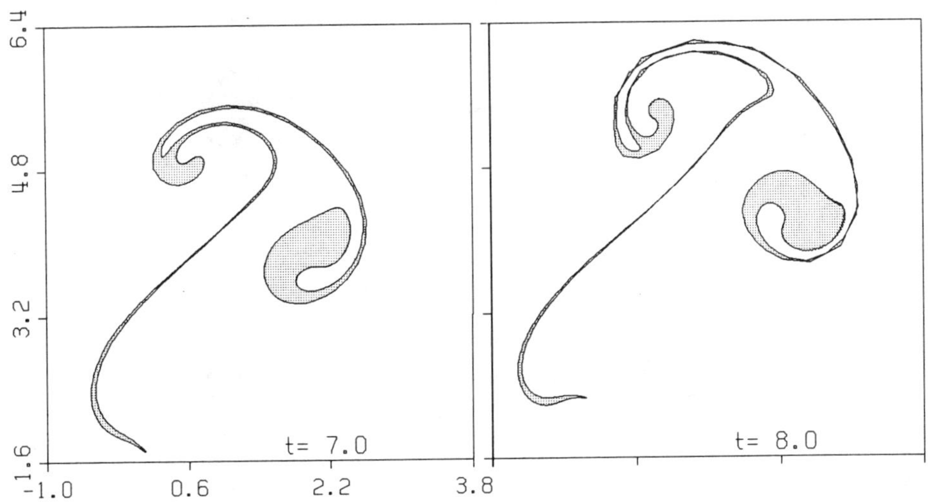

Fig. 14. Close-up portrait of the bubble from Fig.13 for $t = 7$ and $t = 8$. ($\delta^2 = 0.4$, $\theta = 60^o$, $N = 600$)

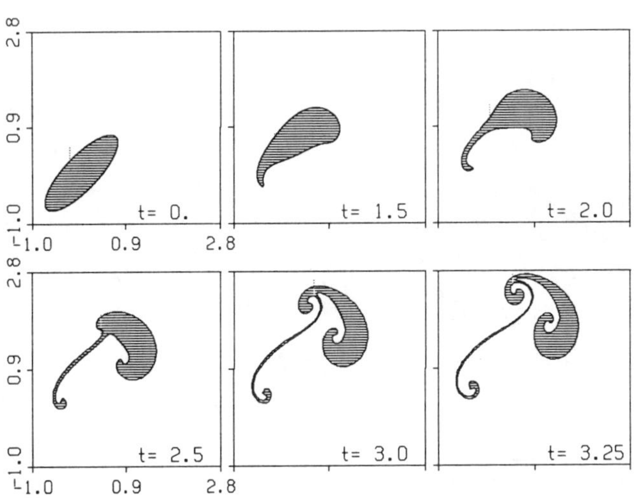

Fig. 15. The successive stages of the bubble in evolution for $\delta^2 = 0.04$, $\theta = 45^o$, $N = 600$.

5. Concluding remarks

This work was inspired mainly by the work of Aref (1987) and Langer (1980) in the context of "dynamical morphology". We were interested in the evolution of the forms that progress in unstable conditions or as one has called it, "far from

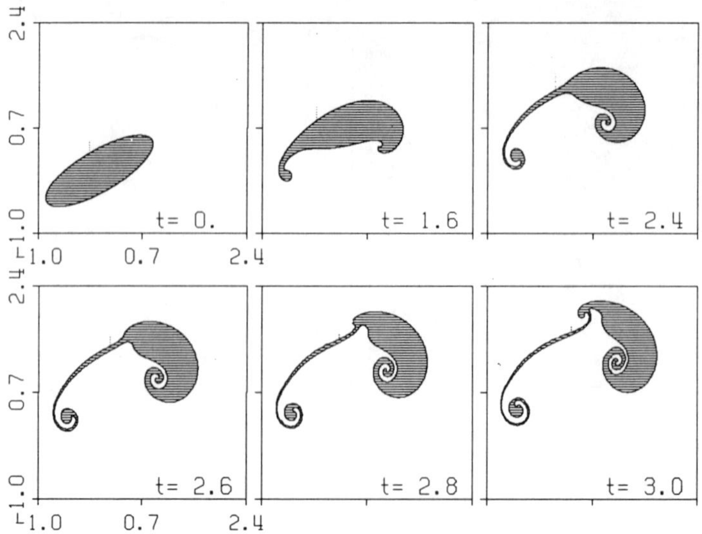

Fig. 16. The successive stages of the bubble in the evolution for $\delta^2 = 0.01$, $\theta = 30^o$, $N = 800$.

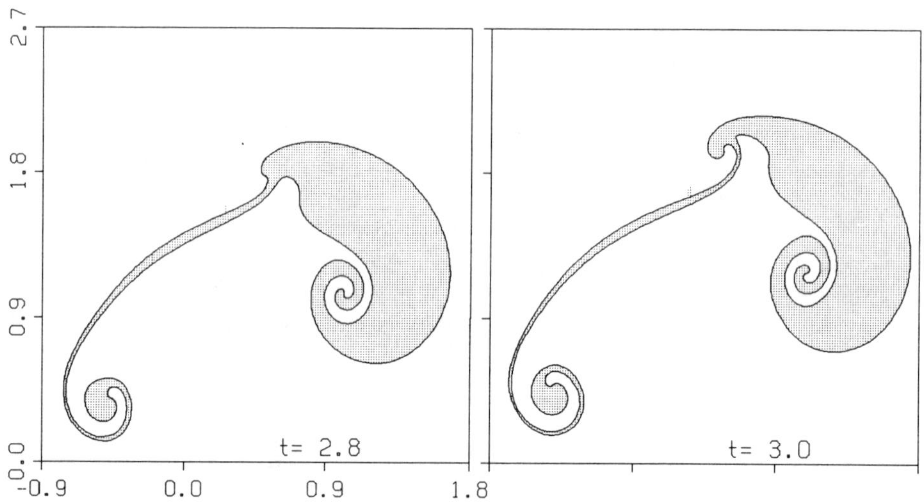

Fig. 17. Close-up of the two portraits from Fig.16 for $t = 2.8$ and $t = 3.0$ ($\delta^2 = 0.01$, $\theta = 30^o$, $N = 800$).

equilibrium". Evolution in such situations inevitably develops strange forms, where in its final stages such entities should emerge as "whorls, tendrils and spikes" [Aref, 1987]. Our results indicate that the tail structure appears due to initially non-circular (elliptical) shapes of the bubble. For the symmetrical case ($\beta = 1/3$, $\theta = 90^o$) our results agree qualitatively, despite the two-dimensionality of our model, with the experimental investigation of Kojima *et al.* (1984) (up to the moment when in the experiment the tail was cut off from the bubble) and the numerical one

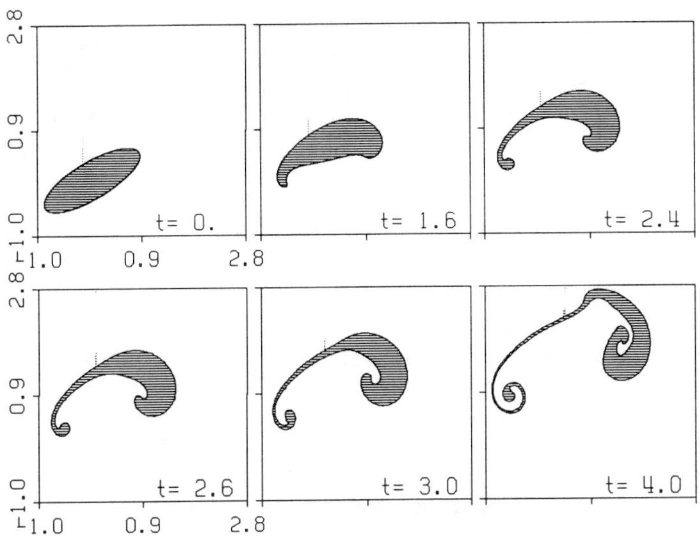

Fig. 18. The successive stages of the bubble in the evolution for $\delta^2 = 0.04$, $\theta = 30^o$, $N = 600$.

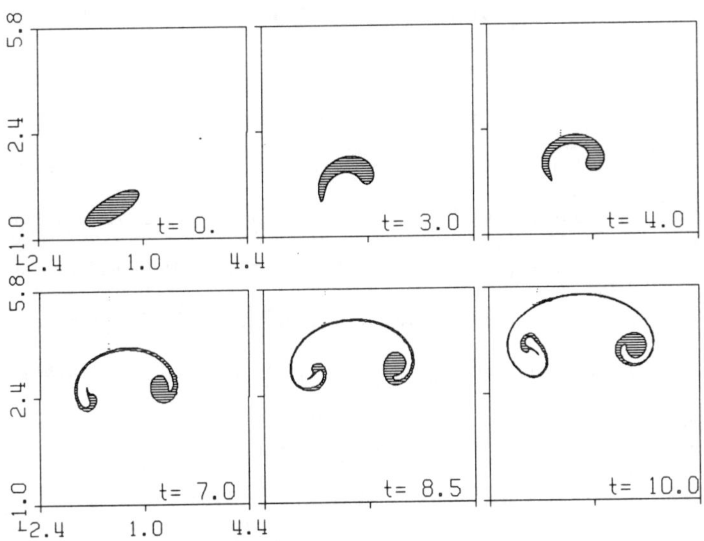

Fig. 19. The successive stages of the bubble in the evolution for $\delta^2 = 0.4$, ($\theta = 30^o$, $N = 600$)

[Pozrikidis, 1990]. As we expected, the shape of the bubble during its evolution is sensitive to the initial condition (the angle θ). We did not refer to the problem of the existence of the limit solution when $\delta \mapsto 0$ (see [Krasny, 1986b]). Our feeling is that such a limit may not exist in the case of Rayleigh-Taylor instability but that some other indications like velocity may converge as $\delta \mapsto 0$. Blob regularization gives results which look more realistic than without it. The δ^2 value may strongly modify the bubble evolution. Due to the simplicity of the mathematical model

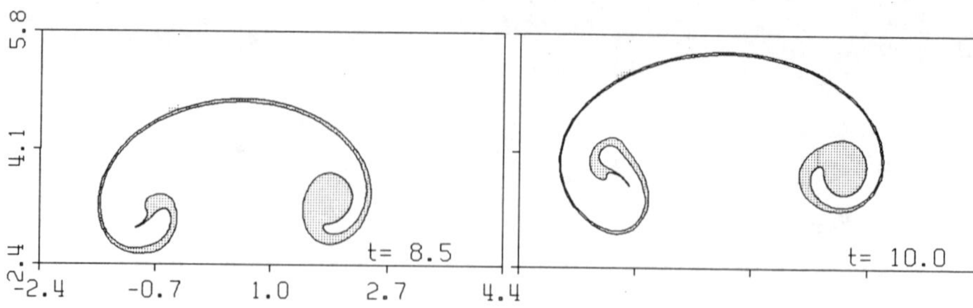

Fig. 20. Close-up of the two portraits from Fig.19 for $t = 8.5$ and $t = 10$. ($\delta^2 = 0.4$, $\theta = 30^o$)

and a not very involved numerical algorithm, the program is very convenient for investigation of diverse aspects of morphological instability.

Acknowledgements

The work was supported by Grant Nr 1542/3/91 from the Polish Committee of Scientific Investigation.

References

Anderson, 1985 Anderson C.R. 1985 *A vortex method for flows with slight density variations.* J. Comput. Phys. **61**, 417-444.

Aref, 1987 Aref H. 1987 *Finger, bubble, tendril, spike.* Fluid Dynamics Transactions, **13**, 25-54, Warszawa PWN.

BMO, 1980 Baker G.R., Merion D.I., Orszag S. A. 1980 *Vortex simulation of the Rayleigh-Taylor instability.* Phys. Fluids, **23**, 1485-1480

BMO, 1982 Baker G.R., Merion D.I., Orszag S.A. 1982 *Generalized vortex methods for free-surface flow problems.* J. Fluid Mech. **123**, 477-501.

Bellman & Pennington, 1954 Bellman R., Pennington R.H. 1954 *Effects of surface tension and viscosity on Taylor instability.* Quart. Appl. Math. **123**, 151-162.

Chorin & Bernard 1973 Chorin A.J., Bernard P.S. 1973 *Discretization of a vortex sheet with an example of roll-up.* J. Comput. Phys. **13**, 423-429.

Griffithes, 1986 Griffithes R.W. 1986 *Thermals in extremely viscous fluids, including the effects of temperature-dependent viscosity.* J. Fluid Mech. **166**, 115-138.

Kojima, Hinch & Acrivos, 1984 Kojima M., Hinch E.J., Acrivos A. 1984 *The formation and expansion of a toroidal drop moving in a viscous fluid.* Phys.Fluids **27**(1), 19-32.

Krasny, 1986a Krasny R. 1986 *A study of singularity formation in a vortex sheet by the point-vortex approximation.* J. Fluid Mech. **167**, 65-93.

Krasny, 1986b Krasny R. 1986 *Desingularization of periodic vortex sheet roll-up.* J. Comput. Phys. **65**, 292-313.

Kudela, 1990a Kudela H. 1990a *The influence of the surface-tension effects on using vortex method in the study of Rayleigh-Taylor instability.* Notes on Numerical Fluid Mechanics **29**, 273-282, ed. P.Wesseling, Vieweg.

Kudela, 1990b Kudela H. 1990b *Numerical studies of Rayleigh-Taylor instability by vortex method.* Arch. Mech. **42**, 493-505.

Kudela, 1992 Kudela H. 1992 *Study of the motion of the two-dimensional bubble by the vortex method.* Z. Angew. Math. Mech. **72**, 5, 369-373.

Langer, 1980 Langer J.S. 1980 *Instability and pattern formation in crystal growth.* Rev. Modern Physics **52**, 1-27.

Meiron, Baker & Orszag, 1982 Meiron D.I., Baker G.R., Orszag S.A. 1982 *Analytical structure of vortex sheet dynamics, Part 1. Kelvin-Helmholtz instability.* J. Fluid Mech. **114**, 283-298.

Meng & Thomson, 1978 Meng J. S., Thomson J.A.L. 1978 *Numerical studies of some nonlinear hydrodynamic problems by discrete vortex element method.* J. Fluid Mech. **84**, 433-453.

Moore, 1979 Moore D.W. 1979 *The spontaneous appearance of singularity in the shape of an evolving vortex sheet* Proc.R.Soc.Lond. A365, 1

Pozrikidis, 1990 Pozrikidis S.C. 1990 *The instability of a moving viscous drop.* J. Fluid Mech. **210**, 1-21.

Scorer, 1978 Scorer R.S. 1978 Environmental Aerodynamics. Halsted Press.

Sparrow, Husar & Goldstein, 1970 Sparrow E.M., Husar R.B, Goldstein R. J. 1970 *Observations and other characteristics of thermals.* J. Fluid Mech. **41**, 793-800.

Sharp, 1984 Sharp D.H. 1984 *An overview of Rayleigh-Taylor instability.* Physica **12D**, 3-18.

Shelley, 1992 Shelley M.J. 1992 *A study of singularity formation in vortex sheet motion by a spectrally accurate vortex method.* to be published in J. Fluid Mech.

Sekerka, 1973 Sekerka R.F. 1973 *Morphological stability.* in Crystal Growth: an Introduction, ed. P. Hartman.

A HYBRID VORTEX METHOD WITH DETERMINISTIC DIFFUSION

Habib N. NAJM
Texas Instruments
Dallas TX
USA

ABSTRACT.

In the general class of particle methods, the vortex method is most convenient for modeling high Reynolds (Re) number vortex flows. This is particularly true because: (1) the Lagrangian solution of the Navier-Stokes equations eliminates the need to discretize the non-linear inertia terms, leading to good numerical stability at high Re, and (2) the restriction of computational elements to the regions of the flow exhibiting shear and finite vorticity leads to significant numerical efficiency.

The present work maintains the above advantages of the vortex method while proposing an accurate scheme for modeling the diffusion component of the flow equations. The method is based on a fractional step solution of the two-dimensional vorticity transport equation. The advection step is implemented using lagrangian propagation of vortex elements, while the diffusion step uses a hybrid lagrangian-eulerian implementation. In this diffusion scheme, both the vorticity field, $\omega(x,y)$, and its Laplacian, $\nabla^2\omega(x,y)$, are evaluated from the summation of the individual analytical elemental fields at any point in the computational domain. In particular, both quantities are computed analytically at the cell centers of a suitable grid overlaid on the computational domain. A second order time integration of the diffusion equation is then utilized to arrive at the new values of the vorticity at the cell centers. These are used to generate a new set of vortex elements at each time step, in the same fashion utilized in the initial discretization of the vorticity field at time $t = 0$. This process of continuous regeneration of vortex elements at each time step, based on the diffusion of the overall vorticity field, allows for creation of new elements where necessary to maintain a time accurate representation of the flow.

The scheme is presented and demonstrated on a 2-dimensional (2D) incompressible flow problem that involves the viscous decay of a vortex tube in an unbounded two-dimensional domain. The analytical solution of this problem is known, and is used to compute the error associated with the numerical solution, to study its convergence characteristics, and to compare the present results against those available from the random vortex method[1], and the deterministic diffusion scheme due to Raviart[2]. Results demonstrate the improvement in accuracy provided by this hybrid scheme, for the selected vortex decay problem, over a wide range of Reynolds number.

1. Introduction

This work is concerned with the modeling of viscous incompressible flow using Lagrangian vortex elements. In the implementation of the vortex element method to two dimensional (2D) incompressible inviscid flow, vortex elements move with the local material velocity in a lagrangian manner. Chorin [1] introduced the "random

J.T. Beale et al. (eds.), Vortex Flows and Related Numerical Methods, 207–222.

vortex method" (RVM), which uses viscous splitting to solve the viscous incompressible 2D vorticity transport equation in a sequence of two steps at each time instant, one convective and one diffusive. The convective step implementation is based on a lagrangian propagation of vortex elements by the material velocity, while the diffusive step is modeled using a random walk algorithm. Properly selected random disturbances to vortex element positions are introduced at each time step, where the size of the disturbance is inversely proportional to the square root of the Reynolds number (Re). This algorithm is quite efficient, but is time-accurate only in a ensemble average sense.

A deterministic particle strength exchange (PSE) diffusion scheme was introduced by Raviart [2] and his coworkers [3; 4; 5]. It is based on the approximation of the Laplace operator ∇^2 by an integral operator, and the application of a quadrature rule to this integral using the locations of the particles as quadrature points. This method has been shown to be conservative, since it does not fundamentally require remeshing the vorticity field. However, it is evident that remeshing is necessary when particle distributions become distorted and overlap is lost, to avoid severe degradation in accuracy [6].

On the other hand, the PSE scheme, while it may accurately describe the decay of vorticity in a flow region where elements exist, has no mechanism for generation of new elements in regions where new vorticity has been created by diffusion. Cottet[7] describes a domain decomposition extension for flow past a cylinder in which the boundary-layer, where diffusion is predominant, is described by means of a finite difference method, while vortex elements take care of the wake, where the flow is convection dominated. Particles are created upstream, which simplifies the interface problem but might not give enough accuracy in the exchange of vorticity between the finite-difference and particle domains.

An alternative hybrid diffusion scheme is proposed that allows for the creation of new vortex elements by viscous diffusion in regions of the flow domain that are initially irrotational. The scheme combines lagrangian and eulerian approaches. Needless to say, an essential advantage of the Lagrangian character of particle/vortex methods is the elimination of the need to discretize the non-linear inertia terms in the Navier-Stokes equations. These terms are implicitly taken into account by the lagrangian motion of the vortex elements. The present scheme is built such as to maintain and use this advantage. The method of fractional steps is used to break the vorticity transport equation into an inviscid convective step where the vortex elements are propagated by the local convective velocity, and a deterministic viscous diffusion step. The diffusion computation does not use discretized spatial derivatives of the vorticity, rather the derivatives are computed analytically from the core functions of the vortex elements used to represent the vorticity distribution. New vortex elements are created at each time step to satisfy the new vorticity distribution over the plane, and are then propagated in a Lagrangian fashion.

This hybrid scheme is non-conservative, because of the remeshing at each time step, but it is shown below, at least in the test case considered, to have superior accuracy compared to other available schemes. In the following, the model is described in detail, and applied to a sample flow case involving the viscous decay of a cylindrical vortex tube in 2D.

2. Numerical Model

Consider viscous incompressible 2D flow in an unbounded simply connected domain, D, and use the vorticity-stream function formulation of the governing equations, with a zero velocity at infinity, and an initial vorticity distribution $\omega_0(x, y)$. The resulting non-dimensional governing equations are:

$$\frac{D\omega}{Dt} = \frac{\partial \omega}{\partial t} + \mathbf{u} \cdot \nabla \omega = \nu \nabla^2 \omega \tag{1}$$

$$\mathbf{u} = \nabla \times \psi \hat{\mathbf{e}}_{\mathbf{z}} \tag{2}$$

$$\nabla^2 \psi = -\omega \tag{3}$$

$$\omega(x, y, 0) = \omega_0(x, y) \tag{4}$$

$$\mathbf{u}(x, y, t) = 0, \quad \text{for}: \quad (x, y) \to \infty \tag{5}$$

where $\mathbf{\Omega} = (0, 0, \omega) = \nabla \times \mathbf{u}$ is the vorticity, $\mathbf{u}$ is the velocity vector, ψ is the stream function, and $\hat{\mathbf{e}}_{\mathbf{z}}$ is a unit vector normal to the (x, y) plane. The dimensional reference velocity and length are U and L respectively, and ν^{-1} is the Reynolds Number, defined as : $Re = \nu^{-1} = UL/\nu^*$, where ν^* is the dimensional kinematic viscosity of the fluid.

The fractional step solution of (1) proceeds at each time step in a sequence of two steps, governed by the following equations:

$$\frac{D\omega}{Dt} = 0 \tag{6}$$

$$\frac{\partial \omega}{\partial t} = \nu \nabla^2 \omega \tag{7}$$

Eq. (6) expresses the convective inviscid motion of vorticity with the local flow velocity, hence the correspondence between vortex lines and material lines. While Eq. (7) describes the diffusion of ω in the (x, y) plane with diffusivity ν.

The convective transport of vorticity (6) is computed using a Lagrangian approach, thus avoiding the discretization of the non-linear inertia terms. The solution of (7) uses a hybrid diffusion scheme, as will be discussed below.

Consider, as a test flow case, the decay of a cylindrical vortex tube. Let the initial condition correspond to a two-dimensional 2nd order gaussian distribution of vorticity centered at (x_0, y_0), given by:

$$\omega(x, y) = \frac{\Gamma_0}{\pi {\delta_0}^2} e^{-{r_0}^2/{\delta_0}^2} \tag{8}$$

where, ${r_0}^2 = (x - x_0)^2 + (y - y_0)^2$, δ_0 is a measure of the width of the distribution, and Γ_0 is the total circulation in the tube, positive counterclockwise. The reference quantities for the non-dimensional equations are chosen to be: $L = \delta_0^*$, and $U = u_\theta^*(r^* = \delta_0^*)$, where $\mathbf{u} = (u, v) = (u_r, u_\theta)$, and δ_0^*, u_θ^*, r^*, are dimensional quantities. Given (8), $u_r = 0$, and u_θ is given by:

$$u_\theta = \frac{\Gamma_0}{2\pi r_0}(1 - e^{-{r_0}^2/{\delta_0}^2}) \tag{9}$$

The following discussion of the model development is in four parts: the discretization of a specified vorticity distribution, the convective step solution of (6), the diffusive step solution of (7), and the overall solution algorithm.

2.1. VORTICITY FIELD DISCRETIZATION

Consider the vorticity distribution of compact support $\omega(x,y)$ defined on D. Overlay a square cartesian grid, with length L_g on a side, of uniform spacing $h = \Delta x = \Delta y$. L_g is chosen to be large enough such that the vorticity is negligible at the grid boundary. The vorticity distribution is discretized at each time step among a number of vortex elements located at the grid cell centers. Each vortex element, by definition, possesses a vorticity distribution of small support. The fundamental attributes defining an element, "j", are: its circulation (strength) Γ_j, its core function f_j which describes the shape of the associated vorticity distribution, and its core radius δ which determines the extent of its support, and hence the smoothness of the distribution. The spatial discretization accuracy depends on the various parameters defined above, namely: h, f_j, δ, as well as on the smoothness of the distribution $\omega(x,y)$[1; 8; 9]. The the choice of (δ, h) for a desired discretization accuracy depends generally on both the vorticity distribution and the core function[10]. Evidently, for smooth $\omega(x,y)$, the choice of higher order f_j generally provides for better accuracy. The choice made here is to use a second order function, f_j, and refine (δ, h) to achieve the desired accuracy. It has been generally demonstrated[1; 10] that a "good" choice of δ requires a significant amount of overlap between adjacent elements, i.e. $\delta/h > 1$.

The vorticity distribution $\delta\omega_j(x,y)$, associated with element "j" is defined to be a 2nd order gaussian distribution given by:

$$\delta\omega_j(x,y) = \Gamma_j f_j(x,y) \tag{10}$$

$$f_j(x,y) = \frac{1}{\pi\delta^2} e^{-r_j{}^2/\delta^2} \tag{11}$$

where, $r_j{}^2 = (x - x_j)^2 + (y - y_j)^2$. The strength Γ_j is the total circulation associated with $\delta\omega_j$, thus:

$$\int_{\mathbf{R2}} \delta\omega_j d\mathbf{x} = \Gamma_j$$

Note that the above $\delta\omega_j$ corresponds to the vorticity distribution associated with a diffused line vortex of strength Γ_j.

Using (10), the cumulative vorticity field $\hat{\omega}(x,y)$ induced by a set of N elements, is given by:

$$\hat{\omega}(x,y) = \sum_{j=1}^{N} \Gamma_j f_j(x,y) \tag{12}$$

Thus, the Γ_j are selected based on the equality:

$$\omega(x_i, y_i) = \hat{\omega}(x_i, y_i) = \sum_{j=1}^{N} \Gamma_j f_j(x_i, y_i), \quad i = 1, 2, \cdots N \tag{13}$$

where (x_i, y_i) is the center of cell i on the grid, and N is the total number of cells. This system of N equations is solved for the unknowns, Γ_j, $j = 1, 2, \cdots N$. The corresponding matrix equation is given by:

$$\mathbf{FG} = \mathbf{H} \tag{14}$$

where, for $i = 1, 2, \cdots N$:

$$F_{i,j} = f_j(x_i, y_i); \quad j = 1, 2, \cdots N \tag{15}$$

$$G_i = \Gamma_i \tag{16}$$

$$H_i = \omega(x_i, y_i) \tag{17}$$

Choquin and Lucquin-Desreux[11] use an iterative solution of the matrix equation (14), after Beale[12], to improve the accuracy of an initial rough approximation given by: $\Gamma_i = \omega(x_i, y_i)/h^2$. In the present work, a direct solution is utilized.

The sparseness of $\mathbf{F}$ depends on the radius of significant interaction for each element. By setting an error threshold ϵ_ω and neglecting the influence of any element "j" beyond a distance $r_{j\epsilon}$ from its center, where $\delta\omega_j(r_{j\epsilon}) = \epsilon_\omega$, the sparseness of the matrix can be specified.

Further, the number of elements actually used in subsequent computation is typically less that N, since many elements may have negligible Γ_i resulting from the solution of (14). The error threshold ϵ_ω is also used in this context. Elements with $\delta\omega_i(r_i = 0) < \epsilon_\omega$ are neglected, leaving in general n_e significant elements.

2.2. ADVECTION STEP

The velocity field induced by element "j" at position (x, y) is found by solving (2) and (3), given (10,11). The resulting expression is:

$$\delta\mathbf{u}_j(x, y) = -\frac{\Gamma_j}{2\pi} \frac{[(y - y_j), -(x - x_j)]}{r_j{}^2} \left(1 - e^{-r_j{}^2/\delta^2}\right) \tag{18}$$

The velocity $\mathbf{u}_i$ induced at the center of vortex element "i" is computed by the sum of the contributions of all elements "j" at that point. Therefore, using (18),

$$\mathbf{u}_i = \sum_{j=1}^{n_e} \delta\mathbf{u}_j(x_i, y_i) \tag{19}$$

The lagrangian equivalent of (6) is given by:

$$\frac{d\mathbf{X}(\alpha, t)}{dt} = \mathbf{u}(\mathbf{X}(\alpha, t), t) \tag{20}$$

$$\mathbf{X}(\alpha, 0) = \alpha$$

where, $\alpha = (\alpha_1, \alpha_2)$, and $\mathbf{X}(\alpha, t)$ is the trajectory of a fluid particle whose location at time $t = 0$ is α.

The integration of (20) at each time step is implemented using a modified Euler method[13], which is a second order Runge-Kutta scheme, with $O(\Delta t^2)$ global integration error.

2.3. DIFFUSION STEP

Eq. (7) describes the diffusion of $\omega(x, y)$ given a diffusivity of ν. A solution that has $O(\Delta t^2)$ global truncation error is presented in the following. Consider the 2D square grid defined above. Let ω_i^n be the value of vorticity at the center of cell "i", i.e. at (x_i, y_i), at time $t^n = n\Delta t$. Then ω_i^{n+1}, using a Taylor Series expansion, is given by:

$$\omega_i^{n+1} = \omega_i^n + \Delta t \left(\frac{\partial \omega}{\partial t}\right)_i^n + \frac{\Delta t^2}{2}\left(\frac{\partial^2 \omega}{\partial t^2}\right)_i^n + O(\Delta t^3). \tag{21}$$

Further, using (7):

$$\frac{\partial \omega}{\partial t} = \nu \nabla^2 \omega \tag{22}$$

$$\frac{\partial^2 \omega}{\partial t^2} = \frac{\partial}{\partial t}\left(\frac{\partial \omega}{\partial t}\right) = \nu^2 \nabla^4 \omega \tag{23}$$

Given that the vorticity field $\omega(x, y)$ is available as the sum of analytical elemental vorticity fields $\delta\omega_j$ as in (13), the spatial gradients of ω are available without grid based differentiation, and $\nabla^2\omega$ and $\nabla^4\omega$ are given by:

$$\nabla^2\omega(x, y) = \frac{4}{\pi}\sum_{i=1}^{n_e}\left[\frac{\Gamma_i}{\delta^4}\left(\frac{r_i{}^2}{\delta^2} - 1\right)e^{-r_i{}^2/\delta^2}\right] \tag{24}$$

$$\nabla^4\omega(x, y) = \frac{16}{\pi}\sum_{i=1}^{n_e}\left[\frac{\Gamma_i}{\delta^6}\left(\frac{r_i{}^4}{\delta^4} - 4\frac{r_i{}^2}{\delta^2} + 2\right)e^{-r_i{}^2/\delta^2}\right] \tag{25}$$

These expressions, substituted in (21) using (22,23) provide the necessary formulation for the diffusive step.

2.4. SOLUTION ALGORITHM

Given the above formulation, the solution proceeds at each time step as follows:

- Discretize the given vorticity field among n_e elements using (14).
- Propagate the elements according to the convection equation (20).
- Find the new values of vorticity at cell centers due to the new distribution of elements using (13).
- Diffuse the vorticity based on the above diffusion scheme using (21).

Recall that a direct solution of (14) is used in step (1). A sparse matrix solver is utilized for this solution. Given that a fixed grid is used, the equation is factored once only, at the first time step. The same LU factorization is used in subsequent time steps. Therefore, a single matrix-vector multiply is required to redistribute vorticity among a new set of elements, also at cell centers, but with new strengths, at each time step.

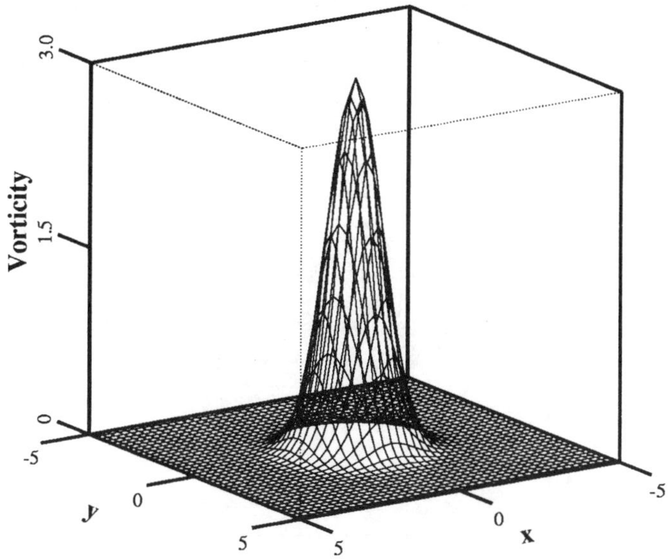

Fig. 1. Initial vorticity discretization, $\delta/h = 1.5$, $\epsilon_\omega = 10^{-6}$, 41x41 grid, $n_e = 593$, $E_\omega^0 = 0.125\text{x}10^{-5}$

3. Results

In this section, computational results are presented. We consider a Reynolds number in the range from 10 to 1000. A fixed square grid with $L_g = 10$ is used. Recall that the reference length is the core radius, δ_0^*, of the initial gaussian vorticity distribution, which, when non-dimensionalized, is given by $\delta_0 = 1$.

The non-dimensional circulation Γ_0 in (9) is specified such that $u_\theta(r_0 = 1, t = 0) = 1$, giving $\Gamma_0 = 2\pi/(1 - e^{-1})$. The initial vorticity distribution is centered at $(x_0 = 0, y_0 = 0)$. Time is non-dimensionalized with respect to $t_c = L/U$, which is the convective time scale. The diffusion time scale is given by $t_v = L^2/4\nu = t_c/4\nu$.

The accuracy of the scheme is determined by comparison to the exact solution of the problem, which is the solution of (7), since $\mathbf{u} \cdot \nabla\omega = 0$ due to symmetry. The exact solution is given by:

$$\omega_e(\mathbf{x}, t) = \frac{\Gamma_0}{\pi({\delta_0}^2 + 4\nu t)} e^{-{r_0}^2/({\delta_0}^2 + 4\nu t)} \tag{26}$$

An integral error norm, E_ω, is used to compare the results to the exact solution, where :

$$E_\omega^2(t) \int_D \omega_e^2(\mathbf{x}, t) dA = \int_D [\omega(\mathbf{x}, t) - \omega_e(\mathbf{x}, t)]^2 dA \tag{27}$$

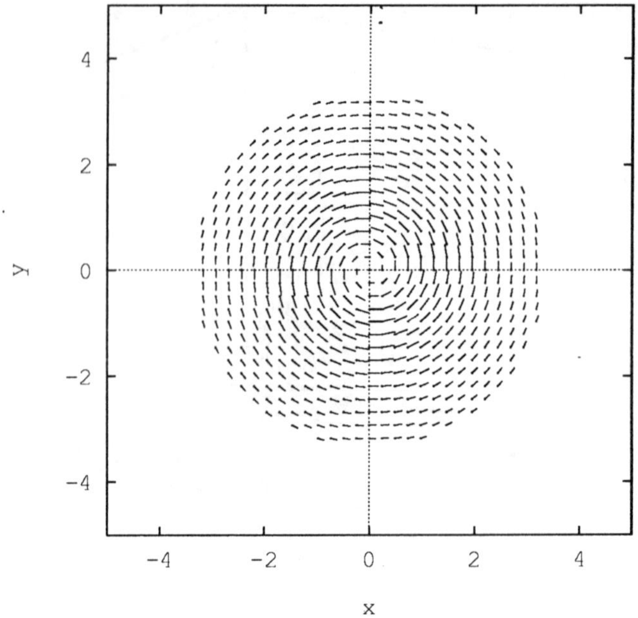

Fig. 2. Initial distribution of vortex elements in the plane, corresponding to the initial vorticity field in Fig. 1. The local vorticity vector is shown originating at the element centers. $\delta/h = 1.5$, $\epsilon_\omega = 10^{-6}$, 41x41 grid, $n_e = 593$, $E_\omega^0 = 0.125\text{x}10^{-5}$.

E_ω is a normalized root mean square norm, evaluated over the computational domain. The integrals are evaluated using a 2D rectangle rule quadrature on a uniform (101x101) cell grid, on the domain D. The results are also compared with those available using other diffusion schemes, namely Chorin's random vortex method[1] and the PSE scheme of Raviart[2].

The initial vorticity distribution at time $t = 0$ is shown in Fig. 1, using the vorticity field values at the cell centers, connected with straight line segments. The figure corresponds to a grid resolution given by $m = 41$ where $h = L_g/m$, or a 41x41 cell grid, and for $\delta/h = 1.5$. The corresponding distribution of vortex elements in the plane is shown in Fig. 2, where the local velocity vector is plotted from the center of each element. The direction of rotation is counter-clockwise. The initial cutoff radius beyond which the vorticity field is neglected, and no elements are generated, is $r_{0i} = 3.32$. The circulation of an element at this radius, normalized with that of the element with highest circulation, is: $3.0\text{x}10^{-6}$. The accuracy of this initial discretization is found to be $E_\omega^0 = 0.125\text{x}10^{-5}$.

The dependence of E_ω^0 on h and δ/h, for the given initial vorticity field, is demonstrated in Fig. 3, for a range of h between 0.2 and 0.5, and for $\delta/h \in [1.1, 1.7]$. It is evident from the figure that, for the given vorticity distribution, E_ω decreases with decreasing h and increasing δ/h. Further, the convergence rate, $\partial E_\omega/\partial h$, is seen to increase with increasing h, and increasing δ/h. The observed order of convergence

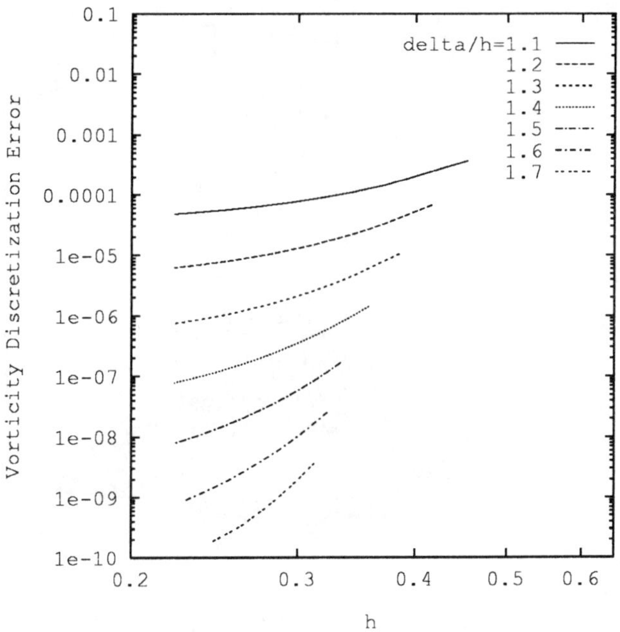

Fig. 3. Dependence of the discretization error at time $t = 0$, E_ω^0, on the grid cell size h and δ/h, for the initial condition shown in Fig. 2, and with $\epsilon_\omega = 10^{-10}$.

with respect to h is clearly higher than the second order convergence expected for the core function chosen here. This is primarily because of the particular initial gaussian vorticity field. Using a less favourable initial profile gives a more realistic convergence rate. Such a profile is shown in Fig. 4. It is a 10th order radial cosine hat function, given by:

$$\omega(r_o) = \begin{Bmatrix} 1 - cos^n[\frac{\pi}{2}(\frac{r_o}{R_o} - 1)] & r_o \leq R_o \\ 0 & r_o > R_o \end{Bmatrix} \tag{28}$$

with $n = 10$, $R_o = 3.0$. The second order convergence rate given this profile is demonstrated in Fig. 5. The figure illustrates also the dependence of the relationship between E_ω and δ/h on the vorticity distribution being discretized. Evidently, $\delta/h = 0.9$ is close to optimum in this case, whereas the above gaussian distribution case exhibited monotonic improvement in E_ω with δ/h. The cosine hat function is presented here for the convergence discussion, and will not be considered further below.

Given an initial distribution of vortex elements, we now consider their propagation in time given the present hybrid scheme. To begin with, and as indicated above, it is expected because of the remeshing at each time step that the scheme is non-conservative. To demonstrate this, consider the results shown in Fig. 6, for an inviscid flow case. The total circulation Γ in the domain was computed at each time step by integrating the computed vorticity field at the cell centers of

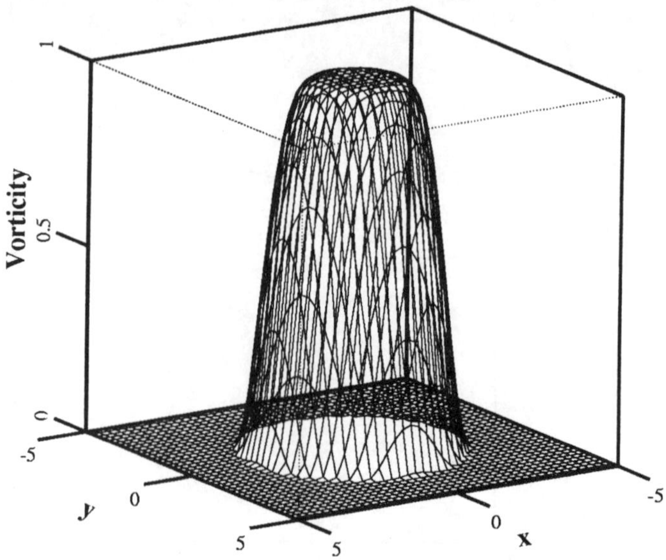

Fig. 4. Initial vorticity discretization for a 10th order cosine hat function, $\delta/h = 0.9$, $\epsilon_\omega = 10^{-10}$, 41x41 grid, $E^0_\omega = 0.81\text{x}10^{-2}$.

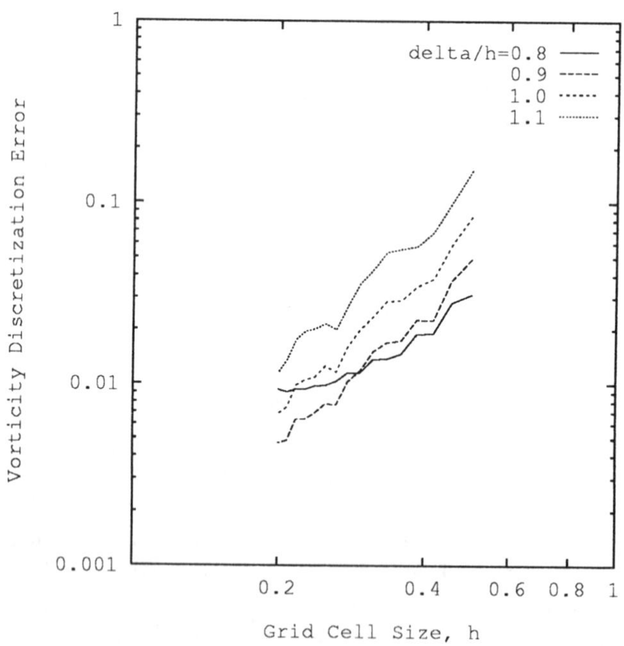

Fig. 5. Dependence of the discretization error at time $t = 0$, E^0_ω, on the grid cell size h and δ/h, for the 10th order cosine hat function initial condition, and with $\epsilon_\omega = 10^{-10}$.

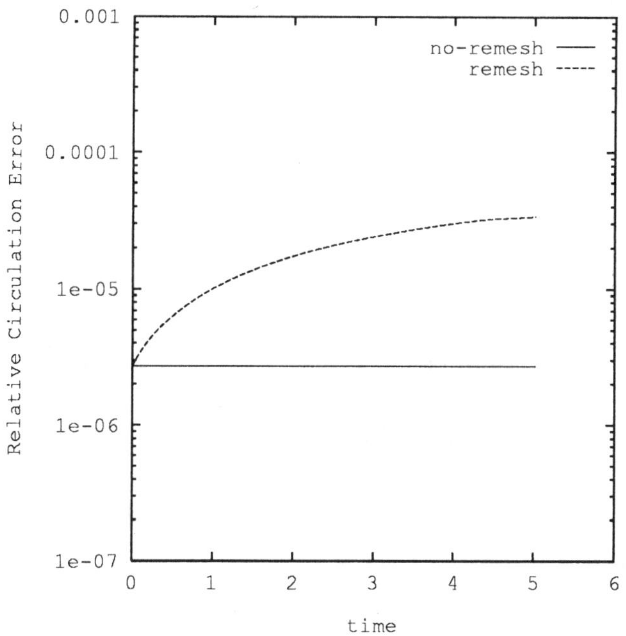

Fig. 6. Growth of the relative error in total circulation, E_Γ, due to remeshing, for an inviscid flow starting with the second order gaussian initial condition, 40x40 grid, $\delta/h = 1.2$, $\Delta t = 0.025$, $\epsilon_\omega = 10^{-10}$.

a uniform (101x101) grid. The relative circulation error is defined as the quantity: $E_\Gamma = |\Gamma - \Gamma_{ex}|/\Gamma_{ex}$. Where Γ_{ex}, a constant, is the total circulation as computed by integrating the exact vorticity distribution over the same above grid. The initial non-zero value of E_Γ is due to the initial discretization of the vorticity field, and is the same for both cases shown in the figure. However, the time evolution of E_Γ is clearly different depending on whether remeshing is done or not. E_Γ is seen to be constant in time if no remeshing is done, while it is monotonically increasing due to remeshing in the present scheme.

The loss of the conservative property is offset, however, by the improved accuracy of the hybrid scheme. This improvement is demonstrated in Fig. 7 where the time evolution of E_ω is plotted for the inviscid case with and without remeshing. Evidently, the non-conservative scheme, at least in this particular case, is superior in terms of the accuracy of the vorticity profile. Further work is required on more realistic vorticity distributions typical of shear flows to support this conclusion more generally.

Consider next the viscous flow case, involving decay of the initial vortex with time. The error norm E_ω is monitored as the vortex decay progresses and is plotted in Fig. 8 vs. time. These results correspond to $h = 0.243$ ($m = 41$), $\delta/h = 1.5$, and $\Delta t = 0.01$. The figure illustrates the error norm evolution for a range of Reynolds number from 10 to 10^3. E_ω is seen to decrease significantly in going from $Re = 10$

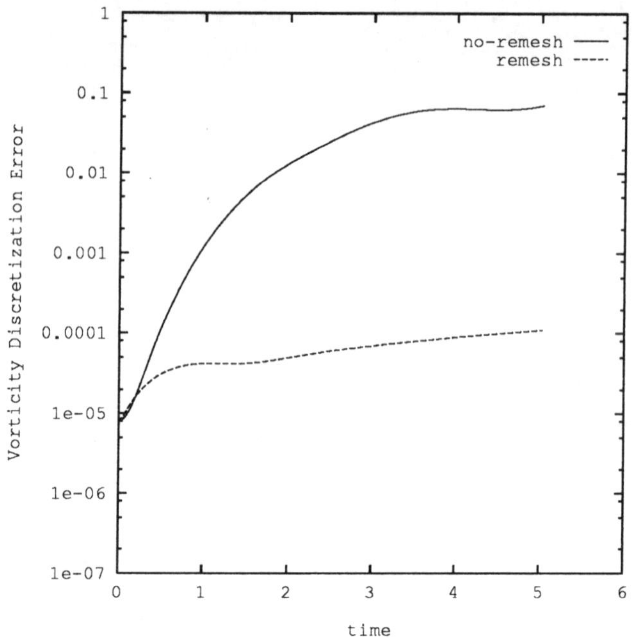

Fig. 7. Growth of E_ω with time, for an inviscid flow starting with the second order gaussian initial condition, 40x40 grid, $\delta/h = 1.2$, $\Delta t = 0.025$, $\epsilon_\omega = 10^{-10}$.

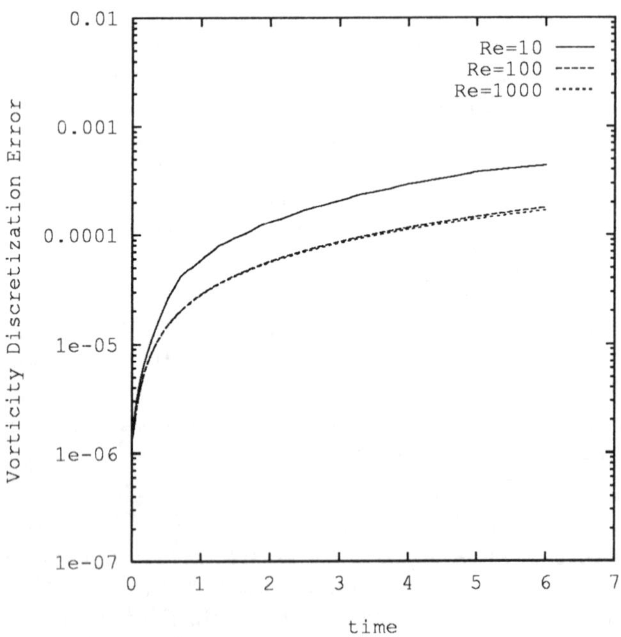

Fig. 8. Growth of E_ω with time for Re range from 10 to 1000, $\delta/h = 1.5$, 41x41 grid, $\epsilon_\omega = 10^{-6}$, using the hybrid scheme.

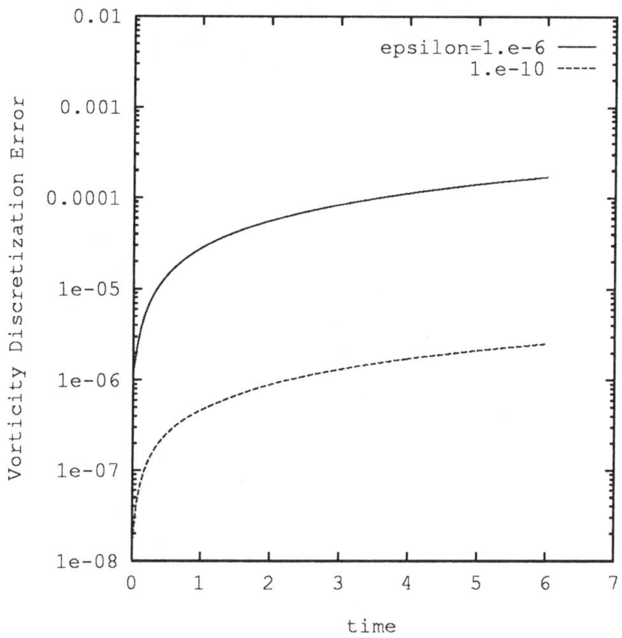

Fig. 9. Figure 26. E_ω growth time, using the hybrid scheme, with Re=1000, $\delta/h = 1.5$, a 41x41 grid, $\Delta t = 0.01$.

to 100, while the next step to $Re = 10^3$ introduces minor improvement. This is expected since, as Re increases, the flow approaches the inviscid limit asymptotically. The significance of the error threshold ϵ_ω is illustrated in Fig. 9, where E_ω is plotted for a flow case with $Re = 1000$ using two values of ϵ_ω. The overall level of E_ω depends on the neglected vorticity in the initial and subsequent discretizations of the vorticity field, and hence on ϵ_ω.

The dependence of E_ω at a given time instant on the time step Δt is illustrated in Fig. 10. The second order convergence rate is evident in the $Re = 10$ case, where diffusion is significant, while 2nd to 3rd order convergence is exhibited in the inviscid limit.

The hybrid scheme is compared against the PSE scheme and the random vortex method using the error norm E_ω. The implementation of these two latter schemes uses the same initial discretization and advection algorithm as the former. The PSE scheme uses the following formulation for updating the strengths of vortex elements at each time step[6],

$$\frac{d\Gamma_j(t)}{dt} = \frac{4\nu h^2}{\pi\delta^4} \sum_{k=1}^{n_e} (\Gamma_k(t) - \Gamma_j(t)) e^{-r_{jk}^2/\delta^2} \quad j = 1, 2, \cdots, n_e. \tag{29}$$

The random vortex method, on the other hand, uses specific random displacements, drawn from a gaussian population with zero mean and a standard deviation given

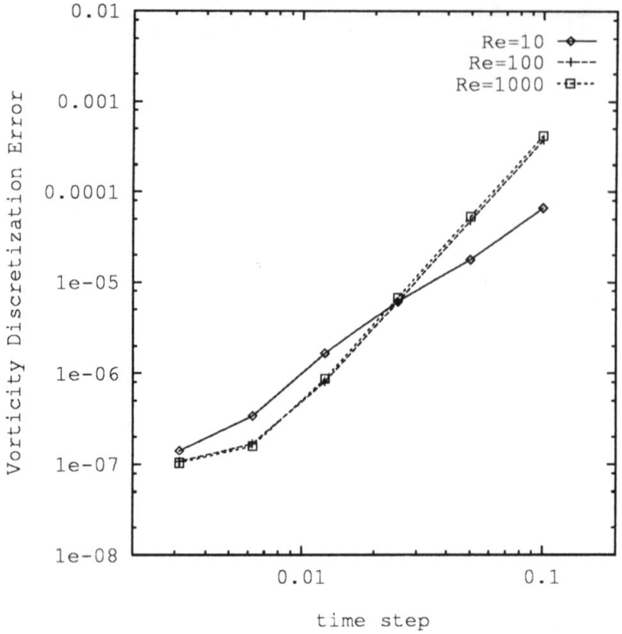

Fig. 10. E_ω at time $t = 1.0$ vs. Δt, for different Re, $\delta/h = 1.5$, 41x41 grid, $\epsilon_\omega = 10^{-10}$, using the hybrid scheme.

by $\sigma = (2\Delta t/Re)^{1/2}$, applied to element locations at each time step.

The error norm E_ω evolution with time is shown in Fig. 11, for a Reynolds number of 1000. Similar E_ω growth trends are observed among the three schemes, however, the amplitude of the error for the PSE and RVM implementations is 4 orders of magnitude higher than the hybrid scheme. The same observation is evident at lower Reynolds number, as shown in Fig. 12 for the $Re = 10$ case.

4. Conclusions

A hybrid diffusion scheme relevant to the vortex method solution of 2D incompressible flow was presented and demonstrated on a vortex decay problem. The accuracy of the numerical solution was computed using the known exact solution of the problem. The accuracy of the initial discretization was shown to be dependent on the initial vorticity distribution as well as the selected attributes of the vortex elements. The growth of the error with time was studied for a range of Reynolds number showing the expected asymptotic approach to the inviscid limit. Spatial and temporal convergence were demonstrated for a range of discretization parameters. The scheme was compared to the random vortex method [1] and the particle strength exchange scheme [2], and demonstrated to achieve superior accuracy for the particular vortex decay problem considered.

A fundamental disadvantage of this hybrid scheme is the requirement for remesh-

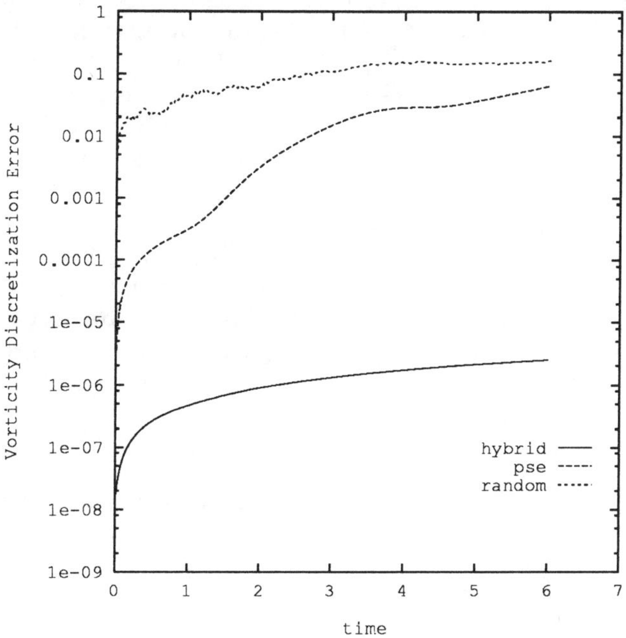

Fig. 11. Growth of E_ω with time, for the hybrid, PSE, and random diffusion schemes, $Re = 1000$, $\delta/h = 1.5$, 41x41 grid, $\epsilon_\omega = 10^{-10}$.

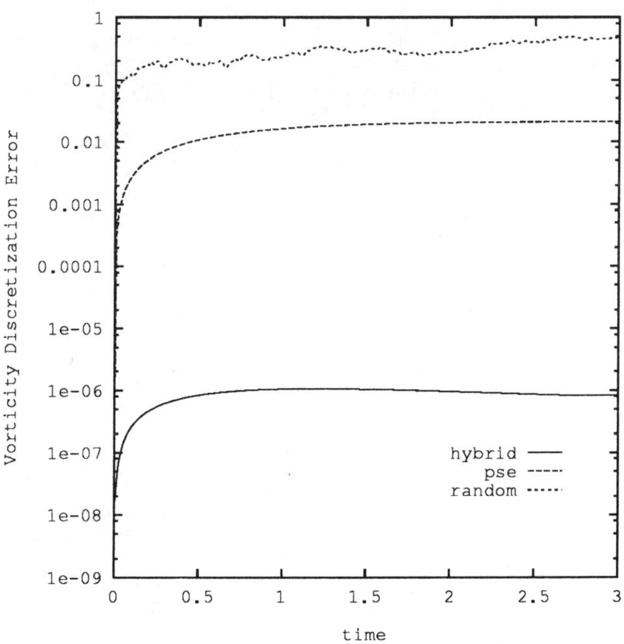

Fig. 12. Growth of E_ω with time, for the hybrid, PSE, and random diffusion schemes, $Re = 10$, $\delta/h = 1.5$, 41x41 grid, $\epsilon_\omega = 10^{-10}$.

ing at each time step, which results in a non-conservative solution. The error in computed total circulation was demonstrated to grow with time. Remeshing may also result in significant grid-related diffusion of high spatial vorticity gradients, a situation which does not arise in the present problem but is of relevance in the computation of shear flows. A possible means to overcome this problem is an adaptive version of the present scheme where the element centers constitute the necessary points for the re-discretization of the vorticity field at each time step, i.e. they form a self-adaptive grid. Initial studies with this adaptive scheme exhibit the tendency of the solution matrix to become ill-conditioned as the flow evolves. This ill-conditioning may be the result of the elements becoming non-equidistant, and may be resolved by using a variable element core size, to be selected for each element based on its proximity to its nearest neighbors. More work is required to fully investigate this adaptive scheme.

Acknowledgements

The author is grateful for the support of Texas Instruments in conducting this investigation.

References

[1] A. Chorin, *J. Fluid Mech.*, **57**, p. 785, (1973).
[2] P-A. Raviart, *Méthodes particulaires*, Lecutre Notes, Ecole d'été d'analyse numérique, Centre d'étude du Bréau-sans-nappe, France (1987).
[3] G-H. Cottet, *Analyse numérique des méthodes particulaires pour certains problémes nonlinéaires*, Thése d'état, Université Pierre et Marie Curie (1987).
[4] S. Huberson, *Modélisation asymptotique et numérique de noyaux tourbillonaires enroulés*, Thése d'état, Université Pierre et Marie Curie (1986).
[5] S. Mas-Gallic, *Contribution à l'analyse Numérique des méthodes particulaires*, Thése d'état, Université Pierre et Marie Curie (1987).
[6] P. Koumotsakos, *Direct numerical simulations of unsteady separated flows using vortex methods*, Ph.D. Thesis, California Institute of Technology, Pasadena, CA, (1992).
[7] G-H. Cottet, *Lectures in Applied Mathematics*, **28**, p. 103, (1991).
[8] L. Rosenhead, *Proc. R. Soc. Lond. A*, **134**, p. 170, (1931).
[9] G. Birkhoff and J. Fisher, *J. Circ. Mat. Palermo*, **8**, p. 77, (1959).
[10] C. Anderson, *J. Comput. Phys.*, **61**, p. 417 (1985).
[11] J.P. Choquin and B. Lucquin-Desreux, *Int. J. Num. Methods in Fluids*, **8**, p. 1439, (1988).
[12] J.T. Beale, 'On the accuracy of vortex methods at large times', *Proc. Workshop on Computational Fluid Dynamics and Reacting Gas Flows*, Institute of Applied Mathematics, Mineapolis, Sep. (1986).
[13] C. Gerald and P. Wheatley, *Applied Numerical Analysis*, p. 305, Addison-Wesley, Reading, MA, (1984).

A SLIGHTLY DIFFUSIVE CONTOUR DYNAMICS

G. RICCARDI and R. PIVA
Dip. di Meccanica e Aeronautica
Universitá di Roma "La Sapienza"
Via Eudossiana 18
00184 Roma ITALY

ABSTRACT.

The Contour Dynamics approach has been revisited to include viscous effects. The vorticity field is approximated by using a piecewise constant distribution, where the jumps between contigous levels are expressed as functions of time. From an integral form of the Helmholtz equation the time derivative of these functions are defined by writing the balances for each region of constant vorticity via a suitable approximation of the exchanged vorticity fluxes. This leads to an ordinary differential system that is solved together with the dynamics of the contours. The calculation of several first integrals of the motion enables to evaluate the accuracy of the integration.

The symmetric vortex merging is studied by using the above technique and a qualitative investigation in terms of relative stream function is conducted. The vorticity structure evolution related to the merging process is observed and, by measuring the time necessary to complete this evolution, an estimate of the merging time is provided. The analysis has been performed for few distances, close to the critical one, between the two vortices and for various Reynolds numbers.

1. Introduction

Contour Dynamics is a very accurate model to study the evolution of vorticity structures in two-dimensional inviscid flows [7], [1]. Actually the use of a piecewise constant approximation of the vorticity field provides a simple and direct approach to the Lagrangian study of the dynamics of vorticity patches. A straightforward generalization of this original idea to a multi-level distribution of vorticity [5] stands on the linearity of the vorticity-velocity relationship which, starting from the Biot-Savart law, may be expressed in terms of the vorticity level contours ∂D_i and the vorticity jumps $[\omega]_i$ as

$$u(x,t) = -\sum_{i=1}^{n}[\omega]_i \int_{\partial D_i(t)} G(x-y)\,dy \qquad (1)$$

where $G(x) = \frac{1}{2\pi} log|x|$. The singularity of the Green's function G, appearing in this formula, is integrable and there is no need of any desingularization procedure, as required in the classical vortex method. As a direct consequence, the most appealing property of this method, in contrast with others, is the absence, at least in principle,

J.T. Beale et al. (eds.), Vortex Flows and Related Numerical Methods, 223–238.

of any form of numerical viscosity, that might enable to follow the small scales of the motion. On the other hand, most of the phenomena of interest, which involve the dynamics of vorticity structures, occur at very large, but *finite*, Reynolds numbers. Therefore the influence of a small vorticity diffusion given by the real viscosity of the fluid may have a paramount importance in understanding the structure evolution in processes like merging or axisymmetrization of vortices.

This suggests to modify the Contour Dynamics technique in order to include diffusive effects by using a suitable approximation of the vorticity fluxes exchanged by adjacent levels. The model that we propose, obtained by an integral form of the Helmholtz equation for the annular regions of the vorticity levels, is described in section 2. The accuracy of the time integration, as well as the effects of the vorticity fluxes approximation on the numerical solution, are evaluated by computing several global quantities of the flow, such as the excess energy or the second moment of the vorticity,that play the role of first integrals of the motion in an inviscid context. They are illustrated in section 3 together with some indications for their analytical calculation.

The primary purpose of the present paper is to analyze by means of this weakly diffusive model of the vorticity dynamics, the process of merging of two vortices and the following axisymmetrization of the vorticity structure, by changing the initial distance of the two vortices and the Reynolds number. As an intrinsic feature of the present model, which is in fact a correction of the inviscid case, the viscosity may be set at very small values without resorting to other concepts, like the hyperviscosity in spectral methods, which might give rise to a non physical behavior of the flow. In the present paper, however, we describe flows at relatively moderate Reynolds numbers in order to enhance the diffusive effects and to understand, in reasonable computational times, their influence on the vorticity structure. The observation of the numerical simulations, illustrated in section 4, is completed by the introduction of a merging time definition, which gives some interesting suggestions about the scale and the similarity of the dynamical evolution.

2. A simple diffusion model for Contour Dynamics

Let us assume a piecewise constant approximation of the vorticity field ω. Hence for an *isolated* vortex the distribution of the vorticity is given by

$$\omega(x,t) \simeq \sum_{i=1}^{n} [\omega]_i(t)\chi_{D_i(t)}(x) \tag{2}$$

where $D_1 \supset D_2 \supset \cdots \supset D_n$. The proposed algorithm consists in following the dynamics of the material domains $\{D_i\}_{i=1,\dots,n}$. Across the boundary of these domains the vorticity field has the jumps $\{[\omega]_i\}_{i=1,\dots,n}$ that are defined as functions of time by solving an integral form of the *Helmholtz* equation

$$\frac{D\omega}{Dt} = \frac{1}{Re}\nabla^2\omega \quad . \tag{3}$$

The vorticity diffusion is simulated by an exchange of vorticity fluxes between a level ∂D_i and the two contiguous ones, namely ∂D_{i-1} (where $D_{i-1} \supset D_i$) and ∂D_{i+1}

(where $D_i \supset D_{i+1}$). The basic approximation on which the model is developed stands in the assumption that the vorticity level $\partial D_i(0)$ (i. e. a curve on which the vorticity is constant at the time 0) becomes the vorticity level $\partial D_i(t)$ at times $t > 0$. In other words, the vorticity included into the curve ∂D_i is spreaded uniformly on the domain D_i and, as a consequence, the vorticity jump is constant along the curve ∂D_i. It is worth to notice that this approximation is caused by the use of a piecewise constant distribution of the vorticity field, inherent to the Contour Dynamics approach and is not related to the model adopted to describe the diffusion process.

2.1. DIFFERENTIAL SYSTEM IN THE VORTICITY JUMPS

The idea is to define the vorticity field ω, in its approximated form (2), by using the integral form of equation (3). The domains $\{D_k\}_{k=1,\ldots,n}$ are material and the left hand side of equation (3) integrated on the annular region $D_i - D_{i+1}$ and on D_n gives respectively

$$\int_{D_i - D_{i+1}} \frac{D\omega}{Dt} dx = \frac{d}{dt}\int_{D_i - D_{i+1}} \omega dx = \frac{d}{dt}(|D_i| - |D_{i+1}|)\sum_{k=1}^{i}[\omega]_k$$

$$= (|D_i| - |D_{i+1}|)\sum_{k=1}^{i}[\dot{\omega}]_k \quad , \tag{4}$$

$$\int_{D_n} \frac{D\omega}{Dt} dx = \frac{d}{dt}\int_{D_n} \omega dx = |D_n|\sum_{k=1}^{n}[\dot{\omega}]_k \quad , \tag{5}$$

where $|D_i|$ is the area of the domain D_i. The boundaries of the domains $\{D_k\}_{k=1,\ldots,n}$ are oriented counter-clockwise and the normal vectors outwards. By introducing an approximated form of the flux

$$\Omega_k := \int_{\partial D_k} \frac{\partial \omega}{\partial \nu} ds \quad , \tag{6}$$

which is illustrated in details below, the integration of the right hand side of equation (3) over $D_i - D_{i+1}$ gives

$$\frac{1}{Re}\int_{D_i - D_{i+1}} \nabla^2 \omega dx = \frac{1}{Re}\left(\int_{\partial D_i} \frac{\partial \omega}{\partial \nu} ds - \int_{\partial D_{i+1}} \frac{\partial \omega}{\partial \nu} ds\right)$$

$$= \frac{1}{Re}(\Omega_i - \Omega_{i+1}) \tag{7}$$

in which there is a balance between the vorticity flux going into the annular region $D_i - D_{i+1}$ through the lower level ∂D_i, that is Ω_i, and the flux going out to the upper level ∂D_{i+1}, that is Ω_{i+1}. For the domain D_n, instead,

$$\frac{1}{Re}\int_{D_n} \nabla^2 \omega dx = \frac{1}{Re}\Omega_n \tag{8}$$

there is only the vorticity flux Ω_n, going into the level through ∂D_n. From equations (4, 7) for the annular region $D_i - D_{i+1}$ and from equations (5, 8) for the domain D_n, it follows the ordinary differential system in the vorticity jumps $\{[\omega]_k\}_{k=1,\ldots,n}$:

$$\begin{cases} \sum_{k=1}^{i} [\dot{\omega}]_k = \dfrac{1}{Re} \dfrac{\Omega_i - \Omega_{i+1}}{|D_i| - |D_{i+1}|}, & i = 1, \ldots, n-1 \\ \sum_{k=1}^{n} [\dot{\omega}]_k = \dfrac{1}{Re} \dfrac{\Omega_n}{|D_n|} \end{cases}$$

that can be easily set in normal form. Actually, taking the equation as numbered from 1 to n, the first equation is already in normal form

$$[\dot{\omega}]_1 = \frac{1}{Re} \frac{\Omega_1 - \Omega_2}{|D_1| - |D_2|} .$$

By subtracting equation 1 from equation 2 we obtain the normal form of the second equation

$$[\dot{\omega}]_2 = \frac{1}{Re} \left(\frac{\Omega_2 - \Omega_3}{|D_2| - |D_3|} - \frac{\Omega_1 - \Omega_2}{|D_1| - |D_2|} \right)$$

and so on, up to equation $(n-1)$. For the n-th equation, the subtraction of the $(n-1)$-th gives

$$[\dot{\omega}]_n = \frac{1}{Re} \left(\frac{\Omega_n}{|D_n|} - \frac{\Omega_{n-1} - \Omega_n}{|D_{n-1}| - |D_n|} \right)$$

and the system is written in normal form as follows

$$\begin{cases} [\dot{\omega}]_1 = \dfrac{1}{Re} \dfrac{\Omega_1 - \Omega_2}{|D_1| - |D_2|} \\ \quad \vdots \\ [\dot{\omega}]_i = \dfrac{1}{Re} \left(\dfrac{\Omega_i - \Omega_{i+1}}{|D_i| - |D_{i+1}|} - \dfrac{\Omega_{i-1} - \Omega_i}{|D_{i-1}| - |D_i|} \right), & i = 2, \ldots, n-1 \\ [\dot{\omega}]_n = \dfrac{1}{Re} \left(\dfrac{\Omega_n}{|D_n|} - \dfrac{\Omega_{n-1} - \Omega_n}{|D_{n-1}| - |D_n|} \right) \\ [\omega]_k(0) \quad \text{given for } k = 1, \ldots, n \end{cases}$$

that must be solved together with the dynamics of the contours $\{\partial D_i\}_{i=1,\ldots,n}$. This system can be integrated with a Reynolds number arbitrarily large, up to $Re \to \infty$ where it reduces to the classical version of Contour Dynamics.

2.2. APPROXIMATION OF THE VORTICITY FLUXES

The evaluation of the integrals (6) is carried out in an approximate way by using a suitable mean value of the gradient $\dfrac{\partial \omega}{\partial \nu}$ on the boundary ∂D_k calculated as a forward derivative. To define this gradient on the lowest level of the vortex (level 1) it is necessary to introduce an extra level ∂D_0 (with $D_0 \supset D_1$) on which the vorticity is assumed always zero. The vorticity flux leaving the first level is going to be lost for the calculation, then if it reaches a sensible value, the circulation around the vortex starts to diminish in an unacceptable way. To avoid this undesired behavior, it is necessary to provide some vorticity levels with $[\omega](0) = 0$ around the last significant level of the initial vorticity field. During the evolution these levels are filled with

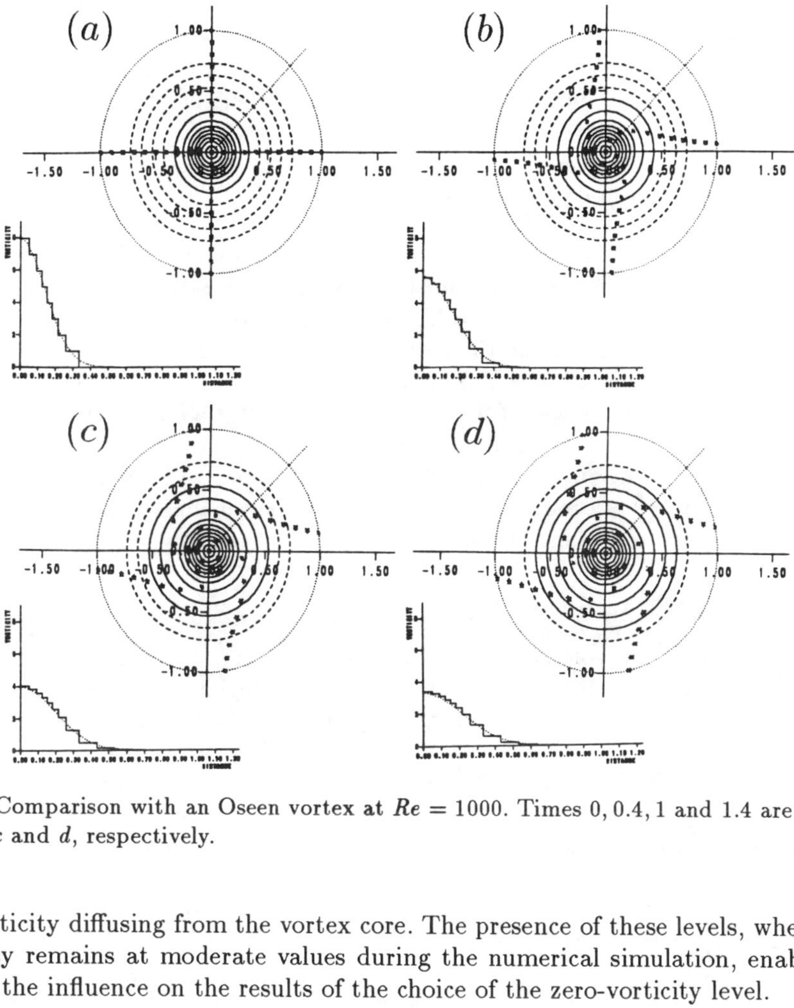

Fig. 1. Comparison with an Oseen vortex at $Re = 1000$. Times $0, 0.4, 1$ and 1.4 are shown in a, b, c and d, respectively.

the vorticity diffusing from the vortex core. The presence of these levels, where the vorticity remains at moderate values during the numerical simulation, enables to reduce the influence on the results of the choice of the zero-vorticity level.

As an alternative way we have considered the concept of "diffusion velocity" introduced by Ogami and Akamatsu [4]. In this context the contours motion is not material and the diffusion velocities are summed to the Biot-Savart ones (1) to keep at a constant value the circulation around each contour. With this approach a problem, still unsolved in a satisfactory way, is the definition of the vorticity gradient along the lowest vorticity level. The mean thickness of the annular region $D_{k-1} - D_k$ is defined by the following relation

$$\bar{s}_{k-1} := \frac{|D_{k-1}| - |D_k|}{\frac{1}{2}(|\partial D_{k-1}| + |\partial D_k|)}$$

in which $|\partial D_i|$ stays for the length of the boundary of the domain D_i, Ω_k is approximated by using the following estimate

$$\Omega_k \simeq -\frac{[\omega]_k}{\bar{s}_{k-1}}|\partial D_k| \ .$$

The adoption of this approximate formula makes simple and not expensive the diffusive correction of the Contour Dynamics algorithm. A comparison between the present model and the motion of an Oseen vortex, i. e. the exact axisymmetrical solution of the Helmholtz equation (2)

$$\omega(r,t) = \frac{Re}{4\pi(t+1)} exp\left[-\frac{Re}{4(t+1)}r^2\right] \quad ,$$

is shown in fig. 1. The zero-vorticity level is represented with a dotted line, whereas the auxiliary levels with vanishing initial jump of vorticity are indicated with dashed lines. The type of the line is changed to a continuos one when the vorticity jump becomes larger than one per cent of the vorticity maximum. A section of the vortex is also represented together with the exact solution (dotted line). Two superimposed sets of particles are placed on the axis at the initial time to compare the analytic integration of their motion with the numerical solution. The approximation given by the present model appears to be satisfactory for an axisymmetric vortex. In more general conditions the reliability of the results may be established by controlling the value of the first integrals.

3. First integrals and accuracy control

The dynamics of the vorticity leads frequently to very complex flow fields with many interacting filaments, high vorticity gradient regions or intricate interfaces between two or more vortices. Consequently these fields are extremely sensible to the discretization (in space and time) employed and the integration accuracy is a relevant question to be posed. The best approach lies in the evaluation of some quantities which are conserved along the motion. By calculating their variations during the numerical simulations, it is understood that their accuracy is under control. The following first integrals hold in the inviscid case

$$\begin{aligned}
\Gamma &= \int_{\mathrm{R}^2} \omega dx && \textit{vorticity flux} \\
M_1 &= \int_{\mathrm{R}^2} x\omega dx && \textit{first order moment} \\
I(t) &= \int_{\mathrm{R}^2} |x|^2\omega dx && \textit{second order moment} \\
E(t) &= \int_{\mathrm{R}^2} \omega^2 dx && \textit{enstrophy} \\
\mathcal{E}(t) &= \frac{1}{2}\int_{\mathrm{R}^2} \psi\omega dx && \textit{excess energy,}
\end{aligned}$$

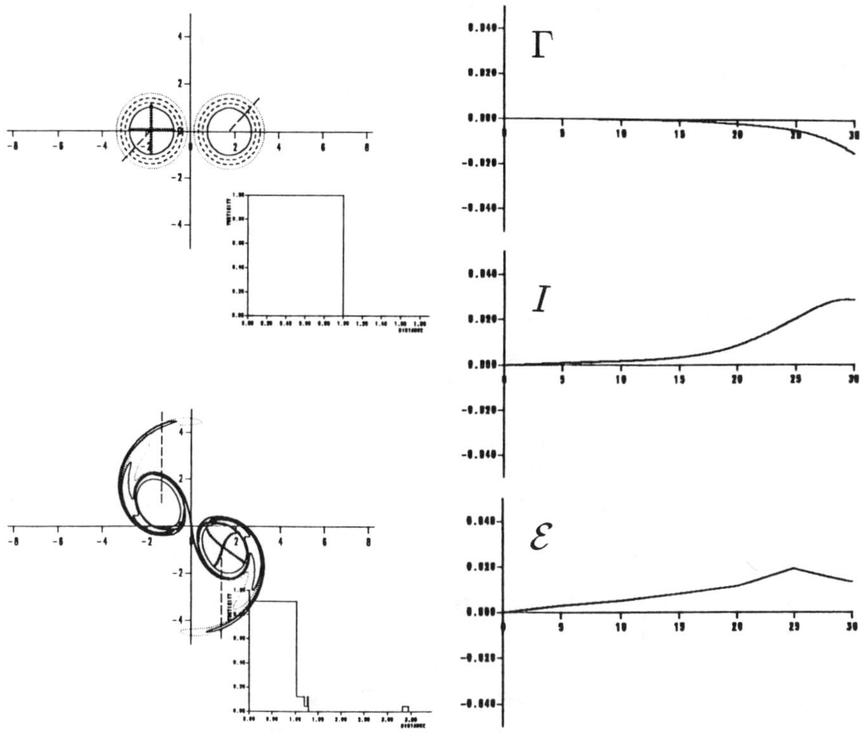

Fig. 2. Initial ($t = 0$) an final ($t = 30$) vorticity fields for the merging at distance $d = 3.5$ and $Re = 5000$ together the accuracy diagnostic (from the top to the bottom: $\Gamma, I, \mathcal{E}$).

where ψ is the stream function, such that $-\nabla^{\perp}\psi = u$. If the vorticity flux $\Gamma \neq 0$ we can't define the kinetic energy of the entire flow field due to the asymptotic behavior of the velocity field. In this case we consider the energy-like quantity "*excess energy*" $\mathcal{E}$ that coincides with the kinetic energy when this can be defined ($\Gamma = 0$). In a viscous context $\mathcal{E}$ satisfies the evolution equation

$$\frac{d\mathcal{E}}{dt} = -\frac{1}{Re}E(t) \ .$$

Also the second order moment is not conserved in these flows, but it changes linearly with time following the law introduced by Poincaré

$$I(t) = I(0) + \frac{4}{Re}\Gamma t \quad .$$

The other two quantities Γ and M_1 are constant also in the viscous case.

The evaluation of the above quantities can be carried out *in analytic way* by calculating the two-dimensional integrals as integrals along the contours $\{D_k\}_{k=1,\dots,n}$. The details of the analytic calculation of the excess energy $\mathcal{E}$, that is rather complex, can be found in [6]. The variations of $\Gamma, I + \frac{4}{Re}\Gamma t, \mathcal{E} + \frac{1}{Re}\int_0^t E(\tau)d\tau$ for the merging at $Re = 5000$ between two patches with respect to their initial value are reported in fig. 2. At the initial time the two patches have a circular configuration with unitary radius and unitary vorticity. The distance between their centers is $d = 3.5$. As shown by the diagrams, the values of these quantities remain confined below a few per cent, at least for the time needed to complete the merging process.

4. Analysis of Axysimmetrization and Symmetric Merging

The study of the motion of elementary vortex structures as the axisymmetrization of a non-uniform elliptical vortex or the amalgamation of two vortices is simplified by analyzing the dynamics of the vorticity structure in a suitable rotating frame of reference. Actually, by choosing the angular velocity of this system with reference to the rotation of the stream function field, i. e. to the bulk motion of the vorticity structure, the resulting flow field in the moving system enables to explain some important aspects of the vortex dynamics.

As shown by Melander at al. [2], for the study of the vortex dynamics the most convenient quantity is the relative stream function field. In the motion of a single isolated elliptical vortex a qualitative description of this field explains the formation of filaments and the axisymmetrization process. As discussed below, the final structure after the axisymmetrization process, at least in an inviscid context, strongly depends on the relative stream function field. The use of this field during the motion of two vortices [3] clarifies the kinematic of the amalgamation process in terms of formation of filaments and exchange of vorticity between the two structures, also when the distance between the vortices doesn't allow the start up of the merging process.

4.1. DIAGNOSTIC ELLIPSE AND STREAM FUNCTION FIELD ANALYSIS

The apparently simple flow generated by the axisymmetrization process is the best context to introduce the numerical tools that are necessary to determine a suitable relative stream function field. For the reader's convenience we recall that an uniform elliptical vortex of vorticity ω is a stationary solution of the Euler equations in a frame of reference rotating with the angular velocity Ω given by

$$\frac{\Omega}{\omega} = \frac{\lambda}{(\lambda+1)^2} \leq \frac{1}{4} \tag{9}$$

where $\lambda \geq 1$ is the ratio between the two axis of the ellipse. A non uniform vortex is not, in general, a stationary solution and its motion can lead, through the formation of filaments, to a configuration having an eccentricity smaller than the initial one. An example is shown in fig. 3, where several neutral particles are placed on the initial core axis to show the core deformations. In this particular case they remain aligned, because the core is always elliptical.

An evaluation of the bulk rotation of the vortex can be obtained by measuring the orientation of a streamline during the vortex motion. Selected a value $\overline{\psi}$ of the stream function, the vortex orientation can be determined by evaluating the second order moments of the domain $D_{\overline{\psi}}$, interior to the closed curve $\psi = \overline{\psi}$. This leads to the definition of an equivalent ellipse, i. e. an ellipse having the same moments

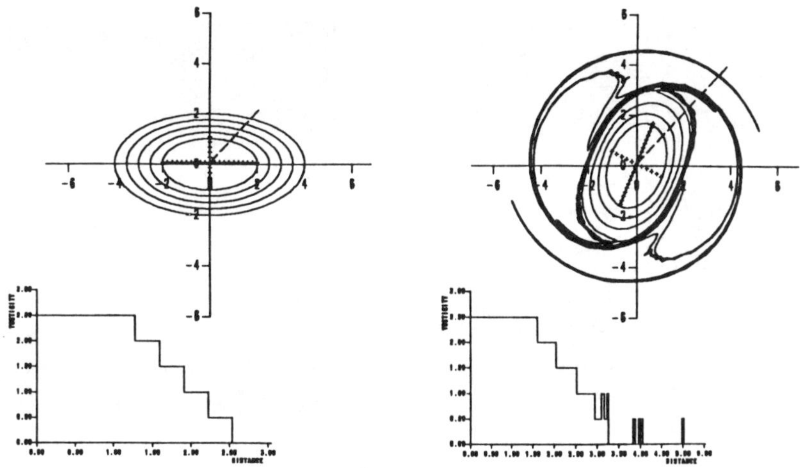

Fig. 3. Initial condition ($t = 0$) and final stage ($t = 27$) of the axisymmetrization of an elliptical vortex having 5 vorticity levels with $[\omega] = \frac{1}{2}$.

of $D_{\overline{\psi}}$ up to the second order, named in [2] "*diagnostic ellipse*", whose orientation gives the vortex orientation χ_ψ. The angular velocity Ω of the rotating frame of reference having its axis superimposed to the diagnostic ellipse axis is calculated numerically by a first order forward derivative

$$\Omega := \dot{\chi}_\psi \quad ,$$

from which the definition (9) is recovered for the uniform vortex. An analogous procedure leads to the definition of a diagnostic ellipse for a vorticity level $\overline{\omega}$.

Finally the relation between the absolute stream function ψ and the stream function in the rotating frame (*relative* stream function ψ_r) follows from the relation between the absolute velocity u and the relative velocity u_r

$$u_r(x) = u(x) - \Omega x^\perp$$

and the definition of the stream function $-\nabla^\perp \psi = u$

$$\psi_r(x) = \psi(x) + \frac{1}{2}\Omega |x|^2 \quad .$$

The above correction of the strem function ψ leads to a nice interpretation of the physical phenomena. As shown in fig. 4 the field of the relative stream function ψ_r can be divided into four parts: a central region, superimposed to the vortex core, in which the flow in the moving frame is co-rotating with the vortex, the outer field and the two (symmetric) regions where the flow is counter-rotating. These latter have been named in [2] "*ghost vortices*" and the vorticity lying into these vortices, being convected along counter-rotating trajectories far from the vortex core, leads to the formation of filaments.

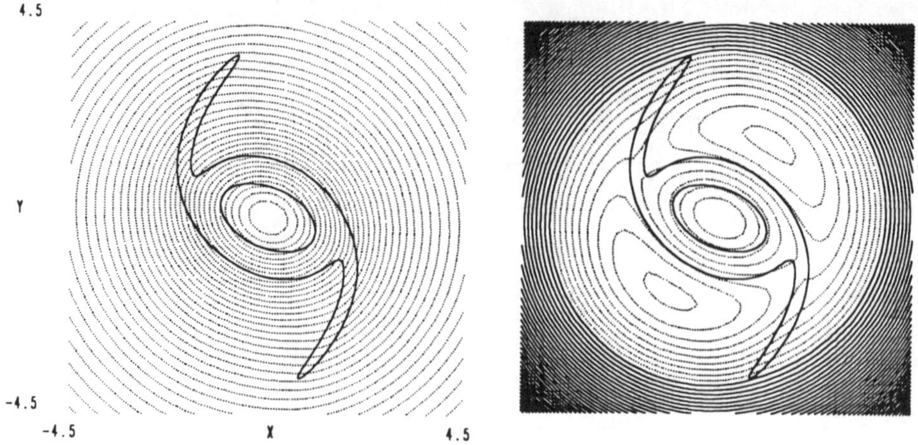

Fig. 4. ψ and ψ_r fields at time $t = 9$ for an elliptical vortex with two vorticity levels with $[\omega]_k = 1$ and initial axis $a_k = \dfrac{3}{2}k, b_k = \dfrac{1}{2}k$ for $k = 1, 2$.

4.2. DEFINITION OF A MERGING TIME

In the same way a rotating frame can be defined for the case of the merging (symmetric, for semplicity) of two vortices by assuming as the bulk orientation of the vortex the orientation of a streamline. The relative stream function in this frame has the structure illustrated in fig. 5 for the case of patches that are initially ($t = 0$) circles of unitary radius with a distance between their centers of 3 radii. The ψ_r field is rather similar to the axisymmetrization case. The relevant difference

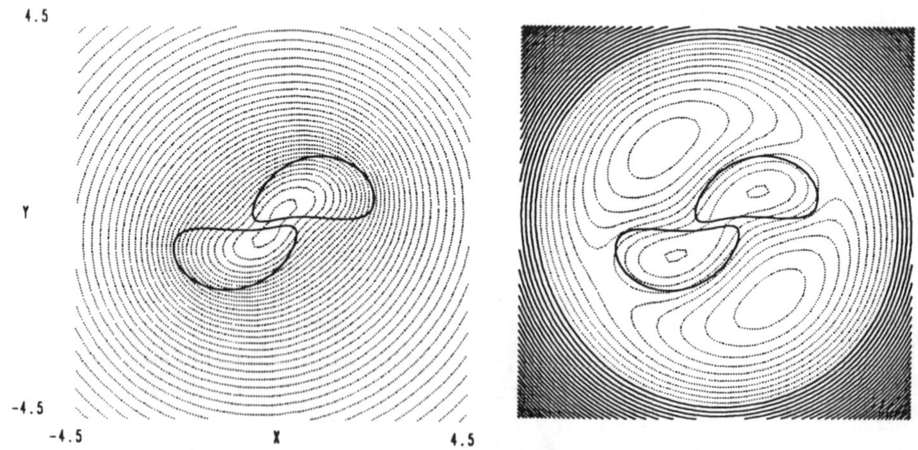

Fig. 5. ψ and ψ_r fields at time $t = 4$ for the merging of two equal patches which initially are two circles of unitary radius and unitary vorticity. The distance between their centers is $d = 2.5$.

is into the core region that appears structurated in three sub-regions: two cores into the physical vortices and an intermediate zone named "*exchange band*" which promotes the exchange of vorticity between the vortices. This leads to another kind of filaments that goes from a vortex to the other, also when the merging doesn't proceed. The amalgamation process implies a change of the structure of the relative stream function field, as discussed in [3]. Evidently, the change is located into the central region that is invaded by the exchange band with a shrinking and a tendential disappearance of the two cores internal to the physical vortices into a small neighbourhood of the center of vorticity of the structure. The complete saturation of the core region is possible only for a viscous fluid. Actually, in this flow the interface between the vortices is spread out by the diffusion process, this leads to an uniform vorticity field corresponding to a ψ_r field without sub-structures into the core region.

The change of the structure of the ψ_r field is the starting point to define a *merging time*, i. e. the time employed by the exchange band to fill the core region. A simple way to calculate this time is to follow the trajectory of a center of one of the two cores during its motion toward the center of vorticity. The center is a maximum of the ψ_r field and some attention has to be paid for its calculation when the field becomes more and more flat around the maximum. The distance between the center of vorticity and a ψ_r maximum vs. time in the inviscid case are shown in fig. 6 for different distances between the two vortices. These results are obtained with two patches of unitary vorticity that are initially circle of unitary radius with a distance d between their center. It appears that the maximum doesn't reach the

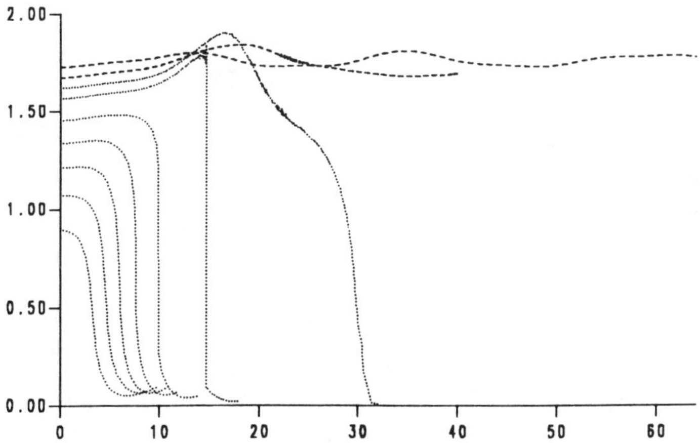

Fig. 6. Distance between a ψ_r center and the center of vorticity vs. time for $d = 2.2, 2.4, 2.6, 2.8, 3, 3.2, 3.3, 3.4$ and 3.5 in the inviscid case.

center of vorticity but it perform a spiral motion around this point. For this reason a definition of the merging time is possible only in a conventional way. We choose to define this time as the intersection between the tangent to the curve of the distance at its flex point and the time axis.

5. Merging time and Reynolds number

A parametrical study of the symmetric merging between two vortices is carried out to determine the dependence of the merging time both on the initial distance between the vortices d and on the Reynolds number Re. In this section we discuss some intersting observations concerning the coupling of the two effects, altought, in our opinion, a deeper understanding of the overall merging dynamics is required. The vortices, initially circular with a unitary radius, are placed at a distance d between the two centers. At the initial time ($t = 0$) the vorticity field is uniformly spread with an unitary value on the two circles. Two auxiliary levels are considered in order to maintain a sufficient accuracy during the merging time. The initial configuration of the ω field is depicted in fig. 2. The simulations are conducted for moderate Re number (1000, 3000, 5000) to find relevant differences between the corresponding behaviors, keeping at the same time the computational effort in a reasonable range.

5.1. ANALYSIS AT DIFFERENT DISTANCES

We consider several values of the separation distance closed to the *critical merging distance* for the inviscid flow, $d \simeq 3.33$. We start with the subcritical distance $d = 3.3$ and we increase it up to the largely supercritical distance $d = 3.5$. For larger distances the start up of the merging process requires a too longe time, hence a more refined discretization of the initial vorticity field.

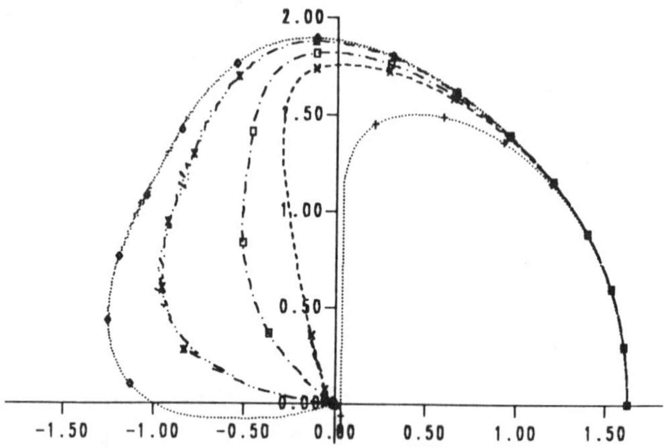

Fig. 7. Trajectories of a center of the ψ_r field for $d = 3.3$ at different Reynolds numbers. From the inner to the outher curve we have $Re = 1000, 3000, 5000, 20000$ and $Re \to \infty$. The time step between two consecutive markers is 2.

We consider first the distance $d = 3.3$ which is slightly less than the inviscid critical one and the merging process is observed for all values of the Reynolds number. In the inviscid case, the merging process occurs past an half turn of the vortices around their center of vorticity, as shown by the ψ_r trajectory. The evo-

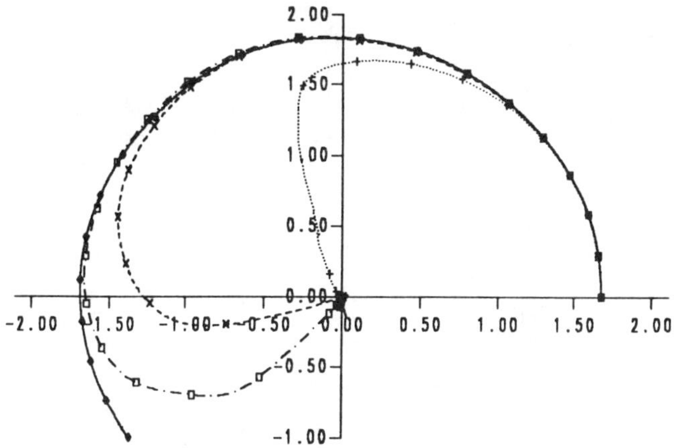

Fig. 8. Trajectories of a center of the ψ_r field for $d = 3.4$ at different Reynolds numbers. From the inner to the outher curve we have $Re = 1000, 3000, 5000$ and $Re \to \infty$. The time step between two consecutive markers is 2.

lution toward an elliptical configuration of the vortex core occurs when the two filaments pertaining to the exchange band reach the opposite extrema of the other vortex. At this stage the convection of these filaments into the vortex core regions leads to a displacement of the bulk of the vorticity toward the center of vorticity. The resulting structure becomes nearly elliptical and the ψ_r-center collapses on the center of vorticity. In presence of a weak diffusion ($Re = 5000$) these filaments are more intense, owing to the transfer of new vorticity diffused from the vortex cores into the exchange band. The convection of the filaments internal to the exchange band it is sufficient to enforce an elliptic core in the resulting structure, hence the merging process is faster: as shown in fig. 7, it is completed in a quarter of the first turn of the vortices. For low Reynolds number ($Re = 1000$) the intensity of the filaments pertaining to the exchange band is so large that the merging process is completed in less than a quarter of the first turn and the elliptical configuration is reached immediately at the entrance of the higher vorticity level into the exchange band. The sudden merging experienced in this case is very close to the case $d = 3.2$ for Reynolds going to infinity.

Let us consider now the super-critical distance $d = 3.4$ for which the merging process doesn't occur in the inviscid case. Nevertheless, the vortices exchange vorticity through the migration of two filaments, internal to the exchange band, from one vortex to the other. They roll up around the vortex cores and a high gradient interface made by both filaments develops on a separatrix line between the core regions and the exchange band, as shown in fig. 9. At the end of this first filamentation the high gradient region is so thin that the vorticity field is practically equivalent to the initial one. The resulting trajectory of the ψ_r center is close to an arc of circle. A slight viscosity ($Re = 5000$) is sufficient for the merging to occur. The two filaments exchanged through the exchange band roll up on the vortex cores. Their

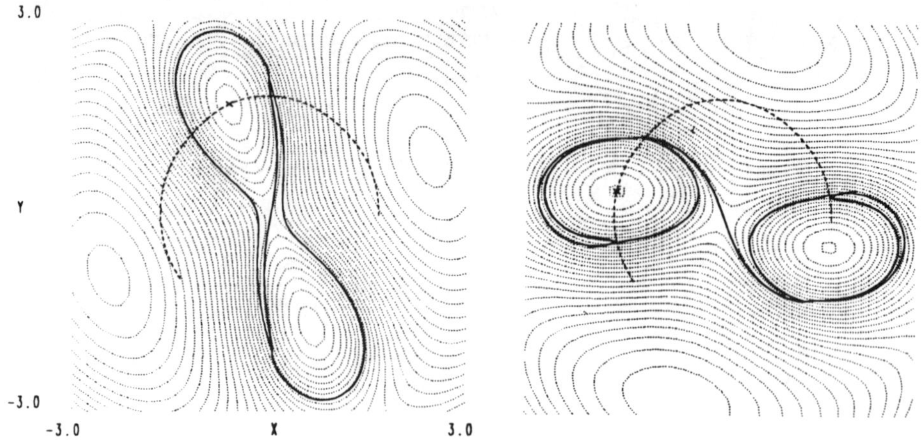

Fig. 9. ω and ψ_r fields for the inviscid motion of two patches at distance $d = 3.4$. Times 20 (left) and 30 (right) are shown.

motion is responsible for the evolution of the core of the resulting vorticity structure toward an elliptical configuration. With a larger diffusion of vorticity ($Re = 1000$) the merging is faster. It is caused by the massive convection of the vorticity internal to the exchange band that enforces the core of the resulting structure toward an elliptical configuration.

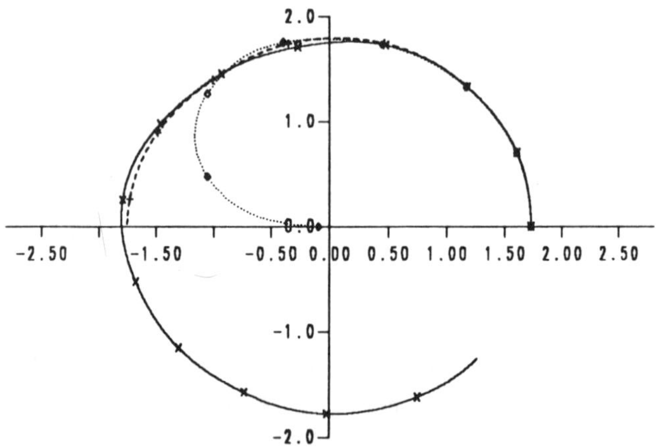

Fig. 10. Trajectories of a center of the ψ_r field for $d = 3.5$ at different Reynolds numbers. From the inner to the outher curve we have $Re = 1000, 3000$ and $Re \rightarrow \infty$. The time step between two consecutive markers is 5.

Finally we consider the even larger distance $d = 3.5$. The flow in the inviscid case is very simple: the two vortices rotate around the center of vorticity with oscillating boundaries, without any exchange of vorticity. In the ψ_r framework, the vorticity is never transferred into the exchange band and the ψ_r center rotates with a nearly circular trajectory. The amount of viscosity, necessary at this distance to start up the merging process in a convective time, is very large: the convective merging is observed only in the case of $Re = 1000$, as shown in fig. 10. For $Re = 5000$ and $Re = 3000$ there are only exchanges of vorticity with the formation of complex filamentary structures that are not going to affect the core configuration (see also fig. 2). By increasing the viscosity ($Re = 1000$) the filaments into the exchange band grow until they enforce the elliptical configuration of the core of the resulting structure. This process is favoured by the motion of the exchanged vorticity into the core regions of both vortices.

5.2. TIME VS. DISTANCE

The above qualitative analysis shows that the transition from the merging to the axisymmetrization stage occurs at a time which is a growing function of both initial distance between the two vortices and Reynolds number. It clearly appears from fig. 11, in which the merging time is represented as a function of the initial distance. We obtain different curves, one for each Reynolds number. For the inviscid flow there is a vertical asymptote near the value of $d \simeq 3.33$, whereas the behavior for higher distance and finite Re is not sufficiently understood. In fact, there is no evidence (at least from the numerical point of view) of the start up of the merging on diffusive time scales, we have only observed the initial stage toward the *metastable* solution described in [3].

An interesting observation suggesting a kind of similarity follows from fig. 11 and from the analysis of the vorticity fields. Actually, a similar behavior corresponds to all cases with different d and Re leading to the same merging time.

For example the two cases $d = 3.4, Re = 1000$ and $d = 3.3, Re = 3000$, the two cases $d = 3.4, Re \to \infty$ and $d = 3.5, Re = 3000$, or the two cases $d = 3.3, Re \to \infty$ and $d = 3.4, Re = 3000$ give similar vorticity dynamics, as it can be inferred from the ψ_r center trajectories of fig. 11.

6. Concluding Remarks

The simple model which has been proposed here to extend the well established Contour Dynamics to weakly viscous flows, has shown to be quite appropriate for the numerical simulation of the dynamical evolution of vorticity structures at large Reynolds numbers. In particular, the analysis of the merging process of two initially circular and uniform vortices has brought to a quantitative evaluation of the time scale dependence on distance and Reynolds number. An interesting similarity, suggested by the observations of the vorticity structures in time, is worth of a further investigation. The simple scheme adopted for the approximation of the diffusive flux (maintained constant along an isovorticity line) and the uniform redistribution of the vorticity to keep the contour an isovorticity line (inherent to the Contour Dynamics approach) make the present model not well suited to represent the filamentation in sufficiently accurate way. A more appropriate model, presently under investigation, may use the concept of diffusion velocity. Another possible improvement could adopt the *surgery technique* [1] to isolate the filaments after their

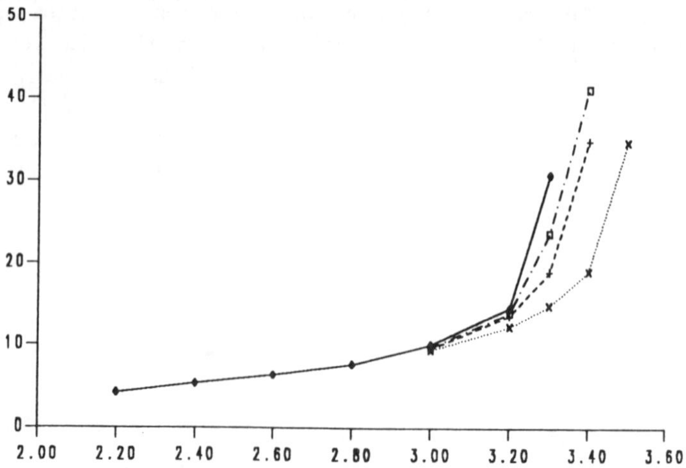

Fig. 11. Merging times vs. initial distance between the vortices for $Re = 1000$ (dotted line), $Re = 3000$ (dashed), $Re = 5000$ (dash-dot) and inviscid (solid).

formation, avoiding thus an excess of vorticity transfer to them during the succesive time evolution. The accuracy control through the computation of the first integrals of motion in viscous flow, sufficiently encouraging for the present simulations, remains the only quantitative tool to evaluate the different approximation schemes for the vorticity diffusion.

References

[1] Dritschel, D. (1989) *Contour Dynamics and Contour Surgery: Numerical Algorithms for extended, high-resolution modelling of Vortex Dynamics in two-dimensional, inviscid, incompressible flows*, Computer Physics Reports **10**, 77 – 146

[2] Melander, M.V.; McWilliams, J.C.; Zabusky, N.J. (1987) *Axisymmetrization and vorticity-gradient intensification of an isolated two-dimensional vortex through filamentation*, J. Fluid Mech. **178**, 137 – 159

[3] Melander, M.V.; Zabusky, N.J.; McWilliams, J.C. (1988) *Symmetric vortex merger in two dimensions: causes and conditions*, J. Fluid Mech. **195**, 303 – 340

[4] Ogami, Y.; Akamatsu, T. (1991) *Viscous flow simulation using the discrete vortex model - The diffusion velocity method*, Computers & Fluids **19** 3/4, 433 – 441

[5] Pullin, D.I. (1992) *Contour Dynamics Methods*, Annu. Rev. Fluid Mech. **24**, 89 – 115

[6] Riccardi, G. *Two-dimensional vorticity dynamics at large Reynolds number*, Ph. D. Thesis, University of Rome "La Sapienza" (1992)

[7] Zabusky, N.J.; Hughes, M.H.; Roberts, K.V. (1979) *Contour Dynamics for the Euler Equations in Two Dimensions*, J. Comput. Phys. **52**, 351 – 373

MODEL COHERENT STRUCTURE DYNAMICS: VORTEX RECONNECTION, CORE DYNAMICS AND INTERACTION WITH TURBULENCE

FAZLE HUSSAIN and MOGENS V. MELANDER
University of Houston
Houston TX 77204

ABSTRACT. Direct numerical simulations of the Navier-Stokes equations for three idealized vortex interactions are analyzed to obtain insights into the dynamics of coherent structures in turbulent flows. We first study the vortex reconnection mechanism which presumably is important in cascade and mixing in turbulent flows. The dynamical significance of axial flow in coherent structures inferred from this interaction is explored further by studying the evolution of an axisymmetric laminar vortex tube with nonuniform core. Evolution of vorticity wavepackets in such a vortex tube is first explained as a coupling between swirl and meridional flow. We show how this evolution is better understood by the motion of polarized vorticity components, obtained via complex helical wave decomposition of the vorticity and velocity fields. Finally, by treating such a vortex tube as a prototypical segment of a coherent structure we study its interaction with background, fine-scale, isotropic turbulence. This study shows polarization and organization of small scales by the coherent structure, and subsequent feedback on the structure from the small scales, thus providing the mechanism for a direct coupling between large and fine scales in turbulent flows.

1. Introduction

Structural approach, distinct from the earlier statistical approach and the more recent dynamical systems approach, to understand turbulence physics started in the early seventies with the discovery of large-scale organized motions – popularly called *coherent structures* (CS). Forerunners among such studies are those by Kline *et al.* (1967), Crow & Champagne (1971), Brown & Roshko (1974). Our long-standing assertion has been that CS should be characterized by *coherent vorticity* – the instantaneous space-correlated vorticity underlying the random field that we call turbulence – and that vortex dynamics is a (perhaps the most) tractable avenue for understanding the evolutionary dynamics of CS, their role in turbulent transport phenomena, their mutual interactions (such as pairing, tearing, and reconnection) and their coupling with fine-scale turbulence. The overwhelming majority of studies in structural turbulence have been based on flow visualization; however, we have also long contended that flow visualization can not only be confusing but also grossly misleading (Hussain 1980, 1986). Even at unity Schmidt number, a passive scalar can faithfully mark vorticity boundaries of structures only in the case of 2D

J.T. Beale et al. (eds.), Vortex Flows and Related Numerical Methods, 239–264.

flow. For the usual 3D flow, and in particular turbulent flows and vortex interactions, visualization may present incorrect information because marker is depleted from the regions of intense vortex stretching but accumulates in regions of minimal stretching. As a result, dynamically dominant events can easily be missed by flow visualization. A highly useful tool for qualitative flow interpretation, flow visualization must be supplemented by quantitative data. Pending the development of an experimental technique (such as Holographic Particle Velocimetry being developed in our laboratory and elsewhere) to obtain well-resolved quantitative data, we must rely on direct numerical simulations DNS) of the Navier-Stokes equations.

The most effective use of DNS is to study idealized vorticity configurations as models of CS interactions in practical flows. Here we discuss an example of such an interaction, viz., vortex reconnection. Our interest in the reconnection process was prompted by our intuitive claim that it is a frequent event in turbulent flows (Hussain 1986), that it plays an important role in cascade and mixing, that it is an important source of aeroacoustic noise (Takaki & Hussain 1985), and also that, while viscosity is crucial to reconnection, the phenomenon occurs at a convective, rather than a diffusive, timescale. Our curiosity was also piqued by our interest in the topological changes of vortex lines during reconnection. For DNS studies, we considered an idealized configuration of two antiparallel vortex tubes which was motivated by our experimental studies of reconnection of elliptic vortex rings in high aspect ratio elliptic jets (Husain 1984, Hussain & Husain 1989).

Analysis of the reconnection data suggested that the often neglected axial flow in coherent structures can be dynamically significant. Thus, we idealized this problem by considering the flow in an axisymmetric vortex column with nonuniform core. To analyze this problem we used the powerful mathematical tool of (complex) Helical Wave Decomposition (HWD). In addition, we have studied the interaction of a nonuniform cross section vortex with background isotropic turbulence as a prototypical interaction scenario between large-scale coherent structures and fine-scale turbulence. Collectively, these studies demonstrate our distinct approach to effectively use DNS, experiments and vortex dynamics to obtain insights into coherent structure dynamics.

2. Simulation Method

All flows are simulated using a spectral method dealiased by a 2/3 k-space truncation and with a fourth-order predictor-corrector algorithm for time stepping. Periodic boundary conditions are appropriate because the interactions studied here are primarily affected by local effects and are insensitive to nonlocal effects such as the influence of adjacent boxes. Tests were undertaken to check that the evolution details remained unchanged when the domain was doubled.

3. Vortex Reconnection

The initial condition consists of two antiparallel vortices with mutually inclined symmetric sinusoidal perturbations (Fig. 1a). The sinusoidal perturbation is provided only to ensure that the two vortex tubes collide so as to induce and sustain reconnection; this perturbation is not central to the flow physics of vortex reconnection. Initially, each vortex has a circular core with a compact support of Gaussian vorticity distribution. We also studied an asymmetric initial condition as a perturbation to Fig. 1a. No change was noticeable in the details of the evolution of the vortex reconnection process, thus confirming that imposition of symmetry (as used in the simple analysis by Takaki & Hussain (1985) and in the DNS study by Kerr & Hussain (1989)) is not a constraint of the simulation (see later). The symmetric initial condition is particularly useful for clearly identifying the effects of viscosity because the two symmetry planes are preserved in time and are material planes in the absence of viscosity. For clarity, the xy-plane is called the 'symmetric' plane or π_s, while the yz-plane is the 'dividing' plane or π_d (see Fig. 1a). In both planes, the normal velocity and the tangential vorticity vanish at all times. Furthermore, $x\omega_z \leq 0$ everywhere in π_s; this eliminates the problem of self-annihilation within each vortex (as well as complicated vorticity peeling which is of no consequence to the reconnection process) as found in the simulation of Meiron *et al.* (1989). Consequently, the circulations Γ_s and Γ_d in the half planes π_s^+ $(x > 0)$ and π_d^- $(z < 0)$ change exclusively by viscous vorticity annihilation. We also know that $d\Gamma_d/dt = -d\Gamma_s/dt$, since an arbitrary vortex line must intercept either π_s or π_d.

4. Three Phases of Reconnection

The evolution of the vortices is studied for the Reynolds number $Re = \Gamma/\nu = 1000$. Fig. 1 shows wire-frame plots of the vorticity norm $|\omega|$ at $t = 0, 2.25, 3.5, 4.75$ and 6; the level surface here represents 30% of the initial peak vorticity $\omega_{\max}(0)$, and t is time, t^*, nondimensionalized by $\omega_{\max}(0)$ such that $t = t^*\omega_{\max}(0)/20$. Fig. 2 shows the evolution of circulation Γ_s. On comparing Fig. 1 with Fig. 2 it is clear how reconnection evolves through three phases. The first phase is governed by inviscid vortex dynamics as almost no circulation is transferred from π_s to π_d. During this phase, while the two vortices move upward by mutual induction as a vortex dipole, self-induction presses the vortices against each other, flattening them and forming a viscous contact zone with steep vorticity gradients. In the second phase, a dramatic change in topology takes place as a rapid transfer of circulation from π_s to π_d (Fig. 2) occurs due to viscous vorticity annihilation (by cross diffusion) in the contact zone. This phase, marked by 'bridging' (growing humps in Figs. 1c,d), is the heart of the reconnection process (discussed later). The growing bridges acquire a curved horseshoe shape, causing them to pull apart by self-induction. Simultaneously, the induced flow by the bridges stretches the dipole in π_s and sustains their annihilation. The third phase is the evolution of the remnants of the original dipole after the rapid circulation transfer. The slender vortices, which we call 'threads', are well defined and sustained because of the stretching by the bridges. These threads arise as a consequence of the incompleteness of the reconnection and are the longest lasting features of the reconnection. As we will see, reconnection must necessarily occur incompletely in the first stage providing the threads and thus will involve a succession of subsequent steps.

The simulation with initial asymmetries produced virtually the same wireframe plots as in Figs. 1(a-e) even at $t = 6$, thus vindicating both the analysis (Takaki &

Hussain 1985) and the high-resolution simulation (Kerr & Hussain 1989) performed by invoking symmetry. The *Re* dependence of the crosslinking process was tested and found to consist of the same sequence of stages with identical successive spatial patterns, except that the corresponding timescales increase with decreasing *Re*.

5. The Bridging Mechanism

The rapid circulation transfer from π_s to π_d begins when the vortices come into contact near $\pi_s \cap \pi_d$ (Figs. 1b,3b) and form a contact zone $C(t)$ with steep vorticity gradients. As the flow is highly viscous around C and nearly inviscid elsewhere, we may think of the vortex lines far away from C, merely for the sake of discussion, as material lines. We use this idea schematically in Figs. 5a-c, where each vortex is represented by three vortex lines; one vortex line from each vortex intercepts C (Fig. 5a), and shortly afterwards these vortex lines annihilate each other at $x = z = 0$ by cross-diffusion. This produces two reconnected lines intercepting π_d at $z = \pm z_0$ (Fig. 5b). The π_d-intercept of z_0 quickly retreats away from $x = z = 0$, because of strong diffusion and nearly antiparallel vorticity lines emerging from cusps.

A cross section AB (Fig. 5c) at the edge of contact zone C reveals the two-dimensional dipole structure of the original vortices. In a co-translating frame the streamline pattern becomes the well-known recirculation bubble of a vortex dipole (Fig. 5d). Although this frame is not precisely defined for an unsteady dipole, the existence of two stagnation points – at the front and back of the dipole – S_F and S_B can be inferred. The large velocity between S_F and S_B advects z_0 towards S_F, ahead of the dipole (Fig. 5c). The relative upward advection of z_0 stops at S_F, and a reconnected vortex line unfolds along the diverging separatrix a-a (Fig. 5d). This mechanism is schematically illustrated in a perspective view of one half of the two vortices in Fig. 5e, which shows a vortex line at five different instants (1-5). Far away from C, a vortex line is simply swirled around the vortex tube. However, in C, vortex lines get reconnected as soon as they come into contact after 1, but before instant 2 and are swept upward by the pumping motion of the dipole. At instant 3, the upward motion of the reconnected vortex line stops at S_F, and it gets stretched and begins to wrap around the dipolar vortex pair near S_F (vortex line at instants 4 and 5). Meanwhile, the circulatory motion in the dipole continuously pumps fresh non-reconnected lines into C near S_B (Fig. 5d) which then in turn go through the same sequence of reconnection. Due to this mechanism, reconnected vortex lines accumulate ahead of the antiparallel vortices and align orthogonal to the latter.

During bridging, a complex interaction of *three* competing effects governs the rate of annihilation of vorticity by changing its gradients in C. These effects are *diffusion*, *stretching* in the z- direction, and local *'self-induction'*. Diffusion counteracts the steepening of vorticity gradients and can only decrease the peak vorticity in the symmetry plane ω_m, while any increase in ω_m is due to stretching (normal to π_s). This 'axial' stretching is generated mainly by the vorticity distribution $\varepsilon(t)$ away from C (ε is shown in Fig. 5a; note that ε includes the bridges), and in part by self-induced lengthening of vortex lines near C (analogous to the axisymmetric collision of vortex rings). Depending on the relative positions of ε and the dipole in π_s, the external stretching is either positive (vortex stretching) or negative (vortex compression): positive if the dipole has fallen behind ε (Fig. 5f), and negative if the dipole has advanced ahead of ε. The self-induction on each side of the sheet-like region C is in π_s, and is directed toward π_d whenever the non-reconnected vortex lines in C curve upward, and away from π_d when they curve downward. The motion toward π_d increases gradients across the y-axis by compressing the vortex cores into a characteristic head-tail structure (Fig. 3d), while motion away from π_d decreases the gradients by separating the vortex core in π_s. The above three effects - diffusion, stretching and self-induction - dominate the evolution in the following order. At first, the vortex lines in C curve upward, and the gradients across π_d increase as self-induction toward π_d presses the dipole cores into a head-tail shape; meanwhile, the external stretching being negative, the increase in ω_m is small. However, as the dipole in π_s falls behind ε (discussed later) such that the vortex lines in C curve downward, the axial stretching picks up and produces a large increase in ω_m, and consequently large gradients. Finally, towards the end of the rapid circulation transfer from π_s to π_d (Fig. 2), the local self-induction away from π_d arrests the cross-diffusion by producing a near balance between axial stretching, diffusion and separation of the dipole cores in π_s, causing the threads to linger.

6. Curvature reversal and reconnection cascade

The curvature reversal, which causes the incomplete reconnection, is a characteristic and crucial event. It occurs because the upward motion of the dipole in π_s slows down relative to that of ε (Figs. 3,4; note that the y-position can be compared using the numbers on the right side of each panel); but why does the dipole in π_s slow down? A downward velocity field in C downwash opposes the dipole's own upward motion in π_s (Fig. 5f). Furthermore, the dipole loses circulation through vorticity annihilation; its upward motion weakens and it relaxes into a head-tail shape. The downwash is induced by the growing bridges as well as by non-reconnected vortex lines in ε, so downwash will be present even at $Re = \infty$. The relaxation into a head-tail shape, an inviscid process, virtually decreases the dipole circulation as the dipole propagation velocity is determined almost exclusively by the head. Although this inviscid effect is weak at $Re = 1000$, it may well be the primary mechanism for curvature reversal at higher Re. An indication of this would be that the area A of the nearly self-similar head decreases with increasing Re, such that $\omega_m A^{1/2}$ (propagation speed of the head to leading order) is bounded as $Re \to \infty$. This indicates the possibility that the amount of circulation transferred from π_s to π_d tends to zero as Re tends to infinity while the transfer is still occurring within a convective timescale.

The above gives us some interesting new results. The vortex reconnection is not complete in a single step; as the bridges move apart by self-induction, the thread-dipole advects forward in the presence of the weakening 'downwash' caused by the bridge-dipole. This way, the threads reverse their curvature again and then collide at the midpoint by self-induction, starting the next burst of reconnection. This is what we call reconnection cascade. Clearly, the frequency of reconnection bursts will increase and circulation transfer will decrease as the Reynolds number increases.

7. Dynamical Interactions between Bridges and Threads

In order to provide further insight into the reconnection mechanism, topology, and dynamics of interacting vortices, we discuss the evolution of helicity, enstrophy production and vortex lines. Even though helicity density (h) is a frame-dependent quantity and hence not a suitable measure of topology (unlike the Helicity integral), in the present idealized configuration it does point out some important features. Helicity in the initial condition and during the first phase of reconnection is very small and vortex lines inside the vortex tubes are almost parallel. However, the helicity distribution after the reconnection (e.g. $t = 4.75$ and $t = 6$) is particularly interesting, as it consists of intertwined regions of positive and negative helicity. Apart from the threads, where the helicity stands out clearly (Figs. 6a,b), the structure is an enigma until one examines bundles of vortex lines. Figs. 7a,b show that vortex lines emerging from the threads do not wrap all the way around the bridges as one would expect from $|\omega|$ surfaces, e.g. expanded views of Figs. 1d,e. Instead, they form asymmetric hairpin-like structures (see A and B, Fig. 7a). The vorticity is high in one hairpin leg and low in the other. Vortical structures of this kind are unfortunately not faithfully reflected in the surface plots of $|\omega$. The intense leg (A) is the extension of the thread (C) and its high vorticity makes its helicity stand out in Fig. 6. The diffuse leg is difficult to identify in the helicity distribution as the $|\omega$ there is low and the angle between vorticity and velocity is larger. The intense leg induces a flow away from π_d on the outside of the bridges causing vortex stretching in the outer part of the bridges (see point A in Figs. 8b,d).

Inside the bridge, the vortex lines have a slow helical twist near π_d (barely discernible in Fig. 7b). Contrary to the intense hairpin leg, the twist produces an axial flow toward π_d at the center of the bridge. Since this axial flow is toward π_d from both sides, it results in a negative enstrophy production, as seen in a cross section of bridges parallel to π_d (dotted lines in Figs. 8b,d). The twisting of vortex lines is the result of a skewed (nonconcentric) vorticity distribution in the bridges (Figs. 4b-d). The peak vorticity in the bridge is not at the location of the geometric centroid of the low level vorticity contours, but is far away from the contact zone, where the vorticity is lower than in the bridges – a consequence of earlier stretching by the dipole during bridging. Figs. 7a,b show that a vortex line which neatly follows the centerline of the vortex far away from π_d ceases to do so in the bridge. This effect clearly results in vortex line twisting and thus in an induced axial flow (Fig. 7c). The direction of the axial flow is determined by the orientation of the twist, which in turn is determined by the highest swirling velocity along the vortex. The bridges, having the highest peak vorticity, also have the highest swirling velocity. Hence the axial flow is toward π_d from both sides, resulting in vortex compression (i.e. negative P_ω in Figs. 8b,d) which decreases the peak vorticity and thereby also the twisting rate. This constitutes a new inviscid mechanism, clearly distinct from diffusion, for smoothing the vorticity intensity along the reconnected vortex lines. We further analyze the dynamics associated with axial flow due to nonuniform vorticity distribution along the bridges in the next section.

Interacting vortex tubes relax to a head-tail structure as shown in Fig. 3d. This is essentially an inviscid mechanism and was also observed in the head-on collision of inviscid vortex rings (Shariff et al. 1988; Stanaway *et al.* 1988). The principal reason for head-tail formation is the strain in π_s due to self-induced motion of the vortex tubes toward π_d. Because of higher vorticity concentration, the annihilation is higher between the heads than between the tails. The tail decays more by viscous diffusion than by annihilation. The higher circulation in the head also causes it to move faster due to mutual induction and around $t = 4.5$, we see a separation of the head from the tail (Fig. 3e). Such deformation of a circular vortex core into a head-tail structure increases the surface area over which vorticity diffuses to make the surrounding fluid rotational, and thus, increases entrainment. This is in addition to the increase in surface area due to the formation of bridges which also contribute to entrainment via strong 'downwash'. In the present simulations, one of the vortices was marked with a scalar with Schmidt number unity. Even under such ideal conditions the scalar does not continue to faithfully mark the vortical fluid (compare Figs. 1 and Fig. 9). At $t = 4.75$, vortex stretching has diluted the scalar in the contact zone in A (Fig. 9d) while the vorticity, being accentuated, remains large, with the result that a key feature of the vortical structure, the threads, is invisible in experimental flow visualization (Fig 9e; also see Schatzle 1987). This serves as another example supporting our long-standing warning against relying too heavily on flow visualization in studying 3D vortex interactions and turbulence dynamics.

8. Vortex Core Dynamics

In order to focus on the flow along the axis of a bridge (see Fig. 7c), we have idealized this problem to that of an axisymmetric vortex with a sinusoidal variation of core size. This way we eliminate the complications of curvature and the flow induced by the threads. Because of the axial variation of the core diameter, there is also an axial variation of swirl and hence twisting of vortex lines soon after $t = 0$. Figure 10 shows a schematic of the vortex at a time $t > 0$ with vortex lines lying on the shaded surface denoting an axisymmetric vortex surface (not a vorticity level surface, but $r\nu =$ constant surface which is tangent to vortex lines); a 90° cutaway is introduced to enhance comprehension of meridional streamlines, and swirl and axial velocity profiles. The vortex evolves as a result of the interaction between meridional flow and swirl ($r\nu$). The velocities u, v, w in the (r, θ, z) coordinates are governed by the following equations

$$D_t\zeta = \text{viscous term; } \zeta \equiv r\nu \quad (1)$$
$$D_t\eta = r^{-4}\partial\zeta^2/\partial z + \text{viscous term; } \eta \equiv \omega_\theta/r, \quad (2)$$

where $D_t \equiv \partial_t + u\partial_r + w\partial_z$ denotes material derivative in the meridional flow, whose streamfunction Ψ is related to u, w and ω_θ as follows: $u = \Psi_z/r, w = -\Psi_r/r$ and $r\omega_\theta = \Psi_{rr} - \Psi_r/r + \Psi_{zz}$.

The vortex evolution in time is depicted by $|\omega|$ contours in figure 11 and meridional streamlines in figure 12. The axial variation of the swirl, ζ, produces coiling of vortex lines and hence axial pumping of fluid in a direction that will tend to smooth out core size variation and hence opposite to the axial variation of $|\omega|$ (i.e., w is opposite to the vorticity flux). Axial variation of swirl is large where core size is small. Also, the twisting is stronger near the axis than at the outer edges. Note that both swirl and meridional flow would have undergone uneventful decay were it not for the coupling term $r^{-4}\partial\zeta^2/\partial z$ between the two fields. Azimuthal vorticity, ω_θ, initially zero, is thus immediately created by axial expansion and contraction of the vortex tube via the coupling term. This is the basic mechanism for motion of waves along the vortex axis. Maximum twisting of the vortex lines occurs at the inflexion points of $\zeta^2(z)$. Creation of ω_θ is necessary to start the meridional flow. Since the meridional flow first forms counter circulating cells (we call them primary cells), the vorticity peak (at the smallest initial core) bifurcates into two which travel in opposite directions. We may also think of this phenomenon as wavepackets which bounce back and forth while undergoing viscous decay. The first rebound induces a meridional flow reversal within each primary cell thereby generating a secondary cell, which soon dominates by ing out the primary cell. The wavepacket behavior of the vorticity is associated with enstrophy production P_ω which can be related to Ψ for inviscid flows as

$$P_\omega \approx -\frac{4\omega_z^2}{r^2}\frac{\partial\Psi}{\partial z}. \tag{3}$$

Thus P_ω can be inferred at any time from the meridional flow streamline pattern (figure 12).

It is important to focus on the generation of meridional flow, hence $\eta = \omega_\theta$. We find that for inviscid flow,

$$D_t\eta = \text{axial vorticity} \times \text{differential rotation along an axisymmetric vortex surface}$$

thus the instantaneous value of η does not affect instantaneous material generation of η, i.e. $D_t\eta$. Because of axial transport of vorticity as wavepackets, η and $\mathrm{Dt}\eta$ also oscillate in time, η leading $\mathrm{Dt}\eta$ by a phase shift of about 90°; this phase shift controls η-transport. The two η-transport mechanisms (convection of η by meridional flow and η-generation by differential rotation along axis) are in opposite directions (hence slower η-transport) in the primary cell, but in the same direction (hence faster η-transport) in the secondary cell.

By investigating this flow for increasingly higher Re we found that the core size variations do not decay in the high Re-limit. Moreover, we find that the wavemotion is more rapid at higher Re, albeit the wavespeed remains finite as $Re \to \infty$. This result has significant implications for computational vortex filament models ;(e.g. Moore & Saffman 1972) where it is assumed that core variations are smoothed out by above wavemotion. This assumption is suspect in the light of our findings.

9. Helicity and Helical Wave Decomposition

The helical nature of flows has become the focus of considerable interest in recent years (Moffatt 1969). Of the three quantities: helicity density $h = \mathbf{u}.\omega$, relative helicity $h_r = h/(|\mathbf{u}||\omega|)$ and helicity integral $H = \int \mathbf{u}.\omega\, dV$, only the last is Galilean invariant and is a conserved quantity for inviscid flows. From $|\omega \wedge \mathbf{u}|^2 = u^2\omega^2(1-h_r^2)$ and the Navier-Stokes equations

$$\mathbf{u}_t + \omega \wedge \mathbf{u} = -\nabla(P/\rho + u^2/2) + \nu\Delta\mathbf{u}, \tag{4}$$

is common to claim that cascade is small if $h_r \approx 1$. But such thinking is fraught with many pitfalls. First, although $h_r \approx \pm 1$ does imply small $|\omega \wedge \mathbf{u}|$, it does not imply small $\nabla \wedge (\omega \wedge \mathbf{u})$ as high frequency oscillations of $\omega \wedge \mathbf{u}$ are possible. In some frames, $\omega \wedge \mathbf{u}$ can be large but purely potential so that vorticity obeys purely diffusion equation and hence there is no cascade. A rectilinear vortex is a glaring example. Thus, for cascade suppression, Beltramization (i.e. $h_r = 1$) is sufficient, but not necessary. An example of cascade suppression for which $\nabla \wedge (\omega \wedge \mathbf{u}) = 0$ is not satisfied is a laminar vortex ring which travels at a constant speed.

Thus, while H is a topological property (discussed later), h and h_r are of questionable use. But recognizing that local helical structure in a flow contains some essential physics, we propose that complex helical wave decomposition (Moses 1971; Lesieur 1990) is the relevant tool for understanding vortex dynamics and interactions. This decomposition in essence expands the field variables (e.g. $\mathbf{u}, \omega$) in terms of the eigenfunctions of the curl operator, which has only real eigenvalues in a periodic box or in infinite space. Moreover, the eigenfunctions can be chosen to form a complete set of orthonormal basis functions. In Fourier space, for each wavevector $\mathbf{k}$ there are two eigenmodes corresponding to the positive and negative eigenvalues k and $-k$. All functions which are linear combinations of eigenfunctions of positive eigenvalues can be called *right-handed* as trajectories or vector lines locally form right- handed helices. Similarly, we call linear combinations of eigenfunctions corresponding to negative eigenvalues *left-handed.* Thus for an incompressible velocity field $\mathbf{u} = \mathbf{u}_R + \mathbf{u}_L + \nabla\phi$. Since $\nabla\phi$ is the projection of $\mathbf{u}$ onto the eigenspace corresponding to the eigenvalue $k = 0$ of the curl operator since $\nabla \mathbf{u} = 0$, we have $\Delta\phi = 0$. If the potential part of the flow is constant at infinity, or for a periodic box, $\nabla\phi$ is a constant. Thus $\mathbf{u}_R$ and $\mathbf{u}_L$ are unique, and both translationally and rotationally invariant. Similarly, $\omega = \omega_R + \omega_L$ (assuming no rotation at infinity). The decomposition of a flow field into polarized (i.e., right- and left-handed) components is rooted in the intrinsic physical nature of the Navier-Stokes equations: the eigenmodes of the curl operator are exact solutions of the Euler equations and constitute a special class of solutions called Beltrami flows with constant abnormality (i.e. $\omega =$ constant $\mathbf{u}$). Because of orthogonality of the eigenfunctions of the curl operator, helicity takes the simple form

$$H = H_R + H_L \tag{5}$$

where

$$H_R = \int \omega_R . \mathbf{u}_R \, dV \geq 0; \; H_L = \int \omega_L . \mathbf{u}_L \, dV \leq 0. \tag{6}$$

The decomposition also provides a clearer insight into the flow physics: cascade is inhibited wherever $(\omega_R + \omega_L) \wedge (\mathbf{u}_R + \mathbf{u}_L)$ and its first spatial derivative are small. One can also easily show that cascade from interactions of modes is small when the modes are of the same polarity (i.e. the same handedness) but large when they are of opposite polarity. One can thus infer that polarized vortices are persistent, hence "coherent", features of the flow.

We apply the decomposition to obtain some understanding of our axisymmetric vortex in the physical space. Remember that vector lines of ω_R (ω_L) are right-handed (left-handed) helices. In general the physical space distributions of ω_R and

ω_L overlap. The ω_R and ω_L distributions of our vortex are shown as functions of time in figure 13 (note identical distributions of $|\omega_R|$ and $|\omega_L|$ at $t = 0$ in figure 13a). This evolution is much more revealing than the evolution of h_r field (not shown). The polarized components move in opposite directions. Aside from the nonlinear interaction between ω_R and ω_L, there is an obvious resemblance to d'Alembert formula for the 1D wave equation.

Note that $|\omega_R|$ is antisymmetric to $|\omega_L|$ in z with respect to $z = l$ at all times, coinciding in frame (a). The helical decomposition thus gives unique separation of ω into ω_R and ω_L components, a feature we think is essential for deeper understanding of coherent structures in turbulent flows. Note that $|\omega_R|$ peak moves to the left instead of to the right as one would expect; this is a consequence of nonlinear interaction (discussed below). In fact, in the case of an isolated vortex with only right-handed polarization, $|\omega_R|$ peak does indeed move to the right (figure 14). In figure 13 we see that as they move, the wavepackets deform: they broaden by diffusion, but also elongate and form a bubble by nonlinear effects. The nonlinearity is also responsible for the breakup of the initial front-back symmetry, i.e. steepening at the front with a tail at the back.

To understand the nonlinear interactions between polarized components, we extract each component by the projection operators (P^+, P^-). Here we discuss only ω_R (by symmetry, ω_L equation is obvious)

$$\begin{aligned} \frac{\partial \omega_R}{\partial t} &= -\nabla \wedge (\omega_R \wedge \mathbf{u}_R) + P^-[\nabla \wedge (\omega_R \wedge \mathbf{u}_R)] - P^+[\mathbf{u}_L . \nabla \omega_R] \\ &\quad + P^+[\omega_R . \nabla \mathbf{u}_L] P^+[\nabla \wedge (\omega_L \wedge \mathbf{u}_L)] + Re^{-1} \Delta \omega_R \\ &= \text{Term I} + \text{Term II} + \text{Term III} + \text{Term IV} + \text{Term V} + \text{Term VI} \qquad (7) \end{aligned}$$

Term I is the inviscid self-evolution; term II is the generation of ω_L by evolution of ω_R; term III is the contribution to ω_R by its advection by $\mathbf{u}_L$; term IV is stretching of ω_R by $\mathbf{u}_L$; term V is the ω_R generation by ω_L, consisting of self-interaction of ω_L and the right-handed part of stretching and advection of ω_L by $\mathbf{u}_R$; term VI is the viscous diffusion term. Terms II-V represent coupling between the left- and right-handed components. We have evaluated these terms for the vortex and find that in general their dominance in decreasing order is as follows: III, I, V, IV, VI and II.

The evolution of ω_R can be largely understood in terms of two effects: self-evolution and advection by the $\mathbf{u}_L$ field. The self-evolution (term I) attempts to move the ω_R-packet to the right by adding vorticity at the front and annihilating it at the back. But because $\mathbf{u}_L$ moves the ω_R-packet to the left, the leftward motion dominates, although it is slower than $\mathbf{u}_L$. Also, because the velocity field has the largest axial component on the axis, it pushes the front of the packet backward at a faster rate near the axis (figure 13g), hence the formation of a low enstrophy bubble (figure 13i). The enhanced elongation is a combined effect of terms III and V.

The above gives a glimpse of how helical wave decomposition provides a new tool for analysis of core dynamics in terms of tractable interaction of polarized velocity and vorticity wave packets (see Melander & Hussain 1992 for details). Such physical insight into the core dynamics is impossible via the usual tools of vortex dynamics such as Biot-Savart law and local induction approximation. We hope many other important vortex phenomena (one is briefly discussed below) can be understood better by this approach. Vortex breakdown is another obvious candidate (subject of a separate study by us).

10. Interaction with Fine-scale Turbulence

In an attempt to understand a possible coupling between large-scale coherent structures and fine-scale turbulence, we have repeated the simulation reported earlier with the axisymmetric sinusoidal coherent vortex (called CV) immersed in a background of isotropic fine-scale incoherent turbulence. Care was taken to assure that the initial rms level was sufficiently high for the fine-scale turbulence to survive long enough to dynamically interact with CV. The fine-scale size was chosen to be the smallest permitted by the resolution, but still with a spectral gap between it and that of CV. We find that the scales in the background incoherent vorticity grow progressively, so that at the end of the simulation the largest incoherent scales are comparable to CV. The spectral gap is gradually filled up not by diffusive mechanisms, but by interaction between incoherent turbulence and CV. While the size of the incoherent turbulence remains uniform in space, the strength is higher at the boundary of CV where it is wrapped around CV and energized by stretching. This is not dissimilar from the growth of axisymmetric vortices on an impulsively rotated rod except that in our case the curved vortices around CV are intrinsically asymmetric, induced by the shear in the boundary of CV.

The spiral structures formed from organization of the incoherent turbulence are indeed quite different from the spiral sheet-like structures proposed by Lundgren (1982). The structures we find are rod-like spiral patterns wrapped around CV with senses aligned or opposed to the swirl of CV (figure 15). The helical wave decomposition is also helpful in understanding the coupling between CV and incoherent turbulence. The spiral structures are found to have a tendency to be highly polarized: they are either predominantly right-handed or predominantly left-handed. We indeed find a strong correspondence between peaks of incoherent turbulence in the boundary of CV and high values of the degree of polarization. The incoherent turbulence spirals undergo growth through merger by (truly inviscid) pairing mechanism and not by fusion through viscous decay and diffusion. In order for the spiral vortices to pair, their transport (axial to CV) is essential. There are two distinct mechanisms for such transport: i) transport of nearby opposite-signed vortices as dipoles, and ii) self-induced transport due to curvature of the azimuthally aligned vortices (not present in the 2D case). Because of the shear of CV, the small-scale vortex spiral undergoes differential stretching which provides the spiral an intrinsic tendency to become polarized.

In virtually all turbulence theory, local isotropy, which also forms the cornerstone of Kolmogroff's hypothesis, is universally assumed and implies decoupling of large and fine scales. By extrapolating our results (figure 15), we conjecture, for the case of very high *Re*, a hierarchy of structures of successively smaller scales, i.e. a fractal cascade as shown in figure 16. This supports our long-held doubt regarding the validity of the concept of local isotropy. Of course local isotropy does not require individual fine scales to be isotropic, but merely the lack of any statistical preference for the fine-scale structure's orientation. We feel that the fine scales retain some preferred orientation with respect to the coherent structures.

The CV-in-turbulence of course decays, and these unforced flows will eventually laminarize, in a timescale depending on both *Re* and the initial rms turbulence level. If the level is too low, CV recovers axisymmetry; if too high, CV is completely disrupted. For intermediate values interactions discussed above occur. However, in this case there is the interesting possibility of excitation of bending waves on CV. Without bending waves, as the spiral structures wrap around CV, they tend to be axisymmetric, thus diminishing their stretching by CV. When bending waves are excited, CV ceases to be axisymmetric and is then in a position to continue the stretching of the spiral incoherent turbulence. This is an example of feedback (or backscatter) from the smaller scales to CV. Thus we have here a mechanism of coherent structure/fine-scale interaction: the former organizes the latter by vortex stretching and the latter, if *Re* is high enough, induces bending waves in the former to generate the mechanism of its own survival.

This progress report is a part of our continuing research supported by the Air Force Office of Scientific Research.

References

[1] Brown, G.L. & Roshko, A., "On density effects an large structure in :turbulent mixing layers", J. Fluid Mech., 64, 1974, p. 775.

[2] Crow, S.C. & Champagne, F.H.,"Orderly structure in jet turbulence", J. Fluid Mech., 48, 1971, p. 547.

[3] Hussain, H.S., "An experimental investigation of unexcited and excited elliptic jets". Ph.D. thesis, Univ. of Houston, 1984.

[4] Hussain, F., "Coherent structures and studies of perturbed and unperturbed jets", in The role of coherent structures in modelling turbulence and mixing, Lect. Notes in Physics, Ed. J. Jimenez, 1980, p. 252.

[5] Hussain, F., "Coherent structures and turbulence", J. Fluid Mech., 173, 1986, p. 303.

[6] Hussain, F. and Husain, H. , "Elliptic jets. Part 1. Characteristics of unexcited and excited jets.",J. Fluid Mech., 208, 1989, p. 257.

[7] Kerr, R.M. and Hussain, F., "Simulation of vortex reconnection", Physica D, 37, 1989, p. 474.

[8] Kline, S.J., Reynolds, W.C., Schraub, F.A., & Runstadler, P.W., "The structure of turbulent boundary layers", J. Fluid Mech., 30, 1967, p. 741.

[9] Lesieur, M., Turbulence in Fluids, Kluwer Acad. Publ., 1990.

[10] Lundgren, T.S., "Strained spiral vortex model for turbulent fine structure", Phys. Fluids, 25, 1982, p. 2193.

[11] Meiron, D.I., Shelley, M.J., Ashurst, W.T. and Orszag, S.A., "Numerical study of vortex reconnection", in Mathematical Aspects of Vortex Dynamics, SIAM, Ed R. Caflisch, 1989, p.183.

[12] Melander, M. and Hussain, F., "Vortex core dynamics, helical waves and organization of fine-scale turbulence", (submitted) 1992.

[13] Moffatt, H.K., "The degree of knottedness of tangled vortex lines", J. Fluid Mech., 35, 1969, p. 117.

[14] Moore, D.W., & Saffman, P.G., "The motion of a vortex filament with axial flow", Phil. Trans. Roy. Soc. A, 272, 1972, p. 403.

[15] Moses, H.E., "Eigenfunction of the curl operator, rotationally invariant Helmholtz theorem and application to electromagnetic theory and fluid mechanics", SIAM J. Appl. Math., 21, 1971, p. 114.

[16] Shariff, K., Leonard, A., Zabusky, N., and Ferziger, J. "Acoustics and dynamics of coaxial interacting vortices", Fluid Dyn. Res. , 3, 1988, p. 337.

[17] Shelley, M.J., Meiron, D.I., and Orszag, S.A., private communication, 1989.
[18] Stanaway, S., Shariff, K. and Hussain, F., "Head-on collision of viscous vortex rings", NASA report CTR S-88, 1988, p. 287.
[19] Schatzle, P.R., "An experimental study of fusion of vortex rings", Ph.D. thesis, Cal. Inst. of Tech., 1987.
[20] Takaki, R. and Hussain, F., "Recombination of vortex filaments and its role in aerodynamic noise",Turb. Shear Flows V, Cornell U., 1985, p. 3.19.

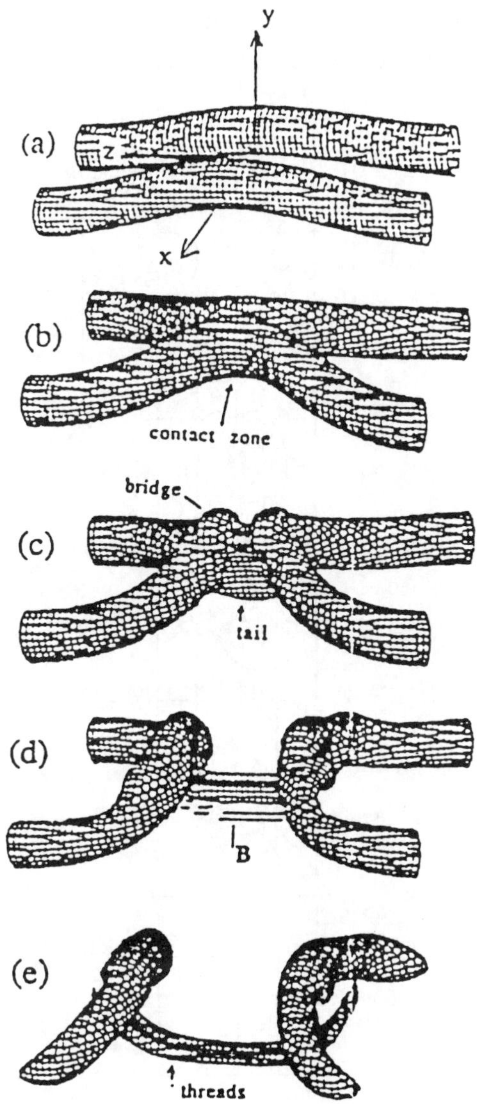

Fig.1 $|\omega|$-surfaces at 30% of the initial peak. (a) $t = 0$; (b) t=2.25; (c) $t = 3.5$; (d) $t = 4.75$; (e) $t = 6.0$.

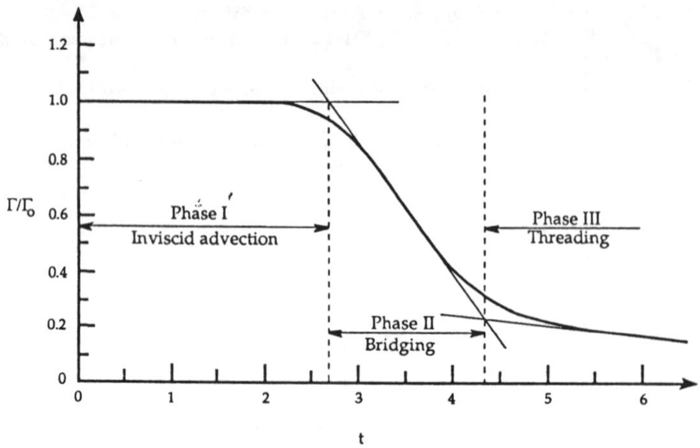

Fig. 2 Circulation in π_s^+ $(x > 0)$. Phases of reconnection.

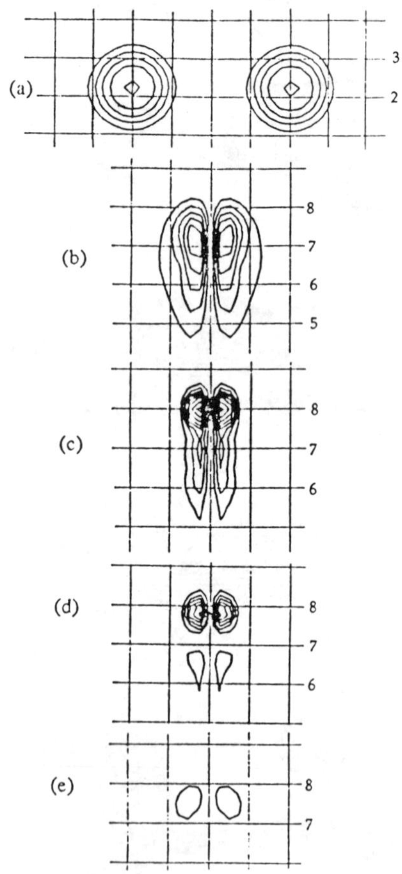

Fig. 3 $|\omega|$ contours in π_s; contour increments $\Delta\omega = 4$. (a) $t = 0$; (b)$t = 3.0$; (c) $t = 3.75$; (d) $t = 4.5$; (e) $t = 6.0$.

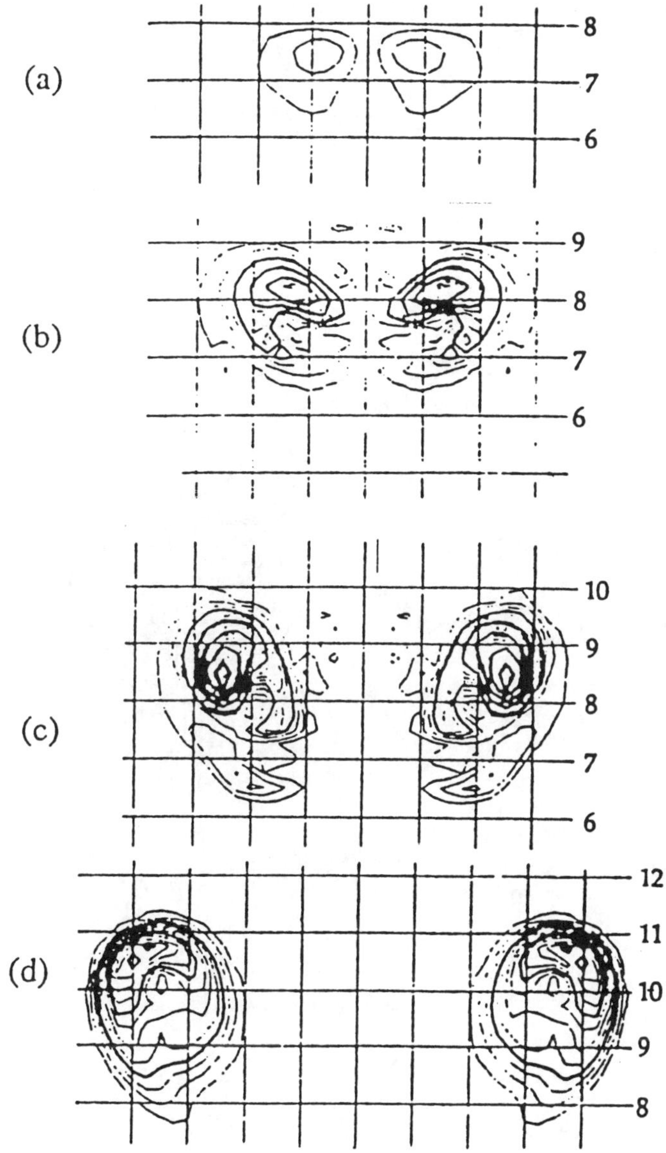

Fig. 4 $|\omega|$ contours in π_d; contour increments $\Delta\omega = 4$. (a) $t = 3.0$; (b) $t = 3.75$; (c) $t = 4.5$; (d) $t = 6.0$.

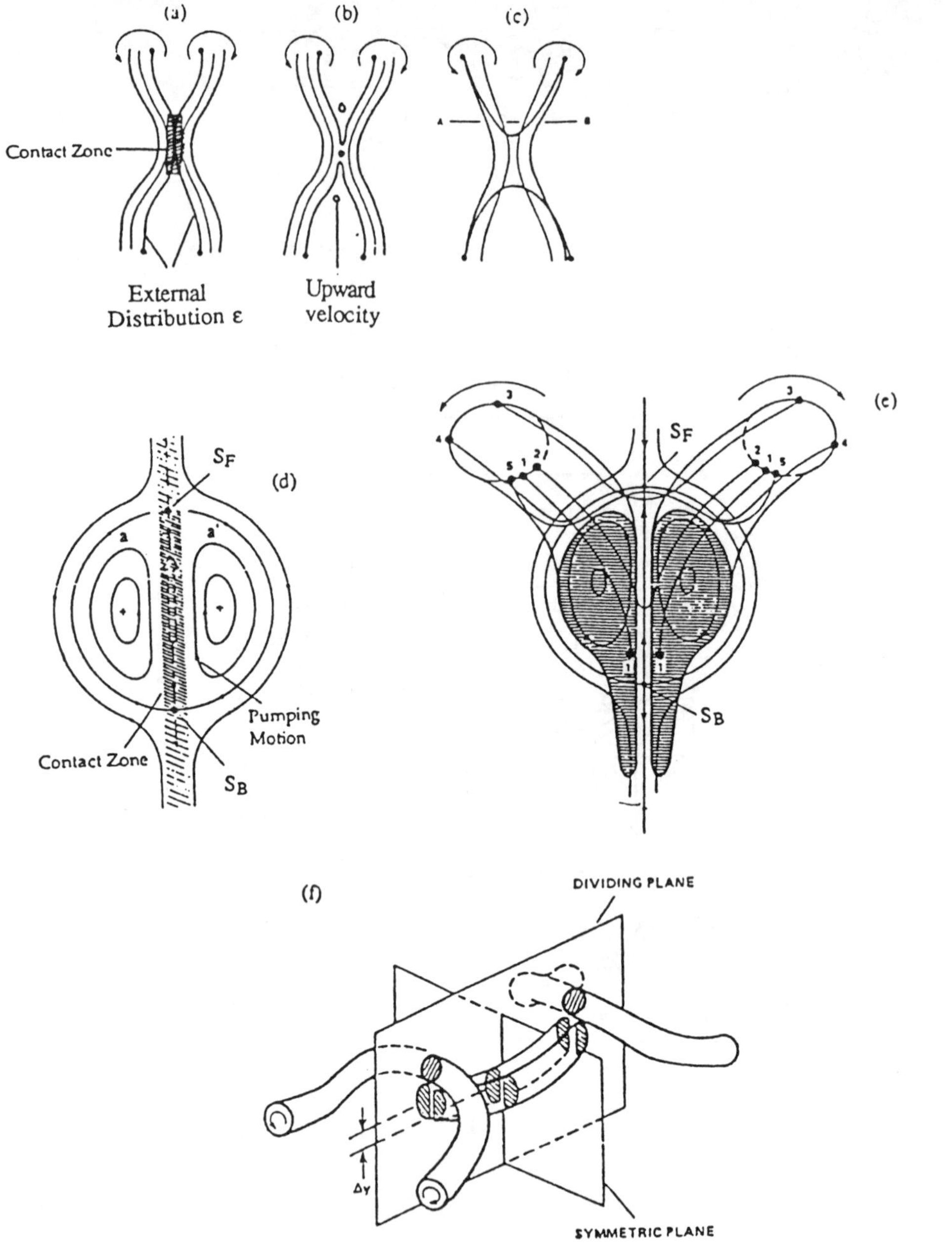

Fig. 5 (a-c) Explanation of bridging mechanism; (d) streamline pattern in π_s; (e) a schematic of the perspective view of one half of the two vortices. A vortex line is sketched at 5 different instants to illustrate the bridging mechanism; (h) downwash.

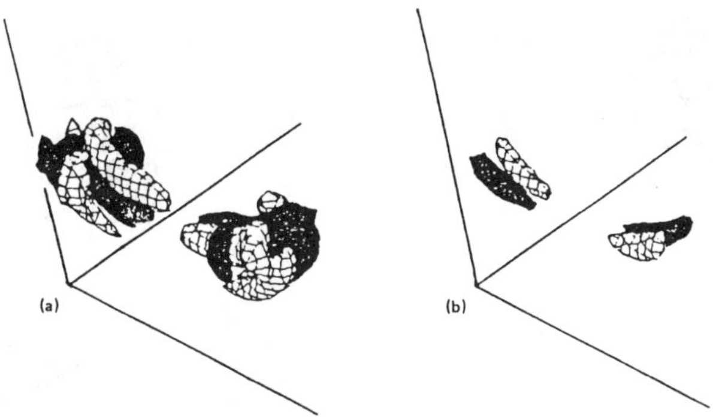

Fig. 6 (a) Helicity density h at t = 4.5, 20% of peak value; (b) h at 50% level. Black regions correspond to negative h, and the white regions to positive h.

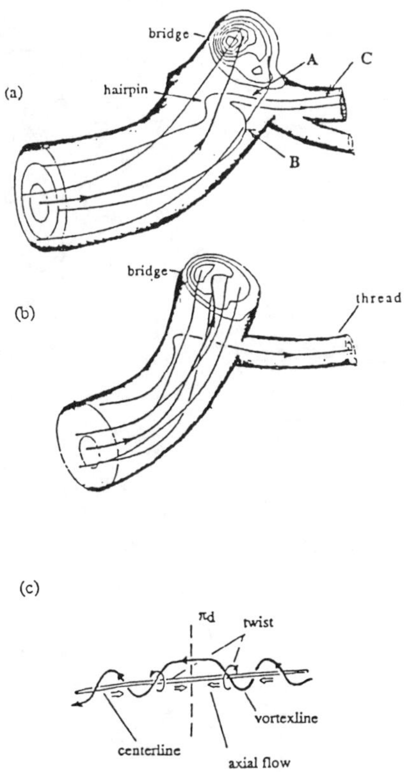

Fig. 7 Vortex lines in a quarter of the computational domain. Also shown are the plane cross sections of the vorticity norms in π_s, π_d and in the box side. A sketch of the vortex has been overlaid to orient the reader. (a) t = 4.75, (b) t = 6.0, (c) axial flow due to vortex line twist.

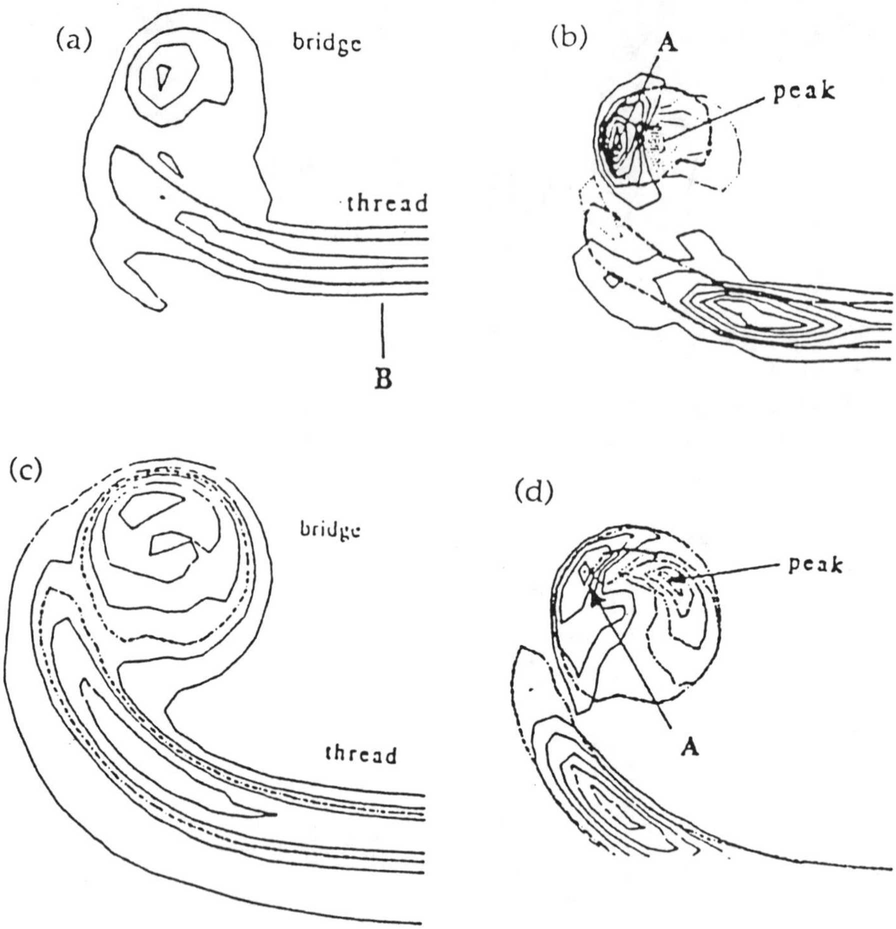

Fig. 8 Plane cross section parallel to π_d but moved $\Delta x = 0.196$. (a) $|\omega|$ at $t = 4.75$; (b) enstrophy production P_ω at $t = 4.75$, chaindotted lines are $|\omega|$ contours; (c) $|\omega|$ at $t = 6.0$; (d) P_ω at $t = 6.0$, chaindotted lines are $|\omega|$.

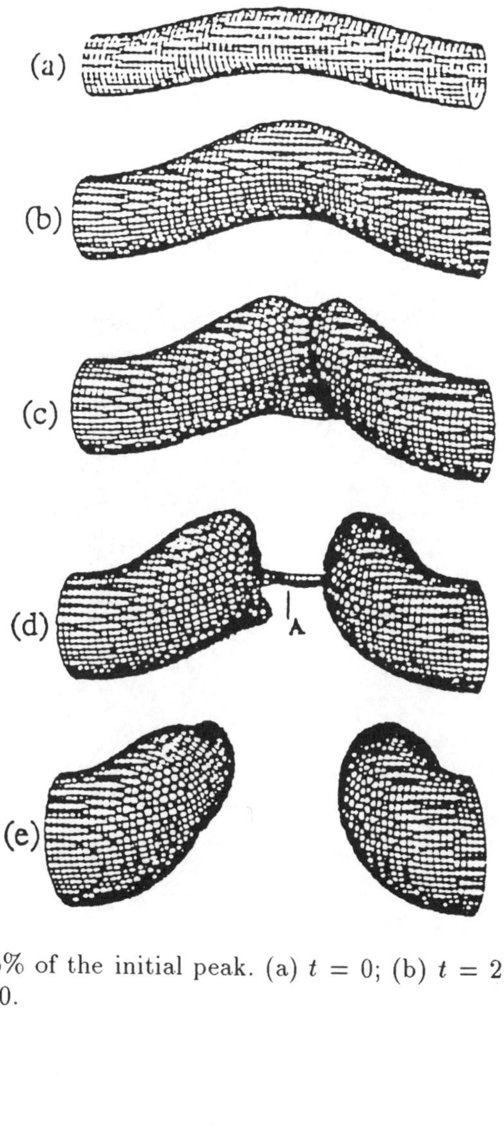

Fig. 9 Scalar at 5% of the initial peak. (a) $t = 0$; (b) $t = 2.25$; (c) $t = 3.5$; (d) $t = 4.75$; (e) $t = 6.0$.

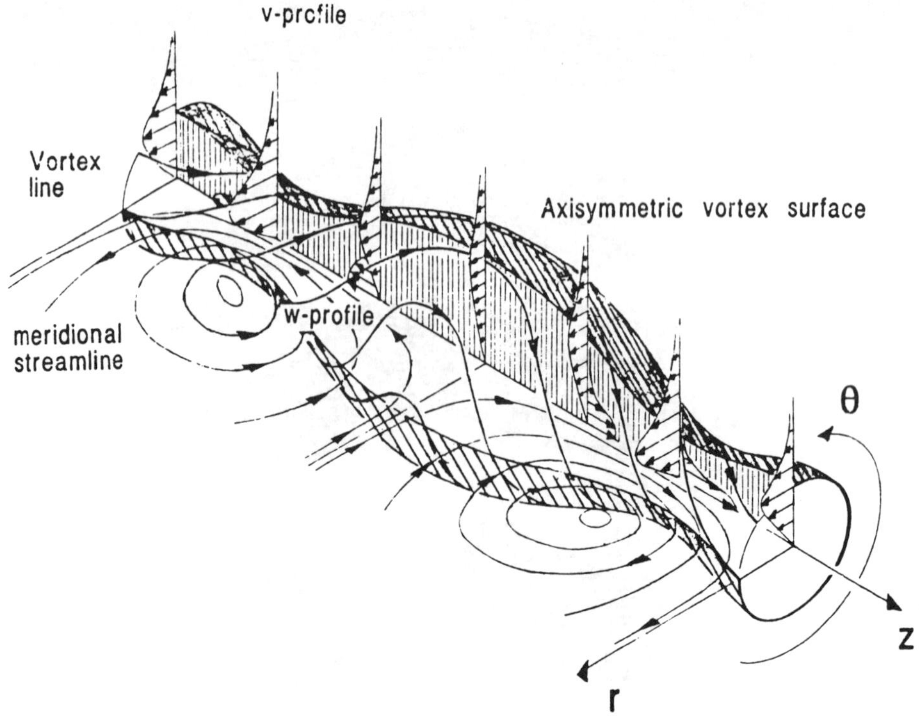

Fig. 10 Axisymmetric vortex surface with a 90° cutaway on which vortex lines lie. Also shown are streamlines in a meridional plane, and $v(r)$ and $w(r)$ profiles.

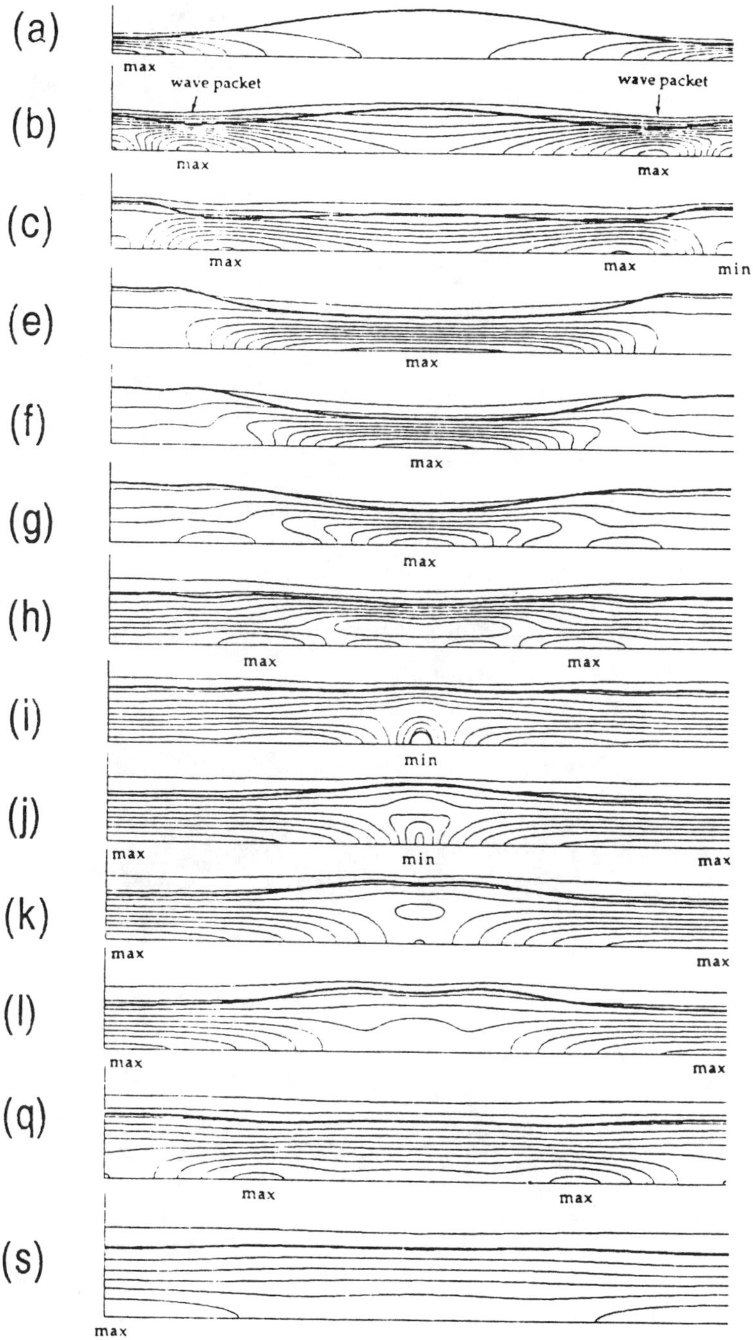

Fig. 11 Time evolution of $|\omega|$ contours. The thick line denotes the contour $N_k = 1$, where $N_k = [(\omega.\omega)/2S_{ij}S_{ij}]^{1/2}/2$ is the kinematic vorticity number.

(b)

(d)

(e)

(f)

(g)

(i)

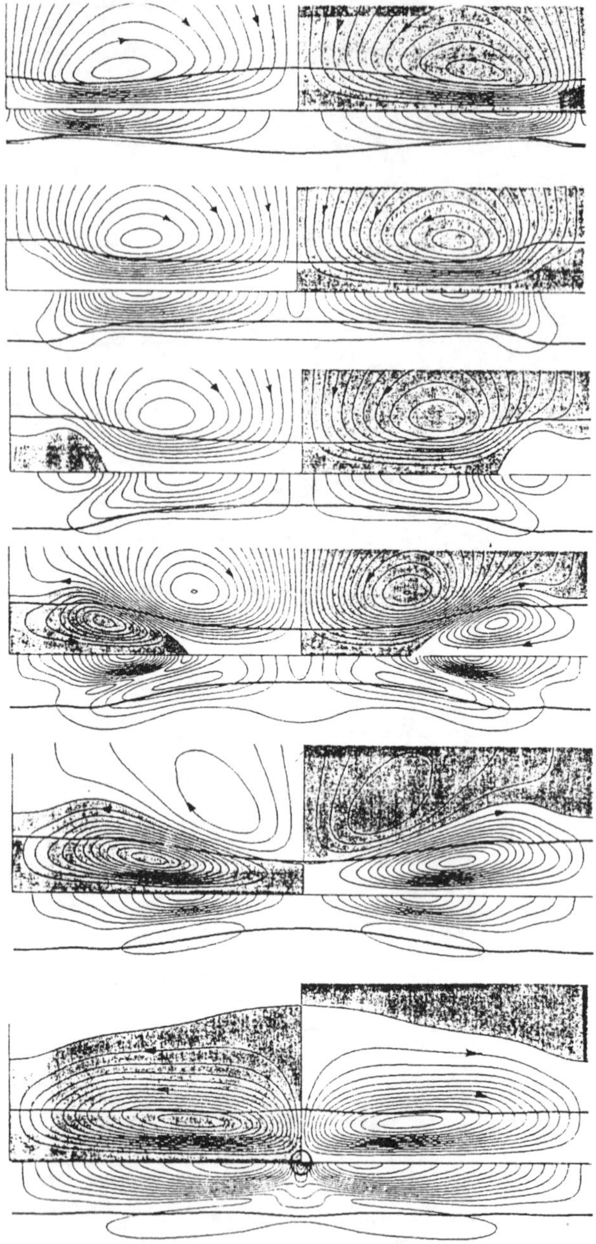

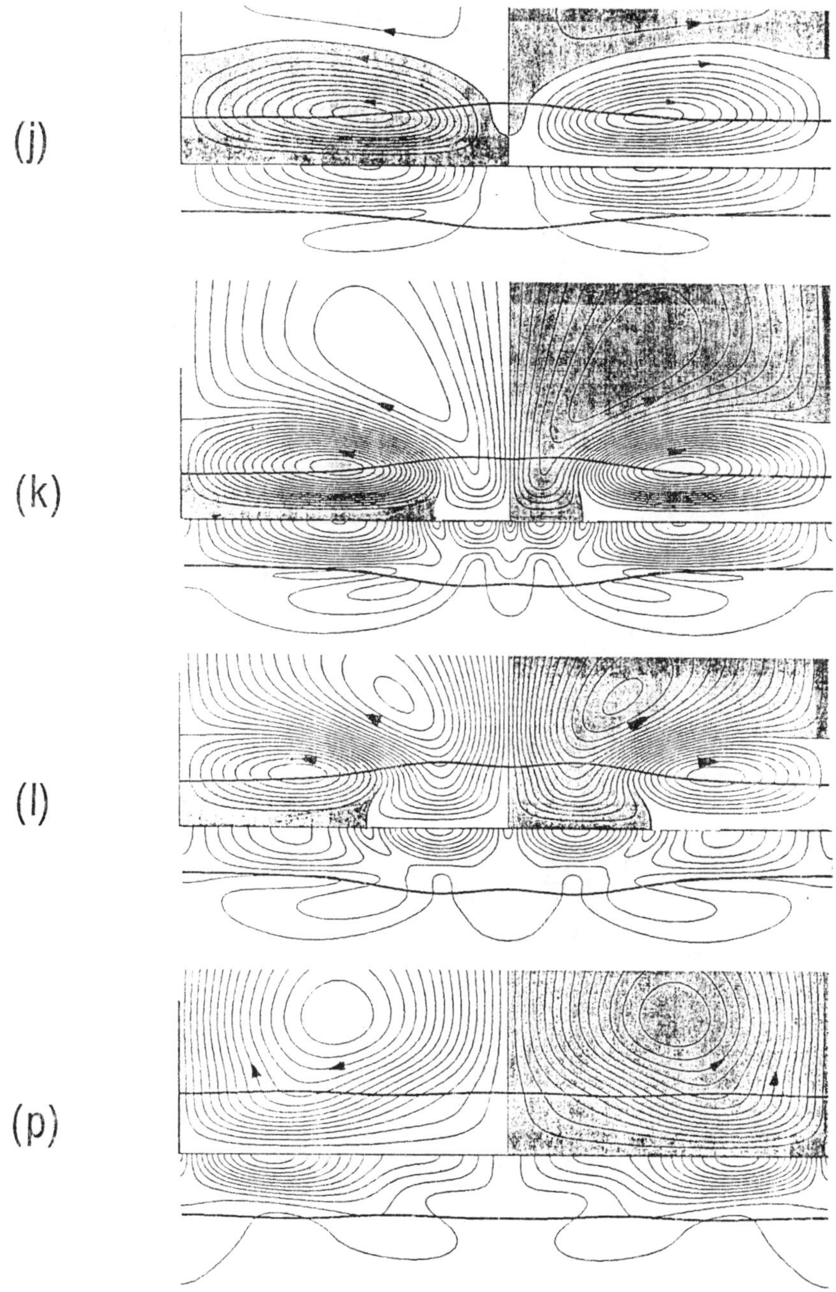

Fig. 12 Time evolution of the meridional streamfunction Ψ (above axis). Meridional plane velocity magnitude $\sqrt{u^2 + w^2}$ is shown below axis. The shaded regions denote negative Ψ. Also shown is the contour $N_k = 1$.

(a)
(b)
(c)
(e)
(g)
(i)
(l)
(q)
(s)

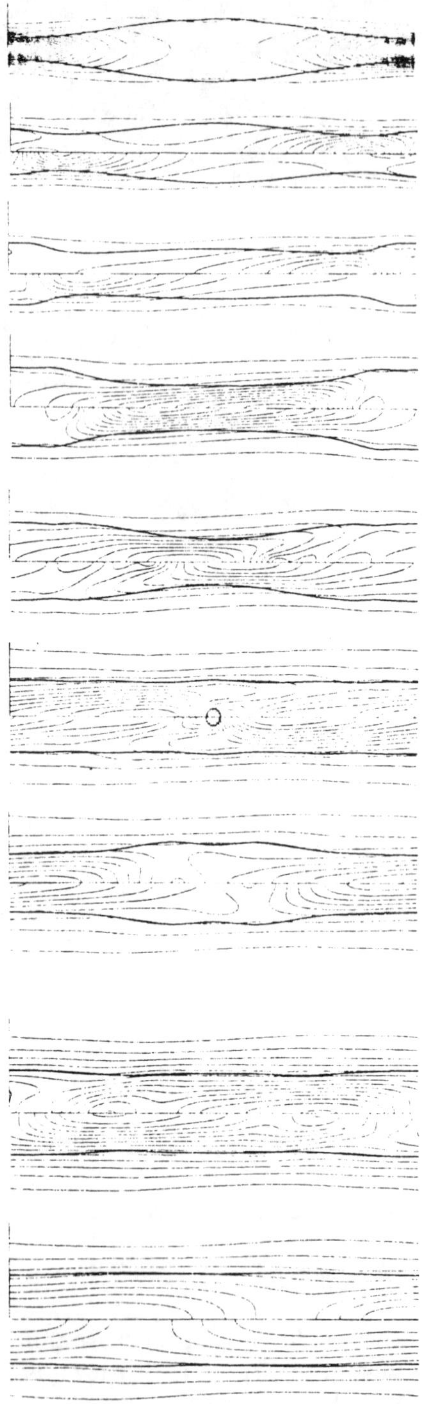

Fig. 13 Evolution of $|\omega_R|$ (above) and $|\omega_L|$ (below) the axis, with the $N_k = 1$ contour overlaid.

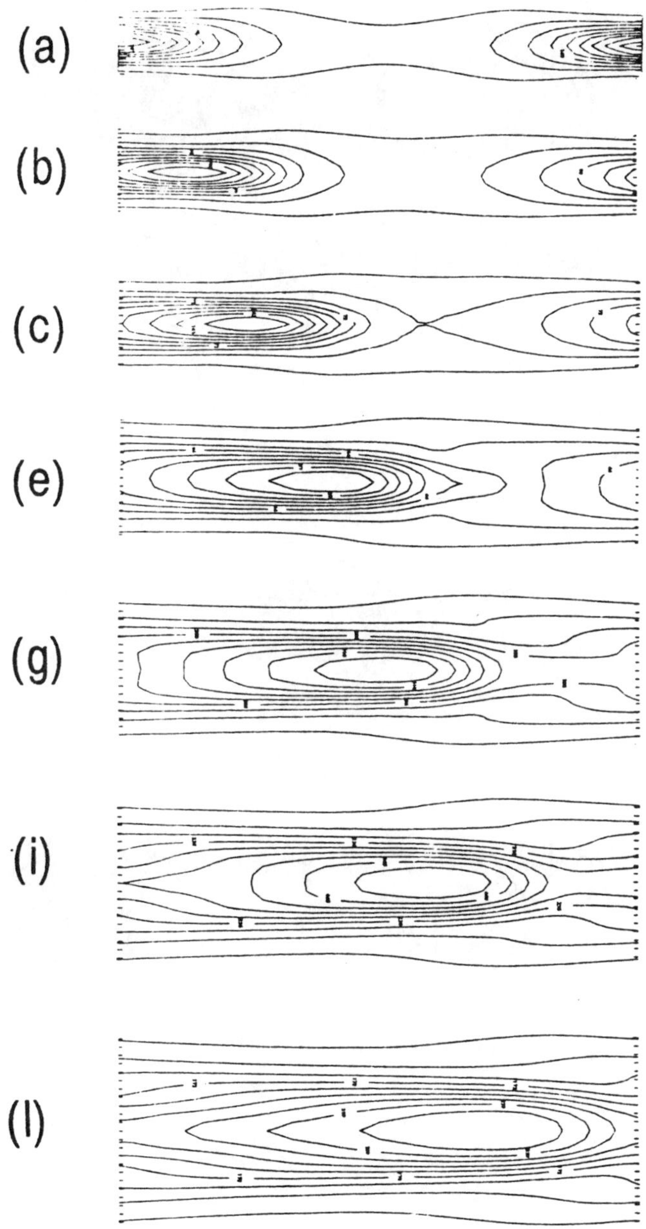

Fig. 14 Evolution of $|\omega_R|$ for an initially right-handed vortex; i.e. the initial condition for this simulation is $\omega = \omega_R$.

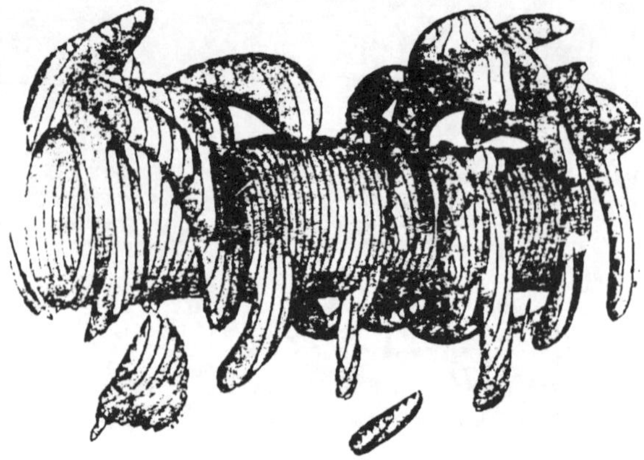

Fig. 15 $|\omega|$ surface at 10% of the peak value.

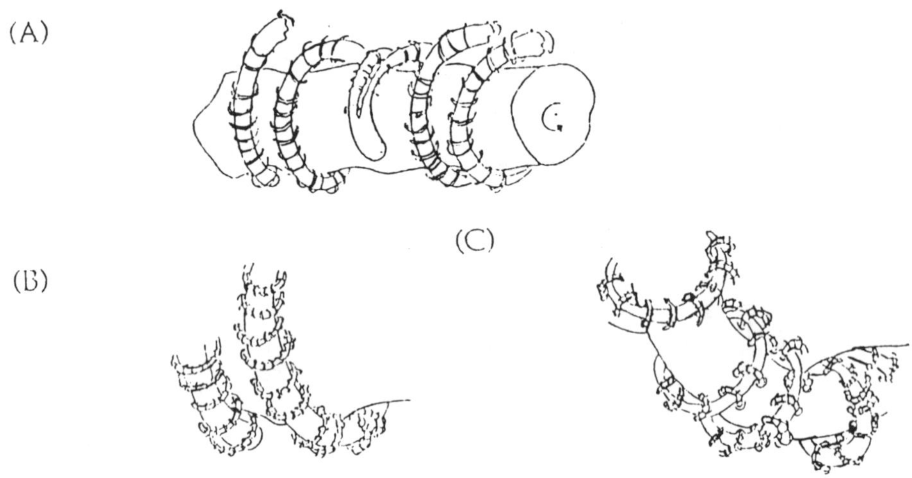

Fig. 16 Schematic of the conjectured fractal vortex with successive magnification.

THE NONLINEAR DYNAMICS OF A JET SHEAR LAYER WITH SWIRL

E. MEIBURG and J.E. MARTIN
Department of Aerospace Engineering
University of Southern California
Los Angeles
CA 90089 USA

ABSTRACT.
The present computational study concerns the three-dimensional evolution of a nominally axisymmetric jet with swirl subject to axisymmetric and azimuthal perturbations. We employ an inviscid vortex filament technique to gain insight into the vorticity dynamics of the flow. As a first step, we consider the effect of streamwise vorticity in the jet shear layer only. For the case of an axisymmetric perturbation, we observe the formation of weakly swirling vortex rings connected by braids in which the effect of the swirl is relatively more important. The streamwise vorticity forming in the braid is of the same sign everywhere, however, due to the axisymmetry, concentrated structures do not form in the braid. By introducing an additional periodic perturbation in the azimuthal direction, the axisymmetry is broken and the emergence of concentrated streamwise braid vortices all of the same sign is observed. A Kelvin-Helmholtz-like instability of the curved braid vorticity layer plays a significant role in their generation.

1. Introduction

Nominally axisymmetric jets have long enjoyed the attention of researchers in the field of fluid mechanics. In part, this is due to their prominence in a variety of technical applications where they are employed, for example, in mixing and combustion processes as well as for propulsion purposes. In addition to their technological importance, however, axisymmetric jets also represent one of the few fundamental classes of free shear flows. Consequently, there has been a considerable effort to study their laminar evolution, the transition to turbulence, and the features of the ensuing fully turbulent flow, including the large scale coherent structures. While most investigations in the past have focused on the case of non-swirling jets (e.g. Batchelor and Gill 1962, Cohen and Wygnanski 1987a,b, Tso and Hussain 1989), swirl is an essential aspect of jet flows in many situations (Wygnanski 1970, Escudier 1987, Mehta, Wood and Clausen 1991). For example, in combustion applications swirl is frequently employed to achieve a large spreading angle of the jet, which leads to an upstream flow of the hot combustion products near the jet centerline and thereby serves to stabilize the flame (Edwards and Rudoff 1990, Edwards, Marx and Ashurst 1991). For the purpose of developing compact, efficient, and non-

J.T. Beale et al. (eds.), Vortex Flows and Related Numerical Methods, 265–278.

polluting combustors, the governing fluid mechanical and chemical processes need to be optimized, which in turn requires a detailed understanding of the dynamical evolution of the swirling flow and its effect on and interaction with the combustion process and the flame structure. The present study is intended as a first step towards an improved understanding and modeling of the effect of swirl in a nominally axisymmetric jet. In particular, we aim at using the insight we have gained from our previous studies of the three-dimensional nonlinear evolution of non-swirling jets (Martin and Meiburg 1991, 1992) towards elucidating the modifications introduced by the presence of swirl. For this purpose, we will in the present investigation only consider the effect of swirl in the jet shear layer. A logical extension of the present study will have to consider jets with swirl distributed across the core.

Section 2 will briefly outline the main features of our vortex filament technique, while sections 3 and 4 address the effects of axisymmetric and azimuthal perturbations on the swirling jet shear layer, respectively. Section 5 will give a short summary of our main findings.

2. Numerical Technique

Since a more detailed account of the numerical method can be found in Ashurst and Meiburg (1988), we limit ourselves here to a description of the main features of our purely Lagrangian vortex method. Using the theorems of Kelvin and Helmholtz for inviscid dynamics and following the general concepts reviewed by Leonard (1985), we employ vortex filaments to construct the vorticity field. While filaments in the form of vortex rings were appropriate to represent a non-swirling axisymmetric shear layer in our earlier studies, we employ helical vortex filaments for the axisymmetric shear layer with swirl. The position of each filament is given by a number of nodes along its centerline, through which a cubic spline is fitted to give it a smooth shape. The three-dimensional Biot-Savart integral is reduced to a line integral by assuming an invariant algebraic vorticity distribution around the filament centerline. The filament core radius is kept uniform along the arclength but changes with time to maintain a constant filament volume.

For the numerical simulation, we limit ourselves to the temporally growing problem, i.e., our flow is periodic in the streamwise direction. Consequently, the Biot-Savart integration has to be carried out over the periodic images of the vortex filaments as well. Based on a series of test calculations, we set the number of periodic images carried along in the streamwise direction equal to three. We take the velocity difference between the centerline and infinity as our characteristic velocity. The thickness of the axisymmetric shear layer, defined as the velocity jump divided by the maximum slope of the velocity, serves as the characteristic length scale, which results in the filament core radius of 0.5. In these units, the radius R of the jet considered here is 5. Hence, the important ratio of jet radius to momentum thickness of the jet shear layer is 22.6 in the calculations to be reported here.

In order to evaluate the Biot-Savart integral with second order spatial accuracy, we represent the filament centerline by a cubic spline using the straight line distance between the node points as the spline parameter. This allows us to determine the tangent vector at each node point. In addition, the use of the spline formulation enables us to obtain a better representation of the locally induced velocity by subdividing filament sections close to a node in the process of calculating the self-induced velocity. Furthermore, the spline representation permits the easy introduction of additional nodes during the calculation in order to maintain a satisfactory resolution even when considerable vortex bending and stretching occurs. The filaments are repeatedly remeshed so that the arc length between successive nodes is always less than one but more than one half of the initial core radius. In addition, the local curvature is determined and nodes are spaced in such a way that, within the above bounds, the product of the arc length between the nodes and the local curvature does not exceed $\pi/8$. These criteria were established by means of test calculations in order to provide an improved resolution of highly curved regions of the filaments without wasting numerical effort on the representation of scales too small to be accurately represented by our filament model.

Both the spatial and the temporal integration are carried out with second order accuracy. The time step is continuously adjusted as local acceleration effects increase. In the course of the calculation to be discussed below, the time step is reduced by approximately one order of magnitude. In the following, we will discuss the nonlinear evolution of the swirling jet shear layer and the resulting large-scale structures for a variety of initial perturbations.

3. The Effect of an Axisymmetric Perturbation

By adding an axisymmetric distribution of streamwise circulation strength γ_x to the unperturbed non-swirling jet shear layer, a circumferential velocity field $v_\theta(r)$ is generated. While v_θ vanishes inside the jet shear layer, i.e. near the jet axis, the v_θ-field outside of the jet is identical to the one we would obtain if all of the streamwise circulation were concentrated on the jet axis. Consequently, if

$$\Gamma_x = 2\pi R \gamma_x$$

we have

$$v_\theta = 0 \quad , \quad r < R$$

$$v_\theta = \frac{\Gamma_x}{2\pi r} = \frac{R}{r}\gamma_x \quad , \quad r > R$$

The jet shear layer itself has a circumferential velocity equal to the average of the inside and outside fluid

$$v_\theta(r = R) = \frac{1}{2}(v_{\theta_-} + v_{\theta_+})$$

It is well known that non-swirling axisymmetric jet shear layers can be unstable with respect to axisymmetric waves. In order to determine how the presence of swirl modifies the stability characteristics of the jet shear layer, the transport equation for the circumferential vorticity in axisymmetric flow with non-vanishing circumferential velocity

$$\frac{\partial \omega_\theta}{\partial t} + v_r\frac{\partial \omega_\theta}{\partial r} + v_\theta\frac{\omega_r}{r} + v_x\frac{\partial \omega_\theta}{\partial x} = \omega_r\frac{\partial v_\theta}{\partial r} + \frac{\omega_\theta v_r}{r} + \omega_x\frac{\partial v_\theta}{\partial x}$$

provides useful information. In particular, it shows that the circumferential velocity component v_θ leads to vortex line deformation by tilting streamwise and radial vorticity into the circumferential direction. By expanding

$$\omega_r = -\frac{\partial v_\theta}{\partial x}$$

$$\omega_x = \frac{1}{r}\frac{\partial}{\partial r}(r v_\theta)$$

we realize that there is a partial canceling of the vortex deformation terms, but the final form

$$\frac{\partial \omega_\theta}{\partial t} + v_r\frac{\partial \omega_\theta}{\partial r} + v_x\frac{\partial \omega_\theta}{\partial x} = \frac{\omega_\theta v_r}{r} + 2\frac{v_\theta}{r}\frac{\partial v_\theta}{\partial x}$$

indicates that the swirl velocity component v_θ does affect the evolution of the circumferential vorticity distribution and thereby modifies the stability characteristics of the non-swirling jet shear layer with respect to axisymmetric waves. Experiments involving non-swirling jets (e.g. Lasheras, Lecuona and Rodriguez 1991) often employ acoustic forcing in order to trigger the axisymmetric instability mode and to study the nonlinear evolution of the jet. In our numerical simulation, we duplicate the effect of such forcing by perturbing the shear layer with an axisymmetric wave that displaces the vortex filaments in the radial direction. Figure 1 shows a side view as well as a streamwise view of the vortex filaments at time $t = 0.325$, i.e. shortly after the start of the calculation. For clarity, the side view displays only those sections of the filaments located at $z > 0$. We recognize the initially helical shape of the individual vortex filament centerlines as well as the displacement by the axisymmetric perturbation wave. The amount of swirl in the flow is linked to the initial streamwise component of the vorticity. For our particular flowfield, in which both streamwise and circumferential vorticity are limited to the jet shear layer, the strength of the swirl can be quantified by means of the angle α that the helical vortex lines initially form with the circumferential direction (figure 2). A suitable swirl number S is hence given by $\tan(\alpha)$, which represents the ratio of streamwise circulation per unit length in the circumferential direction, γ_x, to circumferential circulation per unit length in the streamwise direction, γ_θ. In other words, $S = \tan(\alpha)$ gives the ratio of the circumferential velocity jump Δv_θ to the streamwise velocity jump Δv_x across the jet shear layer

$$S = \tan(\alpha) = \frac{\gamma_x}{\gamma_\theta} = \frac{\Delta v_\theta}{\Delta v_x}$$

The present simulation employs $S = 0.2$. The wavelength λ of the axisymmetric perturbation wave has been set to 2π, which is close to the value at which the non-swirling jet shear layer instability experiences maximum growth.

Figure 3a, which presents a side view of the filament centerlines at $t = 3.281$, shows the subsequent evolution of the swirling jet shear layer. We observe that, as in the non-swirling case, the axisymmetric perturbation leads to a concentration of the circumferential vorticity into a periodic array of emerging vortex rings, whereas the resulting strain field depletes the braid regions in between of circumferential vorticity. As far as the evolution of the streamwise vorticity component is

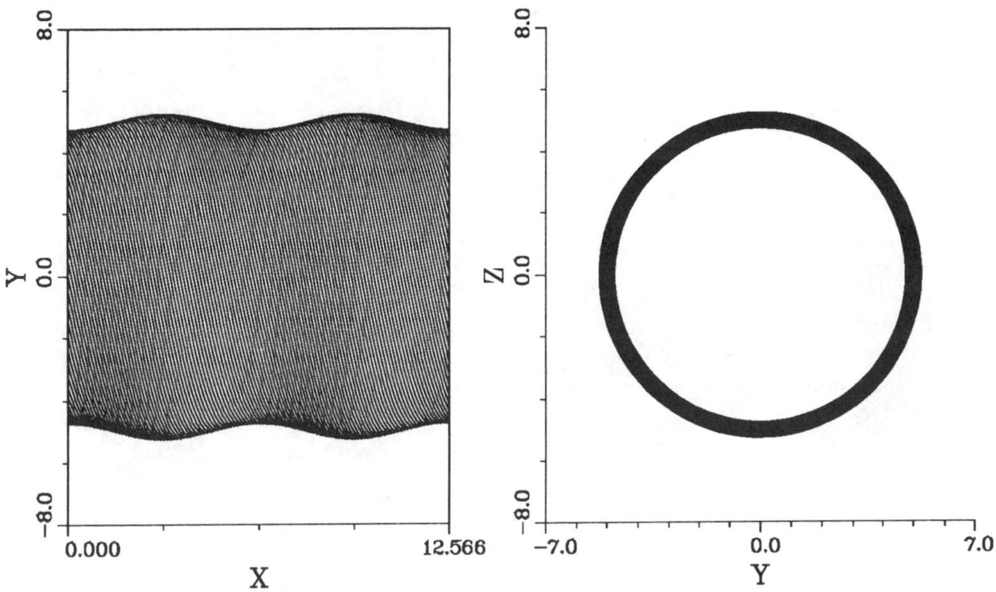

Fig. 1. Side view (a) and streamwise view (b) at time $t = 0.3125$ of the vortex filament centerlines in the swirling jet shear layer perturbed by an axisymmetric wave. The swirl number S has a value of 0.2.

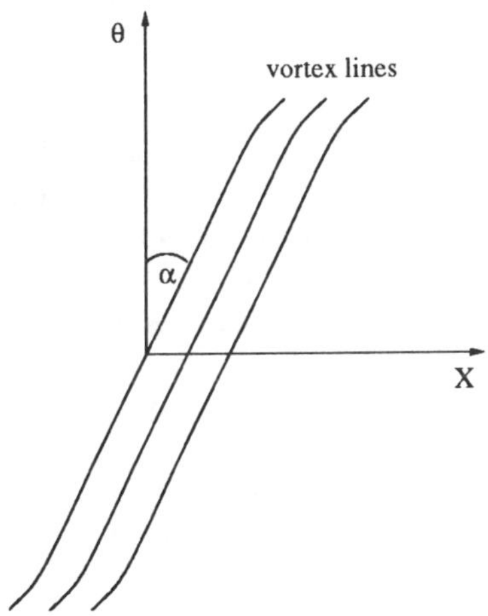

Fig. 2. For our particular flow field, a swirl number can be defined by means of the angle between the vortex line and the circumferential direction: $S = \tan(\alpha)$

concerned, it is important to realize that the imposed axisymmetric perturbation preserves the axisymmetric nature of the initial vorticity field, so that at every x-location the streamwise circulation is evenly distributed around the circumference,

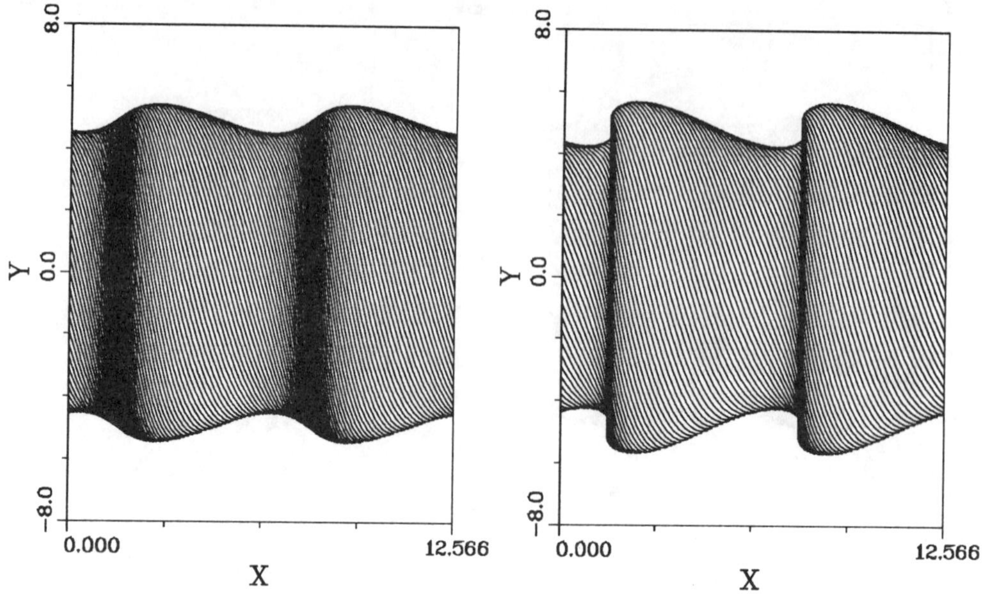

Fig. 3. Side view for the axisymmetrically perturbed swirling jet shear layer at $t = 3.281$ (a) and $t = 4.844$ (b). Notice the roll-up into axisymmetric vortex rings. In the ring region, vortex filaments become more aligned with the circumferential direction, while the braid filaments experience a reorientation into the streamwise direction. Consequently, the swirling component of the motion becomes relatively more important in the braid than in the ring.

and concentrated streamwise braid vortices cannot form. At the same time, the overall streamwise circulation of the jet has to be the same at every x-location, so that the streamwise circulation per unit length in the circumferential direction, γ_x, varies only as a result of a change in the local radius of the jet shear layer. The main effect of the axisymmetric perturbation consequently is to concentrate the circumferential vorticity at periodic streamwise locations, thereby leading to a locally higher γ_θ and a stronger streamwise velocity jump Δv_x at these locations, while reducing γ_θ, i.e. Δv_x, in the braid regions. γ_x and Δv_θ, on the other hand, are only indirectly affected by the axisymmetric perturbation through the change in the local radius of the jet column. This situation is reflected in the continuous deformation of the vortex filaments. The extensional strain field set up in the braid region by the emerging vortex rings leads to a reorientation of the braid filament sections into the streamwise direction, while those sections of the filaments located in the ring experience a reduction of their streamwise component. We thus observe that the local swirl number $S(x) = \tan[\alpha(x)]$ reaches a maximum in the braid region, while it becomes minimal in the ring region. Consequently, we find that even though the amount of streamwise circulation is independent of the x-location, the redistribution of the circumferential circulation caused by the axisymmetric perturbation renders the effect of the swirl proportionately more important in the braid

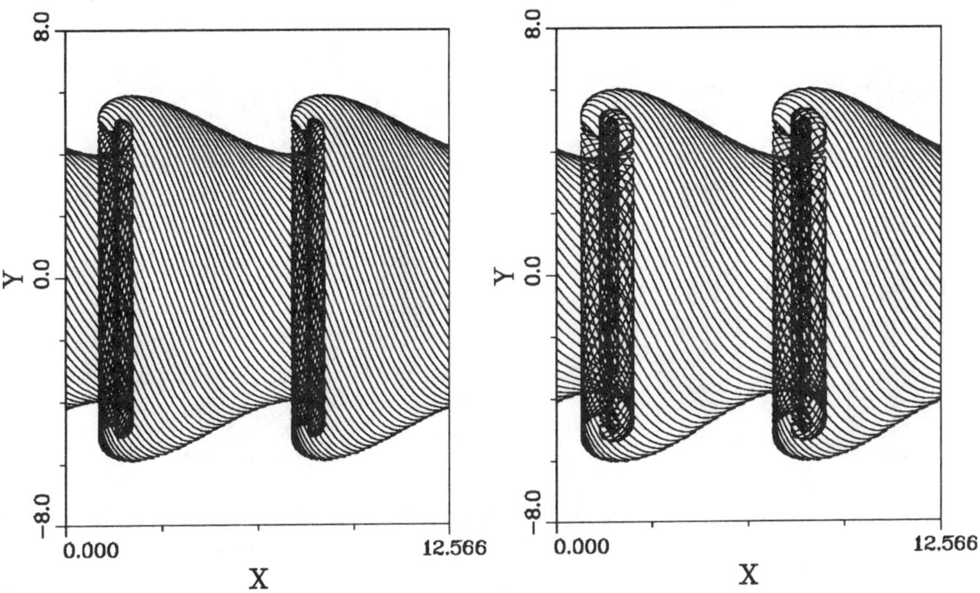

Fig. 4. Side view of the swirling jet shear layer at times 6.406 (a) and 7.969 (b). While the ring entrains progressively more circumferential vorticity, the vortex lines in the braid region become increasingly aligned with the streamwise direction.

region.

At time $t = 4.844$ (figure 3b), we recognize that, as part of the process of forming a vortex ring, the jet shear layer begins to fold over, as it would in the non-swirling case. This mechanism leads to the emergence of a region of the jet shear layer in which the streamwise vorticity component has reversed its direction. As the jet shear layer keeps winding around the growing vortex ring (figure 4), we thus observe the emergence of alternate layers with opposite signs of streamwise vorticity. However, integrating the streamwise vorticity component over the jet's cross-section at any x-location would still show the overall streamwise circulation to be independent of x.

At the final time $t = 11.094$ of our calculation (figure 5), we observe that very little circumferential circulation is left in the braid region, thereby rendering the vortex lines in this region approximately aligned in the streamwise direction.

4. The Effect of an Additional Azimuthal Perturbation

In our earlier studies of non-swirling jets, we investigated the effect of an additional azimuthally periodic perturbation on the three-dimensional evolution of the flow field. For nominally axisymmetric jets dominated by large-scale vortex rings (Martin and Meiburg 1991), we found that azimuthal perturbation waves led to the formation of counterrotating regions of streamwise vorticity which subsequently collapsed into concentrated braid vortices in a way similar to the mechanism described

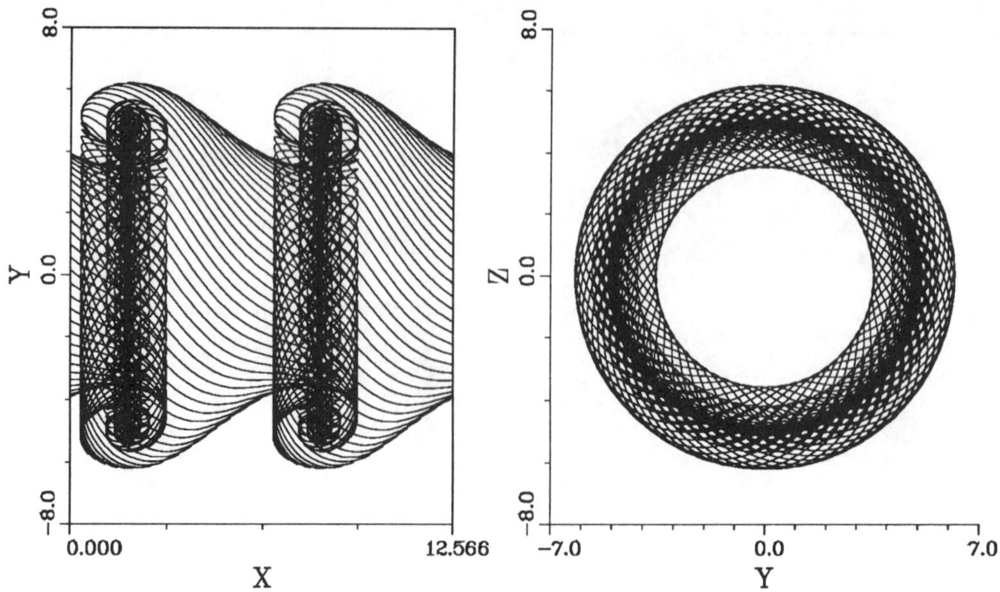

Fig. 5. Side view and streamwise view of the swirling jet shear layer at the final simulation time $t = 11.094$.

by Lin and Corcos (1984) for the plane mixing layer. Jets dominated by large-scale helical structures, on the other hand, tended to develop streamwise braid vortices all of the same sign (Martin and Meiburg 1992). In addition, depending on the azimuthal wavenumber, the large-scale vortex rings and helices can undergo instabilities of their own. In the following, we aim at exploring the effect of azimuthal waves on the swirling jet shear layer.

Depending on how we displace the vortex filaments, it is possible to introduce azimuthal perturbation waves in the form of radial or streamwise perturbation vorticity, respectively, cf. also the experimental forcing of the jet applied by Lasheras, Lecuona and Rodriguez (1991) and the corresponding plane shear layer and plane wake flows (Lasheras and Choi 1988, Meiburg and Lasheras 1988, Lasheras and Meiburg 1990). Some basic insight into the different effects of these two possible forms of azimuthal waves can be obtained from the governing vorticity transport equations. For the non-swirling jet, the streamwise vorticity equation in the absence of a streamwise perturbation, i.e. with $\partial/\partial x = 0$, and for an azimuthal perturbation of radial vorticity only, reads

$$\frac{\partial \omega_x}{\partial t} = \omega_r \frac{\partial v_x}{\partial r} + \frac{\omega_\theta}{r}\frac{\partial v_x}{\partial \theta}$$

With

$$\omega_r = \frac{1}{r}\frac{\partial v_x}{\partial \theta}$$

$$\omega_\theta = -\frac{\partial v_x}{\partial r}$$

we obtain

$$\frac{\partial \omega_x}{\partial t} = 0$$

In other words, an azimuthal wave introducing only radial perturbation vorticity cannot by itself lead to the formation of streamwise vorticity. At each location, the tendency of the radial perturbation vorticity to be tilted into the down(up)stream direction is exactly compensated by the tendency of the circumferential vorticity to be tilted into the up(down)stream direction. It follows immediately that

$$\frac{\partial \omega_r}{\partial t} = 0$$

as well, so that the non-swirling axisymmetric jet shear layer with $\partial/\partial x = 0$ and an azimuthal perturbation that introduces radial vorticity only represents a solution to the steady Euler equations. An additional streamwise perturbation or a streamwise gradient in the base flow is necessary so that the radial perturbation vorticity can lead to the generation of streamwise vorticity as well. One hence has to be careful about assuming in an overly simplistic way that the displacement of vortex lines into the slower outer or the faster inner flow automatically leads to the formation of streamwise vorticity.

If, on the other hand, we perturb the non-swirling jet by an azimuthal wave that introduces streamwise vorticity only, we obtain for the radial vorticity equation with $\partial/\partial x = 0$

$$\frac{\partial \omega_r}{\partial t} - \frac{v_\theta \omega_\theta}{r} = \omega_\theta \left(\frac{1}{r} \frac{\partial v_r}{\partial \theta} - \frac{v_\theta}{r} \right)$$

With

$$\omega_\theta = -\frac{\partial v_x}{\partial r}$$

we obtain

$$\frac{\partial \omega_r}{\partial t} = -\frac{1}{r} \frac{\partial v_x}{\partial r} \frac{\partial v_r}{\partial \theta}$$

It is interesting to note that it is not the streamwise perturbation vorticity that is tilted into the radial direction, but rather the circumferential vorticity. We hence notice that for a non-swirling jet an azimuthal perturbation wave of streamwise vorticity immediately leads to the generation of radial perturbation vorticity as well, whereas an azimuthal wave involving radial vorticity just changes the base state. We will now explore the corresponding situation for the swirling jet shear layer.

For the swirling jet shear layer with $\partial/\partial x = 0$, introducing radial perturbation vorticity by means of an azimuthal wave leads to the following equation for the streamwise vorticity

$$\frac{\partial \omega_x}{\partial t} + v_r \frac{\partial \omega_x}{\partial r} + \frac{v_\theta}{r} \frac{\partial \omega_x}{\partial \theta} = \omega_r \frac{\partial v_x}{\partial r} + \frac{\omega_\theta}{r} \frac{\partial v_x}{\partial \theta}$$

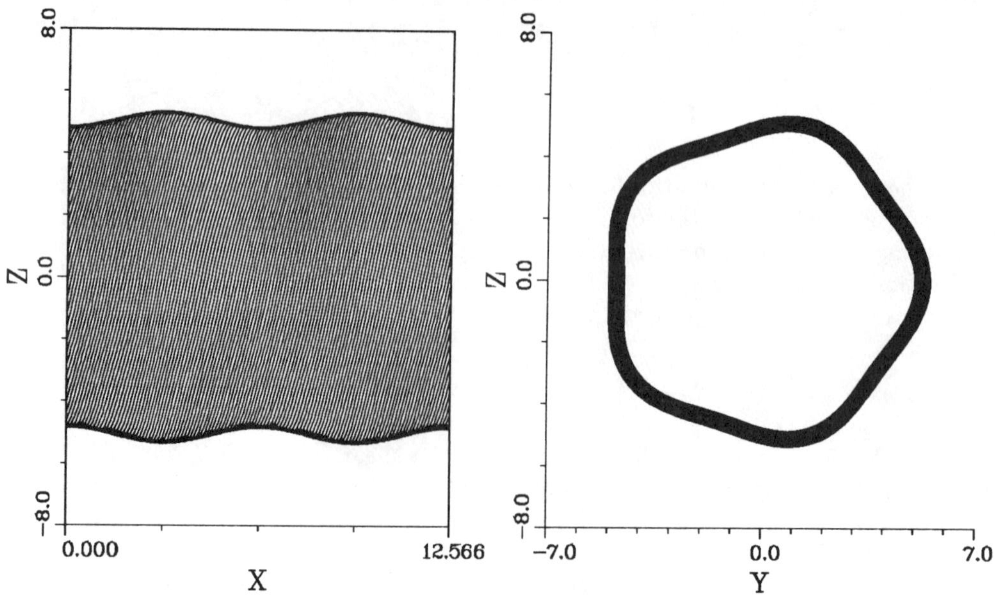

Fig. 6. Side view (a) and streamwise view (b) at $t = 0.3125$ of the swirling jet shear layer perturbed both by an axisymmetric and an azimuthal wave. Both waves displace the vortex filaments in the radial direction.

As for the above non-swirling case, the vortex line deformation terms on the right hand side cancel each other, but this is not the case for the convective terms on the left hand side. Consequently

$$\frac{\partial \omega_x}{\partial t} \neq 0$$

so that, in contrast to the non-swirling case, the introduction of radial perturbation vorticity immediately also leads to the formation of streamwise vorticity in the swirling jet shear layer.

With the above basic considerations as background, we can now discuss a simulation of a swirling jet shear layer perturbed by both an axisymmetric and an azimuthal wave. Each of these two waves displaces the vortex filaments in the radial direction. Figure 6 shows a side view along with a streamwise view shortly after the start of the calculation at time 0.3125. The azimuthal corrugation has an amplitude of 5 per cent of the jet radius. At $t = 4.688$ (figure 7a) we recognize the roll-up of the jet shear layer into vortex rings, in a fashion similar to the axisymmetric case discussed above, but at the same time a modulation in the circumferential direction is becoming visible. By $t = 7.813$ (figure 7b) this modulation has resulted in a slight waviness of the evolving vortex rings, and we furthermore notice the emergence of regions of more concentrated streamwise vorticity in the braids. This is reflected in the streamwise vorticity contours at $x = 5$ (figure 7c), which show the curved shear layer of streamwise vorticity in the braid developing a Kelvin-Helmholtz-like instability due to the presence of the azimuthal perturbation wave. By $t = 10.938$

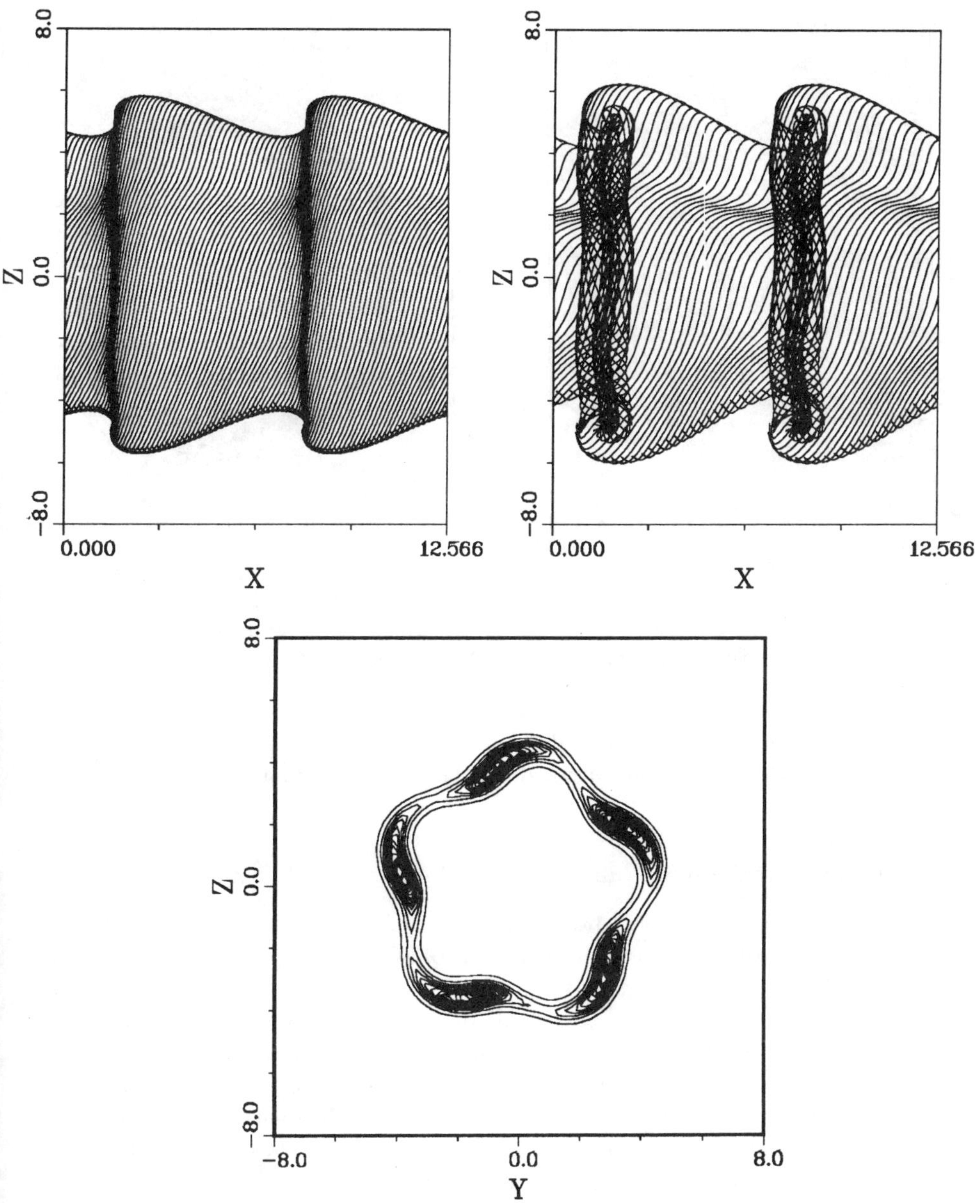

Fig. 7. Side views of the swirling jet shear layer at $t = 4.688$ (a) and $t = 7.813$ (b). By the later time, the azimuthal perturbation has led to a slight waviness along the core of the vortex rings as well as to the emergence of regions of more concentrated streamwise vorticity in the braid. This is reflected in the streamwise vorticity contour plot at $x = 5$ presented in figure 7c.

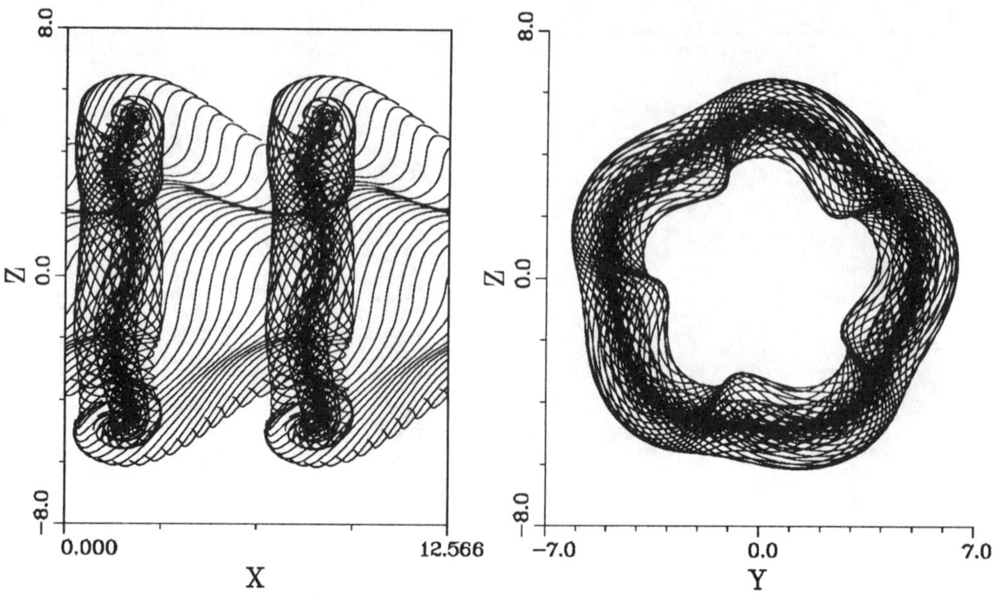

Fig. 8. Side view and streamwise view of the swirling jet shear layer at $t = 10.938$. Concentrated streamwise braid vortices all of the same sign have evolved.

(figure 8), the final time of our simulation, these streamwise vorticity regions have become considerably more concentrated, while the vortex ring waviness has grown as well.

5. Summary and Conclusions

Our aim has been to progress towards a more complete understanding and modeling of the effect of swirl in a three-dimensionally evolving jet. To this end we have extended our previous calculations of non-swirling jets by adding streamwise vorticity to the jet shear layer. These calculations can only represent a first step since the jet core does not contain swirl, but they nevertheless allow us to discuss some of the basic effects caused by the introduction of swirl. For the jet shear layer, the effect of swirl can easily be quantified by the angle formed by the vortex lines and the circumferential direction. This angle is directly related to the ratio of the streamwise and circumferential velocity jumps across the jet shear layer.

The first question concerns the growth of an axisymmetric perturbation wave on the jet shear layer. By inspecting the circumferential vorticity transport equation, it is immediately seen that the presence of swirl leads to a modification of the azimuthal vorticity's evolution. The nonlinear simulation shows the evolution of relatively weakly swirling axisymmetric vortex rings connected by braid regions in which the vortex lines have become nearly aligned with the streamwise direction, so that the swirl is relatively more important in this region. In this way, a curved mixing layer with a primary velocity component in the circumferential direction is formed in the braids. By introducing an additional azimuthal perturbation wave, we can trigger a Kelvin-Helmholtz-like instability of this curved mixing layer, which results in the formation of concentrated streamwise braid vortices as well as in a wavy modulation of the vortex rings. It is important to note that the streamwise vortices in the braid region are all of the same sign, thereby resembling more the situation of a non-swirling jet perturbed by a helical wave rather than an axisymmetric wave. Future work will have to address the quantitative development of the swirling jet in more detail, and it will have to extend the presence of the swirl to the entire jet core as well.

Acknowledgements

We gratefully acknowledge helpful discussions with W.T. Ashurst and C.F. Edwards. This work has been supported by the National Science Foundation under grant CTS-9058065 and by the Electric Power Research Institute under project RP8006. The San Diego Supercomputer Center has been providing computing time on its CRAY-Y/MP.

References

[1] ASHURST, W.T. and MEIBURG, E. 1988 Three-dimensional shear layers via vortex dynamics. *J. Fluid Mech.* **189**, 87.

[2] BATCHELOR, G.K. and GILL, A.E. 1962 Analysis of the stability of axisymmetric jets. *J. Fluid Mech.* **14**, 529.

[3] COHEN, J. and WYGNANSKI, I. 1987a The evolution of instabilities in the axisymmetric jet. Part 1. The linear growth of disturbances near the nozzle. *J. Fluid Mech.* **176**, 191.

[4] COHEN, J. and WYGNANSKI, I. 1987b The evolution of instabilities in the axisymmetric jet. Part 2. The flow resulting from the interaction between two waves. *J. Fluid Mech.* **176**, 221.

[5] EDWARDS, C.F. and RUDOFF, R.C. 1990 Structure of a swirl-stabilized spray flame by imaging, Laser Doppler velocimetry, and Phase Doppler Anemometry. Proceedings of the Twenty-Third International Symposium on Combustion.

[6] EDWARDS, C.F., MARX, K.D., and ASHURST, W.T. 1991 Instantaneous spatial structure of a confined swirling flow after sudden expansion. Preprint.

[7] ESCUDIER, M. 1987 Confined vortices in flow machinery. *Ann. Rev. Fluid Mech.* **19**, 27.

[8] LASHERAS, J.C. and CHOI, H. 1988 Three-dimensional instability of a plane, free shear layer: an experimental study of the formation and evolution of streamwise vortices. *J. Fluid Mech.* **189**, 53.

[9] LASHERAS, J.C., LECUONA, A., and RODRIGUEZ, P. 1991 Three-dimensional vorticity dynamics in the near field of coflowing forced jets. *Lectures in Applied Mathematics* **28**, 403.

[10] LASHERAS, J.C. and MEIBURG, E. 1990 Three-dimensional vorticity modes in the wake of a flat plate. *Phys. Fluids A* **2**, 371.

[11] LEONARD, A. 1985 Computing three-dimensional flows with vortex elements. *Ann. Rev. Fluid Mech.* **17**, 523.

[12] LIN, S.J. and CORCOS, G.M. 1984 The mixing layer: deterministic models of a turbulent flow. Part 3: The effect of plane strain on the dynamics of streamwise vortices. *J. Fluid Mech.* **141**, 139.

[13] MARTIN, J.E. and MEIBURG, E. 1991 Numerical investigation of three-dimensionally evolving jets subject to axisymmetric and azimuthal perturbation. *J. Fluid Mech.* **230**, 271.

[14] MARTIN, J.E. and MEIBURG, E. 1992 Numerical investigation of three-dimensionally evolving jets under helical perturbations. *J. Fluid Mech.* **243**, 457.

[15] MEHTA, R.D., WOOD, D.H., and CLAUSEN, P.D. 1991 Some effects of swirl on turbulent mixing layer development. *Phys. Fluids A* **3**, 2716.

[16] MEIBURG, E. and LASHERAS, J.C. 1988 Experimental and numerical investigation of the three-dimensional transition in plane wakes. *J. Fluid Mech.* **190**, 1.

[17] TSO, J. and HUSSAIN, F. 1989 Organized motions in a fully developed turbulent axisymmetric jet. *J. Fluid Mech.* **203**, 425.

[18] WYGNANSKI, I. 1970 Swirling axisymmetrical laminar jet. *Phys. Fluids* **13**, 2455.

DYNAMICS OF VORTEX TUBES IN THREE-DIMENSIONAL TURBULENCE

MAURICE MENEGUZZI and ALAIN VINCENT
CERFACS
42 avenue Gustave Coriolis
31057 Toulouse
FRANCE

ABSTRACT. Direct numerical resolution of the Navier-Stokes equation is used to obtain a three-dimensional homogeneous turbulent flow at Reynolds numbers $R_\lambda \approx 150$. An inertial subrange appears in the energy spectrum over more than one decade in wavenumbers. Visualization of the vorticity field suggests that the vorticity tubes which are known to be the main structures of the flow, are produced by shear instabilities. This is confirmed by a decay calculation, in which vorticity sheets appear at first, and then roll up to form the first tubes. We conjecture that these are the main mechanisms by which energy is transfered from large to small scales.

1. Statistically steady turbulence calculations

Direct simulations of homogeneous turbulent flow at the highest presently reachable resolution are useful to gather some information on the statistical properties of such a flow and to explore its structure in space. Indeed, these "numerical experiments" allow measurement of many quantities unaccessible in the laboratory and visualization of the small scales structures of the flow. It becomes feasible, on computers of the last generation, to reach Reynolds numbers at which a genuine inertial subrange shows up in the energy spectrum (Kerr 1985, Yamamoto and Hosokawa 1988, She et al. 1988, 1990). The main reason to do a calculation at higher resolution was to obtain a more extended inertial subrange than in previous work, and to study on the inertial domain properties.

We solve the Navier-Stokes equation for incompressible fluids in rotational form

$$\frac{\partial \mathbf{v}}{\partial t} = \mathbf{v} \times \omega - \nabla(p + v^2/2) + \nu \nabla^2 \mathbf{v} + \mathbf{f} \tag{1}$$

with the continuity equation $\nabla \cdot \mathbf{v} = 0$. $\mathbf{v}$ is the velocity field, $\omega = \nabla \times \mathbf{v}$ the vorticity, p the pressure, ν the kinematic viscosity and $\mathbf{f}$ a force field. Since we are interested in (statistically) homogeneous turbulent flows, we take periodic boundary conditions in all directions. In Fourier space the two equations can be combined

J.T. Beale et al. (eds.), Vortex Flows and Related Numerical Methods, 279–291.

to give

$$\frac{\partial \mathbf{v_k}}{\partial t} = \mathbf{P}(\mathbf{k}) \cdot (\mathbf{v} \times \omega)_{\mathbf{k}} - \nu k^2 \mathbf{v_k} + \mathbf{f_k} \tag{2}$$

where the tensor $\mathbf{P}$ is the projector on the space of solenoidal fields, defined as $P_{ij}(\mathbf{k}) = \delta_{ij} - \frac{k_i k_j}{k^2}$. A pseudospectral method is used to compute the right-hand side of this equation (see Gottlieb and Orszag 1977). The time marching is done using a second order finite-difference scheme. An Adams-Bashfort scheme is used for the non linear term while the dissipative term is integrated exactly. The resulting numerical scheme is

$$\frac{\mathbf{v_k}^{n+1} - \mathbf{v_k}^n e^{-\nu k^2 \delta t}}{\delta t} =$$

$$\mathbf{P}(\mathbf{k}) \cdot \left[\frac{3}{2} (\mathbf{v} \times \omega)_{\mathbf{k}}{}^n e^{-\nu k^2 \delta t} - \frac{1}{2} (\mathbf{v} \times \omega)_{\mathbf{k}}{}^{n-1} e^{-2\nu k^2 \delta t} + \mathbf{f_k}{}^n e^{-\nu k^2 \delta t} \right].$$

To start (or restart) the calculation, we use a second order Runge-Kutta scheme.

We force the field at low wavenumbers in a deterministic way. All Fourier modes with $k < 1.5$ are forced with a constant amplitude $\mathbf{f}$ independent of $\mathbf{k}$ (the wavenumber $\mathbf{k}$ has integer components because the space period is 2π). This results in large fluctuations of the energy injection rate $\epsilon_i = < \mathbf{f} \cdot \mathbf{v} >$. One can only hope to reach a steady-state regime in the sense that ϵ_i fluctuates in time around a constant value.

The calculation presented here was done with 240^3 Fourier components, and a viscosity of 10^{-3}. One time-step takes 12 seconds on a Cray-2, using the 4 processors. Integration over one turnover time (defined below) takes of the order of 4 hours of Cray-2.

To estimate the degree of isotropy of the flow, we use a method described in Curry et al. (1984). For each wavenumber $\mathbf{k}$, we define two unit vectors $\mathbf{e_1}(\mathbf{k})$ and $\mathbf{e_2}(\mathbf{k})$ which form with $\mathbf{k}$ an orthogonal reference frame. Since $\mathbf{k} \cdot \mathbf{v} = 0$, each Fourier mode $\mathbf{v}(\mathbf{k})$ is defined by its two components in this frame $v_1(\mathbf{k})$ and $v_2(\mathbf{k})$. We define the isotropy I as $I = \left[\frac{<|v_1|^2>}{<|v_2|^2>} \right]^{1/2}$. In the calculations presented here, I fluctuates by a few percent around 0.95. Therefore, our flow is close to but not statistically isotropic.

Let us recall the definition of some characteristic quantities used in the following. Three characteristic lengths are used, the integral scale $L = \frac{\int_0^\infty k^{-1} E(k) dk}{\int_0^\infty E(k) dk}$, the Taylor microscale $\lambda = \left[\frac{\int_0^\infty E(k) dk}{\int_0^\infty k^2 E(k) dk} \right]^{\frac{1}{2}}$ and the Kolmogorov dissipation scale $l = \left(\frac{\nu^3}{\epsilon} \right)^{\frac{1}{4}}$, where ϵ is the mean energy dissipation rate per unit mass. The two characteristic time-scales of homogeneous turbulence are the eddy turnover time $\tau_0 = L/v_0$ where v_o is the root mean square velocity, and the dissipation time $\tau_\nu = L^2/\nu$. With these quantities one can define two Reynolds numbers, the integral scale Reynolds number $R_L = v_0 L/\nu$ and the Taylor microscale Reynolds number $R_\lambda = v_0 \lambda/\nu$.

2. The spectra of the flow

After integration over several tens of turnover times we obtain a statistically stationary regime in the above sense. The energy spectrum, shown in figure 1, displays a power law range for $k < 30$, with an exponent a little larger than -5/3. The fact that this is an inertial range is confirmed by inspection of the energy flux spectrum $\Phi(k) = \int_k^\infty T(k)dk$ where $T(k)$ is the energy transfer at wavenumber k. $\Phi(k)$ is the energy transfered by unit time from wavenumbers smaller than k to wavenumbers larger than k. This function is found constant for k in the inertial range, as expected. After reaching this stationary regime, we integrate for 30 more turnover times in order to accumulate some statistics. The energy spectrum does not vary significantly during this period.

By fitting the spectrum with the Kolmogorov form $E(k) = C_K\epsilon^{2/3}k^{-5/3}$ we obtain a value of 2 for the Kolmogorov constant C_K, which does not vary by more than 3% in the statistically steady-state period. This value of C_K is a little larger than the experimental value 1.5 (see Monin and Yaglom 1975 for references). Our Reynolds numbers are $R_L \approx 1000$ and $R_\lambda \approx 150$.

3. The statistical distribution of v and its derivatives

Rather than using time distributions of a velocity component v_i or its derivatives $\partial_i v_j$ at a given space location to obtain probability distributions, it is more efficient in our case to use space distributions, for which we have $\approx 1.4 \times 10^7$ grid point values at each time. Using approximately 70 different times, we reach a total of $\approx 10^9$ values.

As an example, figure 2 shows the distribution of $\partial_x v_x$, known from experiments to be strongly non gaussian. Also shown (dashed line) is a gaussian, normalized so that $\sigma = 1$. It is seen that the distribution we obtain is closer to an exponential than to a gaussian, but our statistical sample allows one to see a departure from an exponential law. The wings are indeed not straight lines in this linear-logarithmic plot. The distribution is more intermittent than an exponential distribution (i.e. it has larger wings). Detailed results about the statistics are presented in Vincent and Meneguzzi (1991).

4. The spatial structure of the flow

Figure 3 shows a 3D picture of the vorticity field. The vorticity at each grid point is represented by a vector, here so small that individual vectors can hardly be seen. Vectors are only plotted if their modulus is larger than a given threshold. By varying this threshold and rotating the figure we can explore the structure of the field in detail. One can see that the vorticity is organized in thin elongated tubes, as previously reported by Siggia (1981), Kerr (1985), She et al. (1990). The length of these tubes seems to be of order 1 , i.e. of the same order as the integral scale L (the cube size is 2π). Their thickness is of the order of a few dissipation scales, here a few grid points.

Figure 4 shows a detailed view of a vorticity tube. It displays a sub-cube one sixth the size of the complete one, with 40 grid points on each side. Similar plots for the velocity field show mainly the forcing field if one uses a high threshold. But when the highest velocities are eliminated, the tubes are clearly visible.

From these visualizations, one is lead to the conclusion that these vorticity tubes involve all the scales of the flow. We have done the same kind of pictures of the vorticity field after removing all dissipation range scales. A smooth filter is applied in Fourier space in order to avoid spurious fluctuations. The large scales to which the forcing is applied were also removed. Figure 5 shows the same sub-cube as figure 4 when only inertial range scales are left. One can see the external regions of the tubes. Some helical structure can be seen, as noted by She et al. (1990). This is probably the result of the wrapping up of smaller tubes around the largest.

5. Time evolution of the vorticity structures

When time evolution is added by running a film composed of pictures like figure 3, it strongly suggests shear instabilities as the production mechanism of the voricity tubes. Indeed, the tubes often appear in pairs or in groups of several, approximately parallel, all moving in a direction perpendicular to their axis. One also observes a succession of tube formations with approximately the same alignement and in the same region of the fluid.

The tubes appear with a finite length. During their lifetime, they are often clearly seen to increase in length, but by a factor two or three at most. This is consistent with the shear instability, a two-dimensional instability producing finite length rolls. Random stretching will subsequently tend to increase the length of these vortices.

A precise definition of the typical tube lifetime is difficult because of the necessity to use an amplitude threshold. In figure 3, the threshold vorticity amplitude is 55 , while the maximum vorticity is 223. (Our units are such that the periodic box size is 2π, the rms velocity is ≈ 1, and the turnover time is also ≈ 1.) If this threshold is held fixed in time, the tubes appear with a finite length and can be followed for several turnover times (typically 5), and then disappear.

Figures 6a and b show an example of merging of two parallel worticity tubes, which may have been produced in the same shear layer. Note the antiparallel tube in between the two. Figure 7 displays a different view of the flow detail of figure 6a, a view approximately along the axis of the parallel tubes. One can see the velocity in green and the vorticity in red.

This vortex merging leading to a stronger vorticity structure is generally the origin of the largest observed tubes. Some of them have undergone several mergings. The phenomenon seems much less important for the large number of medium size tubes.

In order to better isolate the mechanism of tube formation, we have done a run at resolution 256^3 without any forcing term in the equation. The initial condition is a random field with only a few small wavenumber components. Initially, the vorticity is very small. When one watches the evolution of the vorticity field above a fixed threshold, the first vorticity structures which appear are pancake-like zones, which flatten in time. The same structures are seen in solutions of the Euler equation (Brachet et al. 1992, Pumir and Siggia 1991, Herring and Kerr 1992). The flattening leads to the appearance of sheets, which tend to bend and roll up, producing the first vorticity tubes. During this process, the vorticity is observed to increase. Therefore, the shear instability is accompanied by vortex stretching. An example of this process is shown in figure 8a and b. The colour scale is kept constant in time, dark blue meaning larger vorticity vectors than light blue, and red even larger. The tip of the vectors is indicated by the white dots. The same conclusion was reached by Brachet et al. (1985) in their study of the solution of the Navier-Stokes equations with a Taylor-Green vortex as initial condition. A self-similar model of the flattening of the vorticity zones leading to sheets is presented in Brachet et al. (1992).

To obtain a more quantitative picture, we have computed the eigenvectors and eigenvalues of the strain tensor around these unstable vorticity sheets. As in paper I, we call λ_1, λ_2 and λ_3 the eigenvalues of the rate of strain tensor $S_{ij} = \frac{1}{2}(\partial_i v_j + \partial_j v_i)$, in ascending order, with λ_1 always negative and λ_3 always positive, and $\mathbf{e_1}, \mathbf{e_2}, \mathbf{e_3}$ the eigenvectors associated with $\lambda_1, \lambda_2, \lambda_3$ respectively, each eigenvector being given the amplitude and sign of its eigenvalue.. ($\nabla \cdot \mathbf{v} = 0$ implies $\lambda_1 + \lambda_2 + \lambda_3 = 0$). In a pure shear, like for instance $v_x = y, v_y = v_z = 0$, one would have $\mathbf{e_1}$ and $\mathbf{e_3}$ at 45 degrees in the plane $x - y$, i.e. $\mathbf{e_1} = (1, -1, 0)$, $\mathbf{e_3} = (1, 1, 0)$, and $\mathbf{e_2} = (0, 0, 1)$. If a straining field is present. like for instance $v_x = ax + y, v_y = -ay, v_z = 0$, for large values of a $\mathbf{e_3}$ will be close to the x axis, and $\mathbf{e_1}$ to the y axis.

What we observe in run B is the following. At early times, $\mathbf{e_1}$ is oriented perpendicularly to the vorticity pancakes, while the vorticity is parallel to the pancakes and to $\mathbf{e_3}$. At a fraction of a turnover time later, $\mathbf{e_1}$ is still perpendicular to the pancake, but the vorticity has become aligned with the intermediate eigenvector $\mathbf{e_2}$. At roughly two turnover times, the pancake has become a thin vorticity sheet, which is beginning to bend, and the vorticity is still aligned with $\mathbf{e_2}$. But we observe that $\mathbf{e_1}$ and $\mathbf{e_3}$ are now approximately at 45 degrees respect to the plane of the sheet. So the picture is that of an initial straining phase which produces a vorticity sheet, with the shear increasing and becoming dominant in a second phase, followed by the rolling up of the sheet.

In this rolling- up phase leading to tube formation, ω is already aligned with the intermediate eigenvector $\mathbf{e_2}$. Therefore, the general alignement of the strain intermediate eigenvector with $\mathbf{e_2}$ found previously by Kerr (1985) and ourselves in Vincent and Meneguzzi (1991) happens long before the vorticity tube production.

A quantitative measure of this alignement was presented in paper I by plotting the distribution of the cosine of the angles between the vorticity ω and the strain eigenvectors $\mathbf{e_i}$. These distribution functions are very similar to the ones found in experiments by Tsinober et al. (1990), and numerically by Ashurst et al.(1987). It was not clear why the distribution of $\omega.\mathbf{e_3}/|\omega||\mathbf{e_3}|$ is so much flatter than the others. In order to clarify this point, we have recomputed these distributions after splitting the set of vorticity values into large and small vorticities, i.e. $\omega > \omega_0$ and $\omega < \omega_0$. The value of ω_0 is 13 while the maximum ω value is 223. The volume in space occupied by the first set is 36% The distributions are shown on figures 9a, b, c The alignement of ω with $\mathbf{e_2}$ is more pronounced in large vorticity regions than in the plot for the general field (see paper I). For small vorticities, the distribution is flat, and there are definitely regions where ω and $\mathbf{e_3}$ are parallel.

Physically, this means that although the vorticity is more often aligned with the intermediate eigenvector $\mathbf{e_2}$, the vortex stretching along the main positive stress direction $\mathbf{e_3}$ produces some vorticity in that direction. To make this more quantitative, we have computed, like it was done by Tsinober (1990), the rate of enstrophy production $\eta = S_{ij}\omega_i\omega_j$. But in addition we have also computed separately the contribution of the three principal strain directions. The production rate in direction $\mathbf{e_i}$ is given by

$$\eta_i = (\omega \cdot \mathbf{e_i})^2/\lambda_i$$

with $\eta_1 + \eta_2 + \eta_3 = \eta$ (remembering that $\mathbf{e_i}$ has the length λ_i). Table 1 gives the values of η_i at two different times. One can see that the enstrophy is produced in approximately equal amounts in direction 2 and 3, despite the fact that the stretching is much larger along direction 3.

Our conclusion is that the main mechanism of vorticity tube production is shear instability, the stress tensor is, when this instability develops, such that ω is oriented in the $\mathbf{e_2}$ direction. In this case, η_3 is not expected to be much larger than η_2, which is what we observe.

6. The role of vorticity tube formation in the energy cascade

It was shown in figures 4 and 5 that, after filtering the vorticity field in Fourier space so as to retain only the inertial range scale, the field in physical space still appears organized in vortex tubes. When we look at the time evolution of the velocity and vorticity fields of run A after this kind of filtering, we essentially observe the evolution of the external part of the vorticity tubes.

To analyse the inertial range velocity field in more detail, we have taken several gaussian filters in Fourier space, separating the inertial domain in three zones, and we have transformed the three fields back to physical space. Figure 10 shows a cut of the velocity field of run A. The low wavenumber part of the inertial domain (k $\approx$ 10)is shown in red, the intermediate part (k $\approx$ 15) in green, and the higher wavenumber part (k $\approx$ 20) in blue. The figure suggest that the rolling up of vorticity sheets plays an essential role in the energy cascade from large to small scales.

This picture is somewhat at variance with the traditional view of a Richardson many step cascade, in which large eddies produce smaller eddies, smaller eddies produce even smaller ones, and so on. On the contrary, what we observe here is a one step process, with a strong correlation of small and large scales. The small scale vorticity constitutes the core of the vorticity structures seen in pictures of the inertial domain. The resulting field is much more intermittent than what would be obtained in a many step process. But an inertial range is definitely present in the energy spectrum, with approximately a $k^{-5/3}$ slope (fig. 1).

To summarize, our visualizations strongly suggest shear instabilities, and more precisely the rolling-up instability of finite extension shear layers, as the mechanism of vortex tube generation in three-dimensional turbulent flow. The alignement of vorticity with the intermediate strain eigenvector is found to exist before this instability develops, and is to be expected in a shear instability.

The enstrophy production rates in the principal strain tensor directions as well as the visualization of the inertial range velocity field in several bandwidths suggest that this rolling-up mechanism plays an important role in the energy cascade from large to small scales. The vortex sheet formation leading to these strong shears is also probably an important contributor.

All our calculations were done on the Cray-2 of the CCVR (Centre de Calcul Vectoriel pour la Recherche), Ecole Polytechnique, Palaiseau, France, thanks to a grant by CNRS (Centre National de la Recherche Scientifique).

References

[1] Ashurst W.T., Kerstein A.R., Kerr R.M., Gibson C.H. (1987), Phys. Fluids, **30**, 2343.
[2] Brachet M.E., Meneguzzi M., Vincent A, .Politano H, Sulem P.L.(1992), Physics of Fluids A **4**, 2845. "Numerical evidence of smooth self-similar dynamics for three-dimensional ideal flows".
[3] Brachet M.E., Meiron D.I., Orszag S.A., Nickel B.G., Morf R.H.,Frisch U.(1983), J. Fluid Mech. **130**, 411.
[4] Curry J.H., Herring J.R., Loncaric J., Orszag S.A. (1984), J. Fluid Mech. **147**,1.
[5] Gottlieb D., Orszag S.A. (1977), CBMS-NSF Regional Conference Series in Aplied Math., **26**.*"Numerical Analysis of Spectral Methods : Theory and Applications."*
[6] Kerr R. (1985), J. Fluid Mech. **153**, 31.
[7] Herring J.R., Kerr R.M. (1992), private communication.
[8] Monin A.S., Yaglom A.M. (1975), Statistical Fluid Mechanics, Vol 2, MIT Press.
[9] Pumir A., Siggia E. (1990), Phys. Fluids A **2**, 220.
[10] She Z.S., Jackson E., Orszag S.A. (1988), J. Sci. Computing, **3**, 407.
[11] She Z.S., Jackson E., Orszag S.A. (1990), Nature, **344**, 226.
[12] Siggia E.D. (1981) J. Fluid Mech. **107**, 375.
[13] Yamamoto K., Hosokawa I. (1988), J. Phys. Soc. Japan **57**, 1532.
[14] Vincent A., Meneguzzi M. (1991), J. Fluid Mech. **225**, 1.

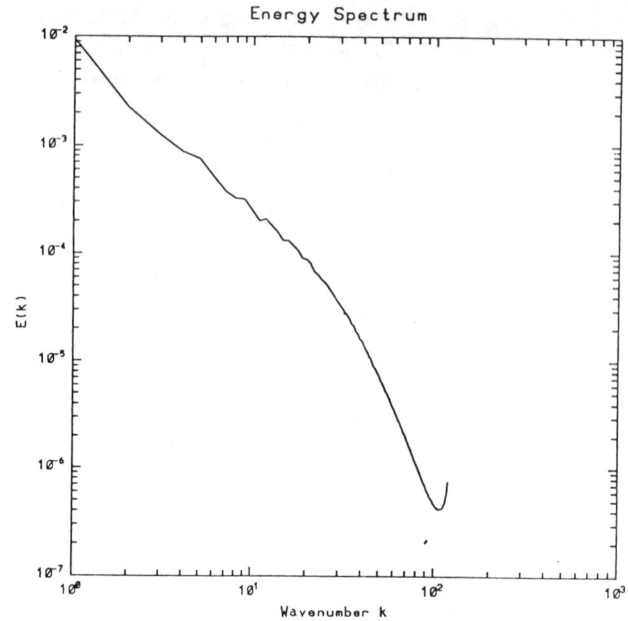

Figure 1. Energy spectrum averaged over 10 different times.

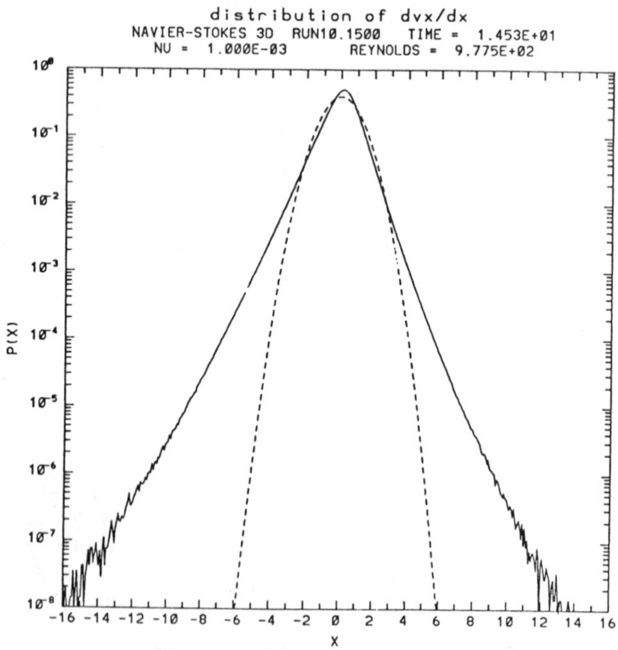

Figure 2. Probability distribution of $\frac{\partial v_x}{\partial x}$ normalized so that $\sigma = 1$, shown together with a gaussian distribution (dashed line).

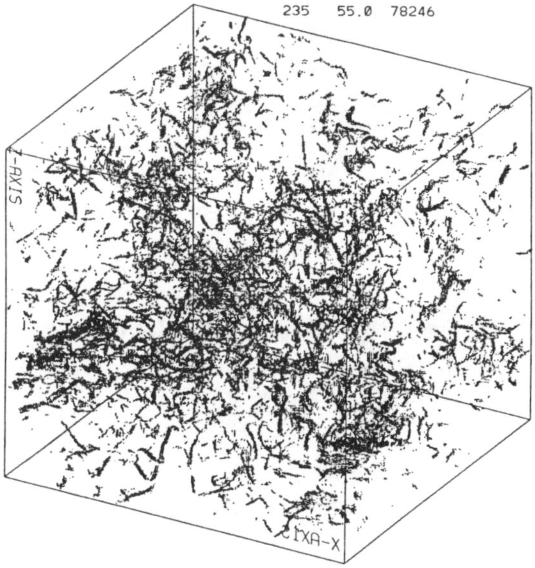

Figure 3.View of the vorticity field, represented by a vector of length proportional to the vorticity amplitude at each grid point. Only vectors larger than a given threshold value are shown.

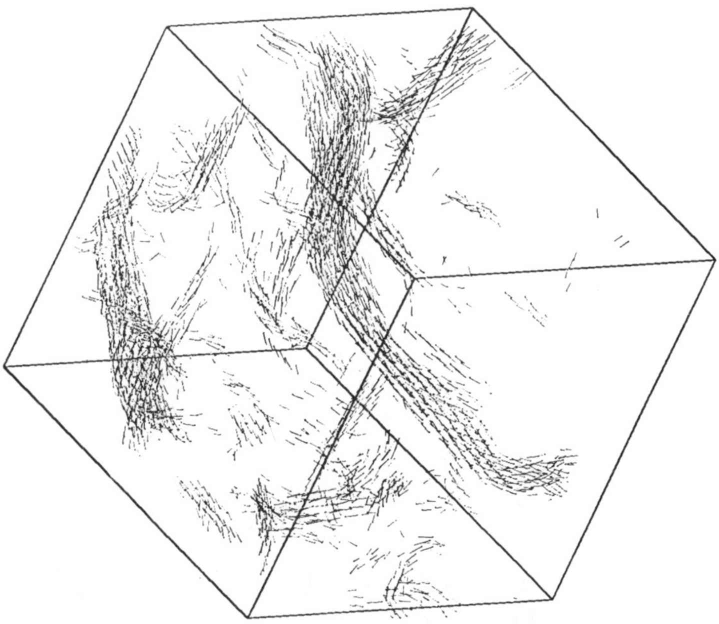

Figure 4. Detail of the vorticity field, showing a sub-cube of size one sixth of the complete cube, i.e. 40 grid points in each direction. The color are associated with vorticity amplitude. The red vectors are larger than the blue ones.

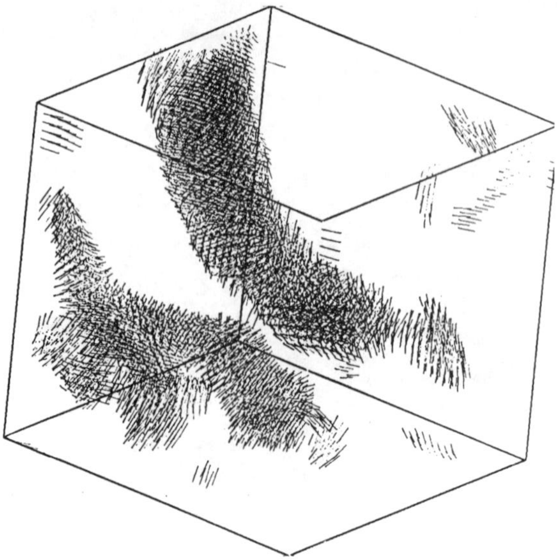

Figure 5. Same vorticity tube as in figure 9, but showing only inertial sub-range scales. Dissipation and energy injection scales have been removed by filtering in Fourier space.

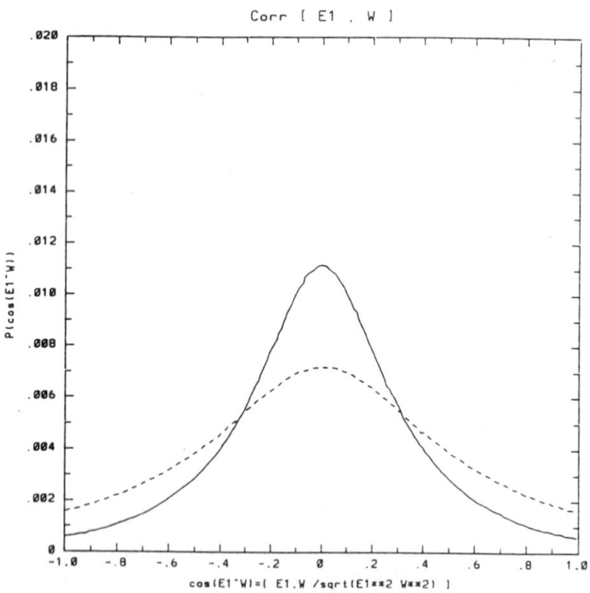

Figure 9a

Figure 9. Distribution of the angle between the vorticity field and the rate of strain eigenvectors. Continuous line: large vorticities; dotted line: small vorticities. (a) case of eigenvector $\mathbf{e}_1$ associated with the smallest eigenvalue λ_1 (always negative). (b) case of eigenvector $\mathbf{e}_2$, associated with the intermediate eigenvalue λ_2. (c) case of eigenvector $\mathbf{e}_3$, associated to the largest eigenvalue λ_3 (always positive).

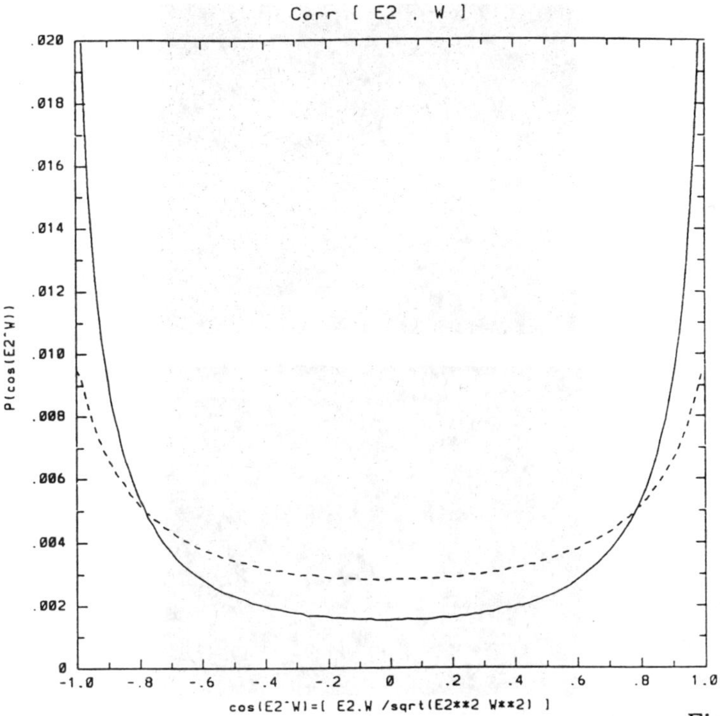

Figure 9b

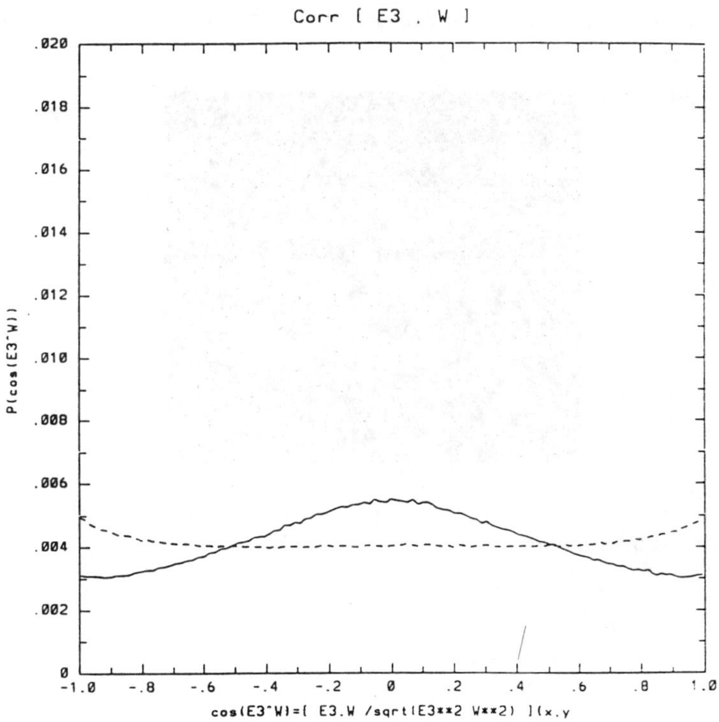

Figure 9c

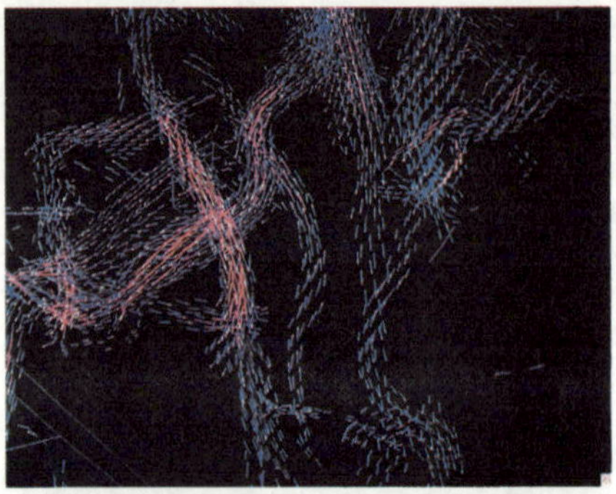

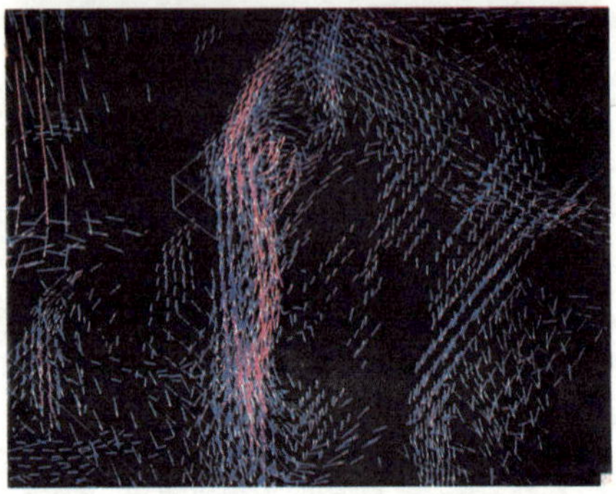

Figure 6. (a)Detail of the vorticity field showing some approximately parallel tubes; (b)The same vorticity region one turnover time later. The three vorticity tubes have merged into one larger tube.

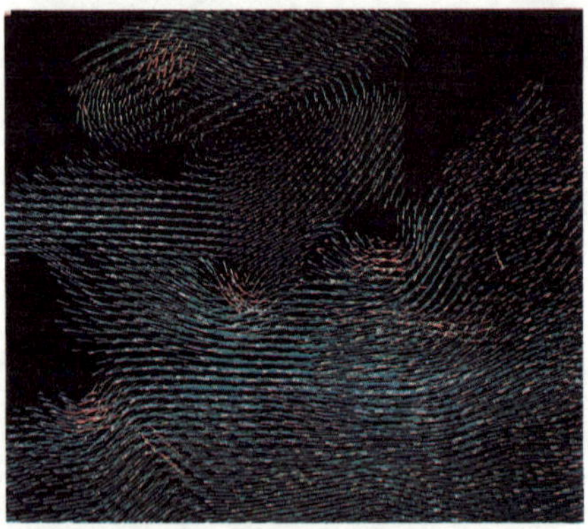

Figure 7. Velocity field, in green, in a plane approximately perpendicular to the vorticity tubes shown in red.

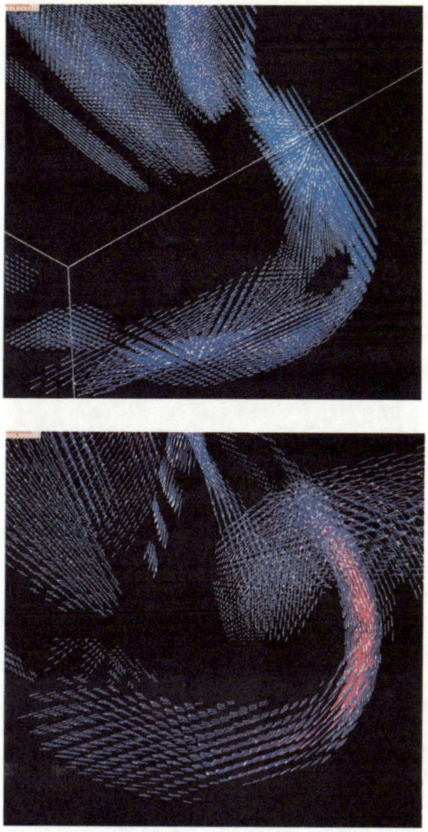

Figure 8. (a)Detail of the vorticity field of run B, at early times, showing several curved vorticity sheets. Dark blue means larger vorticity than light blue, and red even larger; (b)Same a few turnover times later. The first tubes have appeared, as a result of the rolling-up of some of the vorticity sheets, and the vorticity has increased.

Figure 10. Detail of the velocity field after filtering in Fourier space so as to retain a fraction of the inertial domain. In red: low wavenumber range of the inertial domain; in green: intermediate range; in blue: high wavenumber range.

NUMERICAL SIMULATION OF AXISYMMETRIC VORTEX SHEET ROLL-UP

MONIKA NITSCHE
University of Colorado at Boulder

ABSTRACT. The vortex blob method is applied to compute axisymmetric vortex sheet roll-up. An experiment by Didden (1979) of vortex ring formation at the edge of a circular tube is simulated. Numerical results are compared with experimental measurements.

1. Introduction

The vortex sheet is an inviscid model for a free shear layer in which the layer is approximated by a surface. The tangential velocity is discontinuous across the vortex sheet and the flow is irrotational away from the sheet. This article concerns the numerical study of axisymmetric vortex sheet roll-up. As noted in Shariff & Leonard's (1992) review on vortex rings, the axisymmetric roll-up simulates the vortex ring formation process.

The computations are performed using the vortex blob method. The method, which consists of regularizing the Biot-Savart kernel by convolution with a smoothing function, has been applied successfully in 2-d to resolve vortex sheet roll-up (Chorin & Bernard 1973, Anderson 1985, Krasny 1986). The smoothing function depends on a smoothing parameter δ and formally, the Euler equations are recovered in the limit $\delta \rightarrow 0$. It is not clear a priori whether solutions obtained for finite $\delta > 0$ display properties of a real flow.

Several investigators have computed axisymmetric vortex sheet motion using the vortex blob method. Caflisch, Li & Shelley (1992) studied the development of singularities from perturbations of an axisymmetric vortex sheet with swirl. Martin & Meiburg (1991) computed the evolution of ring-like structures from an axial perturbation of a cylindrical vortex sheet. They also observed the growth of azimuthal waves on the rings. The azimuthal instability of a free vortex ring was also studied numerically by Knio & Ghoniem (1990) using a 3-d vortex element method. They obtained good agreement with linear stability theory for the growth rate as a function of the azimuthal wavenumber. Nitsche (1992) computed the roll-up of an initially flat disk of fluid into an axisymmetric vortex ring. Such roll-up is predicted to be self-similar at small times and the computations confirmed this behaviour. Good agreement was also obtained with the theoretical predictions of Taylor (1953) and Saffman (1975) for the vortex ring radius and translation velocity.

J.T. Beale et al. (eds.), Vortex Flows and Related Numerical Methods, 293–301.

The objective of the present work is to investigate by comparison with laboratory experiment whether the vortex blob method applied to axisymmetric flow can recover detailed information about the vortex ring formation process. Didden (1979) reported careful measurements of the vortex ring trajectory and circulation for vortex ring formation at the edge of a circular tube. Didden's (1979) work presented an opportunity to perform detailed comparison of vortex blob computations with experimental measurements.

2. Governing Equations

This section describes the vortex blob method applied to axisymmetric vortex sheet motion. The equations governing axisymmetric vortex sheet motion have also been given by Kaneda (1990).

Let x, r, θ be cylindrical coordinates and u, v, w the corresponding velocity components. For the present computations, we consider axisymmetric flow with zero swirl, $w = 0$. In this case, the fluid velocity is obtained from the stream function ψ by

$$u = \frac{1}{r}\frac{\partial\psi}{\partial r} , \qquad v = -\frac{1}{r}\frac{\partial\psi}{\partial x} . \tag{1}$$

The vortex filaments are circles on the axis $r = 0$ and the vortex sheet is described by its intersection with the x-r plane $(x(\Gamma,t), r(\Gamma,t))$, $0 \leq \Gamma \leq \Gamma_T$. Here Γ is the Lagrangian circulation parameter, Γ_T is the total circulation and t is time. The regularized streamfunction for the flow is

$$\psi_\delta(x,r) = -\frac{1}{4\pi}\int_0^{\Gamma_T}\int_0^{2\pi} \frac{r\tilde{r}\cos\theta}{(\rho^2+\delta^2)^{1/2}}\, d\theta\, d\tilde{\Gamma} , \tag{2}$$

where $\rho^2 = (x-\tilde{x})^2 + r^2 + \tilde{r}^2 - 2r\tilde{r}\cos\theta$, $\tilde{x} = x(\tilde{\Gamma},t)$, $\tilde{r} = r(\tilde{\Gamma},t)$ and δ is the smoothing parameter.

As in 2-d computations, the continuous vortex sheet is replaced by $N+1$ "vortex blob" filaments at $(x_j(t), r_j(t))$, $j = 0,\ldots,N$, corresponding to a discretization $\{\Gamma_j\}$ of $[0,\Gamma_T]$. The streamfunction governing the discrete problem is

$$\psi_\delta(x,r) = -\frac{1}{4\pi}\sum_{k=0}^{N}\int_0^{2\pi} \frac{r r_k\cos\theta}{(\rho_k^2+\delta^2)^{1/2}}\, d\theta\, \Delta\Gamma_k \quad , \tag{3}$$

where $\rho_k^2 = (x-x_k)^2 + r^2 + r_k^2 - 2rr_k\cos\theta$ and $\Delta\Gamma_k$ are the trapezoidal rule weights. The motion of the vortex blobs is governed by the system of ordinary differential equations

$$\frac{dx_j}{dt} = \frac{1}{r_j}\frac{\partial\psi_\delta}{\partial r}(x_j,r_j) , \qquad \frac{dr_j}{dt} = -\frac{1}{r_j}\frac{\partial\psi_\delta}{\partial x}(x_j,r_j) , \tag{4}$$

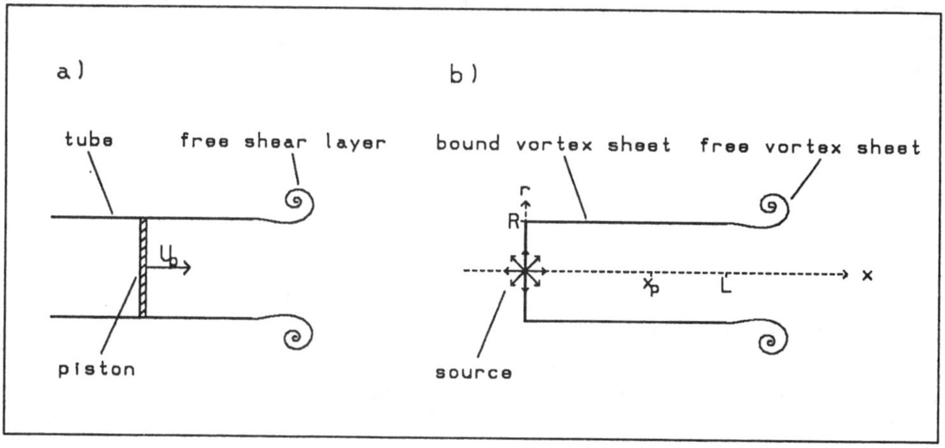

Fig. 1. a) Schematic of Didden's (1979) experiment showing piston, tube and free shear layer. b) Vortex sheet model showing point source, bound and free vortex sheets.

with initial data $x_j(0) = x(\Gamma_j, 0)$, $r_j(0) = r(\Gamma_j, 0)$, $j = 0, \ldots, N$. Following the steps outlined by Lamb (1932) (see also Nitsche 1992), the discrete streamfunction (3) and its derivatives are expressed in terms of complete elliptic integrals. The geometric-algebraic iteration scheme is used to evaluate these integrals. Time integration of (4) is performed using the 4th order Runge-Kutta method.

3. Vortex ring formation at the edge of a circular tube

A schematic diagram of Didden's (1979) experiment is given in Figure 1a. A piston drives fluid out of a tube causing a shear layer to separate at the edge of the tube. The layer rolls up into a vortex ring and propagates away from the edge. Figure 1b shows the vortex sheet model used to simulate the experiment (Nitsche 1992). The separating shear layer is modelled as a free vortex sheet. The tube is modelled as a bound vortex sheet whose strength is determined by ensuring that the fluid velocity is tangent to the wall. The effect of the piston motion is modelled by placing a source at the back of the tube on the symmetry axis. The source strength is chosen so that the fluid velocity equals the driving piston velocity at a location approximately halfway down the tube (at $x = x_p$ in Figure 1b).

Various numerical algorithms have been developed to simulate vortex sheet separation at a sharp edge, as reviewed by Graham (1985). In the present implementation, a vortex blob is released from the edge into the flow at each time step. The vortex blob is released parallel to the edge, with axial velocity equal to the average of the fluid velocities u_- inside and u_+ outside of the tube. The strength of the shed vortex blob is determined by the Kutta condition

$$\frac{d\Gamma}{dt} = -\frac{1}{2}\left(u_+^2 - u_-^2\right) . \tag{5}$$

The values of u_+ and u_- are obtained from the vortex sheet strength at the edge,

$\sigma_e = u_+ - u_-$, and the average axial fluid velocity induced at the edge, $\overline{u} = (u_+ + u_-)/2$. This implementation of the shedding mechanism is due to Krasny (1991).

In Didden's (1979) experiment, the piston moved at constant velocity U_o and stopped at time $t = t_{off}$. Most measurements were taken with $U_o = 4.6$ cm/s and $t_{off} = 1.6$ s. Figure 2 shows a numerical simulation of this experiment. The value $\delta = 0.2$ was used for the smoothing parameter. During the time interval corresponding to nonzero piston velocity, $0 < t < 1.6$, the velocities at the edge satisfy $u_- > u_+$ and a vortex sheet with positive circulation is created. Once the driving velocity is turned off, $t > 1.6$, $u_+ > u_-$ and a secondary vortex ring with negative circulation is formed. For $t > 1.6$, the primary ring travels downstream with essentially constant circulation while the secondary vortex ring slowly propagates upstream inside the tube. Comparison with Didden's experimental visualization showed good agreement in the shape of the rolled-up shear layer. The simulation also captured the ellipticity and tumbling of the outer turns for $t > 1.6$ (see Nitsche 1992).

Figure 3 plots the vortex ring diameter D as a function of axial distance x_c from the tube opening. The experimental measurements from Didden (1979) are denoted by the symbol ($\circ$). Computed results obtained with $\delta = 0.4, 0.2, 0.1$ are shown. Arrows denote the experimental and computed vortex ring position at the time $t = t_{off}$, the left arrow corresponding to the experiment, the right arrow to the computation. For $t < t_{off}$, the vortex ring diameter increases as the ring propagates away from the tube. In this time interval, the computed results are practically independent of the smoothing parameter δ and the agreement with experiment is good. At $t = t_{off}$, the computed ring has travelled further than in the experiment. For $t > t_{off}$, the computed and experimental ring diameter decrease at approximately the same rate.

Figure 4 shows log-log plots of the vortex ring center coordinates (x_c, r_c) vs. time, for $0 < t < t_{off}$. The vertical shift between experimental and numerical data for the axial coordinate $x_c(t)$ (Figure 4a) shows that the numerical ring velocity is higher than the ring velocity in the experiment. This is consistent with the difference in the axial coordinate x_c at $t = t_{off}$ (Figure 3). The radial coordinate $r_c(t)$ of the ring center (Figure 4b) agrees well with experiment. The numerical and experimental ring center coordinates satisfy $x_c \sim t^{3/2}$ and $r_c \sim t^{2/3}$. This causes the curved trajectory for $0 < t < t_{off}$ seen in Figure 3. Such a curved trajectory was also observed by Auerbach (1987) in 2-d experiments of vortex pair formation. However, similarity theory for 2-d edge flow predicts a linear trajectory $x_c \sim t^{2/3}$, $r_c \sim t^{2/3}$ (Pullin 1979).

Various explanations for the discrepancy in x_c between experiment and similarity theory were proposed by Auerbach (1987). Auerbach (1987) suggested that the basic flow may well not be an edge flow. Didden (1982) noted that the similarity theory neglects the effect of the vortex ring's self-induced velocity and predicts an upstream motion of the ring center outside the tube. However, in the experiments and the computations, the vortex ring travels downstream. Thus the discrepancy in the behaviour of the axial displacement x_c can be attributed to the neglect of the ring's self-induced velocity by the similarity theory.

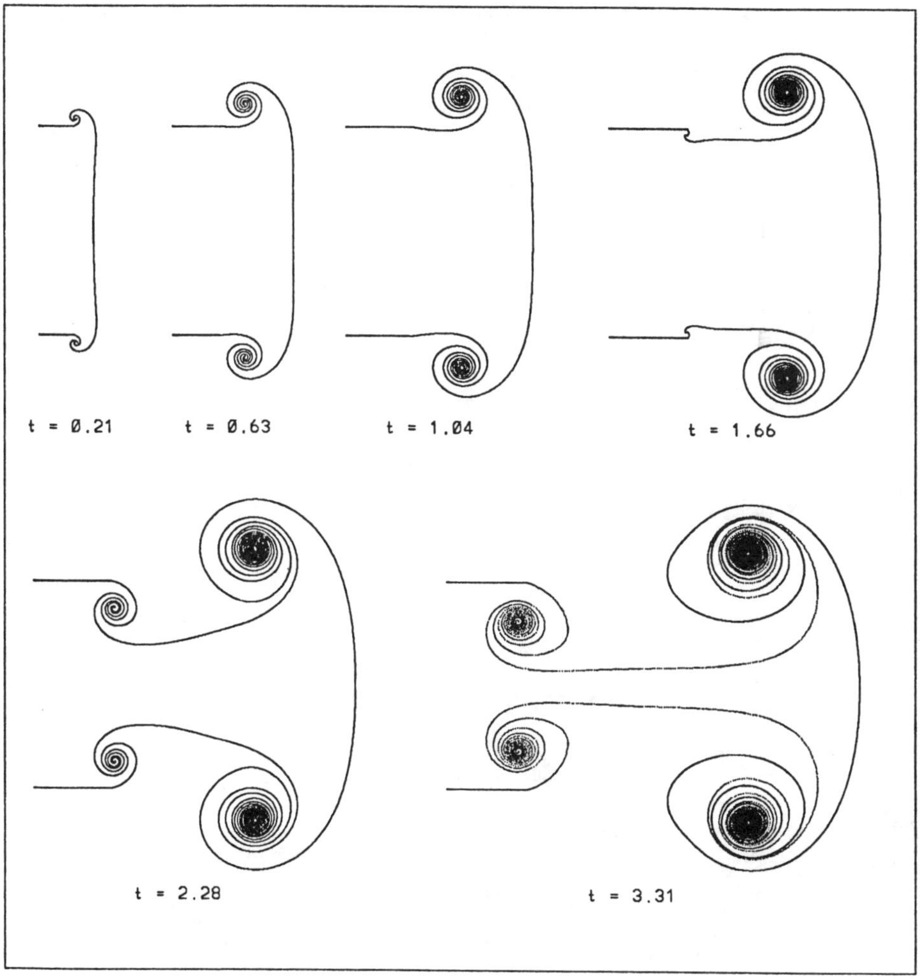

Fig. 2. Numerical simulation of vortex ring formation at the edge of a tube.

Figure 5 plots the total circulation Γ shed as a function of time. Experimental values, computed values and the slug flow model prediction are given. The computed values of Γ decrease with δ but remain significantly larger than the experimental

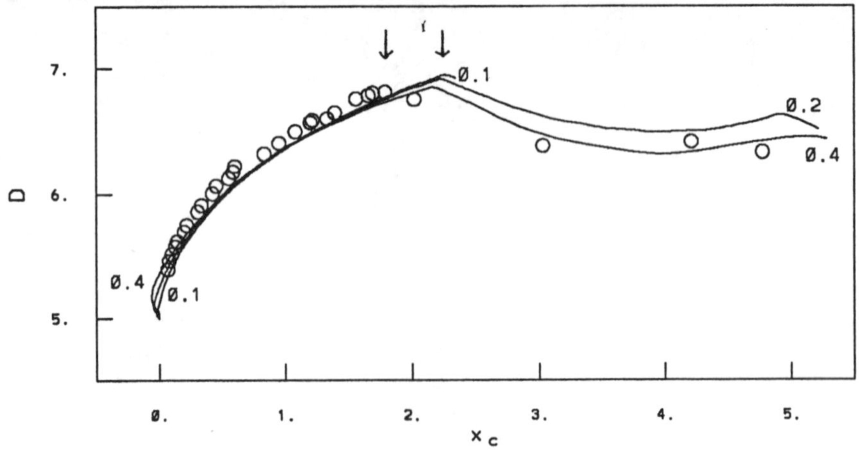

Fig. 3. Vortex ring diameter D vs. distance x_c from the tube opening. Didden's (1979) experimental data (o) and the numerical solution for $\delta = 0.4$, 0.2, 0.1 are shown.

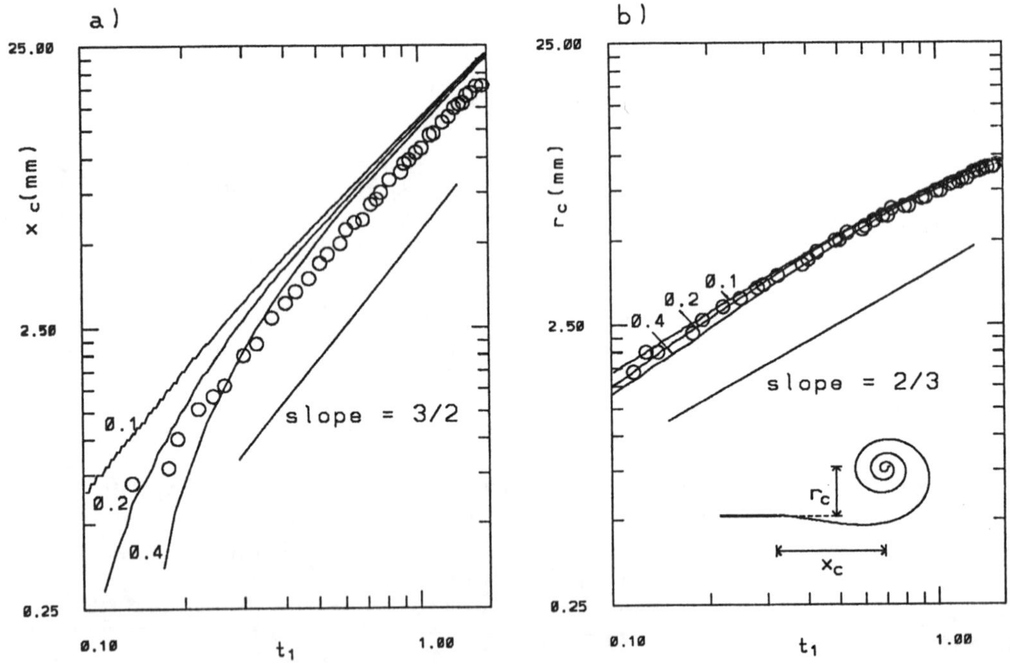

Fig. 4. Log-log plots of the ring center vs. time for $U_o = 4.6$. a) Axial distance x_c, b) radial distance r_c. Didden's (1979) experimental data (o) and the numerical solution for $\delta = 0.4$, 0.2, 0.1 are shown.

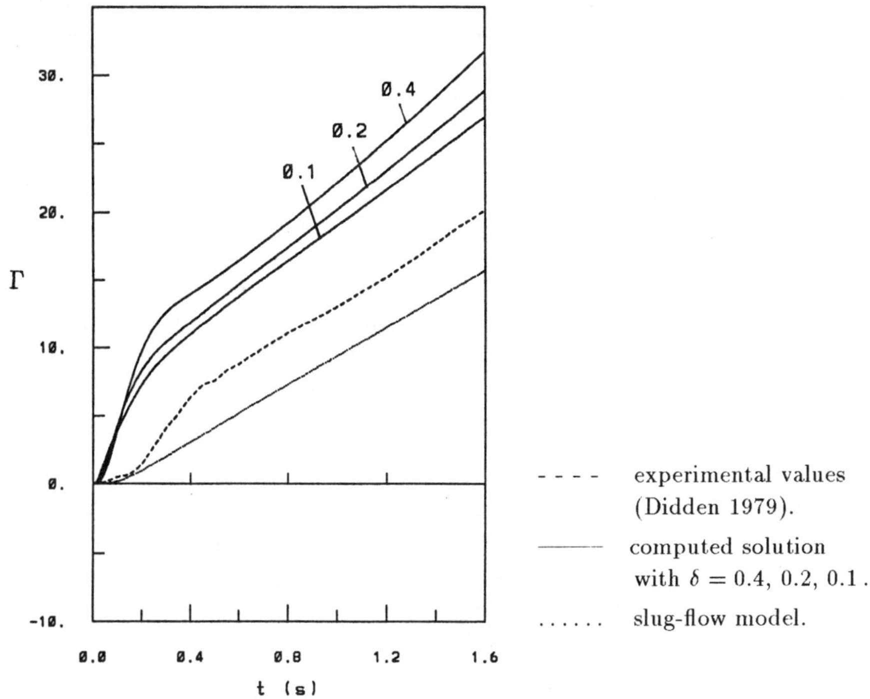

Fig. 5. Total shed circulation Γ vs. time t.

measurements. This is consistent with the observation that the computed ring velocity is greater than in the experiment. Figure 5 shows that the experimental and computed circulation shedding rates $d\Gamma/dt$ agree reasonably well for $t > 0.4$. At smaller times, viscous effects dominate the experimental flow and the computed circulation sheddding rate is higher than the measured rate. The discrepancy between the computed and experimental circulation shedding rates for $0 < t < 0.4$ could be due to the neglect of viscous effects in the vortex sheet model. It may also be noted that the accuracy of the experimentally measured circulation for small times is not known.

4. Conclusions

The vortex blob method has been applied to simulate an experiment by Didden (1979). The comparisons indicate that the method recovers detailed information about the vortex ring formation process. In particular, the shape of the rolled-up shear layer, the vortex ring trajectory, the power laws governing the ring trajectory and the circulation shedding rate at late times are recovered. The computed values of these quantities are practically independent of the smoothing parameter. Considering that the model does not account for viscosity, it may be concluded that these flow quantities do not depend strongly on viscosity. The conclusion that vortex blob simulations recover flow quantities which are independent of viscosity is consistent with recent numerical work by Tryggvason, Dahm & Sbeih (1991). They compared vortex blob and Navier-Stokes solutions, for the case of a periodic 2-d free shear layer, and obtained good agreement between the vortex blob limit of zero smoothing parameter $\delta \to 0$ and the Navier Stokes limit of zero viscosity $\nu \to 0$.

The computed and experimental circulation shedding rates disagree for $0 < t < 0.4$. The experiments were performed with a fixed Reynolds number Re $\approx$ 1500. We conjecture that with increasing Reynolds number the agreement in the circulation shedding rate at small times would improve. Not enough experimental data is available at present to verify this. Further experimental measurements of the dependence of the circulation shedding rate on the Reynolds number would be valuable in helping to develop a more accurate model for the vortex shedding mechanism.

Acknowledgements

I wish to thank my thesis advisor Professor Robert Krasny for guidance in completing this work. This research was partially supported by NSF grant DMS-9003965 and GRI contract 5088-260-1692. The computations were performed at the University of Michigan and the NSF San Diego Supercomputer Center.

References

[1] Anderson, C. 1985 *A vortex method for flows with slight density variations*, J. Comp. Phys. **61**, 417.

[2] Auerbach, D. 1987 *Experiments on the trajectory and circulation of the starting vortex*, J. Fluid Mech. **183**, 185.

[3] Caflisch, R., Li, X. & Shelley, M. 1992 *A vortex stretching instability for axi-symmetric flow with swirl*, submitted for publication.

[4] Chorin, A.J. & Bernard, P.S. 1973 *Discretization of a vortex sheet, with an example of roll-up*, J. Comp. Phys. **13**, 423.

[5] Didden, N. 1979 *On the formation of vortex rings : rolling-up and production of circulation*, J. of Applied Math. and Phys. (Z. Angew. Math. Phys.) **30**, 101.

[6] Didden, N. 1982, *On vortex formation and interaction with solid boundaries*, Vortex Motion, ed. E.-A. Müller, Braunschweig/Wiesbaden: Friedr. Vieweg. & Sohn.

[7] Graham, J.M.R. 1985 *Application of discrete vortex methods to the computation of separated flows*, Numerical Methods for Fluid Dynamics II, eds. K.W. Morton & M.J. Baines, Clarendon Press.

[8] Kaneda, Y., 1990 *A representation of the motion of a vortex sheet in a three-dimensional flow*, Phys. Fluids A **2**, 458.

[9] Knio, O.M. & Ghoniem A.F. 1990 *Numerical study of a three-dimensional vortex method*, J. Comp. Phys. **86**, 75.

[10] Krasny, R. 1986 *Desingularization of periodic vortex sheet roll-up*, J. Comp. Phys. **65**, 292.

[11] Krasny, R. 1991 *Vortex sheet computations: roll-up, wakes, separation*, Lect. Appl. Math. **28**, 385.

[12] Lamb, H. 1932 *Hydrodynamics*, Cambridge University Press. Martin, J.E. & Meiburg, E. 1991 *Numerical investigation of three-dimensionally evolving jets subject to axisymmetric and azimuthal perturbations*, J. Fluid Mech. **230**, 271.

[13] Nitsche, M. 1992 *Axisymmetric vortex sheet roll-up*, Ph. D. Thesis, University of Michigan.

[14] Pullin, D.I. 1979 *Vortex ring formation at tube and orifice openings*, Phys. Fluids **22**, 401.

[15] Saffman, P.G. 1975 *On the formation of vortex rings*, Stud. Appl. Math. **54**, No. 3, 261.

[16] Shariff, K. & Leonard, A. 1992 *Vortex rings*, Ann. Rev. Fluid Mech. **24**, 235.

[17] Taylor, G.I. 1953 *Formation of a vortex ring by giving an impulse to a circular disk and then dissolving it away*, J. of Appl. Phys., Vol 24, 104.
[18] Tryggvason, G., Dahm, W.J.A., & Sbeih, K. 1991 *Fine structure of vortex sheet roll-up by viscous and inviscid simulation*, ASME J. Fluids Engin. **113**, 31.

FREE VORTEX RINGS, ANALOGIES AND DIFFERENCES BETWEEN VORTICITY AND A PASSIVE SCALAR

P. ORLANDI, and R. VERZICCO
Università di Roma
"La Sapienza"
Dipartimento di Meccanica e Aeronautica.

ABSTRACT. In this paper the simulation of a free ring together with a passive scalar has been performed. The axisymmetric simulation has shown that the pseudo-steady solution is obtained when a single functional relationship between $\hat{\omega} = \omega_\theta / r$ and the Stokes streamfunction is achieved. The relationship is linear around the centre of the ring and the slope depends on the initial vorticity distribution. The passive scalar does not correlate with ω_θ and correlates well with $\hat{\omega}$. In the 3D case the simulation has shown completely different distributions between passive scalar and vorticity field. The passive scalar tends to be concentrated in regions with low vorticity and low rate of strain. This is a further proof that flow visualisations can be misleading when conclusions on the vorticity field are withdrawn from markers.

1. Introduction

The vortex ring is one basic structure in many flows, e.g. the round jets can be considered constituted by several rings interacting each other in a complicated manner. Several studies [1] [2] were devoted to explain the formation of vortex rings as a roll-up of a vortex sheet generated at the edge of a nozzle. In the present paper we are interested to analyse, by an axisymmetric simulation, which is the mechanism creating the pseudo-stationary vorticity distribution. The term pseudo-stationary is related to the viscosity causing a non-real steady state. However the numerical simulations, performed at high Re, do not differ from those obtained by the Euler equations. As in two-dimensional flows, it was found that the ring assumes a defined structure producing scatter plots of $\hat{\omega} = \omega_\theta / r$ versus $\hat{\psi}$ not changing in time. $\hat{\psi} = \psi + \int v_z dr$ is the Stokes streamfunction in a frame translating with the ring. The simplest solution in axisymmetric conditions is the Hill vortex, $\hat{\omega} = cons$ in a sphere of radius a. A solution similar to that of Lamb [3], $\omega = k^2 \psi$ in 2D, can not be found analytically [4]. Then to find other classes of solutions the numerical simulation of the Navier-Stokes equations must be used. Different initial conditions produced a linear relationship, at least in the core of the ring, with slopes depending on the ring radius.

In the experiments the direct measurement of vorticity is very cumbersome then its distribution is often derived by flow visualisations. To see whether the passive

J.T. Beale et al. (eds.), Vortex Flows and Related Numerical Methods, 303–313.

scalar does correlate with the vorticity or with $\hat{\omega}$, initially axisymmetric simulations were performed. Moreoveor the influence of the Schmidt number was investigated. The differences between markers and vorticity are mainly due to vortex stretching causing the conservation of $\hat{\omega}$ in the inviscid limit. By a first series of simulations it has been shown that a very good correlation does exist between the passive scalar and $\hat{\omega}$, on the contrary the correlation with the vorticity is very poor. The good correlation has been obtained independently from the Sc number and from initial conditions. In the axisymmetric case a grid refinement is affordable to have a good resolution up to $Sc = 8$. The linear relationship between T and $\hat{\omega}$ at $Sc = 1$ does not hold any more at high Sc, different slopes were found passing from $Sc = 1$ to $Sc = 4$ and these decrease between $Sc = 4$ and $Sc = 8$. The grid refinement check can not be performed in 3D for computer limitations, then the simulation was restricted to $Sc = 1$.

In the three-dimensional case the comparison between the equation of each vorticity component or any combination of them and the equation for a passive scalar, does not show any similarity. Hussain & Melander [5], by direct simulations, reached the conclusion that flow visualisation can not only be confusing but also can be grossly misleading. In fact in 3D flows the vorticity undergoes self-augmentation by vortex stretching and markers are not subjected to this effect. As a consequence the markers are depleted from regions of high vorticity and accumulated in region of low vorticity. In the present paper this behavior has been verified by the direct simulation of a free ring and it has been observed that high gradients of passive scalar are found in regions where the amount of rate of strain is low. A certain amount of markers is concentrated near the ring axis for the effect of axial vorticity.

2. Equations and numerical scheme

2.1. AXISYMMETRIC CASE

In calculations of axisymmetric flows the vorticity streamfunction formulation is convenient and, if the quantity $\hat{\omega}$ is used the non-linear terms attain the form of a Jacobian $J(\hat{\omega}, \psi)$. It is also possible to find discretized forms conserving, in the limit for $\nu = 0$, high order invariants (kinetic energy, $\int \hat{\omega}^2$) and that maintain the skew symmetry of the Jacobian [6]. We wish to emphasise that the fulfilment of the conservation properties avoids the generation of unphysical vorticity.

The adimensionalization, scaling t with respect to Γ/r_0^2 and ω_θ with respect to r_0^2/Γ, defines the Reynolds number as $Re_\Gamma = \Gamma/\nu$ and $Sc = k/\nu$, the Navier-Stokes equations and the equation for a passive scalar become

$$\frac{\partial \omega_\theta}{\partial t} + J(\hat{\omega}, \psi) = \frac{1}{Re}\left[\frac{\partial}{\partial r}\frac{1}{r}\frac{\partial \omega_\theta r}{\partial r} + \frac{\partial^2 \omega_\theta}{\partial z^2}\right] \tag{1}$$

$$\frac{\partial T}{\partial t} + \frac{1}{r}J(\hat{T}, \psi) = \frac{1}{ReSc}\left[\frac{1}{r}\frac{\partial}{\partial r}r\frac{\partial T}{\partial r} + \frac{\partial^2 T}{\partial z^2}\right] \tag{2}$$

with

$$J(\Omega, \psi) \equiv \frac{\partial \Omega}{\partial r}\frac{\partial \psi}{\partial z} - \frac{\partial \Omega}{\partial z}\frac{\partial \psi}{\partial r} \tag{3}$$

Velocity components are related to the streamfunction by $v_z = -\frac{1}{r}\partial\psi/\partial r$, $v_r = \frac{1}{r}\partial\psi/\partial z$ and the streamfunction is related to the vorticity by

$$\omega_\theta = [\frac{\partial}{\partial r}\frac{1}{r}\frac{\partial\psi}{\partial r} + \frac{1}{r}\frac{\partial^2\psi}{\partial z^2}] \tag{4}$$

In the axisymmetric case the solution of the Navier-Stokes equations is performed imposing at $r = 0$ the condition $\omega_\theta = 0$. For the passive scalar, at $r = 0$, the realistic boundary condition of zero flux should be assigned at $r = 0$, but being interested to the differences between vorticity and markers, $T = 0$ has been assumed as for the vorticity. At $r = r_f$, if $r_0(t = 0) \ll r_f$, with r_0 the radius of the torous of the vortex ring, the assumption of $\omega_\theta = 0$ and $\psi = 0$ is satisfactory. Periodicity in the vertical direction has been assumed and this condition causes the wake, shed from the ring during the transient, to impact on the ring itself. At low Re the wake is rapidly dissipated, on the other hand at very high Re the impact does not influence the ring structure. To eliminate this effect the simulations have been performed at an intermediate Re such that the wake is dissipated in few eddy turnover times.

The numerical scheme slightly differs from that described in [7], which we wish to summarise. The viscous terms are discretized by a centered second-order scheme and the advective terms by the Arakawa scheme [6]. To avoid viscous stability conditions, the viscous terms are treated implicitly, and to save CPU time the resulting pentadiagonal matrix has been factorized into two tridiagonal matrices. The homogeneous boundary conditions in the radial directions and the periodicity in the vertical directions avoid the requirement of boundary conditions for the pseudo-vorticity in the first sweep.

A third order Runge-Kutta scheme with two level storage, explicit for the nonlinear terms and implicit for the viscous ones, described by Ref. [8] has been used for time advancement. This scheme allows larger Courant numbers than those permitted by the Adams-Bashforth scheme and it reduces the CPU time requirements. The stream function was calculated by a FFT in the z direction and a tridiagonal solver in the r direction.

2.2. THREE-DIMENSIONAL CASE

In the three-dimensional case the use of the Navier-Stokes equations in primitive variables is a good choice to reduce the number of operations. It is well recognised that it is convenient to stagger the velocity components and to locate the pressure and the passive scalar at the centre of the cell. This choice gives, in a finite difference scheme, the most compact form for the discrete *div* and *grad* operators and well-structured matrices bring to a solenoidal velocity field within round-off errors. In cylindrical coordinates, the continuity equation has a form similar to that in Cartesian coordinates, when the quantities $q_\theta = v_\theta$, $q_r = rv_r$, $q_z = v_z$, are introduced

$$\frac{\partial q_r}{\partial r} + \frac{\partial q_\theta}{\partial \theta} + r\frac{\partial q_z}{\partial z} = 0 \tag{5}$$

as will be discussed later on, these variables simplify the treatment of the region around $r = 0$. The Navier-Stokes equations in terms of q_i and the equation for the passive scalar are

$$\frac{Dq_\theta}{Dt} = -\frac{1}{r}\frac{\partial p}{\partial \theta} + \frac{1}{Re}\left[\frac{\partial}{\partial r}\left(\frac{1}{r}\frac{\partial r q_\theta}{\partial r}\right) + \frac{1}{r^2}\frac{\partial^2 q_\theta}{\partial \theta^2} + \frac{\partial^2 q_\theta}{\partial z^2} + \frac{2}{r^3}\frac{\partial q_r}{\partial \theta}\right]$$

$$\frac{Dq_r}{Dt} = -r\frac{\partial p}{\partial r} + \frac{1}{Re}\left[r\frac{\partial}{\partial r}\left(\frac{1}{r}\frac{\partial q_r}{\partial r}\right) + \frac{1}{r^2}\frac{\partial^2 q_r}{\partial \theta^2} + \frac{\partial^2 q_r}{\partial z^2} - \frac{2}{r}\frac{\partial q_\theta}{\partial \theta}\right]$$

$$\frac{Dq_z}{Dt} = -\frac{\partial p}{\partial z} + \frac{1}{Re}\left[\frac{1}{r}\frac{\partial}{\partial r}\left(r\frac{\partial q_z}{\partial r}\right) + \frac{1}{r^2}\frac{\partial^2 q_z}{\partial \theta^2} + \frac{\partial^2 q_z}{\partial z^2}\right]$$

$$\frac{DT}{Dt} = \frac{1}{ReSc}\left[\frac{1}{r}\frac{\partial}{\partial r}\left(r\frac{T}{\partial r}\right) + \frac{1}{r^2}\frac{\partial^2 T}{\partial \theta^2} + \frac{\partial^2 T}{\partial z^2}\right]$$

with

$$\frac{Dq_\theta}{Dt} = \frac{\partial q_\theta}{\partial t} + \frac{1}{r}\frac{\partial q_\theta q_r}{\partial r} + \frac{1}{r}\frac{\partial q_\theta^2}{\partial \theta} + \frac{\partial q_\theta q_z}{\partial z} + \frac{q_\theta q_r}{r^2}$$

$$\frac{Dq_r}{Dt} = \frac{\partial q_r}{\partial t} + \frac{\partial}{\partial r}\left(\frac{q_r^2}{r}\right) + \frac{\partial}{\partial \theta}\left(\frac{q_\theta q_r}{r}\right) + \frac{\partial q_r q_z}{\partial z} - q_\theta^2$$

$$\frac{Dq_z}{Dt} = \frac{\partial q_z}{\partial t} + \frac{1}{r}\frac{\partial q_r q_z}{\partial r} + \frac{1}{r}\frac{\partial q_\theta q_z}{\partial \theta} + \frac{\partial q_z^2}{\partial z}$$

$$\frac{DT}{Dt} = \frac{\partial T}{\partial t} + \frac{1}{r}\frac{\partial q_r T}{\partial r} + \frac{1}{r}\frac{\partial q_\theta T}{\partial \theta} + \frac{\partial q_z T}{\partial z}$$

The equations have been non-dimensionalized as in the axisymmetric case, and the velocities scale with r_0/Γ and the pressure with $r_0^2/\rho\Gamma^2$. The same considerations made before, about the computational domain and about the boundary conditions, hold in the three-dimensional case.

The numerical scheme is described in Ref. [9]; here we summarise the main features of the method which presents some similarity with those in the vorticity streamfunction formulation, e.g. viscous and advective terms by centered second order schemes. In the limit of $\nu \to 0$, the scheme conserves energy. The system of equations was solved by the fractional step method [8] with the pressure in the step for the non-solenoidal velocities, to simplify the boundary conditions. In this method a scalar quantity Φ, related to the pressure, is introduced to project the non-solenoidal field in a solenoidal one.

The large band matrix associated to the Φ calculation is reduced to a tridiagonal matrix by FFTs in the azimuthal and in the vertical directions. This procedure is very efficient to obtain the solenoidal field within round-off errors. The same third order Runge-Kutta scheme was used to advance in time the non-solenoidal field, it consists of three substeps and at each substep the scalar Φ is calculated.

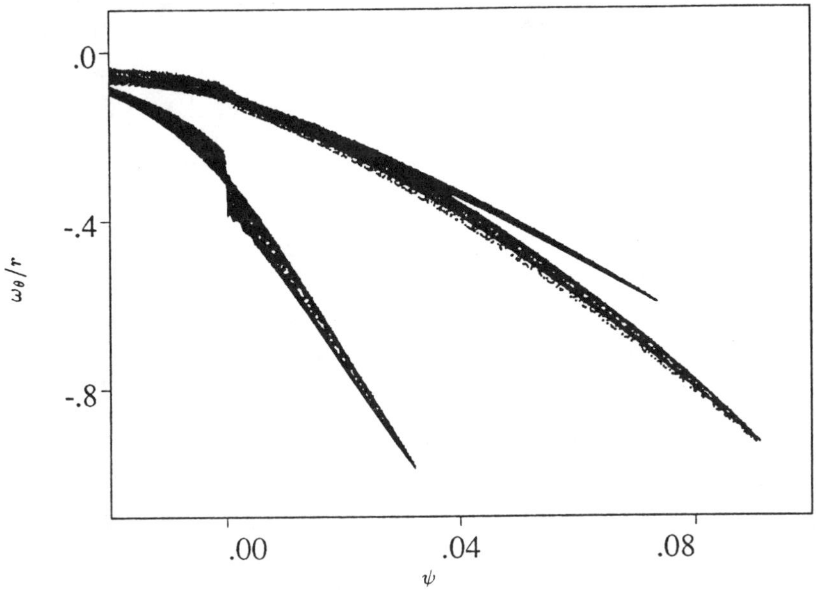

Fig. 1. Scatter plots of $\hat{\omega}$ versus $\hat{\psi}$ with various initial conditions at $Re = 2750$ and $Sc = 4$.

In cylindrical coordinates at $r = 0$ the equations for the velocity components are singular. The advantage of using staggered "fluxes" is that at the centreline $q_r = 0$, thus the q_r equation can be discretized at the grid points $i+\frac{1}{2}, 2, k+\frac{1}{2}$ without any approximation. The viscous terms in the radial direction calculated at the point $i, \frac{3}{2}, k+\frac{1}{2}$ for the q_θ and at the point $i+\frac{1}{2}, \frac{3}{2}, k$ for q_z equations have been expressed by doing approximations, similar to those done in Ref.[8] in wall bounded flows. The accuracy is reduced but in the present case, close to the symmetry axis, the velocity gradients are smaller than those near a solid wall then the reduction of accuracy has a minor effect on the results.

2.3. INITIAL CONDITIONS

The simplest steady solution of the Euler equations for axisymmetric rings is the Hill vortex, $\hat{\omega} = cons$, for $r < a$, with a the radius of the sphere. This distribution is appropriate for the contour dynamic calculations as in Ref. [10], on the contrary it is not appropriate for direct simulations. Stanaway [11] considered a Hill vortex with a smoothing to overcome the abrupt vorticity jump between the vortical and the irrotational regions. The initial vorticity distribution is given by

$$\omega = \frac{3}{2r}\sin\theta \begin{cases} 1 & r \le r_i \\ 1 - f(\xi) & r_i \le r < r_e \\ 0 & r_e < r \end{cases} \tag{6}$$

where $f(\xi) = 0$ for $r = r_i$ and $f(\xi) = 1$ for $r = r_e$, is the smoothing function [12], that smoothly connects the inner with the outer values of the vorticity. A further

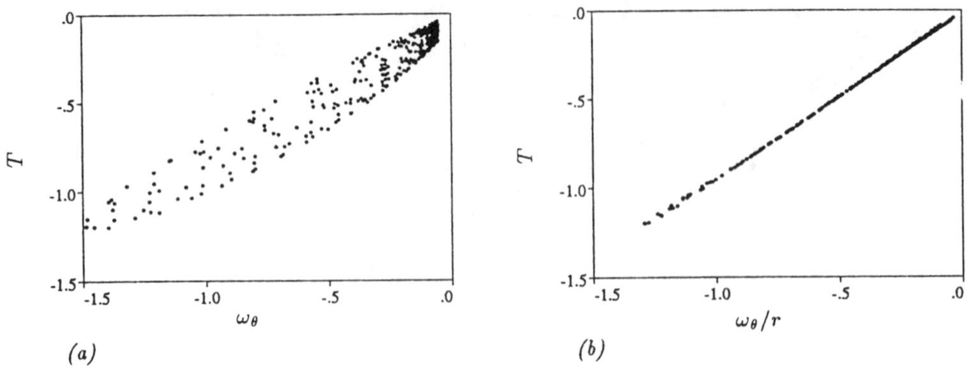

Fig. 2. Scatter plots of T versus a) ω_θ b) $\hat{\omega}$ at $Re = 5500$ and $Sc = 1$.

initial condition has been assigned, as in Ref. [11], by a Gaussian distribution

$$\omega_\theta(t=0) = \frac{\Gamma}{\pi a^2} e^{-s^2}, \qquad s^2 \equiv \frac{(z-z_0)^2}{b^2} + \frac{(r-r_0)^2}{c^2} \tag{7}$$

being (r_0, z_0) the centre of the ring core. To check the effects of the initial conditions the Gaussian profile has been also distributed within an ellipse, with $b = 0.75a$ and $c = 1.25a$.

The equilibrium solution, yielded in the axisymmetric simulation at $Re = 5500$, was assumed as base azimuthal vorticity in the 3-D simulation and superimposed to this a disturbance was introduced by imposing on the axis of the ring core a radial displacement, with amplitude $\epsilon = 0.02$ and wave number $n = 5$. This perturbation is similar to that used by Knio and Goneim [13] in the numerical study of azimuthal perturbations on free evolving rings. This initial condition introduces perturbations in ω_θ and ω_r and not in ω_z, by the kinematic relationships between velocity and vorticity the initial velocity field is evaluated. The initial distribution of the passive scalar coincides exactly with azimuthal vorticity.

3. Results

Simulations of free axisymmetric rings have been performed to study which is the functional relationship the vorticity achieves at the steady state. The impossibility to perform accurate simulations in the inviscid limit, condition at which a real steady state is reached,brought us to perform simulations at a high Re number. This Re is such that the simulation of the passive scalar, at a substantial high $Sc = 8$, can be considered well resolved by a grid with 256×256 points. The simulations were performed at $Re = 2750$ and $Re = 5500$, at these values the 3D calculation was also performed by $64 \times 97 \times 128$ grid but in this case only $Sc = 1$ was considered. The 3D simulation was performed by 64 points in the azimuthal direction in a

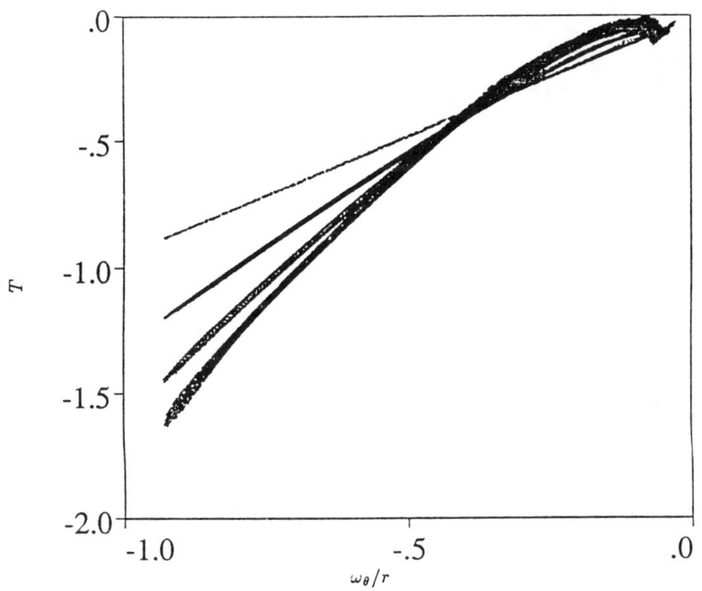

Fig. 3. Scatter plots of T versus $\hat{\omega}$ at $Re = 2750$ and $Sc = 1$, $Sc = 2$, $Sc = 4$ and $Sc = 8$.

domain $2\pi/n$, with n the wave number of the initial azimuthal perturbation. The most unstable wave number obtained by experimental observations [14], $n = 5$, was assumed and 64 points in θ are sufficient to describe both the axial structures generated in the central region and the azimuthal deformations of the core.

3.1. AXISYMMETRIC RINGS

The axisymmetric simulations were performed to study the dependence of the pseudo steady solution from the initial vorticity. The pseudo-steady solution is that at which a well defined functional relationship between $\hat{\omega}$ and $\hat{\psi}$. Simulations, at low Re, produce very smooth solutions, in short times, but these have the disadvantage to furnish scatter plots from which it is difficult to detect single value relationships. Fig.1 shows the functional relationship obtained by four different initial conditions. It results that when the initial distribution is perturbed ($b = 0.75a$, $c = 1.25a$) , without changing σ, the scatter plots are identical to that obtained with $b = c = a$. On the contrary when a larger σ is used the relationship has a reduced slope near the centre of the ring but still linear. When the initial ring has the distribution of Eq.(6) the slope is steeper, and it is related to the vorticity clustered near the axis. We wish to recall that the Hill solution presents a discontinuity between $\hat{\omega} = 0$ and $\hat{\omega} = cons$.

The scalar has been introduced to prove what is evident from the equations, i.e. that ω_θ behaves differently from T. Scatter plots between T and $\hat{\omega}$ and T and ω_θ have been evaluated, showing in Fig.2 that at $t = 100$, a time corresponding to almost 8 eddy turnover times, the pseudo-steady condition is reached. At $Sc = 1$ the

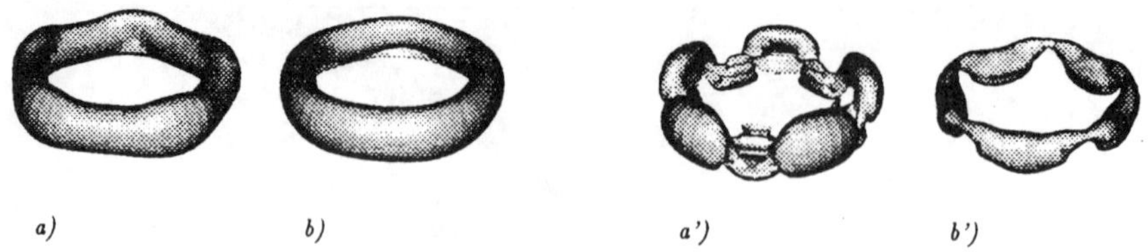

Fig. 4. Surface contours of $\omega = -1.1$ [a),a')] and $T = -1.1$ [b), b')] at $t = 10$ [a), b)] and at $t = 40$ [a'), b')].

vorticity does not correlate with T, on the other hand the correlation between T and $\hat{\omega}$ is perfect. This is not surprising looking at the equations but, to our knowledge, this is the first time that it was proved by a direct simulation. It is interesting to investigate whether the good correlation still holds when the Schmidt number increases. In this case to maintain a good resolution for the passive scalar at $Sc = 8$ the Reynolds number was reduced to 2750. Fig.3 shows that the correlation holds but it lyes on curves of different slopes as a consequence of the reduced dissipation of the passive scalar. By considering that at $k = 0$ ($Sc = \infty$) the T peak remains constant and that at this Re the ω peak is almost halving in 8 eddy turnover times, we expect that an increase, e.g. of four times on Sc from $Sc = 1$, produces more effects than quadruplicating Sc from $Sc = 4$. From these results we can draw a first conclusion that, in a perfect axisymmetric conditions distributions of $\hat{\omega}$ can be extracted from flow visualisations. This is the consequence of vortex stretching, this conclusion can not be extrapolated to 3D flows and this is the reason why a full simulation has been performed.

3.2. THREE-DIMENSIONAL RINGS

The 3D case is very interesting because azimuthal disturbances, at high Re, grow and deform the ring. The knowledge of the mechanism producing these instabilities is important because it characterises the turbulent state. The growth of the kinetic energy at each wave number and its dependence from initial conditions was studied in Ref. [15], on the other hand in this paper we are interested only in the differences or similarities between vorticity and passive scalar. Starting with the equilibrium distributions, obtained by the axisymmetric simulations at $Re = 5500$, Fig.4 shows that at $t = 10$, almost one eddy turnover time, the surface contour plots of ω_θ and T at a value equal to 1.1 do not largely differ. The choice of this high value produces images representing the deformation of the inner part of the ring, which at this time is well described by the marker distribution. Contour plots in horizontal and vertical sections of T and of vorticity magnitude $\omega_M = \sqrt{\omega_\theta^2 + \omega_r^2 + \omega_z^2}$, not reported in the

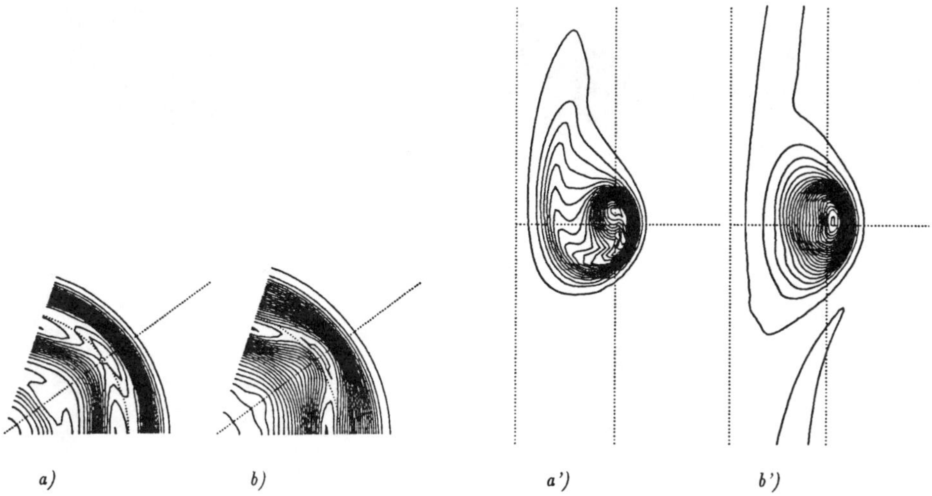

Fig. 5. Contours plots in horizontal and vertical planes of of ω [a),a')] and T [b), b')] at $t = 10$ [a), b)] and at $t = 40$ [a'), b')].

paper, have shown that the passive scalar has been convected near the ring axis in greater amount than the vorticity. At this time the vorticity magnitude almost completely coincides with ω_θ because the time was not sufficient to produces ω_r and ω_z.

Fig.4 shows also that, at a later time $t = 40$, the scalar is depleted from the regions where the amount of vorticity is high, in agreement with the observations by Hussain & Melander [5] in a different vortical structures. The T distribution should be very similar to what obtained by a flow visualisation picture and it does not resemble the vorticity distribution. However we wish to recall that a further reason of large differences between our simulation and a real experiment is due to $Sc >> 1$ in the experiment. Contour plots in horizontal and vertical planes of ω_M and T are easily evaluated in the numerical simulation; a similar picture in an experiment is obtained by light sheets illuminating particles. Also this different and more precise detection method shows that the vorticity and markers behave differently as shown in Fig.5. In fact from this picture it results that the passive scalar does not reproduce the amount of high ω_M near the axis due, for the major part, to the contribution of axial vorticity. Also the vertical sections show that vorticity is distributed differently from T in fact the vorticity presents two maxima on the contrary T presents only one maximum. As was done for the axisymmetric case in the 3D case scatter plots between ω_θ and T and $\hat{\omega}$ and T have been evaluated (Fig.6) and in this case the data show a large scatter for both quantities.

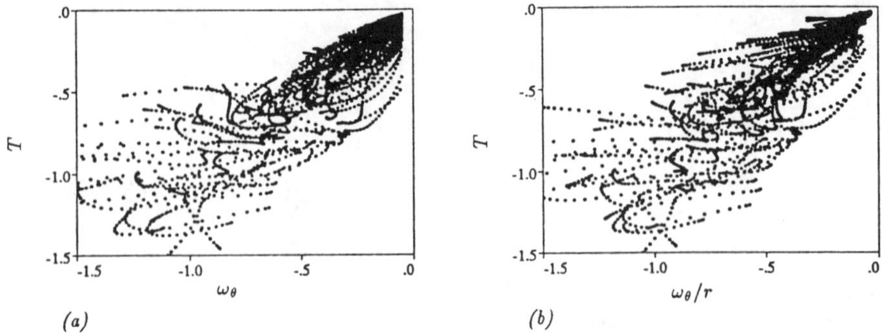

Fig. 6. Scatter plots of T versus a) ω b) $\hat{\omega}$ at $Re = 5500$ and $Sc = 1$ in the 3D case

Conclusions

The present study was devoted to bring a further contribution to the debate whether by flow visualisations informations on vorticity field can be obtained. The aim was pursued considering the free vortex ring, which is a basic structure for several flows. Moreoveor we were interested to analyse the achievement of the steady state. By axisymmetric simulations we concluded that the vortex ring, from an initial out of equilibrium solution, reaches a pseudo-steady solution, which depends on the initial vorticity distribution but has the common characteristic to have a linear relationships between $\hat{\omega}$ and ψ. The slope is as higher as closer to the axis is the centre of the ring.

The axisymmetric calculation has also shown that the passive scalar does not correlate with ω_θ but on the contrary it correlates perfectly with $\hat{\omega}$. This conclusion is built in in the Euler equation and for $Sc = \infty$, but no any information can be derived when viscous and diffusivity effects are included. This is the reason why we decided to perform the direct simulations and from those we concluded that at different Sc the correlation between T and $\hat{\omega}$ is always very good; moreover we understood that initial conditions, are irrelevant.

In the 3D case, as evinced from the equations, the scalar is not correlated with any vorticity components or theirs combination. The full simulation has confirmed these considerations and has proved also for the free ring that T does not correlate neither with ω_θ nor with $\hat{\omega}$. Our simulation confirmed the conclusions by Hussain & Melander [5] obtained in a different vortical structures. The influence of Sc in the 3D case is very interesting but it requires much finer grid, this further study will be performed in the near future.

Acknowledgements

The research was supported by Ministero dell'Universita' e della Ricerca Scientifica e Tecnologica through a grant MURST 40%.

References

[1] Auerbach, D. " Some open questions on the flow of circular vortex rings", Fluid Dynamics Research,3,209-213, 1988.
[2] Glezer, A. " The formation of vortex rings ", Phys. Fluids, 31, pp.3532-42, 1988.
[3] Lamb H. *Hydrodynamics* Cambridge University Press 1932.
[4] Widnall S. E., "The structure and dynamics of vortex filaments" Annual Review of Fluid Mech.,pp.141-166,1975.
[5] Hussain F. and Melander M.V. "Understanding turbulence via vortex dynamics " Studies in Turbulence. Spinger-Verlag 157-178,1991.
[6] Arakawa A.," Computational design for long term numerical integration of the equations of fluid motion: Two-dimensional incompressible flow. Part I", J.of Comp. Physics,1,pp.119,143,1966.
[7] Orlandi P. "Numerical Simulation of Vortices Motion in Presence of Solid Boundaries" Proceedings of 8th GAMM conference on Numerical Methods in Fluid Mechanics Edited by P. Wesseling, (vieweg, Braunschweig) pp. 436-441,1989.
[8] Rai M.M., & Moin P.,"Direct simulations of turbulent flow using finite-difference schemes", J. Comp. Phys.,96,pp.15-53,1991.
[9] Verzicco, R. and Orlandi P., "A finite difference scheme for direct simulation in cylindrical coordinates ", Presented at the ICOSAHOM 92, Montpelier, 1992.
[10] Shariff, K. Leonard,A. and Ferziger,J.H., "Dynamics of a class of vortex rings " NASA TM-102257,1989.
[11] Stanaway S. "A numerical study of viscous vortex rings using a spectral method' 'Ph.D. Thesis Aeronautics and Astronautics Dept.,Stanford University., Also NASA TM 101041,1988.
[12] Melander, M.V., McWilliams J.C. & Zabusky N.J. "Axisymmetrization and vorticity gradient intensification of an isolated two-dimensional vortex through filamentation" J.Fluid Mech.,178,137-159,1987.
[13] Knio O.M., & Ghoniem O.M."Numerical study of three-dimensional vortex method", J. Comp. Phys.,86,75-106,1991.
[14] Sullivan, J.P., Widnall, S.E. & Ezekiel, S. "Study of vortex rings using a laser doppler velocimeter" A.I.A.A., 11, pp.1384-1389,1973.
[15] Verzicco R., Shariff K.R., & Orlandi P., "A numerical simulation of three-dimensional vortex rings instabilities" ,In printing, 1992.

TURBULENT EDDY STRUCTURES, COMBUSTION AND CHEMICAL REACTIONS

J.P. CHOLLET
Laboratoire des Ecoulements Géophysiques et Industriels
Institut de Mécanique de Grenoble

ABSTRACT. Turbulence strongly influences transport and disperson of contaminants advected by velocity fields, especially in the case of fluids made of species which undergo chemical reactions. Large scale coherent vortices take a prominent part in mixing processes as demonstrated here from numerical simulations of two and three dimensional mixing layers. Phenomena are even more intricate for turbulent combustion because of the density variations which are induced by heat released from chemical processes. Moreover, supersonic conditions let pressure waves trigger specific flow instabilities. In order to model all phenomena involved in these gaseous streams, numerical methods have to be carefully selected since flow structures must develop spontaneously whatever they are, vortices or shock waves.

1. Introduction

Turbulent flows, especially when they are submitted to a mean shear, develop eddies which determine transport, mixing, and diffusion of scalar quantities such as temperature T or species concentrations C_i. Such structures generally develop from instabilities and appear as vortices. Each of these vortices keep an identity during some period of time t. The coherence of such flow structures is usually defined referring to the distribution of vorticity field $\boldsymbol{\omega} = \boldsymbol{\nabla} \otimes \mathbf{u}$ in space $\mathbf{x}$. They are mostly produced by shear instabilities such as Kelvin Helmoltz ones which grow from velocity mean profiles as soon as these profiles exhibit inflexion points [5].

Each vortex structure can be characterized by its size which defines a length scale δ_v. As a matter of fact, various values of δ_v coexist in a same flow because instability processes and therefore coherent structures can develop at both large and small scales. Smallest scale structures are not specifically taken into consideration here although they are supposed to significantly influence diffusive and dissipative processes. Interest is mainly focused on those structures which are large enough to depend on boundary or initial conditions and which are the most energetic. These structures strongly determine flow mixing.

The interest for eddy dynamics is not only justified by velocity $\mathbf{u}(\mathbf{x}, t)$ or vorticity $\boldsymbol{\omega}(\mathbf{x}, t)$ field predictions but also by various effects on transport, mixing and diffusion of any property $\theta(\mathbf{x}, t)$ carried by the flow. Fluid composition is affected in this way by turbulent flow structures and its evolution is followed through scalar

J.T. Beale et al. (eds.), Vortex Flows and Related Numerical Methods, 315–324.

quantities which are concentrations of species $C_i(\mathbf{x},t)$ rewritten as non dimensional molar fraction $X_i(\mathbf{x},t)$ or mass fraction $Y_i(\mathbf{x},t)$ for each of the species i. Species i can be subjected to chemical reactions and produced or consumed at a rate $\dot{w}_i(\mathbf{x},t)$. These $\dot{w}_i(\mathbf{x},t)$ can ve viewed as the variables of main interest and are easily accessible in numerical simulations contrary to experiments.

Non compressibility assumption ($\boldsymbol{\nabla}\mathbf{u}(\mathbf{x},t)=0$) allows to restrict the complexity of interacting phenonema (section 2) and, nevertheless, provides realistic models for chemical engineering in liquids, at least when neglecting any heat release. This approximation holds also for strong dilutions of reactive species as in atmospheric pollution problems. Any attempt to model a wider class of reacting flows needs a more realistic account of thermodynamics and energy exchanges under various forms, especially because of heat released from reactions as in combustion processes (section 3) .

In both cases, the basic formalism is built on a continuous fluid medium approach with partial derivative equations whose either elliptic or hyperbolic characters predominate depending on the either nondivergent or compressible frame of approximation in use. Eulerian coordinates with velocity $\mathbf{u}(\mathbf{x},t)$ as the kinematic variables are easy to implement in algorithms and can accomodate a large family of boundary conditions and flow geometries. Moreover, these techniques are very efficient for the calculation of diffusive and viscous mechanisms, the significance of which can be enhanced by chemical reactivity. Lagrangian techniques are used here only to trace particule trajectories computed from time evolving velocity fields. They are also expected to be helpful to derive future subgrid scale models.

Various numerical techniques [6] are appropriate to deal with reacting flow simulations. Usual finite differences and pseudo spectral methods have been used for non compressible flows in section 2. A numerical code has been specifically written for gaseous flow and combustion in section 3, on the basis of Godunov method and PPM approximation [4].

2. Vortices in non compressible chemically reacting flows

A typical picture of reactive shear flow is given under the form of the mixing layer of figure 1. Stream 1 with velocity u_1 is made of species A when stream 2 with velocity u_2 is made of species B.

A typical time to characterize turbulent large scales is δ_i/U where δ_i is the initial thickness of the layer and $U=(u_1-u_2)$. Transition to turbulence through unstable Kelvin Helmoltz modes has been widely documented both experimentally and numerically [5]. Layer thickens through a pairing process which interacts with more intricate dynamics as three-dimensionality gets stronger.

Chemical species A and B react together to produce species P

$$\nu_A A + \nu_B B \longrightarrow \nu_P P \tag{1}$$

with ν_i as stochiometric coefficients. Fluid composition evolves in both time t and space $\mathbf{x}$ and is described with molar $X_i(\mathbf{x},t)$ or mass $Y_i(\mathbf{x},t)$ fractions for each of the species A, B and P.

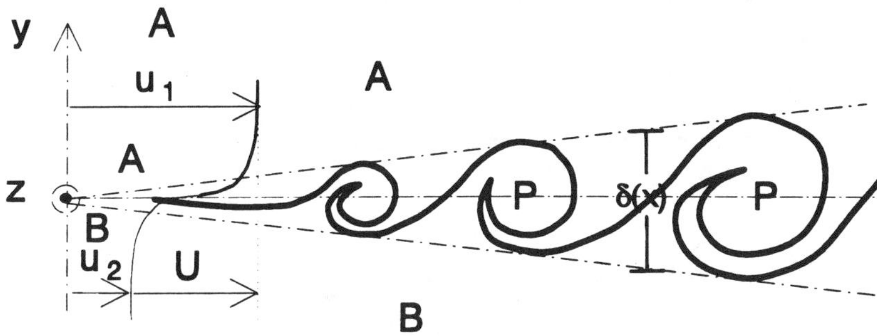

Fig. 1. Reactive mixing layer

Within this first assumption of incompressible flows, reactive scalars are assumed not to influence the kinematic field. Vorticity $\boldsymbol{\omega}(\mathbf{x},t) = \boldsymbol{\nabla} \otimes \mathbf{u}(\mathbf{x},t)$, together with the incompressibility condition $\boldsymbol{\nabla}\mathbf{u}(\mathbf{x},t) = 0$ evolves according to :

$$\frac{\partial \boldsymbol{\omega}}{\partial t} + (\mathbf{u}(\mathbf{x},t)\boldsymbol{\nabla})\boldsymbol{\omega}(\mathbf{x},t) = \nu \boldsymbol{\nabla}^2 \boldsymbol{\omega}(\mathbf{x},t) + (\boldsymbol{\omega}(\mathbf{x},t)\boldsymbol{\nabla})\mathbf{u}(\mathbf{x},t) \tag{2}$$

where ν is the kinematic viscosity. Eq. 2 takes account of transport, diffusion but also production-consumption of vorticity, the latter term being zero in two dimensional approximations only.

Scalar molar fractions X_i evolve also according a transport-diffusion process with sources added, because of chemical reactions

$$\frac{\partial X_i}{\partial t} + (\mathbf{u}(\mathbf{x},t)\boldsymbol{\nabla})X_i(\mathbf{x},t) = D_i \boldsymbol{\nabla}^2 X_i(\mathbf{x},t) + \epsilon_i \nu_i \dot{w}(\mathbf{x},t) \tag{3}$$

with $\epsilon_A = \epsilon_B = -\epsilon_P = -1$, and D_i the diffusivity. The Schmidt number Sc_i is the rate of the kinematic viscosity ν to this scalar diffusivity D_i. Diffusivity and viscosity are, at least in three dimensions, models for physical processes which take place at molecular length scales.

The reaction rate is assumed to be $\dot{w} = (X_A^{\nu_A} X_B^{\nu_B})/\tau_c$ where the time τ_c has a constant value which is typical of the chemical reaction under interest. This formalism stands for a crude model of chemical processes which actually take place at length scales much smaller than any scale explicitely taken into consideration in numerical flow simulations. Chemistry does not introduce any length scale of its own but it provides a time scale τ_c which, in the simplified approach of this section 2, is assumed to remain constant.

Evolutions with time of several vector or scalar fields are computed through numerical simulations : velocity, vorticity $\boldsymbol{\omega}$, molar or mass fractions (X_i or Y_i), and also reaction rates $\dot{w}$. These $\dot{w}$ allow to find out locations and times where most of reactivity takes place. Snapshots of isocontours of these fields can be drawn [12]. Various statistics can be also calculated from these fields for the purpose of a more quantitative approach : for instance, examining spectra shows that chemical reactions affect as much large scales as small ones. Probability density functions plots allow to conclude about the intermittency of reactive patches as reaction gets faster.

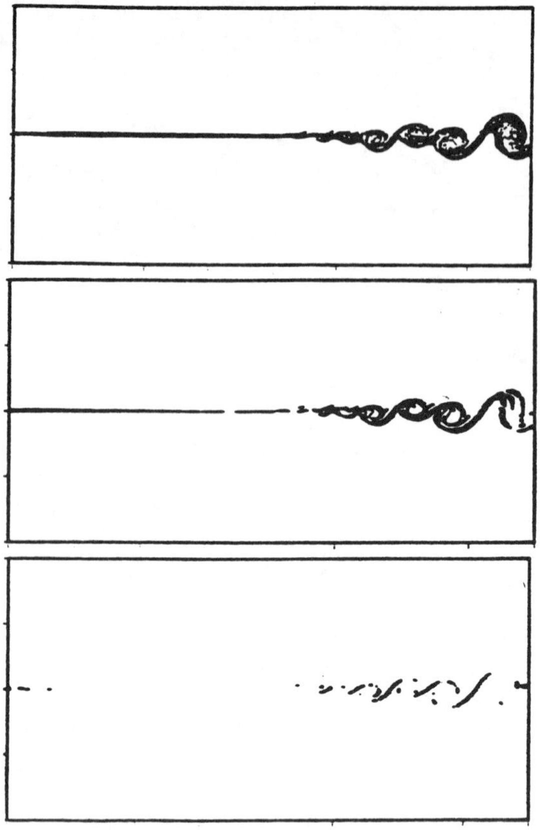

Fig. 2. Two dimensional mixing layer, iso-curves of reaction rate $\dot{w}(\mathbf{x}, t)$ for Da= 0.1, 1., 10 (from top to bottom)

Extensive computations have been performed, for instance in [12], to investigate the effects of parameters such as the Damkhöler number (chemical time), the Schmidt number (scalar diffusivity), the stochiometric coefficients (ν_A, ν_B). Isocontours of reaction rates of fig.2 demonstrates the effect of chemical time scale through the Damkhöler number $Da = \tau_t / \tau_c$ where τ_t characterizes an initial turbulent time scale.

Within this two dimensional approximation, mixing layer evolves through pure interactions between vortices. Eddies in interaction are almost identical one to the other. Reactivity develops on the support made of these eddy structures and more or less spreads out depending on Damkhöler number. As the layer thickness δ increases along the streamwise x direction, the effective Damkhöler number $Da^{(e)}(x) = \delta(x)/U$ also increases and the reaction can be viewed as becoming faster when going downstream.

Three dimensionality is required in order to produce vorticity stetching (Eq. 2) and develop a wider variety of length scale interactions and then give rise to more realistic flow models. A significant drawback of these three dimensional computations is the huge number of computations nodes or, more generally, the many degrees of freedom, to deal with. In this section, where space resolution is restricted to 64^3 computational nodes, a detailed description of eddy interactions has been prefered to an accurate model of layer streamwise evolution. Periodic streamwise boundary conditions are used within the context of the so called temporal approximation which models an evolution along space x streamwise direction through an evolution with time t. This usual approximation yields unaccurate estimations of layer thickening rates, especially for scalars which are passively advected by the velocity field. Reaction rates $\dot{w}$ whose high values can pile up at the periphery of eddies and then along the borders of the mixing region, at least for fast reactions (large Damkhöler number values), are especially sensitive to this underestimation. A quasi spectral method is used in order to take advantage of periodic (in two directions) and free-slip (in the last one) boundary conditions. Such a space scheme provides accurate estimations of space derivatives for both advection term $\boldsymbol{\nabla}\theta$ and diffusive ones $D_i\boldsymbol{\nabla}^2\theta$, whatever salar $\theta(\mathbf{x},t)$ is. Analysis of Schmidt number (Sc) dependency would have been much less sensitive with other numerical methods which generally produce significant numerical diffusivity. Nevertheless, chemical terms add complexity to the calculation since reaction rates $\dot{w}$ need to be estimated in physical space. Moreover errors are induced by truncated Fourier modes and violate accuracy requirements for fast reactions ($Da > 10$). Analysis of $X_i(\mathbf{x},t)$ and $\dot{w}(\mathbf{x},t)$ fields has been carried out through plotting 3D pictures and computing statistics as spectra and probability density functions. Reaction takes place around the surface $\Phi(\mathbf{x},t) = 0$ where $\Phi(\mathbf{x},t) = X_A/\nu_A - X_B/\nu_B$ is a conserved (chemically inert from eq.(3)) scalar but does not fill it up, depending on Da values, as observed in [12].

Lagrangian methods to compute velocity fields are not discussed here. These methods are less easy to implement in three than in two space- dimensions because of the difficulty to track down vorticity filaments or, failing that, vorticity patches. Moreover, an accurate description requires heavy computations with many patches, the number of them guaranteing the accuracy of results only if they are properly distributed. Also lagrangian fluid particles are used here only to analyse the results obtained from numerical simulations, especially for relatively short times. Several particles are introduced in the flow and then tracked down using standard Runge Kutta scheme together with second order interpolation to integrate the velocity field $\mathbf{u}(\mathbf{x},t)$ (figure 3). Each of the particle can record the evolution with time of its composition (through the X_is) and the reaction rates $\dot{w}(t)$ it has experienced.

3. Turbulent gas flow under combustion

Modelling turbulent combustion or any reactive gaseous flow need to take account of the fact that pressure p, temperature T and densiy ρ are also thermodynamics variables. Very small length scale features are modelled through local equilibrium algebraic equations; so is the ideal gas relation $p(\mathbf{x},t) = \rho(\mathbf{x},t)/RT(\mathbf{x},t)$. Chemistry

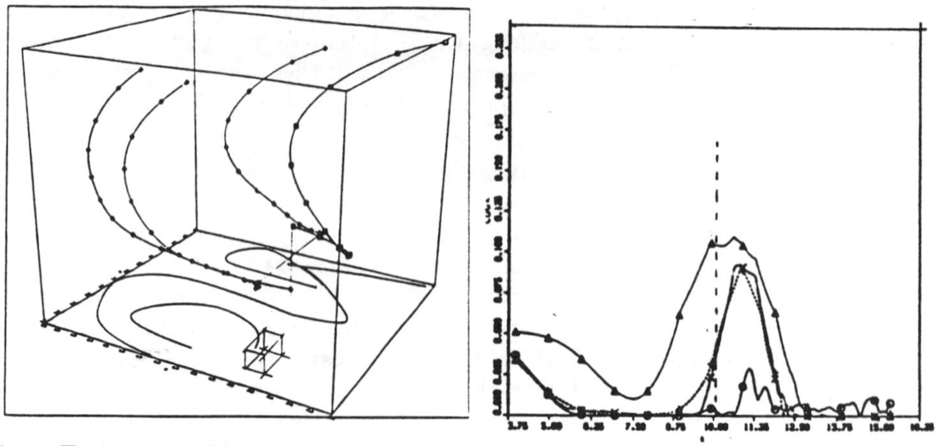

Fig. 3. Trajectories of lagrangian tracors in a mixing layer and reaction $\dot{w}(t)$ recorded by each of these particles

kinetics is also determined by length scales which are so small that the reaction rate $\dot{w}(\mathbf{x},t)$ is written, at every location $\mathbf{x}$ and time step t, according to Arrhenius law :

$$\dot{w} = BT^{\alpha}\exp(-\frac{T_A}{T})\frac{\rho^{(\nu_A+\nu_B)}}{M_A^{\nu_A}M_B^{\nu_B}}Y_A^{\nu_A}Y_B^{\nu_B} \tag{4}$$

where the constants B, α and the activation temperature T_A are specific to the chemistry under consideration. Fluid composition is handled through mass fractions Y_i of species i. M_A and M_B are molar mass of species A and B. Characteristic properties of the gaseous mixture evolve in time and space ; any f (for instance $\bar{c}_v$, $\bar{c}_p$, average molar mass $\bar{M}$), are calculated as : $f(\mathbf{x},t) = \sum_{i=1}^{n} f_i Y_i(\mathbf{x},t)$. The strong dependency of $\dot{w}$ with T often makes the system of equations to be solved very stiff [10; 9], contrary to problems under consideration in section 2.

Gas flow evolves according to continuity equations for *(i)* mass, *(ii)* momentum, *(iii)* energy, written together in a conservative form :

$$\frac{\partial \mathbf{U}}{\partial t} + \frac{\partial \mathbf{F(U)}}{\partial x} + \frac{\partial \mathbf{G(U)}}{\partial y} + \frac{\partial \mathbf{H(U)}}{\partial z} = \mathbf{D} + \mathbf{S} \tag{5}$$

with the unknown functions :

$$\mathbf{U}(\mathbf{x},t) = (\rho,\ \rho u_x,\ \rho u_y,\ \rho u_z,\ \rho E,\ \rho Y_i,\ \rho Y_{i+1}, \ldots) \tag{6}$$

and total energy : $E = \frac{1}{2}({u_x}^2 + {u_y}^2 + {u_z}^2) + \frac{\bar{c}_v}{\bar{M}}$. and with

$$\mathbf{F(U)} = (\rho u_x,\ \rho u_x^2 + p,\ \rho u_x u_y,\ \rho u_x u_z,\ (\rho E + p)u_x,\ \rho Y_i u_x,\ \rho Y_{i+1} u_x,\ \ldots) \tag{7}$$

and similar expressions for $\mathbf{G}$ and $\mathbf{H}$. As chemical reactions can build up strong gradients (∇T or ∇Y_i), $\mathbf{D}$ diffusive terms need to be accurately taken into consideration. $\mathbf{S}$ are computed from (4). Formulation is detailed in [7]. The numerical scheme is splitted into basic operators so that solutions at time $n+2$ are computed from solutions at time n through :

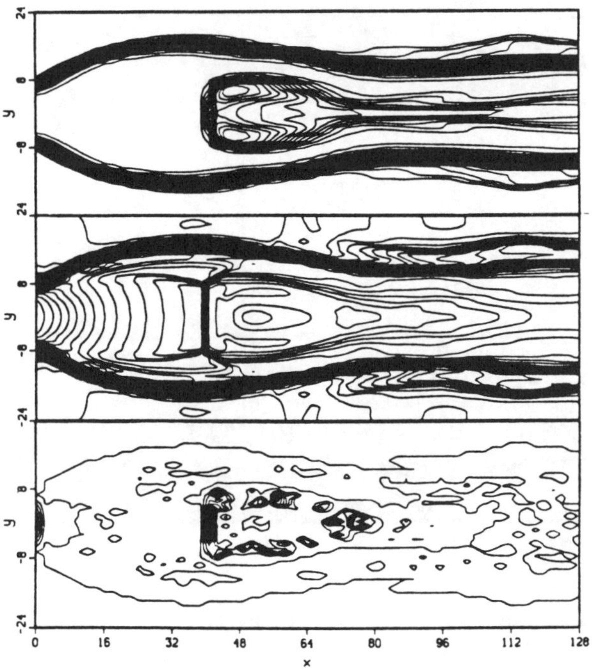

Fig. 4. Numerical three dimensional simulation of an underexpanded round jet : isocontours of NO mass fraction, temperature, NO production rate (from top to bottom) in a cross section

$$\mathbf{U}^{(n+2)} = \Psi^{\frac{1}{2}} \mathcal{L} \Psi \mathcal{L}^{-1} \Psi^{\frac{1}{2}} \mathbf{U}^{(n)} \tag{8}$$

$\mathcal{L}$ stands for the hyperbolic operator which is, itself, splitted into $\mathcal{L} = L_x L_y L_z$ ($\mathcal{L}^{-1} = L_z L_y L_x$) with basic operators L_a for each space direction a. Operator L_a is made from a Riemann solver and a PPM (parabolic piecewise method) interpolation [4].

Turbulent combustion, especially in the cases of supersonic flows presented hereafter, leads to consider two completely different types of flow structures : eddy vortices and pressure waves. These structures of course interact. Such waves when they grow stiff enough act as strong vorticity sources, for instance in cellular detonations [2] or even in high speed jets or shear layers. Chemical systems under consideration are the NO thermal production and $H_2 - O_2$ combustion.

Mixing layers between rigid flat walls and underexpanded jets have been considered because of their relevance to supersonic combustion engines [13]. Behind a strong shock, (fig. 4), high values of vorticity develop, which improves species mixing ant can then trigger chemical reactions. Two regions are of special interest : *(i)* the periphery of the jet where small scale structures develop with vorticity aligned along streamwise direction x *(ii)* larger scale structures just behind the main shocks with high temperatures and recirculating eddies which favour the production of NO species (figure 4 from [8]).

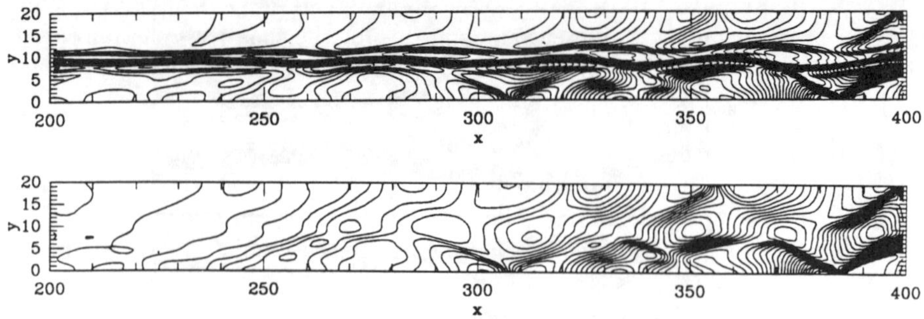

Fig. 5. Two dimensional mixing layer between walls : density and pressure isocontours

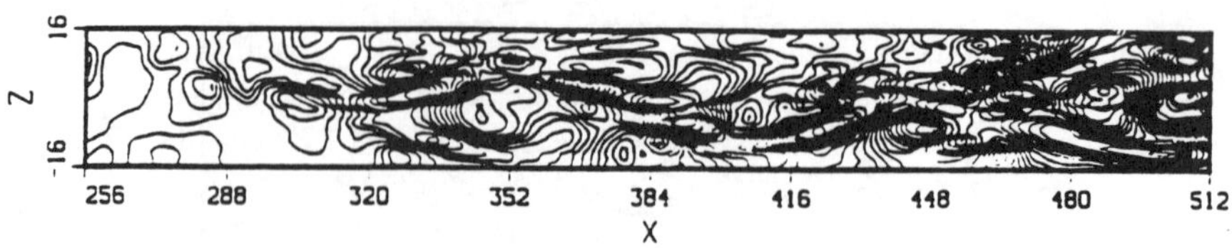

Fig. 6. Three dimensional simulation of a mixing layer between walls : cross section of pressure field in spanwise-streamwise directions

Mixing layers between rigid walls also demonstrate what numerical simulations can do when pressure waves triggered by high speed flows and reflected by walls interact with shears (fig.5). Frequency analysis of numerical results shows that observed unstable modes are in good agreement with classical results from linear theory. Moreover, real dynamics seem to be monitored mainly by oblique and therefore three dimensional unstable modes (fig.6).

4. Conclusions and prospects

Numerical simulations which are based on eulerian formalisms are very efficient to model large scale flows since large scales strongly depend on boundary conditions. Moreover so many numerial techniques have been developed to discretize partial derivative equations that the efficicency of eulerian methods is constantly improving, even when flow exhibits discontinuities such that shocks in high speed compressible flows, which nevertheless requires appropriate schemes.

The main weakness of numerical simulations remains the incapability to deal simultaneously with both large and small scales, at least in three dimensions which is the only case relevant for real turbulence. Therefore any model which aims at taking explicit account of viscous-diffusive scales, is restricted to rather moderate values for Reynolds and Peclet numbers. In order to simulate the evolving fluid composition, several scalar fields have to be considered. Depending on the reactions under interest and the level of complexity of chemical schemes, combustion in gas requires to manipulate more than 10 scalar fields. Numerical difficulties can arise because of the stiffness of differential equations for the chemical reaction rate (eq.4) with possibly very small chemical times τ_c especially when temperature T increases.

Eddy structures build up from instability processes, so the only eddies or vortices to be produced are those whose characteristic length scales are about the same as the scales excited by unstable modes. Therefore numerical simulations are suitable only for unstabilites whose wave lengths can be computed explicitely, small ones in full (or direct) simulation, large ones (for instance Kelvin Helmoltz eddies) in large eddy simulations. Optimizing numerical simulations of chemically reacting flows needs to deal with two conflicting requirements : *(i)* increasing space resolutions, that is the number of computation nodes, in order to widen the range of explicitely computed turbulent length scales and then allow higher Reynolds and Peclet numbers in direct simulations (simulations up to Kolmogorov and molecular diffusive scales) *(ii)* detailing chemistry kinetics in order to get more realistic chemical processes although those reaction steps (j) which are characterised by very short time scales $\tau_{c(j)}$ cannot be taken into explicit account. Parallel computers are giving a chance to improve the range of Reynolds numbers or the details of the chemical schemes, or both of them depending on computational techniques to be available in future.

Promising prospects seem to be offered by appropriate combination between eulerian and lagrangian methods. Because of fluid continuity, it does not seem there is more arbitrariness in eulerian formulation than in lagragian ones and fluid particles whose trajectory can be traced are defined as fluid particle whose properties depend more or less on the model in use. The two major restrictions of lagrangian techniques seem to be : *(i)* the need for a huge number of fluid elements to be followed, especially if large scales are expected to be calculated since many of the particles can be lost through boundaries or can possibly pack together letting wide space domains without any tracor, *(ii)* the difficulty of modelling diffusive processes which are of special interest especially in this problem of scalar mixing and reactive flows. On the opposite, eulerian methods were observed to conveniently reproduce phenomena strongly determined by boundary conditions, or initial ones in temporal approximation. A wide variety of numerical scheme is available and, especially when dealing with simple geometries and structured space mesh, efficient algorithms are known. The main restriction of these eulerian methods is their incapability of dealing with both largest and finest scales at the same time. Fully direct numerical simulations are relevant to basic studies but simulation of real industrial or natural flows correspond almost always to relatively large length scales problems. Consequently, future development can combine : *(i)* a description of large scales in eulerian formulation calculating their evolutions in both time and space ; temporal approximation can bring information for relatively small scales phenomena but must not be used whenever the flow depends on boundaries ; *(ii)* a lagrangian description of small scale phenomena which requires to follow trajectories for relatively short paths. The evolution of variables related to the fluid composition can be recorded on each of the particles and then locally modified by the reaction activity.

Acknowledgements

R.J. GATHMANN, M.R. VALLCORBA, F.MATHEY are acknowledged for their contribution, DRET (Ministry of defence) and SNECMA for their support, respectively in incompressible and gaseous flows. Simulations were performed on the CRAY II thanks to the scientific committee of the CCVR (Centre de Calcul Vectoriel pour la Recherche), Palaiseau, France. The Laboratoire des Ecoulements Géophysiques et Industriels is associated to CNRS and is part of Université J.Fourier and Institut National Polytechnique de Grenoble.

References

[1] Chollet, J.P., Gathmann, R., Vallcorba, M.R., 1991, Reactive mixing layers and Turbulent combustion, *Third European Turbulence Conference*, Stockholm ,july 3-6, 1990 , Advances in Turbulence 3, Springer Verlag

[2] Chollet, J.P., Gathmann, R.J., 1991, Reactive mixing layer between pressure reflecting boundaries, *8th symp. on turbulent shear flows*, Munich, sept. 9-11

[3] Chollet, J.P., Vallcorba, M.R., Gathmann, R. , 1990, Reactive scalar as a tracor of organized structures in turbulent mixing layers, *Organized structures and turbulence in fluid mechanics*, Kluwer Acad. publ.,

[4] Colella, P., Woodward, P.R., 1984, The piecewise parabolic method (PPM) for gas-dynamical simulations, *J. Comp. Phys.*, **54**, 174-201.

[5] Drazin, P.G., Reid, W.H., 1981, Hydrodynamic stability, *Cambridge University Press*

[6] Fletcher, C.A.J., 19898, Computational techniques for fluid dynamics, *Springer Verlag.*

[7] Gathmann, R.J., 1991, Direct numerical simulations of supersonic turbulent combustion, *4th symp. on comp.fluid dynamics*, Davis, California, 8-12 sept.

[8] Gathmann, R.J., Baudoin C., Chollet J.P, 1992, Direct numerical simulation of nitric oxyde evolution in underexpanded jets, *ASME- paper*, 92-GT-372

[9] Gathmann,R.J, Hebeker F.-K., 1989, Modelling a one-dimensional convection-diffusion problem with an Arrhenius source term, *IBM Technical Report*, TR 89.11.019

[10] Gathmann, R.J., Hebeker, F.K., 1991, On the Numerical Treatment of a One-Dimensional Convection-Diffusion Problem with an Arrhenius Source Term, *ZAMM*, Z.angew.Math.Mech. 71 (1991) 6, T 668- T 670.

[11] Rudman, M., Gathmann, R.J., Lesieur, M., 1992, Instability in variable density round jets, *Proceedings of 11th Australien Fluid Mech. Conf.*, Hobart, Aus, Dec 14-18

[12] Vallcorba, M.R., 1992, Ecoulements cisaillés réactifs : simulations numériques de couche de mélange, *thèse de l'Université J.Fourier- Grenoble 1.*

[13] Zapryagaev, V.I., Solotchin, A.V., 1990, Three dimensional structures of the flow in supersonic underexpanded jet, *Separated Flows and Jets*, IUTAM Symposium Novosibirsk, Springer Verlag

VORTEX GENERATION AND EVOLUTION IN NUMERICAL SIMULATION OF TRANSITIONAL SHEAR FLOWS

P. Comte, E. David, F. Ducros & Y. Fouillet
Laboratoire des Ecoulements Géophysique et Industriels
Institut de Mécanique de Grenoble
*CNRS - UJF - INPG**
B.P. 53X
F38041 Grenoble-Cedex
Tel: (33) 76 82 51 21; Fax : (33) 76 82 52 71; E-Mail: comte@imga.img.fr

ABSTRACT. Vortex generation is investigated by means of three-dimensional numerical simulations with intensive use of visualizations. The following cases are considered:

- direct simulations of temporally-growing (streamwise-periodic) incompressible mixing layers for 2 different types of initial perturbations
- direct simulations of temporally- and spatially-growing compressible mixing layers for different convective Mach numbers
- a large-eddy simulation of a weakly-compressible spatially-growing boundary layer over an adiabatic flat plate
- a large-eddy simulation of the flow over a heated wedge
- a direct simulation of a high-Mach number temporally-growing boundary layer

Strong analogies are found between the routes of transition observed, with a predominance of Λ-vortices except in one case (quasi-two-dimensionally-forced incompressible mixing layer). This suggests a possible universality.

1. Introduction

The experiments of mixing layers performed at CalTech by Roshko and his group (see Brown and Roshko, 1974, for example) showed the presence of large-scale organized vortices even extremely far downstream of their origin. It turned out that these vortices resulted from the growth of hydrodynamic instabilities yielding roll-up of fundamental vortices and their successive pairings. Such a causal relationship between instabilities and coherent structures is so often encountered in numerical simulations of turbulent flows that it might be a general feature of turbulence: remember for example the video cassette presented by Meneguzzi at the workshop (Meneguzzi, 1992),

* Centre National de la Recherche Scientifique; Université Joseph Fourier de Grenoble; Institut National Polytechnique de Grenoble

J.T. Beale et al. (eds.), Vortex Flows and Related Numerical Methods, 325–340.

showing the formation of long thin vortices in a simulation of isotropic turbulence. The vortices were visibly the result of the roll-up of vortex sheets forming between fluid blobs pressed together. This was again evidence of Kelvin-Helmholz instability acting in local shear zones.

This is one more reason to invite numerical simulations of the transition to turbulence of shear flows such as mixing layers or boundary layers in which shear is prescribed either in the initial or the boundary conditions. In the next two sections, we will present simulations of mixing layers and consider the effect of compressibility on the coherent structures. Sections 4, 5 and 6 will be devoted to boundary layers and feature both low and high Mach number simulations of their transition to turbulence.

2. Incompressible Mixing Layers

We will first assume the flow periodic in the streamwise and spanwise directions. The latter assumption is a commonly-accepted modelization of an infinitely-wide wind tunnel. The former is much more questionable : if, in Brown and Roshko's experimental facility, we follow the vortices in their downstream motion with a video camera, we can have the impression that nothing comes into nor gets out of the field of vision, which would justify the assumption of streamwise periodicity. When we do that in practice, we turn a spatially-growing problem into a temporally-growing one and speak of *temporal approximation.* By means of a Galilean translation, it is possible to relate the temporal growth-rate of the periodic layer to the spatial spreading of the experimental flow: we then find values underestimated by about 50%, which shows the limit of the temporal approximation (Lesieur *et al.*, 1988). However, the dynamics of the vortices seem to be the same and the saving in computational cost due to the restriction of the domain and the simplicity of the boundary conditions is extremely appealing. Besides, linear stability analysis is much easier to perform in the temporal case, which is therefore better documented than the spatial case.

Figure 1 shows the vortical structure of a temporal incompressible mixing layer, visualized through vortex lines and an isobaric surface of level lower than the mean pressure [1]. The initial conditions which led to this vortex lattice consisted of a laminar flow $\bar{u}(y) = U \tanh(2y/\delta_i)$ of vorticity thickness δ_i, perturbed by small-amplitude random noise with a spectrum peaking in the large scales. The instable modes which responded the most to this forcing were a pair of oblique Kelvin-Helmholtz modes of streamwise period equal to twice the most amplified wavelength λ_a predicited by the linear stability theory (Michalke, 1964). They yielded the vortex lattice shown in Figure 1, formed by 4 oblique Kelvin-Helmholtz vortices undergoing helical pairings.

[1] it is indeed possible to show by means of simple analytical arguments that coherent vortices induce pressure troughs in their cores (see for example Lesieur, 1992)

Notice also a hairpin vortex stretched in the streamwise direction between the large vortices. Its strongest leg induces a depression strong enough to appear on the isobaric surface.

By contrast, Figure 2 shows a quasi-two-dimensional vortex structure obtained in the same conditions as before, but with a quasi-two-dimensional initial forcing (see Comte *et al.* 1992a for details). This discrepancy is not confined to the shape and arrangement of the large-scale coherent vortices, but it also affects the dynamics of all the turbulent scales: in the first case, short $k^{-5/3}$ sub-ranges are found in the kinetic enery and passive-scalar variance spectra, whereas the second case features spectra with slopes comprised between -3 and -4 for the kinetic energy and between -1 and -2 for the variance of the passive scalar (see Figure 3). This strong difference in dynamics is also evident when looking at the peak value of the vorticity modulus. In the second case, it remains bounded by the initial value ω_i, whereas in the first case, it grows up to 4 ω_i, due to intense vortex stretching.

3. Compressible Mixing Layers

Helical pairings are also found in temporal simulations of weakly-compressible mixing layers: see for example Figure 4, obtained from a simulation at a convective Mach number $M_c = 0.3$, with a white-noise initial forcing. The corresponding spatial simulation is presented in Figure 5: it shows quasi-two-dimensional Kelvin-Helmholtz billows undergoing branchings in some locations of the span, as observed experimentally by Browand and Ho (1983) in the incompressible regime. It thus seems evident that these branchings are nothing but helical pairings.

When repeating the simulation at higher M_c, (Figures 6 and 7), one finds that the helical pairings are inhibitted for $M_c \leq 0.6 \approx 0.7$. The flow then becomes dominated by Λ-shaped vortices facing each other (as sketched in Figure 8), analogous to those found by Sandham and Reynolds (1991) in similar calculations, but with deterministic initial forcing at the fundamental mode of instability instead of white noise. Figure 9 shows the spatial counterpart of Figure 6. For higher Mach numbers, the reader is referred to the extensive study of Gathmann *et al.* (1992).

4. Weakly-Compressible Boundary Layer Over An Adiabatic Flat Plate

Figure 10 shows iso-surfaces of streamwise vorticity and low pressure in a large-eddy simulation of a spatially-developing boundary layer at Mach 0.5, for a Reynolds number, based upon the upstream displacement thickness, of 1000. The resolution is of $320 \times 32 \times 20$ grid points. A laminar profile is imposed at the inflow boundary, with two types of perturbations: a deterministic forcing at the fundamental mode (2D Tollmien-Schlichting waves of amplitude 8 10^{-3} U_∞) plus three-dimensional white noise of amplitude 5 10^{-3} U_∞. The development of the T.S waves is visible on the isobaric surfaces. Further downstream, three-dimensionality appears in the form of Λ-shaped vortices. Their staggered arrangement[2] proves that they result from the growth of oblique waves of streamwise period double of that of the T.S. modes, *viz.*, a sub-harmonic mode of secondary instability as investigated by Herbert (1988) from the point of view of Floquet's stability analysis. Small-scale turbulence appears first at the tip of the Λ-vortices, where the streamwise velocity profile is locally inflexional. This causes the development of a Kelvin-Helmholtz-type instability of growth-rate much larger than that of the T.S waves, which yields very fast breakdown into turbulence. A subgrid-scale model is then needed, otherwise the calculation would blow-up at this point. The one used in the present simulation is derived from the *structure-function model* of Métais and Lesieur (1992) and adapted to the boundary layer's case by Normand and Lesieur (1992).

5. Low-Mach-Number Boundary Layer Over A Heated Wedge

Even more than in mixing layers, transition in boundary layers is known to be extremely sensitive to the initial conditions (shape of the velocity and temperature profiles; nature of the perturbations...). In order to prove that the simulation presented in the previous section is relevant to aeronautical or aerospatial applications, we now include the leading edge of the wall into the computational domain. The inflow boundary condition is now well upstream of this leading edge, and the flow imposed there is uniform in the mean (velocity U_∞), with white noise perturbations of amplitude 10^{-3} U_∞. In fact, what is really simulated is the upper-half of the flow over a wedge of angle 20°. This new geometry requires extra computational effort (the code has been re-written in orthogonal curvilinear co-ordinates and the mesh refinment near the leading edge implies smaller time-steps than in the previous section). In order to have a fairly rapid transition, we consider the case of an isothermal wedge at a temperature T_w equal to 2.5 T_{ad}, the adiabatic recovery temperature. Thus, although the Mach number is still small (0.5), the compressibility effects due to the wall heating accelerate the transition: the time- and spanwise-averaged velocity profile $\bar{u}(x,y)$ which establishes before transition presents a generalized inflexion point, *i.e.* a distance y_S to the wall where $\partial/\partial y(\rho \partial \bar{u}/\partial y)$ goes to zero and changes sign. This ensures that the dominant instabilities are of inviscid nature (Lees and Lin, 1946), like the Kelvin-Helmholtz instability of mixing layers. They thus have growth rates much larger than the Tollmien-Schlichting waves encountered in the previous section, which are of viscous origin. Consequently, with a resolution

[2] Periodic boundary conditions are prescribed in the spanwise direction

of $237 \times 30 \times 32$ grid points and a cost of about 10 C.P.U. hours of cray 2, it was possible to reach a turbulent regime before the downstream boundary of the domain, with the aid of a sub-grid scale model similar to that used for the flat-plate boundary layer presented before.

Figure 11 shows a top view of the whole computational domain. The surface represented is a surface iso-norm of the vorticity $||\boldsymbol{\omega}|| = 0.09\ \text{Max}||\boldsymbol{\omega}||$. This threshold is low so that the surface visualizes the outer region of the boundary layer. The jagged edge of the surface (on the left part of the figure) corresponds to the leading edge of the wedge. Figure 12 is a blow-up of the downstream half of the domain, showing the isosurface $||\boldsymbol{\omega}|| = 0.30\ \text{Max}||\boldsymbol{\omega}||$, which lays deeper inside the boundary layer. One can see an array of staggered Λ-vortices which break down into turbulence farther downstream. The flow pattern then becomes very complicated but, despite the evident lack of resolution, one can make out streamwise streaks and tips of horseshoe vortices. Figure 13 shows the downstream half of the isothermal surface $T = 1.4\ T_\infty$ which lies, on average, at a height $\approx 0.9\ \delta_1(x)$, the local displacement thickness of the boundary layer. This figure shows, on the left, quasi-two-dimensional corrugations upstream of the Λ-vortices, which confirms that a two-dimensional unstable mode has grown and reached a certain finite amplitude before the onset of the three-dimensional modes. We are thus well within the framework of secondary instability as in the previous section, by contrast with the *helical pairing* case reported in the first section, in which the primary instability was oblique. Figure 14 focusses on this region of secondary instability, showing low-pressure isosurfaces, the most upstream of them being indeed quasi two-dimensional. Notice also the rapid growth of three-dimensionality in the form of a pair of oblique sub-harmonic modes (it is easy to check visually that the streamwise period of the oblique vortex pattern is approximately double the spacing between the quasi-2D vortices.

6. High-Mach-Number Boundary Layer Over An Adiabatic Flat Plate

When the external Mach number M_∞ is high enough, the flow solution of the laminar similarity equations presents a generalized inflexion point y_S, even when the wall is insulated. For $M_\infty = 4.5$, a case which is well documented from the point of view of the stability theory (see Ng and Erlebacher, 1992, for example), the dominant instability is inviscid and two-dimensional (Mack's second mode, Mack, 1969). It is thus natural to expect the formation of the same kind of vortices as in high-Mach number mixing layers or heated boundary layers, *viz.* Λ-vortices stretched by the mean flow. This point of view is confirmed by the large-eddy simulations of a spatially-growing boundary layer at Mach 5 performed by Normand and Lesieur (1992). We will here present a brief extract of the continuation of this work, aiming at a clearer identification of the first coherent structures to form. In order to make the results more convincing, we decided to work in the same conditions as Ng and Erlebacher (1992), *i.e.* in a temporally-evolving configuration and without a subgrid-scale model, at Reynolds number (based upon the initial displacement thickness) $R = 10\,000$. The external temperature is taken equal to 180 K.

For such a Mach number, y_S is above y_1, the height of the sonic line in a frame of reference moving at the phase speed of Mack's second mode, which is very close to $\bar{u}(y_S)$ (see Figure 15). Consequently, this mode, which develops about y_S, is of a subsonic nature. However, its wavelength is determined by the reflexion of acoustic waves between the sonic line and the wall (Mach waves, see Morkovin, 1987). Since this forcing is two-dimensional, it is not surprising that Mack's second mode is also two-dimensional. In order to work out the eigenfunctions of this mode, we ran a two-dimensional simulation in which the laminar flow was perturbed by white noise of small amplitude (10^{-4} that of the laminar flow). Figure 16 shows the vorticity map for $t = 400\ \delta_i/U_\infty$, which corresponds to the end of the linear regime. The extremal values for the vorticity are reached near the wall, with both signs alternatively. This is due to reflections of acoustic waves between the wall and the sonic line, coupled to baroclinic effects. Vorticity is also concentrated in 4 structures located at a height y_S and reminiscent of the Kelvin-Helmholtz vortices observed in mixing layers. The spacing of these structures is equal to λ_a, the most amplified wavelength of the second mode.

The fluctuations obtained at the end of the linear stage of the calculation are used as initial perturbations for the three-dimensional simulation, after rescaling to an amplitude of $A = 8\ \%$ of that of the laminar flow. Secondary instabilities are triggered by three-dimensional white noise of amplitude $5\ 10^{-4}$. The resolution is now of $40 \times 70 \times 36$ grid points, for a domain length of $2\ \lambda_a$ and width equal to the spanwise period of the most amplified mode of secondary instability found by Ng and Erlebacher. Figure 17 shows two isobaric surfaces at $t = 350\ \delta_i/U$: the dark one corresponds to a pressure high ($1.004\ p_\infty$), the light one to a pressure trough ($0.995\ p_\infty$). This picture shows the emergence of an oblique subharmonic of Mack's second mode, with a streamwise period $2\ \lambda_a$ and the spanwise period $L_z = 6.28\ \delta_i$ found by Ng and Erlebacher (1992). The associated vortical structure is materialized in Figure 18 through isosurfaces of the streamwise vorticity component, with thresholds equal to $0.1\ U/\delta_i$ (light surface) and $-0.1\ U/\delta_i$ (dark surface) respectively. This succession of thin tongues carrying streamwise vorticity of both signs, overlapping like fish scales and stretched longitudinally is characteristic of couples of Λ-vortices facing each other, relatively well materialized by surfaces iso-vorticity-magnitude (Figure 19). Such an arrangement is strongly reminiscent (see Figure 8) of the vortical structure of compressible mixing layers at high enough a convective Mach number ($M_c > 0.7$) so that analogous Λ-vortices slide on top of each other and do not undergo helical pairings, as they do at lower Mach numbers (Comte *et al.* 1992b) and 1993). This is again evidence that transition of boundary layers at high Mach numbers is dominated by the development of inflexional shear instabilities.

7. Conclusion

The simulations presented here have shown some strong analogies between high-Mach number or heated boundary layers and mixing layers (see also Gathmann *et al.* 1993), with a transitional vortex structure dominated by Λ-vortices stretched in the streamwise direction at the level of the generalized inflection point. As suggested in particular in Comte *et al.* (1992a), this analogy can be extended to low-Mach number boundary layers dominated by viscous instabilities, keeping in mind the large difference in growth rates. It thus seems that, despite a great sensitivity to initial conditions (impredictability), the processes (in the most general sense of the term) leading to developed turbulence are always the same. The ones we have distinctively seen at work so far are:

- bundling of the vortex lines of the laminar flow (which can be interpreted as the roll-up of an initial vortex sheet (or, rather, slab) into vortex tubes)
- streching by the mean flow of vortex lines or tubes perturbed three-dimensionally (by means of oblique modes for example). This makes Λ-vortices if the level of 3D upstream or initial disturbances is high enough.
- pairings (inverse cascade), which can be helical if the merging vortices are not two-dimensional
- direct cascade due to stretching of the remaining vortex lines by the strain field generated by the larger vortices (3D mechanism yielding $k^{-5/3}$ kinetic energy spectra at least in the incompressible regime)
- another mechanism of direct cascade due to the lamination of fluid elements (the baker's transform). It is a scenario of 2D enstrophy cascade yielding k^{-3} - k^{-4} spectra and thus bound to be overwhelmed by the former mechanism of direct cascade if the initial conditions are sufficiently three-dimensional.

Of course, all these processes do not necessarily occur in a given configuration: in particular, forced incompressible mixing layers can keep a quasi-2D appearance at least in their largest scales if the forcing is mostly two dimensional; on the other hand, in compressible mixing layers, helical pairings are inhibitted by compressibility for $M_c > 0.7$ (Comte *et al.*, 1993). Furthermore, these processes are not independent from each other and do not necessarily correspond to well-separated and identifiable stages of transition; for example, in the low-Mach-number simulations of boundary layers presented here, spanwise vortices (or Tollmien-Schlichting wave packets) roll-up in a quasi-two-dimensional fashion before being stretched into Λ-vortices. In some other cases (helical or compressible mixing layers), the unstable modes which emerge the first are oblique and the resulting vortices make an array following these modes; which requires stretching by the mean flow.

Anyway, the similarities between the manners transition was achieved in the different cases suggest a possible universality. In a recent paper, Lesieur (1992) proposed a *universal description of three-dimensional turbulence in terms of five fundamental vortex interactions*: a *sheet mode* responsible for the formation of vortex sheets when there is no mean shear; *spiral* and *pairing* modes, and two modes called *even longitudinal* and *odd longitudinal*, representing the formation and stretching of hairpin vortices in an aligned or staggered arrangement respectively.

One can of course question the fundamental character of these interactions and the exhaustiveness of the list, but this idea of a finite (and small) number of interactions to explain transition and turbulence is a puff of oxygen which is worth gulping with the aid of numerical simulations, supercomputers, 3D interactive graphics and subgrid-scale models.

Acknowledgements

Part of the calculations were done on a grant from the Centre de Calcul Vectoriel pour la Recherche.

References

Browand, F.K. & Ho, C.M. 1983 "The mixing layer: an example of quasi two-dimensional turbulence", in *Two-dimensional turbulence*, J. Mec. Théor. Appl., Suppl., R. Moreau ed., p. 99.

Brown, G.L. & Roshko, A. 1974 "On density effects and large structure in turbulent mixing layers", J. Fluid Mech., **64**, p. 775.

Comte, P., Lesieur, M. & Lamballais, E. 1992a, "Large- and small-scale stirring of vorticity and passive scalar in a three-dimensional temporal mixing layer", Phys. Fluids A, **4**, (12), pp. 2761-2778.

Comte, P., Fouillet, Y. & Lesieur, M. 1992b, "Simulation numérique des zones de mélange compressibles", Revue Scientifique et Technique de la Défense, 3ème trimestre 1992, pp 43-63.

Comte, P., Fouillet, Y. & Lesieur, M. 1993, "The effect of compressibility upon helical pairings in a mixing layer", preprint I.M.G., in preparation for submission to *Eur. J. Mech.*

David, E.,"Simulation numérique de la transition à la turbulence sur un cône en régime haut-supersonique", Doctorat de l'Institut National Polytechnique de Grenoble, soutenance prévue en 1993.

Gathmann, R.J., Si-Ameur, M. & Mathey, F. 1993, "numerical simulation of three-dimensional natural transition in the compressible confined shear layer", submitted to *Phys. Fluids* A.

Herbert, T. 1988, "Secondary instability of boundary layers", Ann. Rev. Fluid Mech., **20**, p. 487.

Lees L. & Lin, C.C. 1946, "Investigation of the stability of the laminar boundary layer in a compressible fluid", NACA TN 1115.

Lesieur, M., 1992, "Five elementary vortex interactions in turbulence", in *Points de vue sur la Physique et les Mathématiques*, J. Horowitz and J.L. Lions eds, Masson.

Lesieur, M., Staquet, C., Le Roy, P. & Comte, P. 1988, "The mixing layer and its coherence examined from the point of view of two-dimensional turbulence, J. Fluid Mech., **192**, p. 511.

Mack, L.M. 1969, *Boundary-layer stability theory.* Jet Propulsion Lab., Pasadena, Calif., rep. 900-277.

Meneguzzi, M., 1992, "Dynamics of vortex tubes in 3D turbulence", in the present proceedings.

Métais, O. & Lesieur, M. 1992, "Spectral large-eddy simulation of isotropic and stably-stratified turbulence", *J. Fluid Mech.*, **239**, pp. 157-194.

Michalke, A. 1964, "On the inviscid instability of the hyperbolic tangent velocity profile", J. Fluid Mech., **19**, p. 543.

Morkovin, M.V. 1987, *ICASE interim report 1*, NASA contr. rep. 178315.

Normand, X. & Lesieur, M. 1992, "Direct and large-eddy simulations of transition in the compressible boundary Layer", *Theor. and Comp. Fluid Dynamics*, **3**, p. 231.

Ng, L.L. & Erlebacher, G., 1992, "Secondary instability in compressible boundary layers", *Phys. Fluids A*, **4**, p. 710.

Pruett, C.D. & Zang, T.A. 1992, Direct numerical simulation of laminar breakdown in high-speed, axisymmetric boundary layers, *Theor. and Comp. Fluid Dyn.*, **3** pp. 345-367.

Sandham, N.D. & Reynolds, W.C. 1991, "Three-dimensional simulations of large eddies in the compressible mixing layer", *J. Fluid Mech.*, **224**, p. 133.

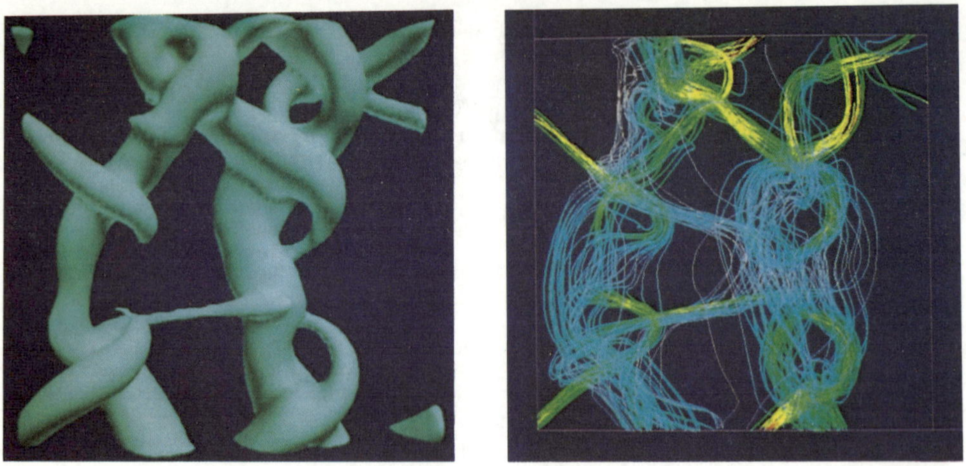

Figure 1: helical pairings in a temporal incompressible mixing layer forced initially by 3D noise, viewed through a low-pressure isosurface (left) and some vortex lines (right). (pseudo-spectral methods, resolution 128^3, initial Reynolds number 100).

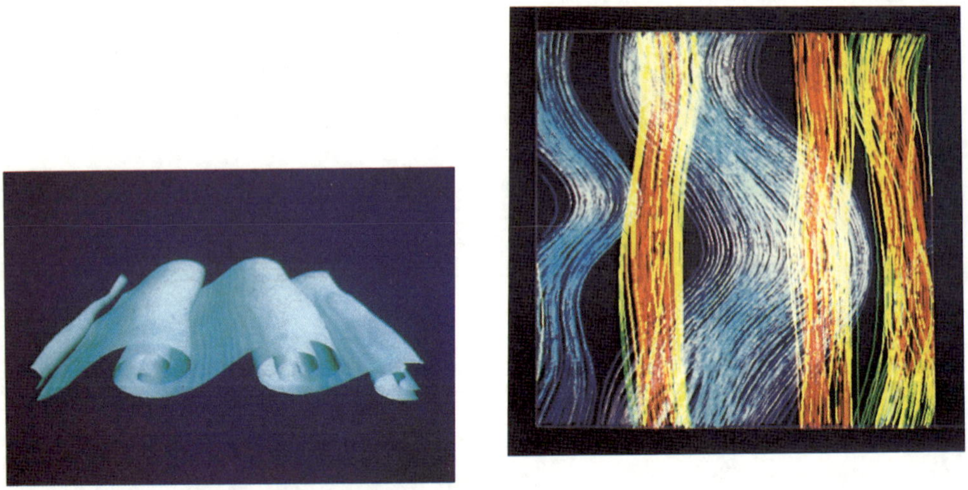

Figure 2: temporal incompressible mixing layer for the same parameters as before, but forced initially by quasi-2D noise. Passive scalar interface between the two streams (left) and some vortex lines at the same instant (right).

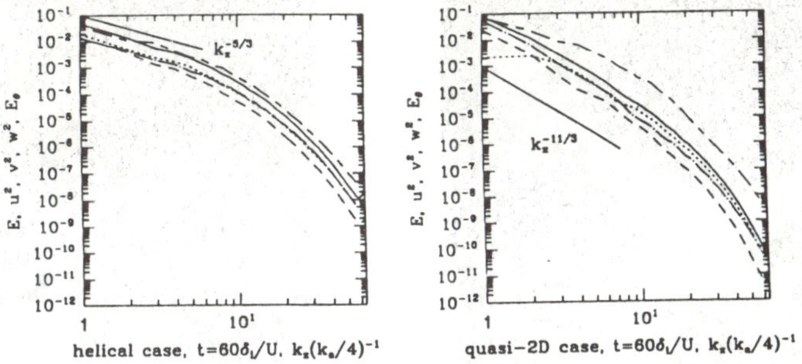

Figure 3: streamwise spectra taken at the end of the simulations corresponding to Figures 1 (on the left) and 2 (on the right): (solid line: kinetic energy $E(k_x)$; u', v' and w' variance: dashed, chain-dotted and dotted lines respectively; passive-scalar variance $E_\theta(k_x)$: chain-dashed line).

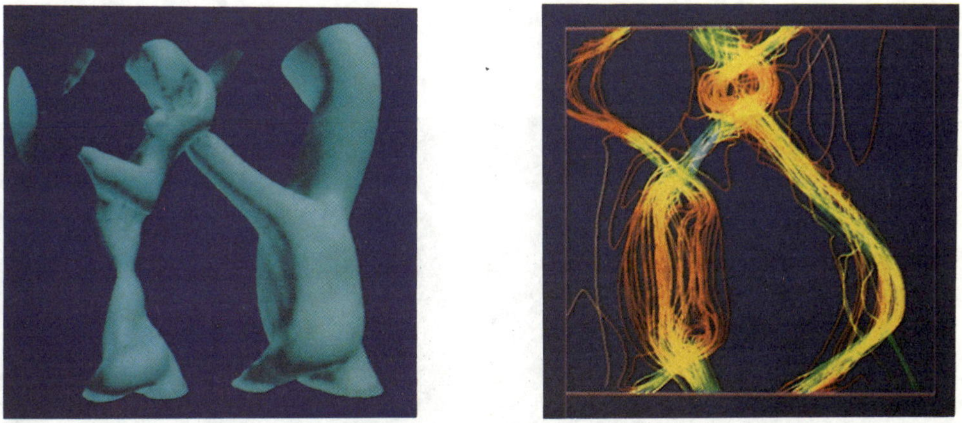

Figure 4: compressible temporal mixing layer at Mach 0.3. Low-pressure isosurface on the left, vortex lines on the right (resolution 48^3, initial Reynolds number 1000)

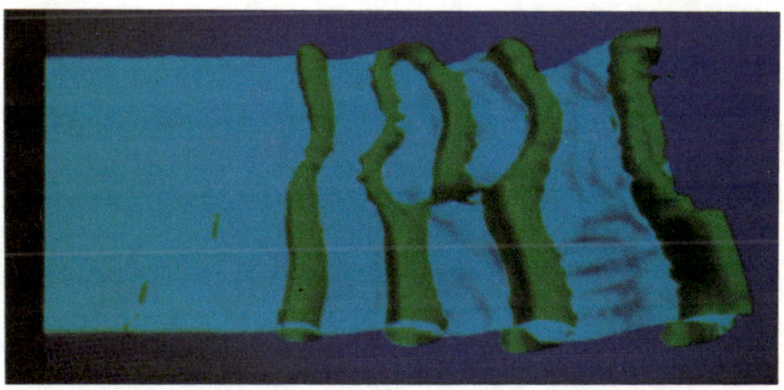

Figure 5: spatially-growing compressible mixing layer at Mach 0.3. Low-pressure isosurface and density interface (resolution $80 \times 40 \times 30$, upstream Reynolds number 1000.)

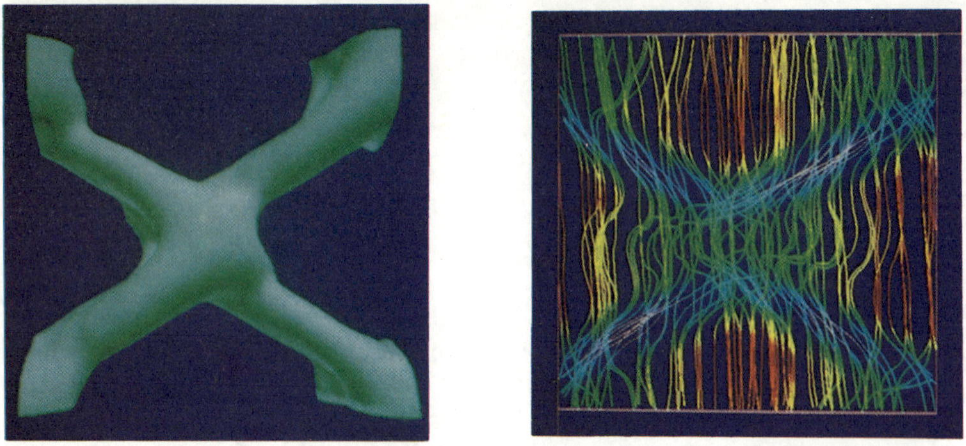

Figure 6: compressible temporal mixing layer at Mach 0.8. Low-pressure iso-surface on the left, vortex lines on the right (resolution 48^3, initial Reynolds number 1000). By contrast with Figure 4, there is no helical pairing between the two Λ-vortices.

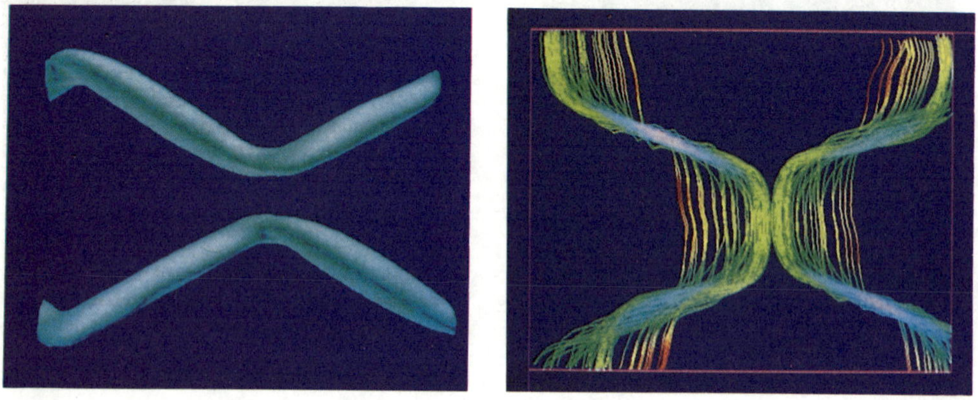

Figure 7: same as Figure 6, but for $M_c = 1$. Notice that the low pressures do not follow the vortex tubes anymore

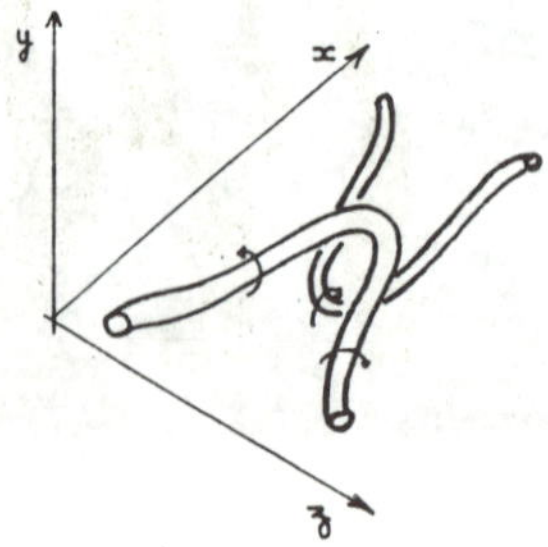

Figure 8: sketch of the vortex structure in a mixing layer at $M_c \approx 1$.

Figure 9: spatially-growing compressible mixing layer at Mach 0.7. Low-pressure isosurface and density interface (same resolution and Reynolds number as for Figure 5).

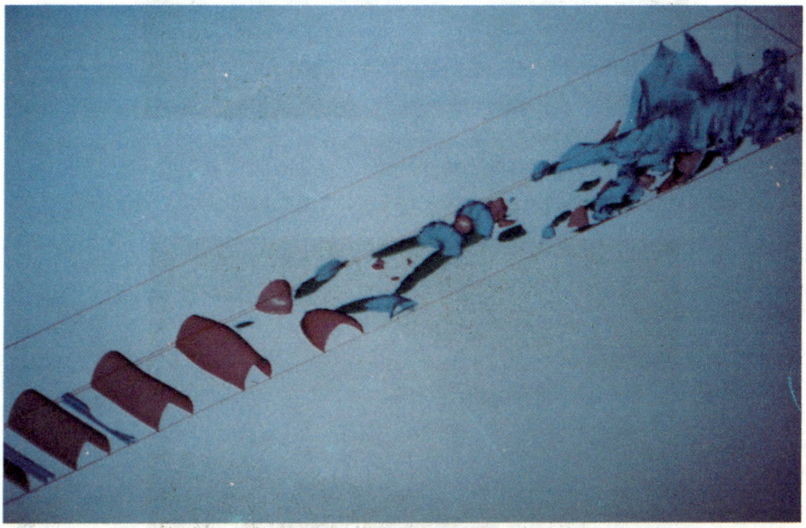

Figure 10: spatially-growing boundary layer over an adiabatic flat plate (resolution $320 \times 32 \times 20$, upstream Reynolds number 1000). The green surface is an iso-streamwise-vorticity surface, for a threshold of 1/7 the maximal vorticity norm at the upstream boundary. The quasi-2D structures seen before transition correspond to the high pressures generated by the 2D determinstic forcing at the upstream boundary.

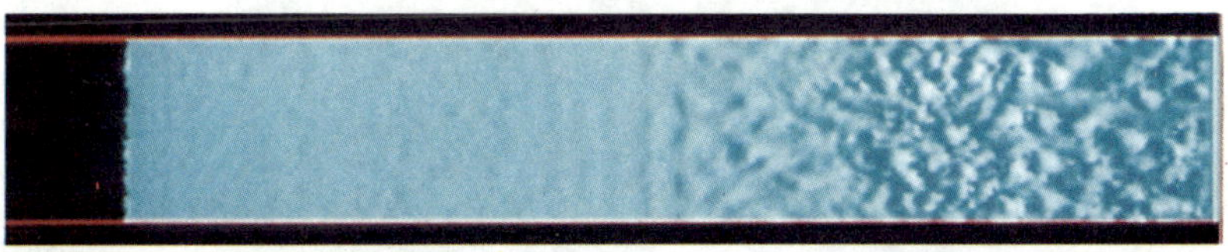

Figure 11: flow over a heated wedge (resolution of $237 \times 30 \times 32$, Reynolds number based upon the length L of the wedge $Re_L = 3.910^5$). Iso-vorticity-norm surface $||\boldsymbol{\omega}|| = 0.09\ \mathrm{Max}||\boldsymbol{\omega}||$ seen from the top. The leading edge of the wedge is located at about 1/8 of the the domain's length.

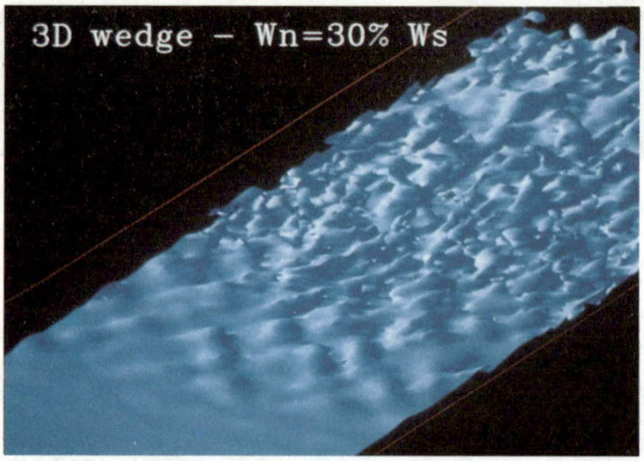

Figure 12: zoom of the downstream half of the isosurface $||\boldsymbol{\omega}|| = 0.30 \text{ Max}||\boldsymbol{\omega}||$ showing staggered Λ-vortices. Same calculation as for Figure 11.

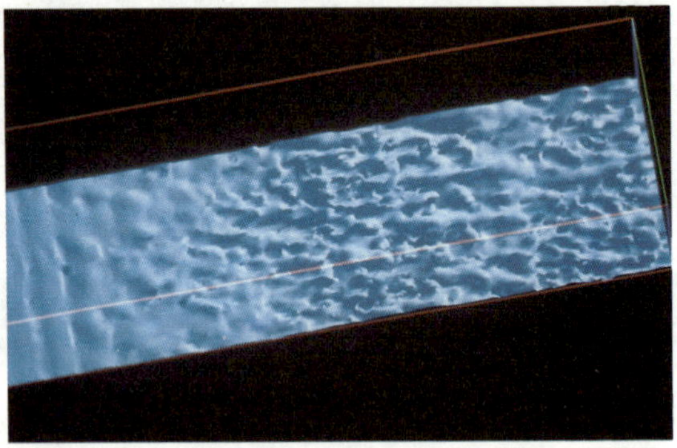

Figure 13: downstream half of the isosurface $T = 1.4\ T_\infty$ showing the rapid transition from a quasi-2D flow to a highly-3D flow with streamwise streaks. Same calculation as before.

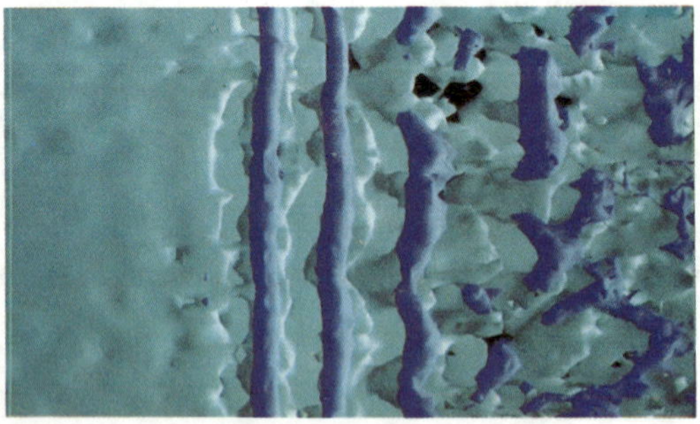

Figure 14: blow-up of the transition region showing a low-pressure isosurface (dark) and the iso-vorticity-norm $||\boldsymbol{\omega}|| = 0.8 \text{ Max}||\boldsymbol{\omega}||$ (light). Notice the strong resemblance with the low-Mach number spatially-growing mixing layer shown in Figure 5.

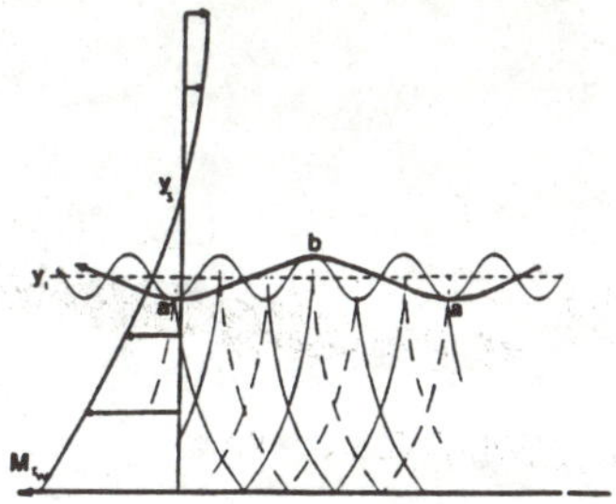

Figure 15: scenario of subsonic excitation of inflexional instabilities in a high-Mach number boundary layer over a flat plate. Excitation occurs about the generalized inflexion point y_S, viewed in a frame of reference traveling at $\bar{u}(y_S)$. It is ascribed to the perturbation of the sonic line (at y_1) by reflections of acoustic waves in the supersonic part of the flow ($y \leq y_1$). The flow is subsonic above, and the perturbations radiate toward y_S. (*Courtesy M.V. Morkovin, 1987*).

Figure 16: 2D simulation of a temporally-growing boundary layer over a flat plate at Mach 4.5. Map of the vorticity field showing vortices developping about y_S.

Figure 17: 3D simulation of a temporally-growing boundary layer over a flat plate at Mach 4.5 (resolution $40 \times 70 \times 36$, initial Reynolds number 1000). Low- and high-pressure surfaces (in light and dark respectively) at $t = 350\ \delta_i/U$. They show the emergence of an oblique mode of streamwise and spanwise periods equal to the box dimensions $L_x = 5.5\ \delta_i = 2\ \lambda_a$ and $L_z = 6.28\delta_i$.

Figure 18: isosurfaces $\omega_x = 0.1U/\delta_i$ (in pale) and $-0.1U/\delta_i$ (in dark) corresponding to Figure 17, still at $t = 350\ \delta_i/U$.

Figure 19: the corresponding isosurface $\|\boldsymbol{\omega}\| = 0.85\ U/\delta_i$, still at $t = 350\ \delta_i/U$. After re-centering by translation of the domain along x and z, the vortex structure appears to be very close to that sketched in Figure 8, which corresponded to the case of a mixing layer.

STABILITY ANALYSIS OF DIFFERENTIALLY-HEATED ASYMMETRIC VORTICITY LAYERS

OMAR M. KNIO
Department of Mechanical Engineering
The Johns Hopkins University
Baltimore MD 21218
USA

and

AHMED F. GHONIEM
Department of Mechanical Engineering
Massachusetts Institute of Technology
Cambridge MA 02139
USA

ABSTRACT. The linear instability of a family of inviscid, two-dimensional, variable-density shear layers and wakes is investigated. Vorticity profiles corresponding to a monotonically increasing velocity profile are first examined. A larger family of initial vorticity distributions which model the merger of two unequal vorticity layers of opposite sign is then considered. The latter is obtained by superimposing on the former a wake component, characterized by a spread, δ, and a velocity deficit, W. The initial density distribution resembles a temperature spike and is described by a thickness, σ, and a temperature ratio, T_r. The stability properties of the layers are interpreted in terms of a four-dimensional parameter space (W, δ, T_r, σ). The non-linear evolution of the flowfield is illustrated using the transport element method.

Flowfield stability exhibits strong sensitivity to the details of the density distribution. In the absence of the wake component, the stability properties of the heated layer are divided into three categories according to the thickness of the density profile, σ, and the vorticity thickness, δ_w. For $\sigma >> \delta_w$, instability of the Kelvin-Helmholtz mode in a uniform-density flow is recovered. When $\sigma \sim \delta_w$, the shear layer mode is inhibited; while this trend persists for $\sigma < \delta_w$, the layer becomes characterized by the appearance of additional short-wavelength unstable modes which become dominant as σ decreases and T_r increases. Addition of a wake component is shown to alter this behavior, and to oppose the stabilizing effects of heat release. In this case, the shear layer mode always dominates the wake mode, and the presence of heated sublayer has a weak effect on the instability of the vorticity layer when δ is large, but may influence the phase speed of unstable waves whenever the zones of high vorticity and high density gradient coincide.

1. Introduction

The evolution of high Reynolds number chemically-reacting free shear flows is governed by a large number of fundamental processes. These processes can be described in terms of the dynamic effects of combustion, which leads to the establishment of

J.T. Beale et al. (eds.), Vortex Flows and Related Numerical Methods, 341–372.

an expansion flowfield and zones of sharp density variation, and in terms of intrinsic instabilities of the underlying shear flow which shape the evolution of the vorticity field. In most cases of practical interest, these effects are closely coupled, as the vorticity-induced convective field governs the molecular mixing processes and hence modulates local reaction rates, while the evolution of the chemical reaction affects the vorticity field through flow divergence field and baroclinic vorticity.

The description of such flows is complicated, due to nonlinear flow-combustion interactions and due to the large number of parameters which govern the relevant molecular and chemical processes. This complexity often necessitates the construction of simplified fundamental models which facilitate the isolation of particular interaction modes, and reduce in the number of governing parameters.

Linear stability theory has proven to be an important tool in theoretical studies of reacting and heterogeneous shear flows [1-12]. In most of these studies, heat release mechanisms associated with the mixing of initially separated reacting species are modelled by corresponding temperature and density profiles which are imposed on an otherwise homogeneous shear flow. Thus, while dynamic effects associated with the presence of an expansion field and the details of the chemical reaction process are omitted, spatial density (and temperature) variation resulting from heat release mechanisms is retained. Using this approach, the essential stability properties of the flowfield have been predicted. In particular, linear stability studies have shown that the presence of two or more zones of different density significantly affects the development of the flow. For instance, it has long been observed that a non-unity density ratio alters the growth of shear flows and influences the entrainment induced by the vortical structures embedded therein [1-4], even when gravity effects are weak [5-6]. In chemically-reacting shear layers, the effects of density variation are equally pronounced; stability results indicate that flowfield stabilization or destabilization may occur, depending on the details of the density and vorticity distributions [7-9]. Furthermore, linear stability results show that the presence of zones of large density variation may also affect the nature of flow instabilities (e.g., by altering the boundaries separating absolute and convective instability modes), and may result in reshaping the global features of the flow [10-12].

Unfortunately, the application of linear stability results to predict the behavior or reacting flows has been complicated due to the large sensitivity of the results to the initial vorticity and density profiles. In this work, this issue is tackled by analyzing the stability of heterogeneous flows for a wide range of initial conditions. Initial density (and temperature) profiles are assumed which model the development of nonpremixed combustion. In addition, initial vorticity profiles corresponding to symmetric and asymmetric shear layers and wakes are considered. A large number of initial conditions is thus constructed, and flowfield stability is examined in a four-dimensional parameter space which models a large family of reacting layers and wakes.

The stability problem is based on linearization of the inviscid heterogeneous flow equations. The formulation of this fluid flow problem is described in Section 2, and computed results are discussed in Section 3. Trends in the behavior of the flowfields are established and further examined by performing non-linear simulation of selected cases using the transport element method (e.g. [13-14]). Major conclusions are given in Section 4.

2. Formulation

2.1. THE STABILITY PROBLEM

In a right-handed coordinate system (x, y), the initial flow field is given in terms of the steady flow velocity, $U(y)$, and the imposed density profile, $\varrho(y)$. The flow is assumed two-dimensional and inviscid, and both the perfect gas and the low Mach number approximations of the governing equations are employed [15-16]. The mean density profile is assumed to be the result of heat deposition by the reaction, while diffusion effects are neglected. Under these assumptions, the fluid flow is governed by the momentum and continuity equations, respectively written as:

$$\varrho \frac{D\mathbf{u}}{Dt} = -\nabla p \tag{1}$$

$$\frac{D\varrho}{Dt} = 0 \tag{2}$$

$$\nabla \cdot \mathbf{u} = 0 \tag{3}$$

where $\mathbf{u} = (u, v)$ is the velocity, t is time, $\nabla = (\partial/\partial x, \partial/\partial y)$ is the gradient operator, D/Dt is the material derivative, and p is pressure. The governing equations may be recast in vorticity form by replacing Eqs. (1) and (3) by:

$$\frac{D\omega}{Dt} = \frac{\nabla \varrho}{\varrho^2} \times \nabla p \tag{4}$$

where ω is the vorticity. This formulation will be later used in the simulation of the nonlinear evolution of the flow.

The stability properties of the variable-density shear flow are studied by determining the temporal behavior of small amplitude disturbances imposed on the initial steady flow. These properties are analyzed in terms of the disturbance cross-stream velocity component v_0 which is first written in the form:

$$v_0(y, t) = v(y) \exp\left(i\alpha(x - ct)\right) \tag{5}$$

whereα is the normalized wavenumber, taken to be real, and $c = c_r + ic_i$ is the complex wave speed. The cross-stream component $v(y)$ obeys the modified Rayleigh equation:

$$v'' = -\frac{\varrho'}{\varrho} v' + \left(\frac{U''}{U - c} + \frac{\varrho' U'}{\varrho(U - c)} + \alpha^2 \right) v \tag{6}$$

with boundary conditions, $v(y \to +\infty) \sim \exp(-\alpha y)$ and $v(y \to -\infty) \sim \exp(\alpha y)$. In Eq. (6) and the following, primed quantities indicate differentiation with respect to y.

We are interested in determining whether waves of the form expressed by Eq. (5) are unstable, i.e. whether their growth rate $\alpha c_i > 0$. To this end, the eigenvalue problem is solved using a shooting technique, in which Eq. (6) is integrated from one side of the layer to the other using a fourth-order Runge-Kutta predictor-corrector scheme [17]. We keep iterating in the complex eigenvalue space c using a secant algorithm until the absolute change in c is reduced below 10^{-7}. This procedure yields dispersion relations which relate the growth rate αc_i and the phase speed c_r to the disturbance wavenumber α.

2.2. INITIAL CONDITIONS

The formulation stability of the stability problem is completed by specifying the steady, parallel shear flow velocity and density profiles. A family of shear layers and density profiles is constructed to model the physical processes shown schematically in Fig. 1. We consider chemically-reacting layers formed by merging oxidizer and fuel streams downstream of a thin splitter plate or a bluff body. Immediately following the tip of the plate, the velocity distribution may be modelled as the superposition of two Blasius profiles, starting from a vanishing streamwise velocity and increasing to the free stream velocities $U1$ and $U2$ (Fig. 1b). Further downstream, the velocity profile approaches that of a shear layer, in which the velocity increases monotonically from one side of the splitter plate to the other (Fig. 1a). In the third flow configuration, the incoming streams are separated by a bluff body (Fig. 1c). In this case, the velocity field is regarded as the superposition of two shear layers, each resembling that shown in Fig. 1a. In all cases, the heat deposited by the initial development of the chemical reaction is modelled by a spike in the temperature profile.

In order to model the various flow configuartions anticipated in Fig. 1, we start with the experimentally fitted velocity profile for large distances downstream of the splitter plate,

$$U(y) = U_m + \frac{U_1 - U_2}{2} \tanh\left(\frac{y}{\delta_\omega}\right) \tag{7}$$

where U_m is the mean flow velocity and δ_ω is the local vorticity thickness. The local vorticity thickness is chosen as characteristic length scale, and the velocity is normalized such that the reduced expression,

$$U(y) = \tanh(y) \tag{8}$$

replaces Eq. (7). As suggested by Koch [18], a continuous family of velocity profiles, which approximates all of the distributions shown in Fig. 1, can be obtained by modifying Eq. (9) by letting:

$$U(y) = (1 + W)\tanh(y - \delta) - W\tanh(y + \delta) \tag{9}$$

where W is the wake-deficit, and δ as the displacement of the vorticity layer. In Eq. (10), W and δ are restricted such that $W \leq 0$, and $\delta \geq 0$. When $W = 0$, Eq. (8) is recovered for $\delta = 0$, and increasing the value of $\delta > 0$ results in a pure translation of the *tanh* profile. On the other hand, when $W < 0$, near wake velocity profiles are approximated for large δ, and the double Blasius profile is imitated when δ is small.

The free stream density and temperature are chosen as a reference density and temperature scales. Accordingly, the normalized initial density profile is taken as

$\delta = 1$	$W = 0.2$	$W = 0.4$	$W = 0.8$	$W = 1.2$
$\delta = 2$	$W = 0.10299$	$W = 0.24709$	$W = 0.55080$	$W = 0.860922$
$\delta = 3$	$W = 0.09261$	$W = 0.22988$	$W = 0.52204$	$W = 0.82149$

TABLE I

a Gaussian profile with standard variation σ, and is expressed in terms of the temperature ratio T_r as follows:

$$\varrho(y) = 1 - \frac{T_r - 1}{T_r} \exp\left(\frac{-y^2}{\sigma^2}\right) \tag{10}$$

In most shear flow applications, including shear layers, the vorticity thickness is usually larger than the product thickness, so that $\sigma < 1$. However, we do not enforce this restriction in order to account for fluid flows characterized by high mass diffusivities, or bluff body flows. When $\delta \simeq \sigma$, the flowfield approximates the merger of two layers of unequal density, a flow configuration that has already been analyzed (see, e.g. [13,19]).

The temporal stability of the family of variable-density layers described above is investigated in the four-dimensional parameter space (W, δ, T_r, σ). We consider four values of the temperature ratio, $T_r = 1, 2, 4$, and 8.0, i.e. we start with a uniform-density field and then vary the temperature ratio in a range that is representative of most chemically-reacting flows. For each of these cases, the effects of the wake component deficit and thickness are investigated by varying W and δ in such a way as to approximate all the flow configurations shown in Fig. 1. In order to separate the effects of the strength of the wake component, W, from those associated with its thickness, δ, we alter the values of W and δ simultaneously so that velocity profiles having having the same maxima and minima are obtained for all values of δ. In addition to the *tanh* profile having $W = \delta = 0$, we consider 12 (W, d) combinations which are described in Table I. The corresponding velocity profiles are plotted in Fig. 2 for all thirteen cases, the shear layer profile being included with the set of profiles having $\delta = 1$.

Thus, W varies in a wide range of wake deficits, while increasing δ from 1 to 3 represents a migration from an asymmetric shear layer velocity distribution to an asymmetric wake profile, in which vorticity layers are well separated. The stability properties of variable-density layers are first determined assuming equal density and vorticity thicknesses, $\sigma = 1$. The effect of the thickness of the density profile is then investigated by repeating the analysis at $T_r = 4.0$ for four additionalσ values, $s = 1.5, 0.75, 0.5$, and 0.25.

2.3. BRIEF THEORETICAL REVIEW

The linear stability problem of inviscid incompressible parallel shear flow has been studied extensively (e.g. [10,20-21]). In this section, we summarize aspects of the theory which directly affect our search for unstable eigenfunctions. We first note that the classical stability results expressed by Rayleigh s theorem [20-21],

A necessary condition for instability is that the profile $U(y)$ admits an inflection point

and by Fjortoft s extension,

A necessary condition for instability is that $U''(U - U_s) \leq 0$ somewhere the flow, where U_s is the velocity at the inflection point

have to be modified when considering a variable density flow. It may easiliy be shown that the appropriate generalizations of the above results may be respectively expressed as:

A necessary condition for instability is that $U'' + (\rho'/\rho)U'$ admits a sign change

and

A necessary condition for instability is that $U'' + (\rho'/\rho)U'(U - U_s) < 0$ somewhere in the flow.

Thus, in a variable-density field, the behavior of the quantity $U''+(\rho'/\rho)U'$ replaces that of U'' in the determination of the stability of flow. As indicated in Refs. 1 and 8, this leads us to expect strong interactions between the density variation and the shear flow, whenever zones of high vorticity and density gradient coexist.

The behavior of the curves of $U''+(\rho'/\rho)U'$ for the flow configurations of Table I is plotted in Fig. 3 for $\delta = 1, \sigma = 1$, and $T_r = 1, 2, 4$, and 8, and in Fig. 4 for $\delta = 1, T_r = 4.0$ and $\sigma = 1.5, 0.75, 0.5$, and 0.25. By the preceding, we are led to expect unstable modes whenever the curves intersect the zero axis. In the uniform-density case, a single intersection point is observed for the *tanh* profile, and two for the asymmetric layers. This is not surprising, since we expect to be observe one unstable shear layer mode, and one unstable wake mode. However, for large T_r, several intersection points appear for the *tanh* profile, and may yield additional instability modes. This unexpected result is further investigated in the following section where the behavior of these modes, whose appearance depends on the vorticity-density configuration, is computed.

We conclude this section by extending to variable-density flow the analysis of Drazin and Howard [22] who studied the the long wave behavior of unstable modes, i.e. the limiting behavior of unstable eigenvalues as $\alpha \to 0$. Our analysis is reduced to a form similar to the incompressible case by rewriting the modified Rayleigh equation as:

$$(Z^2F')' = \alpha^2 Z^2 F \tag{11}$$

where

$$Z^2 \equiv \varrho(U - c)^2 \; ; \; F \equiv \frac{\phi}{U - c} \tag{12}$$

and ϕ is the perturbation potential. Noting that Eq. (12) is identical to Eq. (1.6) of Drazin and Howard [22], we are able to carry out a similar analysis to the one performed there. (Details will be presented elsewhere). In the limit $\alpha \to 0$, Eq. (14) implies that Z^2F' is constant and that this constant is zero in order to satisfy the boundary conditions. Hence F is constant in intervals where Z does not vanish, but may have jumps when $Z = 0$. Assuming F has no jumps, a similar argument to that presented in [22] shows that the limiting eigenvalues satisfy:

$$Z_{\infty}^2 + Z_{-\infty}^2 = 0 \tag{13}$$

For equal free stream densities, Eq. (14) yields the unstable eigenvalue $c = i$ which is recognized as the limiting eigenvalue of the shear layer mode. Density variation does not seem to affect the asymptotic behavior of these long waves.

A similar result is reached if F admits jumps. While the details of the algebraic manipulations are more involved than in the uniform-density case, we are still able to show that if F has a jump at y_0, then $U'(y_0) = 0$. Thus, velocity maxima and minima are expected to represent limiting values of unstable eigenfunctions. For $W < 0$, the profiles considered in this study admit a velocity minimum, which is recognized as the limiting eigenvalue of the (unstable) wake mode.

The long wave approximation estimates are used in the following section to initialize the search for unstable eigenfunctions and to characterize the additional inflection points which appear for the *tanh* profile at high temperature ratio (Fig. 3). Should these additional inflection points correspond to unstable modes, then the above argument shows that the associated instability mode affects shorter wavelength instability, since the asymptotic behavior of long waves is solely dependent on the details of the velocity profile.

3. Results

3.1. STABILITY OF VARIABLE-DENSITY LAYERS AND WAKES

Stability analysis of the variable-density parallel asymmetric shear flow is first conducted for the velocity profiles of Table I, and density profiles having $\sigma = 1$, and $T_r = 1, 2, 4$, and 8. Since a sharp estimate of an upper bound on unstable wavenumbers was not sought, a complete search for the unstable eigenfunctions over a bounded wavenumber-eigenvalue region cannot be easily conducted. Instead, the search for unstable waves is initiated by concentrating on the behavior of long waves, and extrapolating the asymptotic behavior of unstable shear layer and wake modes, as determined by the theoretical predictions. To this end, the wavenumber is increased incrementally with small step size $\Delta\alpha = 0.01$, until the iterations stop to converge. In all cases, the modified Rayleigh equation is integrated over a mesh of 4000 grid points, equally distributed over the interval $-4 - \delta \leq y \leq 4 + \delta$. Thus, the expected short wave behavior of the additional instability modes is only determined in cases where their instability band extends that of the other modes.

Figures 5-8 show the growth rate and phase speed of unstable modes, computed for $T_r = 1, 2, 4$, and 8, respectively. For brevity, results obtained for the intermediate value $\delta = 2$ are not illustrated. In all cases, the thickness of the heated layer coincides with that of the vorticity layer, $\sigma = 1$. We start with the uniform-density flow (Fig. 5), which is later used as reference to quantify the effects induced by the density variation. Cold flow calculations are summarized as follows:

(1) The growth rate of the shear layer mode increases with increasing wake deficit. This result is expected since higher wake deficits correspond to higher vorticity values.

(2) The highest increase is achieved for the smallest separation distance, $\delta = 1$. This is not surprising and is due to the construction of the family of velocity profiles and the selection of (W, δ) pairs so that the same velocity extrema are obtained for different separation distances. Hence, smaller δ values correspond to higher vorticity concentrations.

(3) The wavenumber of the most unstable shear layer mode exhibits a small increase with increasing wake deficits at $\delta = 1$, but is almost independent of W at higher separation distances. Thus, the wake deficit does not affect the frequency selection of unstable shear layer mode, which is closely approximated by estimates based on the *tanh* profile.

(4) While the growth rate of the most amplified wake mode increases by increasing W, its wavenumber of is almost independent of W. The growth rate maxima for both the shear layer and wake modes are reached for $\alpha \approx 0.6$.

(5) The phase speed of most unstable mode vanishes for the *tanh* hyperbolic velocity profile but increases in the direction of wake deficit with increasing W.

(6) The maximum growth rate of wake mode increases by increasing the separation between the positive and negative vorticity layers. However, in all cases considered, the shear layer mode always dominates the wake mode. This result is best interpreted by focusing on the positive and negative vorticity layers separately, which may be used to distinguish between the symmetric shear layer and asymmetric wake-like profiles. For the latter, the negative vorticity layer, whose inflection point is associated with the unstable shear layer mode, has considerably higher strength and thus dominates wake component.

The impact of heat release on flowfield stability is examined in Figs. 6-8, which show dispersion relations for $T_r = 2, 4$, and 8. The analysis is divided into two sections; results for the symmetric *tanh* profile are discussed first, and then contrasted with corresponding results for the asymmetric layer and wake profiles. As before, dispersion relations for the symmetric *tanh* profile are lumped with those of the asymmetric shear layer mode with $\delta = 1$, and identified by a wake deficit $W = 0$. Examination of the dispersion relations reveals:

(1) By increasing T_r, stabilization of the Kelvin-Helmholtz mode is gradually achieved. As noted by McMurtry et al. [9], who studied the stability of a heated layer idealized by broken-line vorticity and density profiles, heat release inhibits the growth of the most unstable mode, whose amplification rate decreases to a small fraction of the uniform-density maximum as temperature ratio becomes high. The results also indicate that the wavenumbers of the most amplified mode and neutrally stable modes decrease as T_r increases. Thus, the density variation alters the features of the instability in such a way as to stabilize short wavelength disturbances and collapse the instability to a thin band affecting long wave perturbations. On the other hand, the vanishing phase speed property of shear layer mode persists.

(2) The stability properties of the layer are significantly altered for $T_r \geq 4$.

As predicted in the previous section, the computed results show that the flow field now admits three unstable modes. The additional pair of unstable modes are associated with the outer zeros of $(rU')'$ and are thus called outer modes [8]. Dispersion relations of these modes appear as extensions of that of the Kelvin-Helmholtz mode.

(3) The outer modes have identical growth rates and equal but opposite phase speeds. The phase speeds of the neutrally stable solutions coincide with the flow velocity at the outer inflection points. Meanwhile, at high T_r the outer mode dominates the shear layer mode which is almost stabilized by the density variation. The wavenumber of the most amplified outer modes increases with increasing temperature ratio; thus, the associated mechanism promotes short-wavelength instability.

The stability properties of asymmetric shear layers and wake profiles differ significantly from those of the symmetric *tanh* profile, as the heat release has a less significant impact than in the former case. The stability results for these profiles are divided into two groups, according to the separation distance between the layers of opposite vorticity. For a large separation distance ($\delta = 3$), the amplification curves for the shear layer mode are weakly affected by the presence of a heated region within the vorticity layer. In particular, all of the stability properties of the unstable modes are unaffected, as neither the instability bandwidth nor the wave speed of unstable modes are affected by the temperature ratio. This result is expected, since the density is constant and equal to unity except in the region separating the distinct vorticity layers. In this region, the velocity profile is almost constant, so that $(\rho')'$ is closely approximated by the vorticity derivative U'' in the entire flow. Thus, the vorticity maxima do not lie with regions of high density gradient, and results for uniform-density flow are recovered.

For small separation distances, the presence of a heated fluid layer results in appreciable changes in the stability of the flow, and has a different influence on the behavior of the wake and shear layer modes. By increasing the temperature ratio, instability of the wake mode is promoted, as the corresponding amplification curves admit higher maxima. While the phase speeds of the unstable wake modes is weakly affected, the corresponding instability bandwidth is increased, and the wavenumber of the most unstable wake mode favors shorter wavelength instability. This result should be favorably contrasted with the results of Koochesfahani and Frieler who showed that the spatially-developing wake mode may become dominant when the asymmetric vorticity layer is subjected to a severe monotonic density difference [1]. However, in all cases considered in our study, the highest amplification rate of the shear layer mode is always considerably higher than for the corresponding wake mode. Therefore, the effect of heat release is not expected to lead to a qualitative change in the behavior of the perturbed reacting layer, which remains dominated by the growth of unstable shear layer waves.

Shear layer modes exhibit a different response to the imposed density variation. This response resembles the behavior of shear layer modes in heterogeneous layers for which the density increases monotonically from one side of the layer to the other. In both instances, density variation has minimal influence on either the growth rate of unstable Kelvin-Helmholtz modes, or on their stability bandwidth. However, it significantly affects the phase velocity of the waves. In the heterogeneous shear layer [19], unstable shear layer modes acquire an additional convective velocity component of the same direction as the high density stream. The analogy between the two cases can be established by inspecting the behavior of the density field in the neighborhood of the inflection point of the velocity profile. Here, the "unstable shear layer mode" is associated with the "upper" inflection point of the velocity profile. The density profile has positive derivative in the neighborhood of this inflection point, so that the density increases as we move towards the top free stream. Thus, we are led to expect an increase in the phase velocity of unstable shear layer mode, since the top stream velocity is positive. This expectation is reflected in the stability calculations, which show that the phase velocity of the most unstable shear layer mode increases with increasing temperature ratios and becomes positive for small wake deficits.

3.2. EFFECT OF DENSITY PROFILE THICKNESS

The stability results of the previous section are reexamined for different thicknesses of the density profile. For laminar nonpremixed flames, the thickness of the low-density zone depends on both the thermal and mass diffusivities. Thus, accurate estimates of this thickness relative to the vorticity thickness require the solution of the boundary layer equations for reacting flow, and the results will depend on the Prandtl and Lewis numbers, and on the details of the chemical reaction. In general, the product zone will be embedded within the vorticity layer, since near unity Prandtl and Lewis numbers generally prevail, and initial conditions describe a finite thickness vorticity layer and a sharp interface separating the oxidizer and fuel streams.

In this study, however, such a detailed study is replaced by the simplified approach of considering different values for the density thickness, which are selected in a wide parameter range in order to cover most situations of interest. To this end, the stability of all the velocity profiles of Table I is investigated for a variable density field specified by a fixed temperature ratio $T_r = 4$, which is characteristic of a large number of combustion applications. Meanwhile, the density thickness is gradually varied; the values $\sigma = 1.5, 0.75, 0.5$, and 0.25 are considered. The results of the computations are shown in Figs. 9-12, in terms of the growth rate and phase speed of unstable modes. As before, results for the symmetric *tanh* velocity profile are lumped with those of shear layer modes having $\delta = 1$.

The discussion of the results distinguishes between the stability properties of the shear layer and wake modes, and those obtained for the symmetric *tanh* profile. The latter case is discussed first and is summarized as follows. The behavior of the *tanh* shear layer mode is non-monotonic with respect to variation of the density thickness. The results are best interpreted by first considering the limiting case of very large density thickness. For this density configuration, the vorticity field lies in a zone of almost constant density, so that the uniform-density results are recovered. As σ is decreased and becomes close to the vorticity thickness, stabilization of the shear layer mode is observed. The maximum growth rate, the most amplified wavenumber and the stability bandwidth are all reduced. On the other hand, the nature of the instability is not altered, as all unstable modes have zero phase velocity. As σ is further decreased, stabilization of the shear layer mode continues but the flow field acquires an additional pair of unstable modes. Due to the symmetry of the density and velocity profile, these additional "outer modes" have identical instability bandwidths and amplification curves, with equal magnitude but opposite-sign phase velocities. In the range of density thicknesses considered, the instability of the additional modes is promoted by decreasing σ, as the corresponding most-amplified modes admit higher growth rates and wavenumbers. This is accompanied by an increase in the stability bandwidth and the phase velocity magnitude. Thus, *tanh* shear layers which accommodate high heat release slow reactions are more susceptible to such short wavelength instabilities. However, we do not expect this mechanism to persist continuously as σ is further decreased, since in the limit $\sigma \downarrow 0$ the deposited energy vanishes so that results for uniform-density flow should be approached.

The effect of the density density thickness is greatly attenuated in the presence of a wake deficit. In all cases, weak variations of the stability properties of the flow are recorded as the value σ is altered. The response of the wake mode to changes in the density thickness σ is extremely weak, as the results exhibit almost insignificant changes in the stability bandwidth, amplification curves, and phase relationships. These changes are of little importance since the shear layer mode dominates the initial evolution of the flow. The latter is weakly affected by changes in the density thickness, which result in small modulation of the phase speed of unstable modes. As previously mentioned, this result is expected, based on the similarity between the behavior of the vorticity and density profiles around the unstable inflection point and that described in monotonic variable-density shear layers [19].

3.3. VISUALIZATION OF UNSTABLE EIGENFUNCTIONS

Finally, the evolution of asymmetric shear layers is numerically computed. Attention is focused on the late stages of flowfield evolution, which witness the formation of vortical structures due to the non-linear breaking of unstable waves. The results are used to examine the validity of extending the trends established in the linear stability analysis. This exercise is limited to a visualization of the effect of these structures on the deformation of the flow and the evolution of the vorticity. A detailed investigation of the dynamics of the flow is not attempted in this work, as detailed reacting flow computations will be discussed in a subsequent study.

Numerical simulation of the variable-density flowfield is performed using the two-dimensional transport element method. The numerical scheme, which belongs to an adaptive class of Lagrangian field methods, is based on the discretization of the vorticity and density gradient fields into a number of transport elements of finite overlapping circular cores. Accordingly, the velocity field is given by a discrete, desingularized convolution over the induced vorticity field of the transport element. A similar convolution yields the density field. Once the velocity field is computed at the element centers, a second-order predictor corrector integration scheme is used to track their motion and to advance the numerical solution. Discrete vorticity and density values evolve according to Eqs. (4) and (2), respectively. Meanwhile, discrete density gradients are updated by relating their evolution to the material

TR	σ	W	δ	NR	Core size	Wavenumber	Wavelength	Growth rate
1	-	0	0	11	0.72	0.45	13.9	0.19
1	-	1.2	1	17	0.70	0.57	11	0.40
1	-	.8215	3	25	0.70	0.45	13.9	0.35
4	1	0	0	17	0.35	0.41	15.3	0.06
4	1	1.2	1	23	0.35	0.66	9.52	0.42
4	1	.8215	3	33	0.35	0.47	13.4	0.37
4	0.5	0	0	33	0.175	0.7	8.98	0.13
4	0.5	1.2	1	45	0.175	0.68	9.24	0.44
4	1.5	1.2	1	25	0.35	0.59	10.65	0.37

TABLE II

deformation of the Lagrangian mesh. Details of the formulation and construction of this numerical scheme, which has been extensively employed in the simulation of variable-density and reacting flows, can be found elsewhere [13-14,19]. Thus, only a brief account of the computational parameters used in the calculations is given.

The visualization of unstable modes is performed using the temporal model of the vorticity layers. In this model, the velocity and density fields are spatially periodic in the streamwise direction. The periodicity length, λ, is close to the wavelength of the most unstable mode, as determined above. The region of finite vorticity and density gradient is initially discretized on a square mesh, having NR and NS grid points in the cross-stream and streamwise directions, respectively. The core size of the transport elements, δ, and the discrete values of vorticity and density gradient are found by minimizing the error between the numerical and initial fields [13].

Nine cases of differenitally-heated layers and wakes are numerically investigated. The layers are identified by values of the temperature ratio, the thickness of the density layer, the wake deficit and the displacement of the wake profile, which are listed in Table II alongside the wavenumber, wavelength, and growth rate of the corresponding most unstable mode. Table II also shows the number of computational grids in the cross-stream direction, NR, and the core radius of the transport elements. In all calculations, the time step $\Delta t = 0.02$.

The first three cases correspond to uniform-density flow, and are selected in order to examine the effects of the wake deficit and the displacement of the wake profile. The numerical experiments are repeated by keeping the same (initial) vorticity field parameters and altering the initial density field by letting $T_r = 4$, and $\delta = 1$. Thus, for these initial flow configurations (cases 4-6), the initial vorticity and density thicknesses coincide. Finally, the dependency of the evolution of the flow field on the thickness of the initial density profile is examined in cases 7-9. These cases correspond to vorticity field configurations for which the linear stability analysis predicts a strong (case 7) or weak (cases 8 and 9) response to changes in δ.

The computations are initialized by introducing a perturbation in the flow field using sinewaves having the same periodicity wavelength as the computational domain, λ, and amplitude 0.01λ. The perturbation is applied by displacing the location of the transport elements in the cross-stream direction according to the sinewaves. The calculations are extended in order to observe the linear amplification of the unstable modes, and the early non-linear stages of their evolution. For cases 1, 4, and 7, this objective is achieved by carrying out the computations until $t = 12$. In the remaining cases, calculations are stopped at $t = 9$. Results of the simulation of the flowfield starting from the flow configurations of cases 1-9 are shown in Figs. 13-21, respectively. The figures are generated by plotting the location and instantaneous position and velocity vector of the transport elements. The development of the vorticity field thus illustrated is discussed below.

Figure 13 shows that the evolution uniform-density shear layer initially described by the symmetric *tanh* velocity profile does not destroy the symmetry of the vorticity field. The vanishing phase speed of the instability wave, predicted by the linear theory for the linear Kelvin-Helmholtz wave, persists as the waves undergo a non-linear growth regime, and roll to form a concentrated core of vorticity. Detailed computations of similar flow fields have been performed previously [13], and have shown that, when pairing is disabled, the late stages are characterized by a maturation of the vortex cores and the continuous entrainment of the vorticity from the braids into the cores.

The superposition of wake deficit on the symmetric, monotonic layer profile results in a significant departure from the previously discussed behavior. The evolution of the asymmetric layers of Figs. 14 and 15 is characterized by the finite wave speed of the linear instability modes, and by the convective motion of the vortical structures which form following their nonlinear evolution. As previously discussed, the direction of motion of the linear instability waves can be determined from the the linear stability analysis and by inspection of the initial velocity profile. The computations are in agreement with the results of the linear theory, which predicts almost equal phase speeds for the most unstable modes, and a slightly larger growth rate for the layer with the smaller separation distance. Alternatively, this behavior can be qualitatively predicted by considering the contribution of the "weaker" vorticity layer, whose induced flow field leads to the motion of the linear waves and of the vortices.

The consideration of the effects of opposite regions of vorticity is easier for the larger displacement parameter ($\delta = 3$, Fig. 15), since the corresponding vorticity profiles are formed of well-separated strips of opposite sign of vorticity. The computed results indicate that the evolution of the flowfield is dominated by the stronger (negative) vorticity layer. The latter appears to develop independently of the weaker vorticity layer, which does not exhibit significant deformation during the period of the simulation. This assessment no longer holds in the case $\delta = 1$, which shows that both regions of vorticity deform simultaneously (Fig. 14). This results in a greater deformation of the vorticity field and the formation of a substantially larger vortex core.

The impact of the density variation on the development of the flow is illustrated in Figs. 16-18, which show the evolution of shear layers characterized by the same initial vorticity distributions previously considered, and a density profile having $T_r = 4$, and $\sigma = 1$. When starting with the symmetric *tanh* velocity profile, the growth of Kelvin-Helmholtz waves is significantly suppressed. Moreover, the nonlinear wavebreaking of these modes results in the formation of weak vortex cores which are less coherent than in the uniform-flow case (Fig. 16). These results are in agreement with the predictions of the linear stability theory and with the computations of McMurtry et al. [9] who considered the evolution of symmetric reacting shear layers.

Density variation effects are considerably less pronounced in the presence of a wake component. For large separation distance, $\delta = 3$, comparison of the uniform- and variable-density results (Figs. 15 and 18, respectively) shows that the presence of a heated layer has almost no influence on the development of the flow. These results extend those of the linear stability theory, which predicts little changes in the wavelength, growth rate, and phase speed of the most unstable Kelvin-Helmholtz waves with the variation of the density profile for this initial vorticity configuration. On the other hand, the interactions of the density and vorticity field cannot be neglected for small separation distance, $\delta = 1$. As indicated in Table II, the wavelength of the most-unstable mode is decreased as the temperature ratio is increased to $T_r = 4$. Moreover, while the growth rate of this mode is not significantly altered by the presence of the heated layer, its phase speed is noticeably reduced. This observation also holds when considering the convective motion of the vortices which form following the nonlinear evolution of the unstable waves (Figs. 14 and 17). However, in both cases, the vorticity-density interactions do not suppress linear growth and do not inhibit the formation of large coherent vortex cores.

The sensitivity of the flow-field to the details of the initial density profile is examined in Figs. 19-21. In particular, the stabilization of the symmetric layer by heat release and the weak dependence of the asymmetric layer on the presence of the heated layer is investigated. As previously mentioned, when the regions of high vorticity and density gradient are well separated, the initial development flow is approximated by the uniform-density equations, so that layers characterized by a large separation distance will not be further considered. For the symmetric case, Fig. 19 indicates that by decreasing the thickness of the density profile to $\sigma = 0.5$, the behavior of the shear layer undergoes an additional transition. The results reflect the prediction of the stability theory, which indicates that in this regime the most unstable mode consists of a pair of travelling waves of equal growth rates, and equal but opposite phase speeds. These waves amplify as they move away from the middle of the computational domain. The rollup of the waves occurs as the two trains of periodic waves meet at the boundaries of the computational domain, and is followed by the formation of vortices whose cores are smaller and less coherent than their uniform-density counterparts. This mechanism differs significantly from the rollup of the linear Kelvin-Helmholtz waves in uniform-density flow, which, as indicated by the theory, have appreciably larger growth rates. Thus, the stabilizing effects of heat release are expected to persist for this profile, though the details of the density distribution may lead to radical changes in the development of the flow and in the structure of the associated vorticity field.

On the other hand, the addition of a wake deficit greatly diminishes the "stabilizing" impact of the density variation. As predicted by the linear theory and observed in the computations, the early evolution of the asymmetric layer having a large wake deficit is almost insensitive to the presence of the heated layer. For such initial vorticity configurations, the effects of density variation are limited to a small modulation of the wavelength, growth rate, and phase speed of the most unstable mode. Moreover, as illustrated in Figs. 14, 17, 20 and 21, similar vortical structures are obtained as a result of the non-linear wavebreaking of the unstable modes. This observation holds for all the asymmetric layers considered, despite the fact that the details of the vorticity distribution within the vortex cores and in the braids joining neighboring vortices and the convective motion of the eddies are strongly affected by baroclinic vorticity generation in the later stages of evolution of the flow [13,19].

4. Conclusions

In this work, stability of heterogeneous shear flows is investigated using linear stability analysis and numerical simulations. A large number of initial flow configurations which model the development of nonpremixed reacting shear flow are analyzed. The initial conditions are used to examine the effects of heat release associated with the development of a nonpremixed flame on the stability of asymmetric shear layers and wakes. The latter belong to a continuous family of velocity profiles which is constructed by deforming the symmetric *tanh* shear layer profile using a wake component characterized by characterized by a spread, δ, and a velocity deficit, W. Density variation is used to model the effects of heat release. The initial density distribution corresponds to a temperature spike and is described by a thickness, σ, and a temperature ratio, T_r. Stability curves are obtained in this four-dimensional parameter space and numerically visualized using the transport element method. The numerical simulations are extended into the nonlinear stages of flowfield evolution in order to examine the validity of extrapolating the linear stability results.

Stability of the *tanh* reacting shear layer exhibits strong sensitivity to the details of the density distribution. When the density and vorticity thicknesses are close, stabilization of the shear layer occurs as the temperature ratio increases. This effect is manifested by a sharp decrease in the instability growth rate; the nonlinear evolution of unstable eigenfunctions yields weaker less-coherent vortex structures than those observed in uniform-density flow. By decreasing the thickness of the density profile, additional inflection points and instability modes are observed. These instability modes dominate the shear layer mode which is almost stabilized by the heat release. However, while the associated growth rates increase with decreasing density thickness and appear to approach the maximum growth-rate values obtained for uniform-density flow, the nonlinear evolution of these instability modes does not result in the formation of concentrated vortices or in substantial deformation of the flow. Thus, for the *tanh* profile, heat release tends to stabilize of the flow.

Addition of a wake deficit significantly alters the stability of the flow. In the parameter range considered, two flow configurations are distinguished. When the zones of high density gradients and vorticity magnitude are well separated, heat release and density variation have almost no impact on flowfield stability. In this case, the growth rate and phase speed of unstable waves are weakly affected, and their non-linear evolution results in the formation of large concentrated asymmetric vortices. On the other hand, when regions of high vorticity and density gradients are close or coincide, heat release has a weak effect on the development of the flow. While the growth rates of unstable waves are almost unaffected, their phase speeds depend on the details of the density variation. This mechanism is also observed in the later stages of evolution of the flow, which indicate that the convective motion of the large vortices is modulated in the same manner as the phase velocity of the corresponding linear waves. Thus, heat release is not expected to inhibit the growth of unstable modes in asymmetric layers and wakes, but may however influence the global features of the flow.

The correspondence between computed stability results and initial flow configurations indicates that the effects of heat release may be intuitively predicted by simple examination of the behavior of the density profile in the neighborhood of the inflection point of the velocity profile. If the density profile reaches a minimum in the neighborhood of the inflection point of the velocity profile, stabilization of the corresponding unstable mode is expected. If the density gradient does not vanish in this neighborhood, we expect a minor variation in the growth rate of the unstable mode, and a modulation of its phase speed which depends on the sign of the density gradient. In this case, the results exhibit analogous trends to those established for symmetric shear layers separating streams of unequal density, where unstable waves are observed to acquire a streamwise convection component in the direction of the high-density stream [13].

The large predicted differences in the stability properties for symmetric and asymmetric reacting layers lead us to expect significant sensitivity of developing reacting shear layers to initial disturbances. In particular, if unstable modes are excited at short distances downstream of the splitter plate, i.e. before viscous diffusion leads to the destruction of the wake deficit associated with the merger of the Blasius profiles, significant stabilization by heat release is not expected. Otherwise, a sharp decrease in shear layer growth, mixing and burning efficiency is anticipated.

We finally note that, in the parameter range considered, shear layer modes were always found to dominate the wake modes. Thus, heat release and density variation are not expected to produce a significant change in the shape of instability [8]. Our results should be contrasted with those of Koochesfahani and Frieler [1] who showed that for large wake deficits and density differences, the initial development of the layer can be dominated by the amplification of unstable wake modes. A stability analysis in this flow regime was omitted, since the corresponding flow configurations are not representative of typical reacting shear flow applications. Extension of the parameter range to include these initial conditions is postponed to a subsequent study which will focus on direct simulation of the developing reacting flow.

Acknowledgements

This work was partially supported by the Air Force Office of Scientific Research under Grant AFOSR 89-0491, and the Gas Research Institute. Linear stability calculations were performed by Ms. Xiyan Shi. Transport element computations were performed at the Illinois National Supercomputer Center.

References

[1] Koochesfahani, M.M., and Frieler, C.E., "Instability of Nonuniform Density Free Shear Layers with a Wake Profile," AIAA Journal, Vol. 27, No. 12, 1989, pp. 1735-1740.

[2] Yu, M.-H., and Monkewitz, P.A., "The Effect of Nonuniform Density on the Absolute Instability of Two-Dimensional Inertial Jets and Wakes," Physics of Fluids A, Vol. 2, No. 7, 1990, pp. 1175-1181.

[3] Dimotakis, P.E., "Turbulent free shear layer mixing," AIAA 27th Aerospace Sciences Meeting, AIAA Paper 89-0262, 1989.

[4] Brown, G.L. and Roshko, A., "On Density Effects and Large Structure in Turbulent Mixing Layers," Journal of Fluid Mechanics, Vol. 64, July 1974, pp. 775-816.

[5] Maslowe, S.A., and Kelly, R.E., "Inviscid Instability of an Unbounded Heterogeneous Shear Layer," Journal of Fluid Mechanics, Vol. 48, No. 2, 1971, pp. 405-415.

[6] Koop, C.G., and Browand, F.K., "Instability and Turbulence in Stratified Fluid with Shear," Journal of Fluid Mechanics, Vol. 93, 1979, pp. 135-159.

[7] Hegde, U.G., and Zinn, B.T., "Vortical Mode Instability of Shear Layers with Temperature and Density Gradients," AIAA Journal, Vol. 28, No. 8, 1990, pp. 1389-1396.

[8] Shin, D.S., and Ferziger, J.H., "Linear Stability of the Reacting Mixing Layer," AIAA Journal, Vol. 29, No. 10, October 1991, pp. 1634-1642.

[9] McMurtry, P.A., Riley, J.J., and Metcalfe, R.W., "Effects of Heat Release on the Large-Scale Structure in Turbulent Mixing Layers," Journal of Fluid Mechanics, Vol. 199, 1989, pp. 291-332.

[10] Huerre, P., and Monkewitz, P.A., "Local and Global Instabilities in Spatially Developing Flows," Annual Review of Fluid Mechanics, Vol. 22, 1990, pp. 473-537.

[11] Monkewitz, P.A., and Sohn, K.D., "Absolute Instability in Hot Jets," AIAA Journal, Vol. 26, No. 8, 1988, pp. 911-916.

[12] Huerre, P., and Monkewitz, P.A., "Absolute and Convective Instabilities in Free Shear Layers," Journal of Fluid Mechanics, Vol. 159, 1985, pp. 151-168.

[13] Krishnan, A., Numerical Study of Vorticity-Combustion Interactions in Shear Flow, Ph.D. thesis, M.I.T., 1989.

[14] Ghoniem, A.F. and Krishnan, A., "Origin and Manifestation of Flow-Combustion Interactions in a Premixed Shear Layer," Twenty-Second Symposium (International) on Combustion, The Combustion Institute, 665, 1988.

[15] Rehm, R.G. and Baum, H.R., "The equations of motion for thermally driven, buoyant flows," J. Res. N.B.S. 83, 297, 1978.

[16] Majda, A. and Sethian, J.A., "The derivation and numerical solution of the equations for zero Mach number combustion," Comb. Sci. Tech. 42, 185, 1987.

[17] Betchov, R., and Criminale, W.O., Stability of Parallel Flows, Academic Press, 1967.

[18] Koch, W., "Local Instability Characteristics and Frequency Determination of Self-Excited Wake Flows," Journal of Sound and Vibration, Vol. 99, 1985, pp. 53-83.

[19] Soteriou, M.C., Knio, O.M., and Ghoniem, A.F., "Manipulation of the Growth Rate of a Variable Density, Spatially Developing Mixing Layer via External Modulation," AIAA 29th Aerospace Sciences Meeting, January 7-10, 1991, Reno. Nevada, AIAA Paper 91-0081.

[20] Drazin, P.G., and Howard, L.N., "Hydrodynamic Stability of Parallel Flow of Inviscid Fluid," Advances in Applied Mechanics, Vol. 9, 1966, Academic Press, pp. 1-89.

[21] Drazin, P.G., and Reid, W.H., Hydrodynamic Stability, Cambridge University Press, 1981.

[22] Drazin, P.G., and Howard L.N., "The Instability to Long Waves of Unbounded Parallel Inviscid Flow," Journal of Fluid Mechanics, Vol. 14, 1962, pp. 257-283.

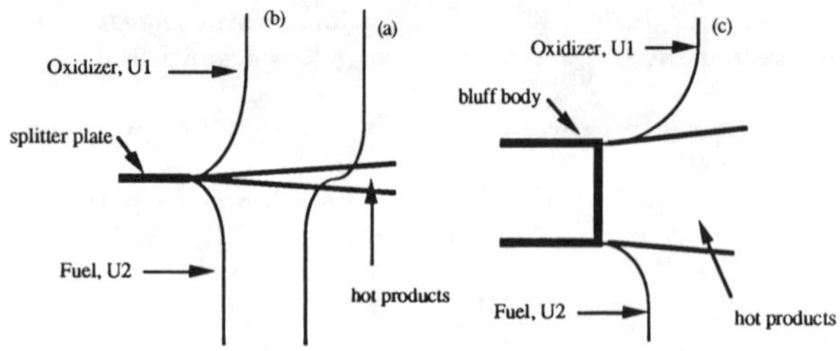

Figure 1. Schematic sketch of a shear layer (left) and a bluff-body wake flow (right). The self-similar monotonic shear layer profile is shown in curve (a), while curves (b) and (c) respectively illustrate asymmetric shear layer and wake profiles. The top (bottom) row corresponds to $\delta = 1$ ($\delta = 3$).

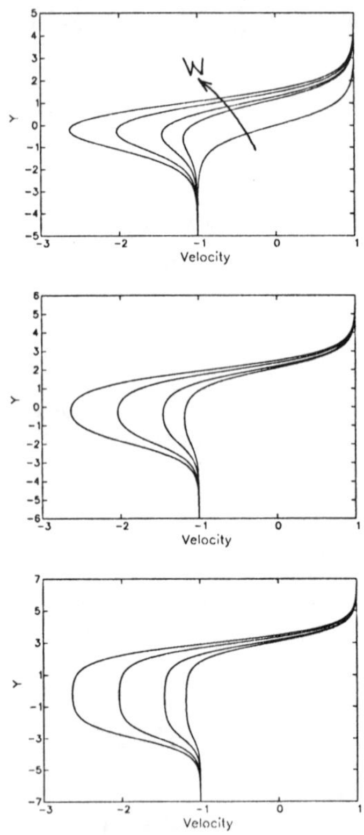

Figure 2. Velocity profiles for the family of shear layers given by Eq. (10), with (W, d) combinations of Table I. The plots are respectively arranged from top for increasing wake increasing thickness, $\delta = 1, 2$, and 3. The *tanh* profile is shown with the $\delta = 1$ subcollection. The top (bottom) row corresponds to $\delta = 1$ ($\delta = 3$).

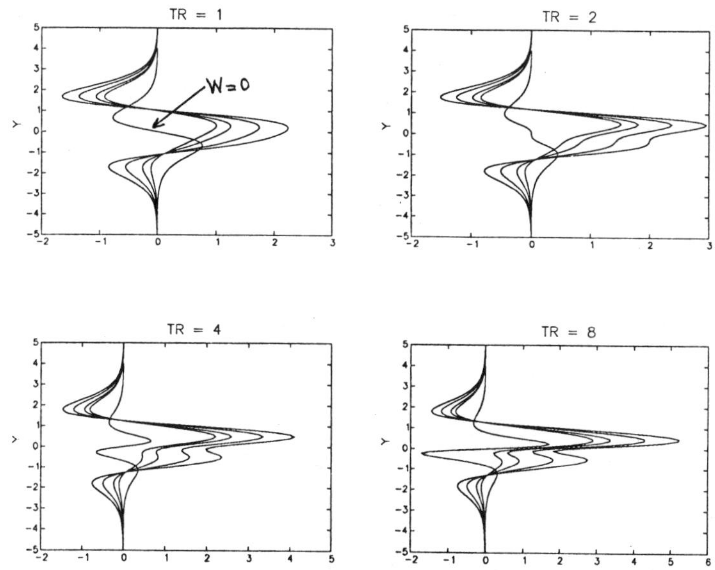

Figure 3. Profiles of $U'' + \rho' U'/\rho$ for layers described by $\sigma = 1$, $\delta = 1$ and (a) $T_r = 1$ (uniform-density flow); (b) $T_r = 2$; (c) $T_r = 4$; and (d) $T_r = 8$. Similar behavior is obtained for the larger separation disctances, $\delta = 2$ and 3. The top (bottom) row corresponds to $\delta = 1$ ($\delta = 3$).

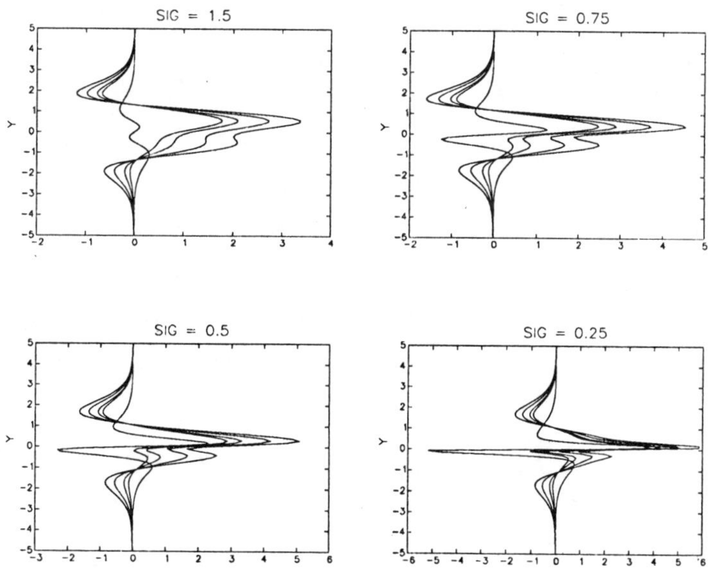

Figure 4. Behavior of $U'' + \rho' U'/\rho$ for the velocity profiles of Fig. 2, with density distribution given by $T_r = 4$, and (a) $\sigma = 1.5$; (b) $\sigma = 0.75$; (c) $\sigma = 0.5$; and (d) $\sigma = 0.25$. The top (bottom) row corresponds to $\delta = 1$ ($\delta = 3$).

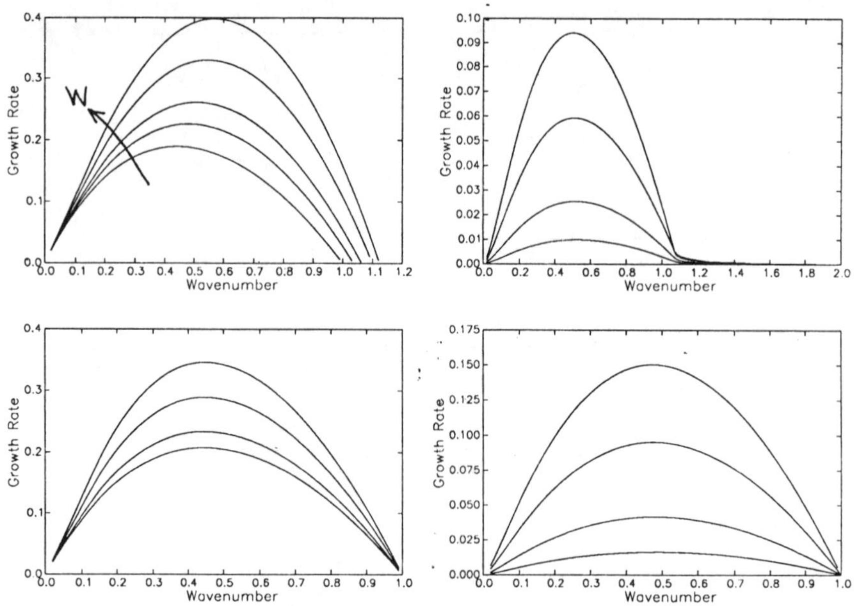

Figure 5a. Growth rate vs. wavenumber of the shear layer (left) and wake (right) modes for the velocity profiles of Fig. 2 and uniform-density profile $T_r = 1$. The top (bottom) row corresponds to $\delta = 1$ ($\delta = 3$).

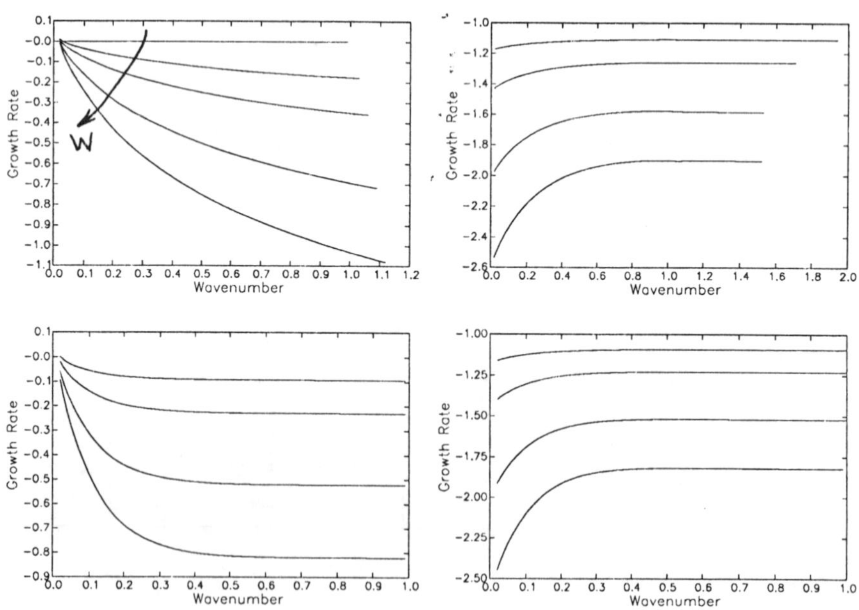

Figure 5b. Phase speed vs. wavenumber for the shear layer (left) and wake (right) modes for the velocity profiles of Fig. 2 and uniform-density profile $T_r = 1$. The top (bottom) row corresponds to $\delta = 1$ ($\delta = 3$).

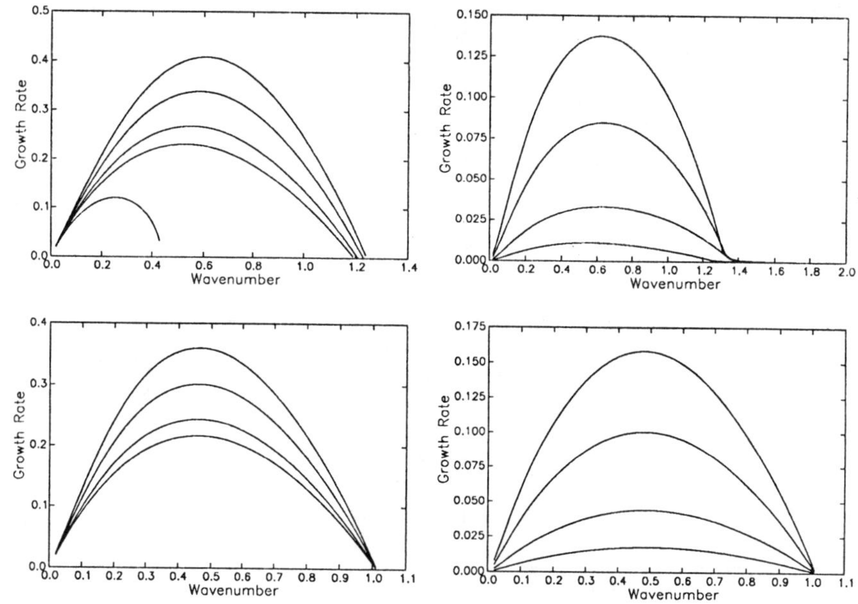

Figure 6a. Growth rate vs. wavenumber of the shear layer (left) and wake (right) modes for the velocity profiles of Fig. 2 and density profile given by $T_r = 2$ and $\sigma = 1$. The top (bottom) row corresponds to $\delta = 1$ ($\delta = 3$).

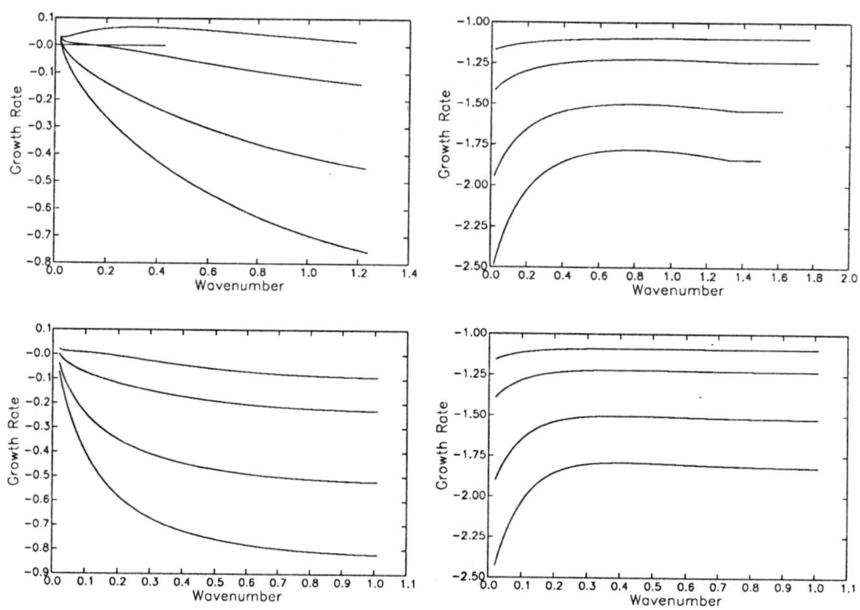

Figure 6b. Phase speed vs. wavenumber of the shear layer (left) and wake (right) modes for the velocity profiles of Fig. 2 and density profile given by $T_r = 2$ and $\sigma = 1$. The top (bottom) row corresponds to $\delta = 1$ ($\delta = 3$).

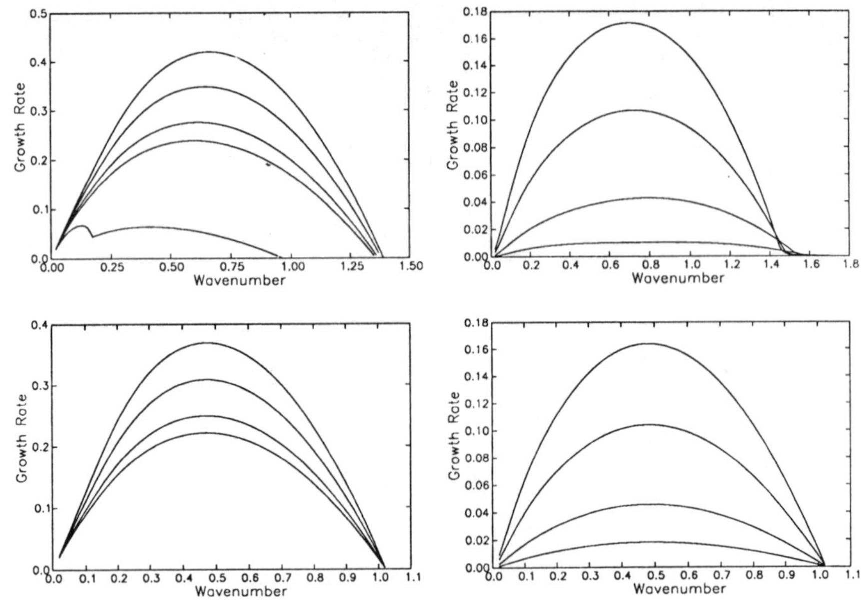

Figure 7a. Growth rate vs. wavenumber of the shear layer (left) and wake (right) modes for the velocity profiles of Fig. 2 and density profile given by $T_r = 4$ and $\sigma = 1$. The top (bottom) row corresponds to $\delta = 1$ ($\delta = 3$).

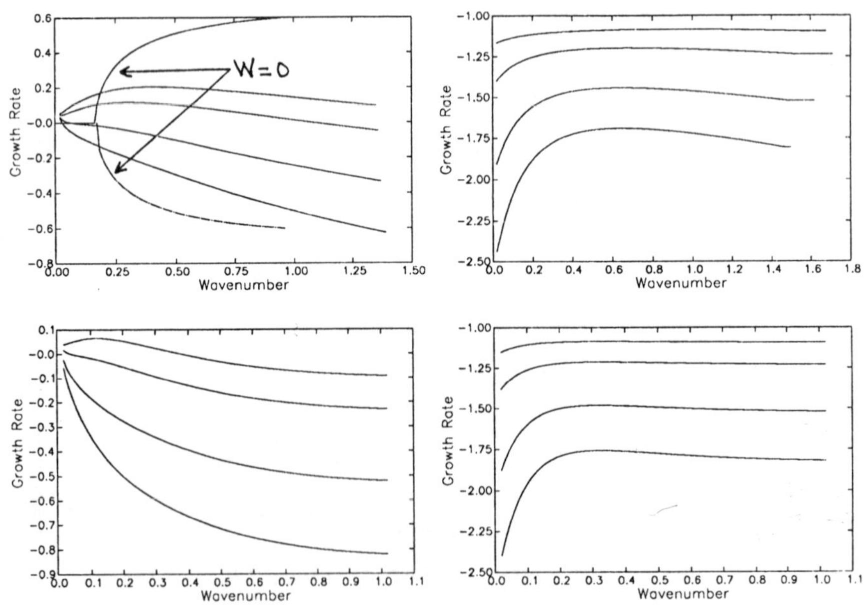

Figure 7b. Phase speed vs. wavenumber of the shear layer (left) and wake (right) modes for the velocity profiles of Fig. 2 and density profile given by $T_r = 4$ and $\sigma = 1$. The top (bottom) row corresponds to $\delta = 1$ ($\delta = 3$).

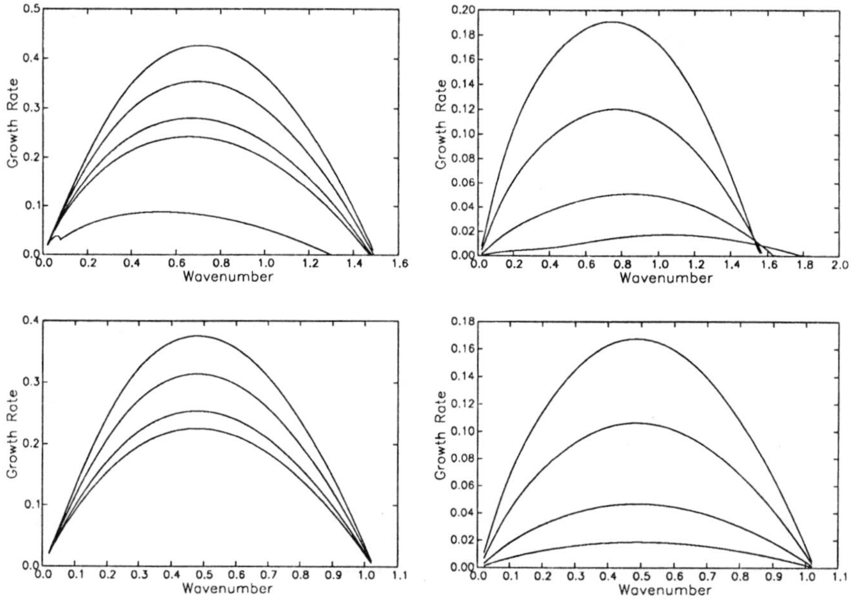

Figure 8a. Growth rate vs. wavenumber of the shear layer (left) and wake (right) modes for the velocity profiles of Fig. 2 and density profile given by $T_r = 8$ and $\sigma = 1$. The top (bottom) row corresponds to $\delta = 1$ ($\delta = 3$).

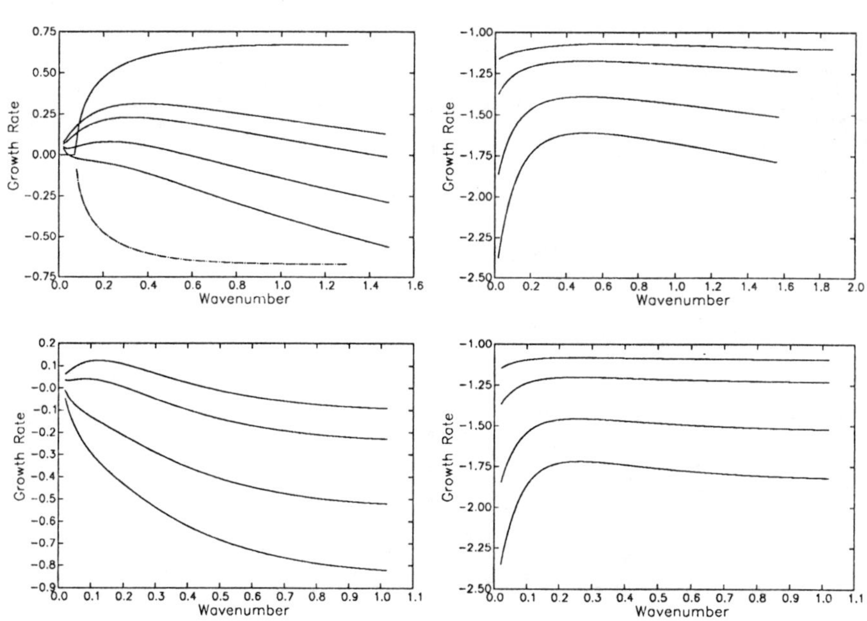

Figure 8b. Phase speed vs. wavenumber of the shear layer (left) and wake (right) modes for the velocity profiles of Fig. 2 and density profile given by $T_r = 8$ and $\sigma = 1$. The top (bottom) row corresponds to $\delta = 1$ ($\delta = 3$).

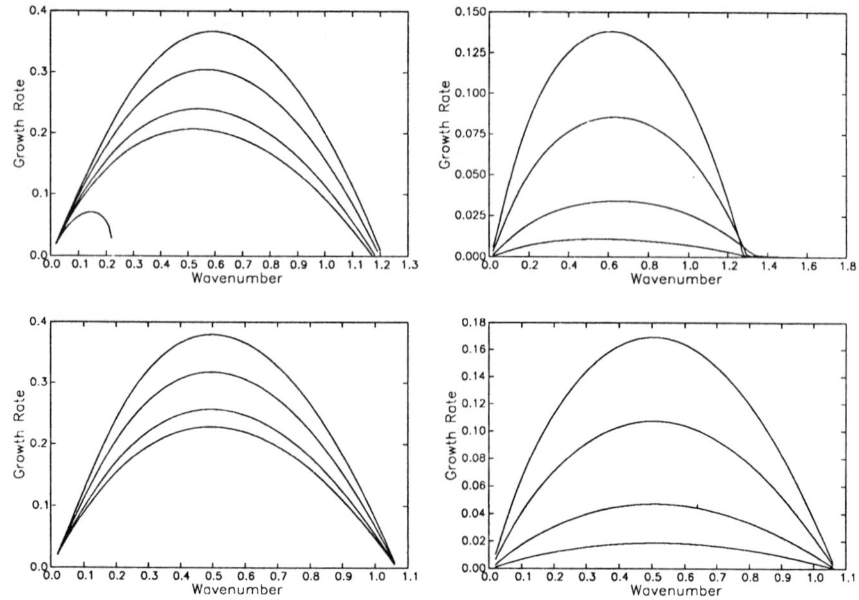

Figure 9a. Growth rate vs. wavenumber of the shear layer (left) and wake (right) modes for the velocity profiles of Fig. 2 and density profile given by $T_r = 4$ and $\sigma = 1.5$. The top (bottom) row corresponds to $\delta = 1$ ($\delta = 3$).

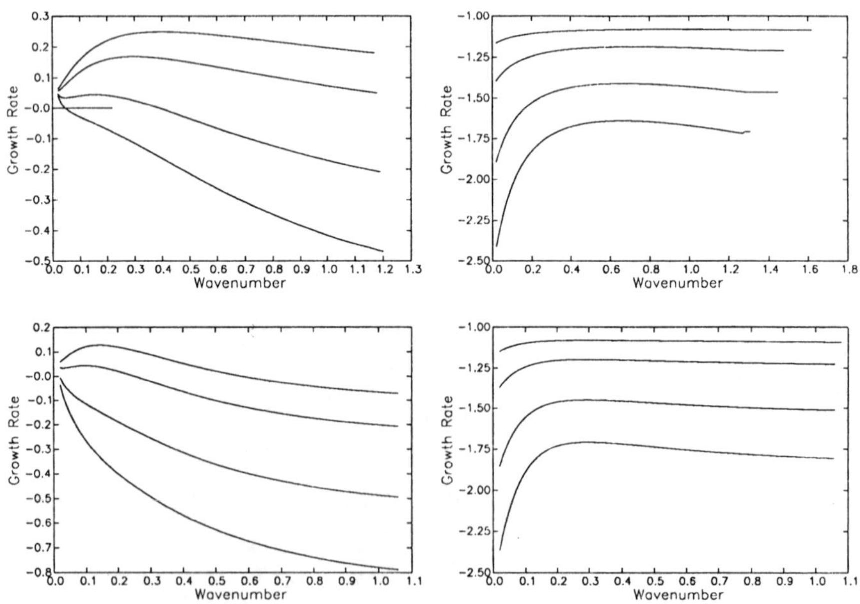

Figure 9b. Phase speed vs. wavenumber of the shear layer (left) and wake (right) modes for the velocity profiles of Fig. 2 and density profile given by $T_r = 4$ and $\sigma = 1.5$. The top (bottom) row corresponds to $\delta = 1$ ($\delta = 3$).

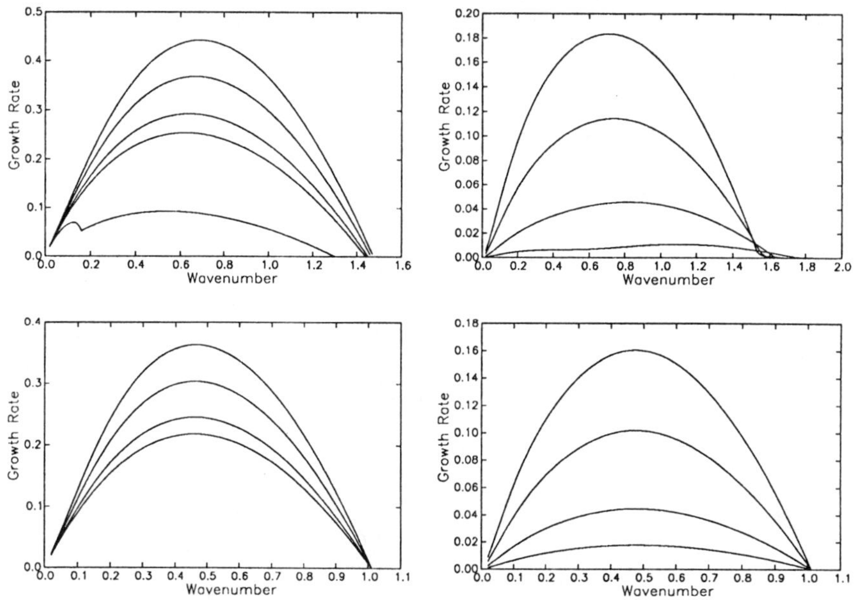

Figure 10a. Growth rate vs. wavenumber of the shear layer (left) and wake (right) modes for the velocity profiles of Fig. 2 and density profile given by $T_r = 4$ and $\sigma = 0.75$.

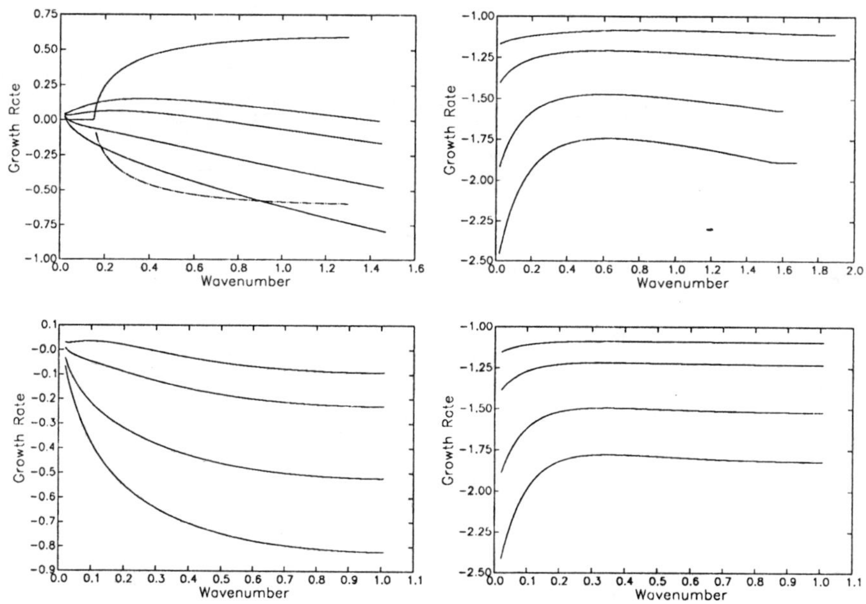

Figure 10b. Phase speed vs. wavenumber of the shear layer (left) and wake (right) modes for the velocity profiles of Fig. 2 and density profile given by $T_r = 4$ and $\sigma = 0.75$.

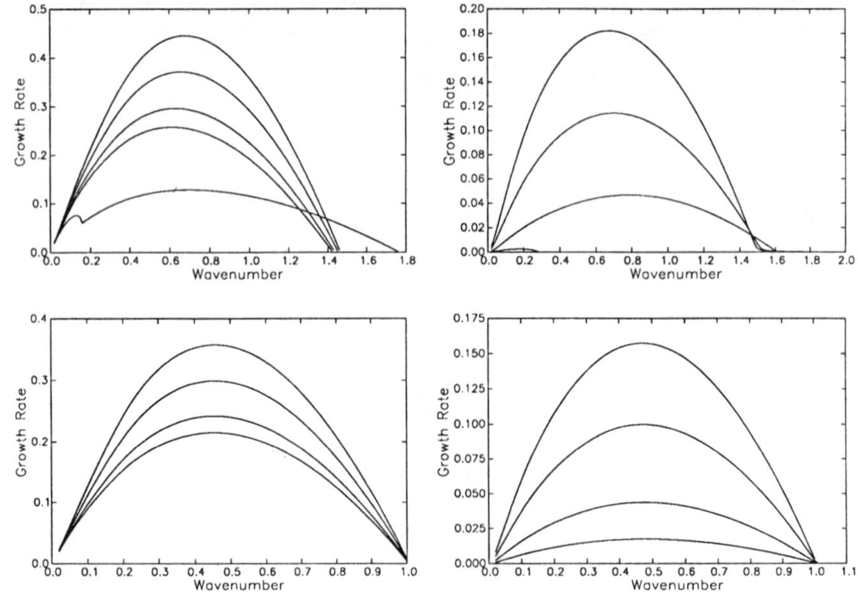

Figure 11a. Growth rate vs. wavenumber of the shear layer (left) and wake (right) modes for the velocity profiles of Fig. 2 and density profile given by $T_r = 4$ and $\sigma = 0.5$.

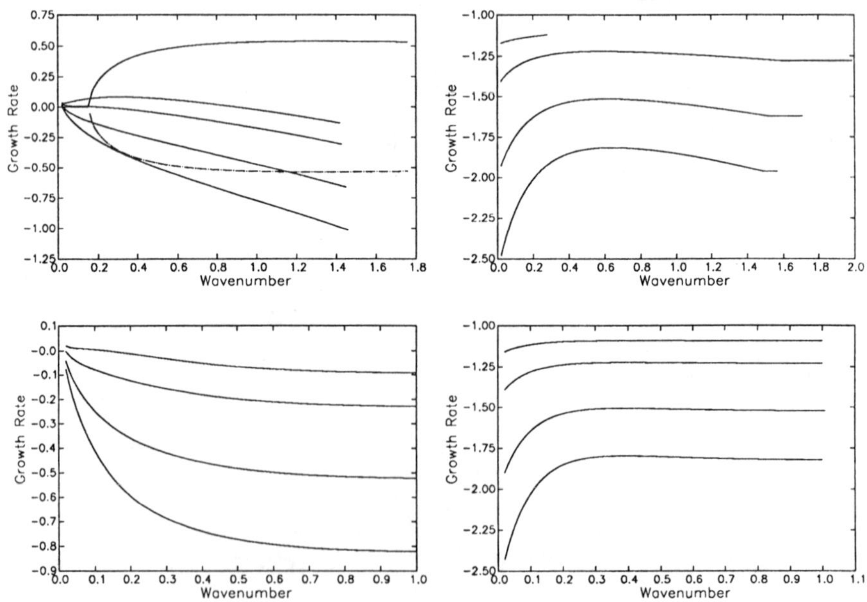

Figure 11b. Phase speed vs. wavenumber of the shear layer (left) and wake (right) modes for the velocity profiles of Fig. 2 and density profile given by $T_r = 4$ and $\sigma = 0.5$.

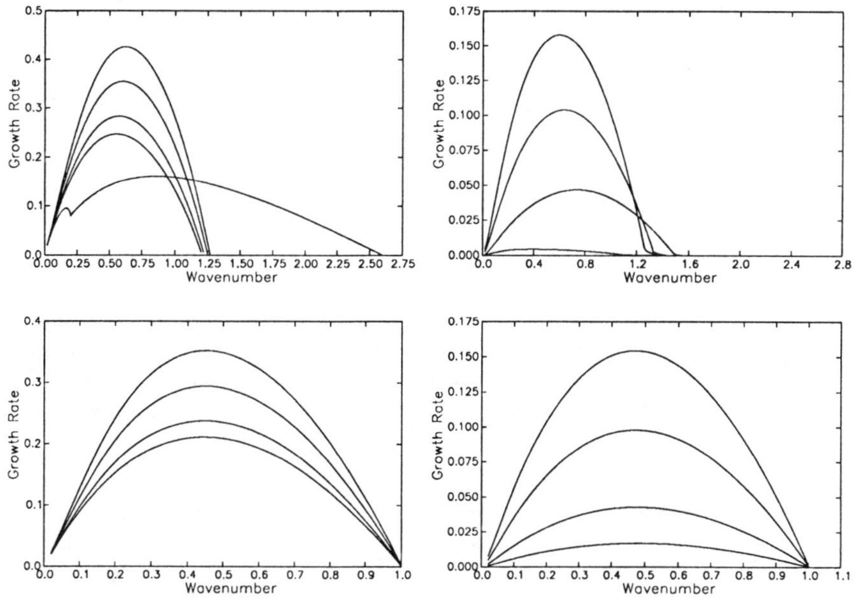

Figure 12a. Growth rate vs. wavenumber of the shear layer (left) and wake (right) modes for the velocity profiles of Fig. 2 and density profile given by $T_r = 4$ and $\sigma = 0.25$.

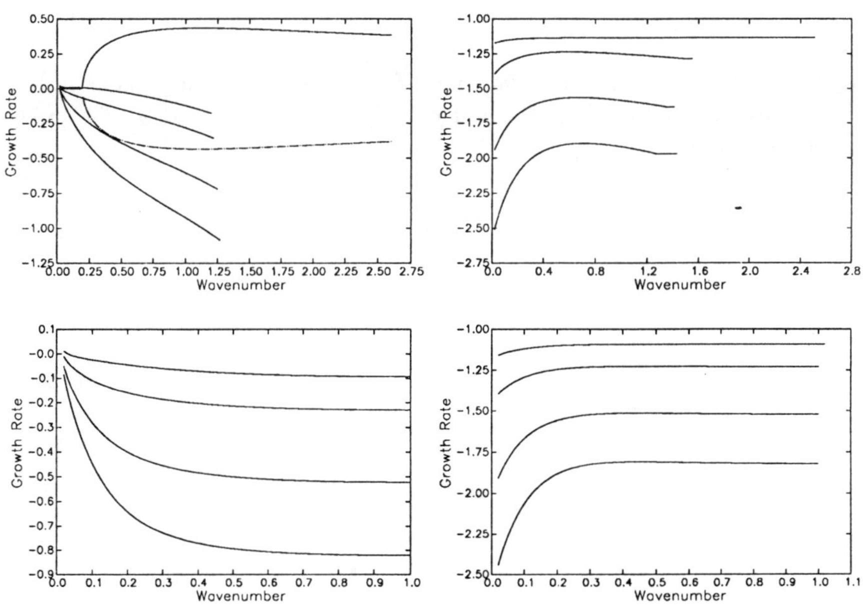

Figure 12b. Phase speed vs. wavenumber of the shear layer (left) and wake (right) modes for the velocity profiles of Fig. 2 and density profile given by $T_r = 4$ and $\sigma = 0.25$.

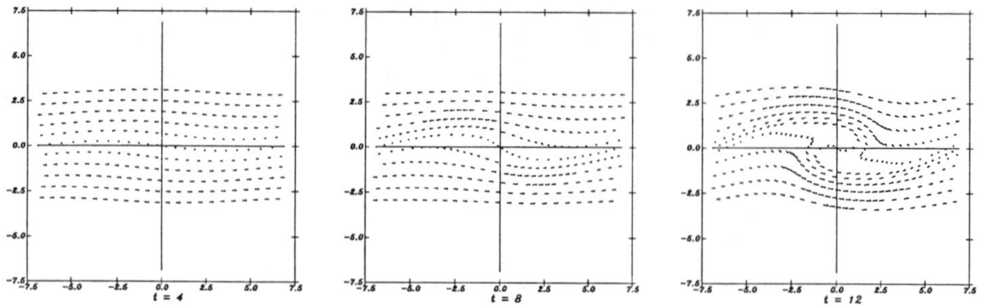

Figure 13. Evolution of the uniform-density shear layer with velocity profile given by $W = \delta = 0$, illustrated in terms of the vortex elements. The plots are generated by drawing the location and instantaneous velocity vector of the elements at $t = 4, 8$, and 12.

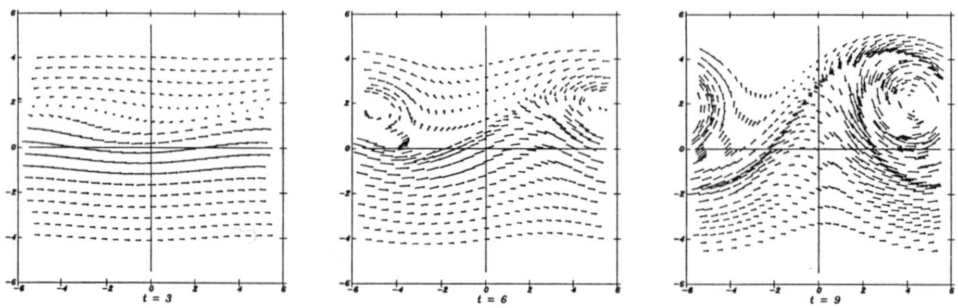

Figure 14. Vortex element representation of the uniform-density shear layer with velocity profile given by $W = 1.2$ and $\delta = 1$, at $t = 3, 6$, and 9. The plots are generated as in Fig. 13.

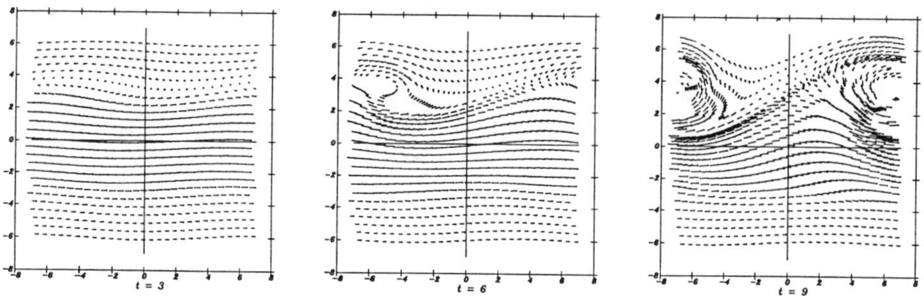

Figure 15. Vortex element representation of the uniform-density shear layer with velocity profile given by $W = 0.82149$ and $\delta = 3$, at $t = 3, 6$, and 9. The plots are generated as in Fig. 13.

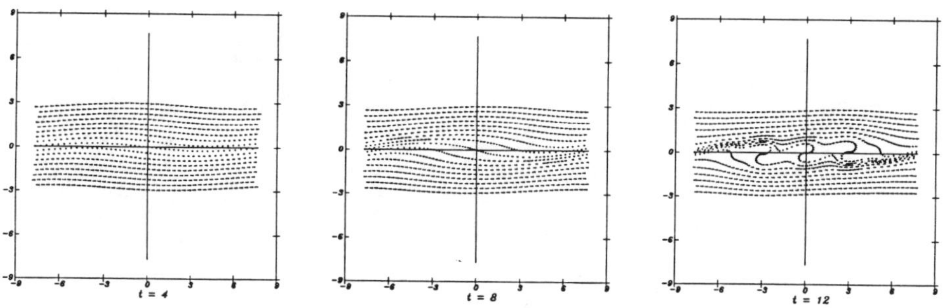

Figure 16. Vortex element representation of the shear layer with velocity profile given by $W = \delta = 0$ and density profile having $T_r = 4$ and $\sigma = 1$. The plots are generated as in Fig. 13.

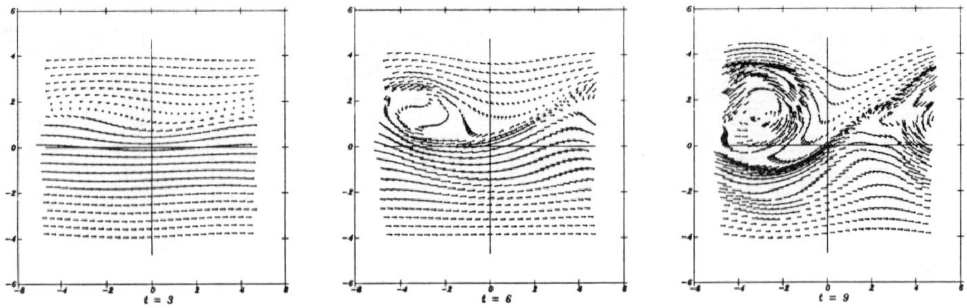

Figure 17. Vortex element representation of the shear layer with velocity profile given by $W = 1.2$ and $\delta = 1$ and density profile having $T_r = 4$ and $\sigma = 1$. The plots are generated as in Fig. 13.

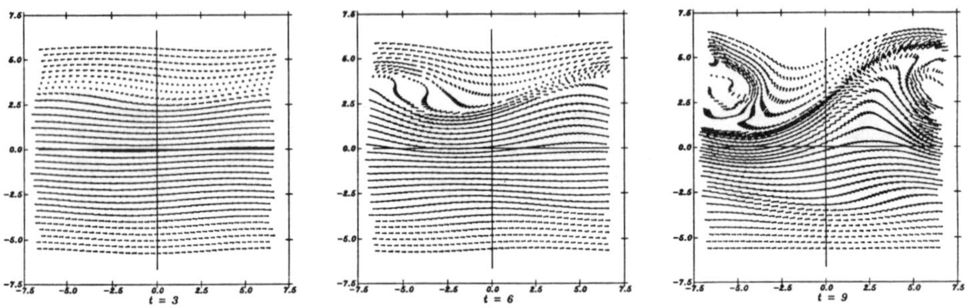

Figure 18. Vortex element representation of the shear layer with velocity profile given by $W = 0.82149$ and $\delta = 3$ and density profile having $T_r = 4$ and $\sigma = 1$. The plots are generated as in Fig. 13.

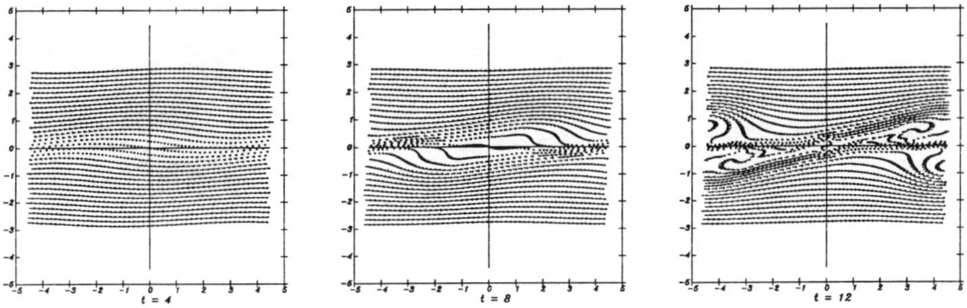

Figure 19. Vortex element representation of the shear layer with velocity profile given by $W = \delta = 0$ and density profile having $T_r = 4$ and $\sigma = 0.5$. The plots are generated as in Fig. 13.

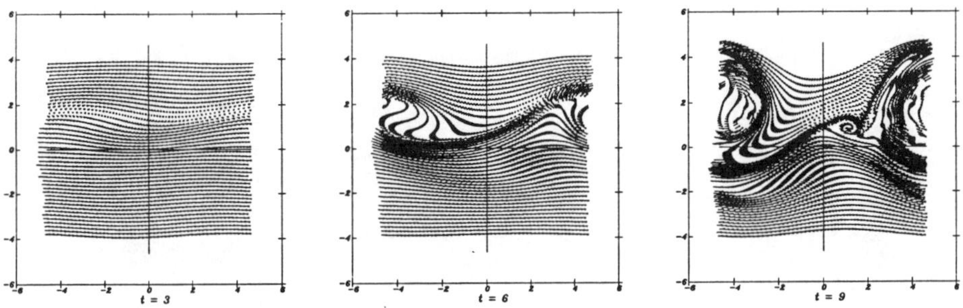

Figure 20. Vortex element representation of the shear layer with velocity profile given by $W = 1.2$ and $\delta = 1$ and density profile having $T_r = 4$ and $\sigma = 0.5$. The plots are generated as in Fig. 13.

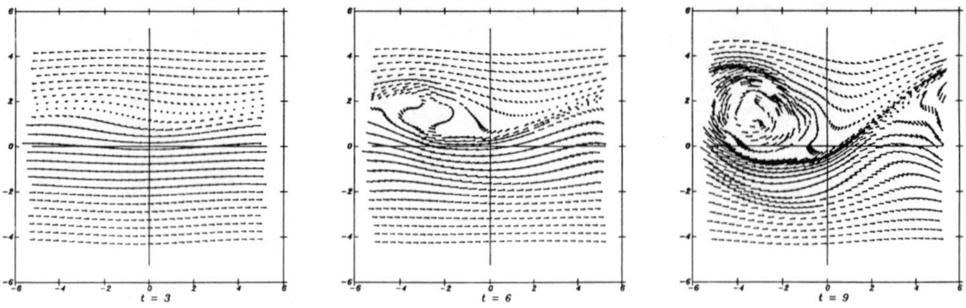

Figure 21. Vortex element representation of the shear layer with velocity profile given by $W = 1.2$ and $\delta = 1$ and density profile having $T_r = 4$ and $\sigma = 1.5$. The plots are generated as in Fig. 13.

A PARTICLE IN CELL METHOD FOR THE 2-D COMPRESSIBLE EULER EQUATIONS

S. MAS-GALLIC
Centre de Mathématiques Appliquées
Ecole Polytechnique
91128 Palaiseau Cedex FRANCE

and

M. LOUAKED and O. PIRONNEAU
Laboratoire d'Analyse Numérique
Université Pierre et Marie Curie
4 Place Jussieu
75252 Paris Cedex 05 FRANCE

1. Introduction

The method presented here is an extension to the case of an isentropic compressible flud of the Vortex In Cell method classically used to compute incompressible inviscid fluid flows (see [4;5], we also refer to [9;10] for pioneering Particle In Cell methods in compressible fluid flow calculations). To begin with, the Euler equation are written in the velocity-vorticity formulation

$$\begin{aligned} \frac{\partial \omega}{\partial t} + u.\nabla \omega &= 0 \\ \nabla . u &= 0 \\ \nabla \times u &= \omega . \end{aligned}$$

or in (ψ, ω) formulation, where the stream function ψ is defined by

$$u = \nabla \times \psi .$$

The method is semi-lagrangian ; it consists in discretizing the vorticity into vorticity elements which are convected by the flow and follow the characteristic curves of the velocity field, while the computation of the velocity is done on a fixed grid. In order to mimic the different steps of the method when applied to the case of an incompressible fluid, we introduce the Helmholtz decomposition of the velocity u

$$u = v + \nabla \phi, \quad \nabla v = 0.$$

J.T. Beale et al. (eds.), Vortex Flows and Related Numerical Methods, 373–387.

This decomposition separates the compressible part, $\nabla\phi$, of the velocity from the incompressible part, v. Taking advantage of this natural splitting, we end up with the following formulation of Euler equations,

$$\frac{\partial \rho}{\partial t} + \nabla.(\rho(\nabla\phi + v)) = 0 \tag{1}$$

$$\frac{\partial \omega}{\partial t} + \nabla.(\omega(\nabla\phi + v)) = 0 \tag{2}$$

$$\nabla.v = 0 \tag{3}$$

$$\nabla \times v = \omega \tag{4}$$

$$\frac{\partial \phi}{\partial t} + \frac{1}{2}|v + \nabla\phi|^2 + k\frac{\gamma}{\gamma - 1}\rho^{\gamma-1} = q \tag{5}$$

$$\Delta q = \omega^2 + \nabla\omega \times (v + \nabla\phi). \tag{6}$$

As in the case of incompressible fluids, the method is semi-lagrangian in the sense that the convection equations (1) and (2) are solved by a particle method whereas the remaining equations are solved by finite difference schemes. The Poisson equations (3)-(4) and (6) are solved by a solver introduced by R. W. Hockney in [11] and rewriting equation (5) under the form

$$\frac{\partial \phi}{\partial t} + \frac{1}{2}|\nabla\phi|^2 + v.\nabla\phi = f,$$

yields a Hamilton-Jacobi equation with a convection term. This equation will be numerically solved by a E.N.O. type scheme introduced by S. Osher and J. Sethian in [14] for Hamilton-Jacobi equations.

The assignment operator of lagrangian values to the grid is the one introduced by J. U. Brackbill and H. M. Ruppel in [1]. Finally the introduction of a cut-off allows to convert grid quantities such as the velocity into particle values. Notice that, if at time zero the vorticity is equal to zero, it remains so for all time t and the flow is potential. This case has been studied in a previous work [13] and we present the general case here.

An outline of the paper is as follows. The formulation of the problem is detailed in Section 2. Then in Section 3 the schemes and the projection operator used are presented. Section 4 is devoted to the presentation of numerical tests in one and two dimensions.

2. Potential-vorticity-density formulation of the problem

We consider a two dimensional compressible fluid. The evolution of the flow is governed by Euler equations which written in density, velocity and pressure formulation are

$$\frac{\partial u}{\partial t} + u\nabla u + \frac{1}{\rho}\nabla p = 0 \tag{7}$$

$$\frac{\partial \rho}{\partial t} + \nabla.(\rho u) = 0 \tag{8}$$

$$p = k\rho^{\gamma}. \tag{9}$$

We have assumed in equation (9) that the fluid is a perfect gas and the constant γ is the specific heat ratio. The Helmholtz decomposition of the velocity field u,

$$u = \nabla\phi + v \quad \text{where } \nabla.v = 0 \tag{10}$$

introduces the rotational part v and the potential part $\nabla\phi$ of the velocity. Then, a little algebra and the vector identity

$$(u.\nabla)u = -u \times (\nabla \times u) + \frac{1}{2}\nabla(|u|^2) \tag{11}$$

lead to a natural separation between v and ϕ. We obtain an evolution equation for each of them and we end up with the new following formulation of problem (7)-(9)

$$\frac{\partial v}{\partial t} - (v + \nabla\phi) \times \nabla \times v + \nabla q = 0 \tag{12}$$

$$\nabla.v = 0 \tag{13}$$

$$\frac{\partial \phi}{\partial t} + \frac{1}{2} \mid v + \nabla\phi \mid^2 + k_1\rho^{\gamma-1} = q \tag{14}$$

$$\frac{\partial \rho}{\partial t} + \nabla.(\rho(v + \nabla\phi)) = 0. \tag{15}$$

This system is a hyperbolic system in ϕ and ρ coupled with a Euler type equation in v. Let us now apply the curl operator to equation (12). We then obtain the equation which governs the evolution of the vorticity, $\omega = \nabla \times v$ of the flow :

$$\frac{\partial \omega}{\partial t} + \nabla.(\omega(v + \nabla\phi)) = 0.$$

Now, we need an equation to determine the unknown function q which appears in the evolution equation for the potential and has the dimension of a pressure. Applying the divergence operator to equation (12), we find

$$\Delta q = \nabla.(u \times \omega) = \omega^2 + u.(\nabla \times \omega).$$

The resulting system of equations is

$$\frac{\partial \rho}{\partial t} + \nabla.(\rho(\nabla\phi + v)) = 0 \tag{16}$$

$$\frac{\partial \omega}{\partial t} + \nabla.(\omega(\nabla\phi + v)) = 0 \tag{17}$$

$$\nabla.v = 0 \tag{18}$$

$$\nabla \times v = \omega \tag{19}$$

$$\frac{\partial \phi}{\partial t} + \frac{1}{2}|v + \nabla\phi|^2 + k\frac{\gamma}{\gamma - 1}\rho^{\gamma-1} = q \tag{20}$$

$$\Delta q = \omega^2 + \nabla\omega \times (v + \nabla\phi). \tag{21}$$

We notice at first glance that the evolution equation for the vorticity is identical to the one obtained in the incompressible case which justifies *a posteriori* the idea to extend the classical vortex method to the case of a compressible fluid. We also notice that the equations which govern the evolution of the density and the vorticity are identical and will thus be numerically solved by the same method.

Remarks:

a) The vector identity

$$\nabla \times (\nabla \times v) = -\Delta v + \nabla(\nabla . v)$$

gives the following equation for the rotational part of the velocity

$$-\Delta v = \nabla \times \omega \tag{22}$$

b) In the case of periodic boundary conditions, we have uniqueness of v and q when adding the following additional equations

$$< q >= 0, \tag{23}$$

$$\begin{cases} \frac{d}{dt} < v_1 > = - < \omega u_2 > \\ \frac{d}{dt} < v_2 > = < \omega u_1 > \end{cases} \tag{24}$$

where, for a periodic function f, the notation $< f >$ stands for the mean value of f on a period. Observe that equation (24) is a consequence of (12).

3. Numerical resolution

The numerical method chosen is a semi-lagrangian method. The purely convective equations (16) and (17) are solved by a particle method whereas equations (18)-(21) are solved by a finite difference scheme. The fluid is represented by finite size particles to which are associated a position x_p, a velocity u_p, a vorticity ω_p and a density ρ_p. Due to the fact that we use a particle in cell method, the quantities carried by the particles (vorticity and density) need to be defined on the grid while the velocity which is computed on the grid has also to be assigned to the particles. The first part is called the particle-grid assignment and the second is the grid-particle interpolation. In order to define a conservative method, when associating a lagrangian quantity to a grid point, the method defined by Brackbill-Ruppel [1] is used. In this method, the volume of a grid point depends on the number of particles which surround it.

The next sections are devoted to the presentation of the schemes and the particle-grid exchange operators.

3.1. FINITE DIFFERENCE METHOD

Let us first focus our attention on equation (20). Setting

$$H(\nabla\phi) = H_1(\frac{\partial\phi}{\partial x}) + H_1(\frac{\partial\phi}{\partial y}) \tag{25}$$

and

$$H_1(\psi) = \frac{1}{2}\psi^2, \tag{26}$$

the equation for the potential ϕ is rewritten under the following form

$$\frac{\partial \phi}{\partial t} + H(\nabla\phi) + v.\nabla\phi = f \tag{27}$$

where we have set

$$f = q - k\frac{\gamma}{\gamma-1}\rho^{\gamma-1} - \frac{1}{2}|v|^2. \tag{28}$$

In equation (27), we recognize a Hamilton-Jacobi equation with an advection term. The chosen finite difference scheme is a E.N.O. type scheme which is derived for the approximation of hyperbolic conservation laws as follows. Let us consider the following one dimensional Hamilton-Jacobi equation

$$\frac{\partial \phi}{\partial t} + H_1(\frac{\partial \phi}{\partial x}) = 0.$$

We notice that the derivative u of ϕ satisfies the following conservation law

$$\frac{\partial u}{\partial t} + \frac{\partial}{\partial x}H_1(u) = 0$$

and we choose a monotone scheme. In the conservative form, the scheme is written as follows

$$u_j^{n+1} = u_j^n - \frac{\Delta t}{\Delta x}(g_{j+\frac{1}{2}}^n - g_{j-\frac{1}{2}}^n)$$

where the numerical flux $g_{j+\frac{1}{2}}^n$ is defined by

$$g_{j+\frac{1}{2}}^n = g(u_{j-p+1}^n, \ldots, u_{j+q+1}^n).$$

After integration on a cell, we obtain the following numerical scheme

$$\phi_j^{n+1} = \phi_j^n - \Delta t\; g(\Delta_-\phi_{j-p+1}^n, \ldots, \Delta_+\phi_{j+q}^n)$$

which gives an approximation of the Hamilton-Jacobi equation. This method is now readily extended to the Hamilton-Jacobi equation with an advection term: we set

$$D_+^x\phi_{ij} = \frac{\phi_{i+1,j} - \phi_{ij}}{\Delta x}, \quad D_-^x\phi_{ij} = \frac{\phi_{ij} - \phi_{i-1,j}}{\Delta x}.$$

The discrete Hamiltonian H_{H-J} is defined by

$$H_{H-J}(\bar{\phi})_{ij} = [|D_+^x\phi_{ij}^n|_+^2 + |D_-^x\phi_{ij}^n|_-^2] + [|D_+^y\phi_{ij}^n|_+^2 + |D_-^y\phi_{ij}^n|_-^2] \tag{29}$$

where $\bar{\phi} = (\phi_{ij})$. We set

$$f_{ij} = q_{ij} - k\frac{\gamma}{\gamma-1}\rho_{ij}^{\gamma-1} - \frac{1}{2}|v_{ij}|^2 \tag{30}$$

and the finite difference scheme is then (see [14]) :

$$\frac{\phi_{ij}^{n+1} - \phi_{ij}^n}{\Delta t} = H_{H-J}(\bar{\phi})_{ij} + v_1^+ D_-^x\phi_{ij}^n + v_1^- D_+^x\phi_{ij}^n + v_2^+ D_-^y\phi_{ij}^n + v_2^- D_+^y\phi_{ij}^n + f_{ij}^n \tag{31}$$

Remark :
1.- Under the following C.F.L. condition

$$(\frac{\Delta t}{\Delta x}+\frac{\Delta t}{\Delta y}) \mid H_1^{'} \mid \leq \frac{1}{2} \tag{32}$$

the scheme is monotone and stable (see [15]).
2.- A second order scheme in space and time is obtained by E.N.O. interpolation in space and Runge-Kutta discretisation in time.
Equations (21)-(22) are Poisson equations and their numerical resolution is achieved by Hockney's method (see [11]). The right hand side of equation (21) is rewritten in the following form

$$\omega^2+\nabla\omega\times(v+\nabla\phi)=\omega^2+(v+\nabla\phi).(\nabla\times\omega) \tag{33}$$

and the derivatives which appear in the right hand side of equations (21) and (22) are discretized by a centered scheme.

3.2. PARTICLE METHOD

The equations to be solved by the particle method are equations (16) and (17) which are purely convective and the method reduces then to a motion of the particles. To be more precise, we define the following measures

$$\omega^h=\sum_p W_p\omega_p\delta(x-x_p(t)), \ ; \ \rho^h=\sum_p W_p\rho_p\delta(x-x_p(t)), \tag{34}$$

where in the weights of the particle p we dissociate the volume $W_p>0$ and the local values of the vorticity $\omega(x_p,t)$ and of the density $\rho(x_p,t)$. These two quantities evolve with the flow but it must be observed that, due to the conservative form of the advection equations, the products $W_p\omega_p$ and $W_p\rho_p$ remain constant. The particle positions and volumes are obtained through the following system of ODE's

$$\frac{dx_p}{dt}(t)=u_h(x_p(t),t); \ \frac{dW_p}{dt}=\mathrm{div}u_h(x_p,t)W_p. \tag{35}$$

Equation (35) is numerically solved by a second order Adams-Bashforth scheme, giving in particular.

$$x_p^{n+1}=x_p^n+\Delta t(\frac{3}{2}u_p^n-\frac{1}{2}u_p^{n-1}). \tag{36}$$

3.3. PARTICLE-GRID EXCHANGE OPERATORS

Let us now present the operators which allow the exchange of information between the particles and the grid. We refer to [12] for a VIC method using similar ideas in the aprroximation of incompressible Navier-Stokes equations.

3.3.1 Particle-grid assignment

Let us recall Brackbill-Ruppel's method of assignment of particle quantities to a grid. We consider a structured mesh on a given domain Ω and denote by S_{ij}, $(i,j)\in$

I the vertices of the grid and ψ_{ij} a basis of interpolation such that $\sum_{ij}\psi_{ij}\equiv 1$ and for $(i,j),(k,l)\in I$, $\psi_{ij}(S_{kl})=\delta_{ij}^{kl}$. Let v^h be the following measure

$$v^h=\sum_p W_p v_p \delta(x-x_p)$$

where as before the coefficient $W_p>0$ is the volume associated to the particle x_p. The grid approximation $\bar{v}$ of v^h is given by

$$\bar{v}=\sum_{ij} v_{ij}\psi_{ij}, \tag{37}$$

where for $(i,j)\in I$, the coefficients v_{ij} are defined as follows

$$v_{ij}=\frac{< v^h,\psi_{ij} >}{< 1,\psi_{ij} >}=\frac{\sum_p W_p v_p \psi_{ij}(x_p)}{\sum_p W_p \psi_{ij}(x_p)}. \tag{38}$$

Appluting this formula to the particle approximations (34) of the density and the vorticity, we obtain for each grid point S_{ij},

$$\rho_{ij}=\frac{\sum_p W_p\rho_p\psi_{ij}(S_{ij}-x_p)}{\sum_p W_p\psi_{ij}(S_{ij}-x_p)}\simeq\rho(S_{ij}) \tag{39}$$

$$\omega_{ij}=\frac{\sum_p W_p\omega_p\psi_{ij}(S_{ij}-x_p)}{\sum_p W_p\psi_{ij}(S_{ij}-x_p)}\simeq\omega(S_{ij}) \tag{40}$$

Remark :
The assignment operator defined is conservative in the following sense : if $v_p\equiv 1$, then $v_j\equiv 1$. We set

$$J_{ij}=\sum_p W_p\psi_{ij}(x_p).$$

This quantity depends on the number of particles which are in the neighbourhood of the grid point S_{ij} and can be seen as a lagrangian volume assigned to the grid point.

We now recall the two different types of interpolation we have used in our numerical tests, the N.G.P. interpolation (nearest grid point) and the bilinear interpolation.

The basis function ψ for the N.G.P. interpolation is defined by

$$\begin{cases}\hat{\psi}(\xi-i,\eta-j)=1, & \text{if } 0\le|\xi-i|,|\eta-j|<\frac{1}{2}\\ \hat{\psi}(\xi-i,\eta-j)=0, & \text{if not}\end{cases}$$

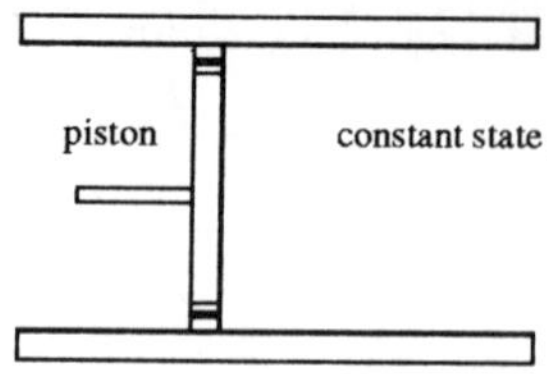

Fig. 1. Monodimensional shock tube

This interpolation distributes the particle quantities among the four grid points which surround the particle.

3.3.2 Grid-particle interpolation

In order to make the particle move, we need to assign a velocity to each of them. This is done in two steps, first, we define a velocity for each grid point, then we interpolate the grid quantity obtained to get a particle velocity. The first step is defined as follows: the value of the velocity $u = v + \nabla\phi$ at the grid point S_{ij} is approximated by

$$u_{ij} = \nabla_h \phi_{ij} + v_{ij}$$

where $\nabla_h \phi_{ij}$ is the approximation of $\nabla\phi$ obtained by upwind divided difference. Then the velocity of the particle x_p is given by

$$u_p = \sum_{ij} u_{ij} \psi_{ij}(S_{ij} - x_p) \tag{41}$$

where ψ_{ij} is one of the interpolation functions above defined.

4. Numerical tests

We present two monodimensional tests in which the P.I.C. method is compared to a classical finite difference scheme and two bidimensional tests.

Test 1 *Monodimensional infinite tube*
We consider an infinite tube along the x axis (see Figure 1). At time 0, we assume that the gas fills half of the tube and that the piston is located at $x = 0$. We assume moreover that the density ρ_0 is constant when the gas is at rest.
At time $t = 0$, we start pulling the piston from $x = 0$ in the direction $x < 0$ with a motion $x = x(t)$. noindent Since at time 0 the gas was at rest with a uniform density

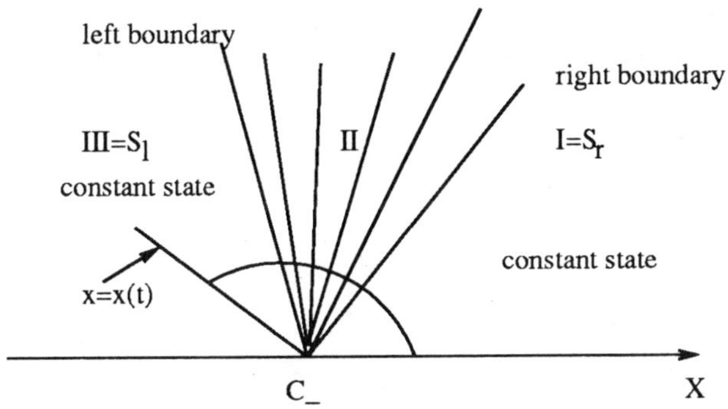

Fig. 2. Phase portrait for the shock tube problem

ρ_0, there exists a constant state I in the plan (x, t) with characteristic curves $C_\pm$ and slopes $\pm c(\rho_0)$. The Riemann invariants are :

$$\Gamma_\pm^0 = \pm \int_a^{\rho_0} \frac{c(s)}{s} ds = \text{constant}.$$

The boundaries of state I are characteristic lines C_+ issued from the origine. The characteristic curve C_- enters in the next region II. If the piston is pulled with a uniform velocity U, $x(t) = Ut$, state II is characterized by a self similar solution which is a rarefaction wave and there exists a constant state III bordering the piston path. In the case of an isentropic flow, we have

$$p = A\rho^\gamma$$

where A and γ are constants such that $\gamma > 1$ and we can do explicit calculations. The sound speed in the gas is given by

$$c = \frac{dp}{d\rho} = \sqrt{A\gamma\rho^{\gamma-1}} = \sqrt{\frac{\gamma p}{\rho}}, \tag{42}$$

and the Riemann invariants are

$$\Gamma_\pm = u \pm \int \frac{c(\rho)}{\rho} d\rho = u \pm \frac{2}{\gamma - 1} c.$$

The constant state I, S_r is such that

$$u_r = 0,\ \rho_r = \rho_0,\ p_r = A\rho_0^\gamma,\ c_r = \sqrt{\frac{\gamma p_r}{\rho_0}}.$$

The constant state II, S_l is defined by (u_l, ρ_l, p_l, c_l). Since $u_l = U$, and

$$\Gamma_- = U - \frac{2}{\gamma - 1} c_l = u_r - \frac{2}{\gamma - 1} c_r = -\frac{2}{\gamma - 1} c_r,$$

we determine the value of c_l and then of (u_l, ρ_l, p_l, c_l). The left boundary of the rarefaction wave is given by $x = (U + c_l)t$. Finally, if (x,t) is in region II, we have

$$u + c = \frac{x}{t}$$

and

$$\Gamma_- = u - \frac{2}{\gamma - 1}c = u_r - \frac{2}{\gamma - 1}c_r,$$

which leads to the following expressions

$$u(x,t) = \frac{2}{\gamma+1}(\frac{x}{t} - c_r),$$

$$c(x,t) = \frac{\gamma-1}{\gamma+1}(\frac{x}{t} + \frac{2}{\gamma-1}c_r).$$

The functions $\rho(x,t)$ and $p(x,t)$ are determined thanks to (42).

The results of this test are illustrated in Figure 4 (left picture for the velocity and right picture for the density). The velocity and density computed with the PIC method are compared with the result's obtained by Glaister's solver [7] at time $t = 1$ for $\Delta x = .01$ and $CFL = .1$.

Test 2 [6]
The initial conditions for this test are

$$\rho(x,0) = \rho(x), \qquad u(x,0) = u(x).$$

The boundary conditions are supersonic at the entry, $(x = 0)$ and subsonic at the exit, $(x = 1)$. The numerical values are

$$A = \frac{1}{(1.4 \times 1.2^2)} \simeq 0.5 \qquad \rho_L(t) \equiv u_L(t) \equiv 1.$$

For $u_R(t) = 0.8$, the computed solution is a 1-shock, it is characterized by a right state $\rho_R(t) = 1.256$ and a shock velocity $V_s = 0.0447$. The value of the density $\rho_R(t)$ is given as boundary condition for the numerical method.

The results of this test are illustrated by Figure 5, where again the PIC method is compared to Glaister's method, at time $t = 7.93$ with $\Delta x = .01$ and $CFL = .1$. The results are in good agreement although the PIC method is slightly more diffusive for the computation of the velocity. The reason is that our PIC method is based on a Lax-Friedrichs scheme for the computation of the velocity. By contrast, the density-shock is better captured by the PIC method, illustrating the fact that particle methods are well suited for the resolution of transport equations.

Test 3 *Simulation of an oblique stationary shock, with a 45° orientation* ([15])
The speed of the shock is 1.7 and the boundary conditions are determined by the exact solution of the continuous problem

$$\begin{aligned} \rho_{/\partial\Omega}(t,x,y) &= \rho_0(x - 2.5t, y) && \text{if } x > 2.5t \\ \rho_{/\partial\Omega}(t,x,y) &= \rho(t_s = \frac{x}{2.5}, 0, y) && \text{otherwise} \end{aligned}$$

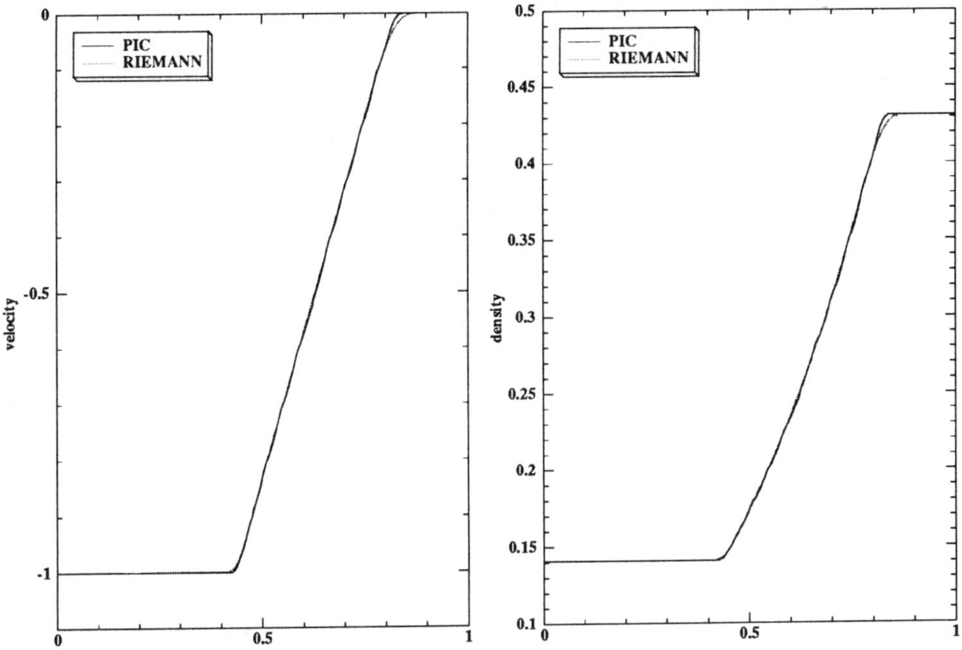

Fig. 3. Test 1: density and velocity profiles for PIC and Glaister's method at $t = 1$ for $\Delta x = .01$ and $CFL = .1$

with $\rho(t_s, 0, y) = \rho_0(2.5t_s, y)$.
Some results of the test at time $t = .1$ for $\Delta x = .05$ are illustrated by Figure 6. The first component of the velocity and the density computed with $CFL = .25$ by the N.G.P. interpolation and the bilinear interpolaton are respectively plotted in the top-left and top-right pictures. The bottom pictures present the results of the same calculations with $CFL = .1$, showing the attenuation of the oscillations.

Test 4 *Kelvin-Helmholtz instabilities* [2;3;16]
The fluid is initially confined in a strip of width d. The initial conditions are
1) *within the strip*

$$\phi_0(x,y) = xy, \quad \rho_0 \equiv 1 \quad \omega_0 \equiv -1 \qquad q(x,y) = \frac{1}{2}y^2 + 1$$

$$v_0(x,y) = \begin{bmatrix} 0 \\ -x \end{bmatrix}, \quad u_0(x,y) = \begin{bmatrix} y \\ 0 \end{bmatrix}$$

2) *outside the strip*

$$\phi_0(x,y) = \pm\frac{1}{2}xd, \quad \rho_0 \equiv 1 \quad \omega_0 \equiv 0 \qquad q \equiv \frac{1}{8}d^2 + 1$$

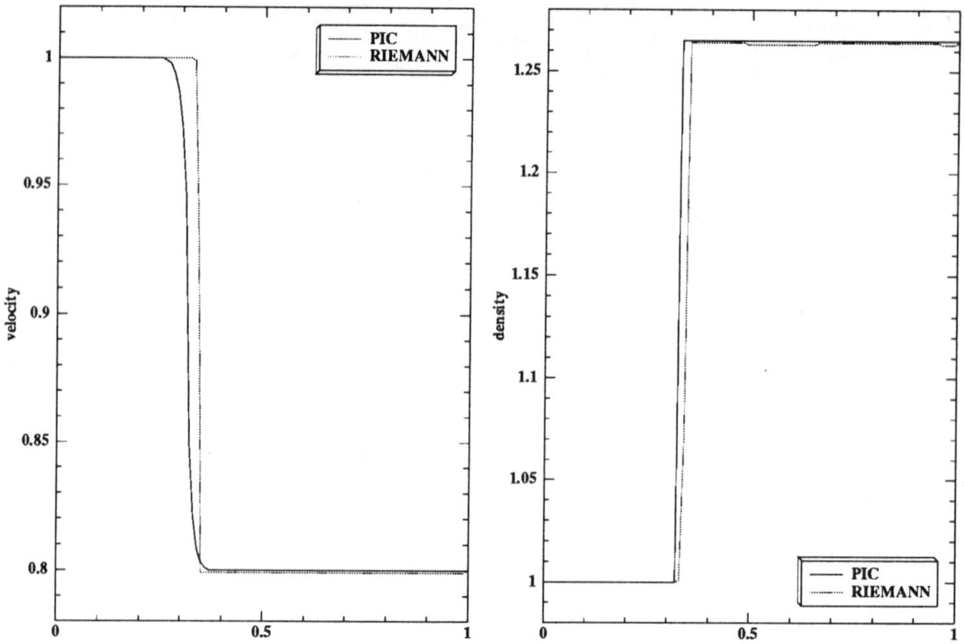

Fig. 4. Test 2: density and velocity profiles for PIC and Glaister's method at $t = 7.93$ for $\Delta x = .01$ and $CFL = .1$

$$v_0(x, y) = \begin{bmatrix} 0 \\ 0 \end{bmatrix}, \quad u_0(x, y) = \begin{bmatrix} \pm 0.5d \\ 0 \end{bmatrix}$$

Some results of this test for $CFL = .1$ and $\Delta x = \Delta y = 1/128$ are illustrated in Figures 7 and 8. Figure 7 shows the time evolution of the velocity between times .0013 and .134 (top pictures) and compares the potential and rotational parts of the velocity fields (bottom pictures), from which it is clear that the flow is mostly rotational. Finally Figure 8 gives the density isovalues. It is interesting to notice that the highest density values remain on the initial separation line.

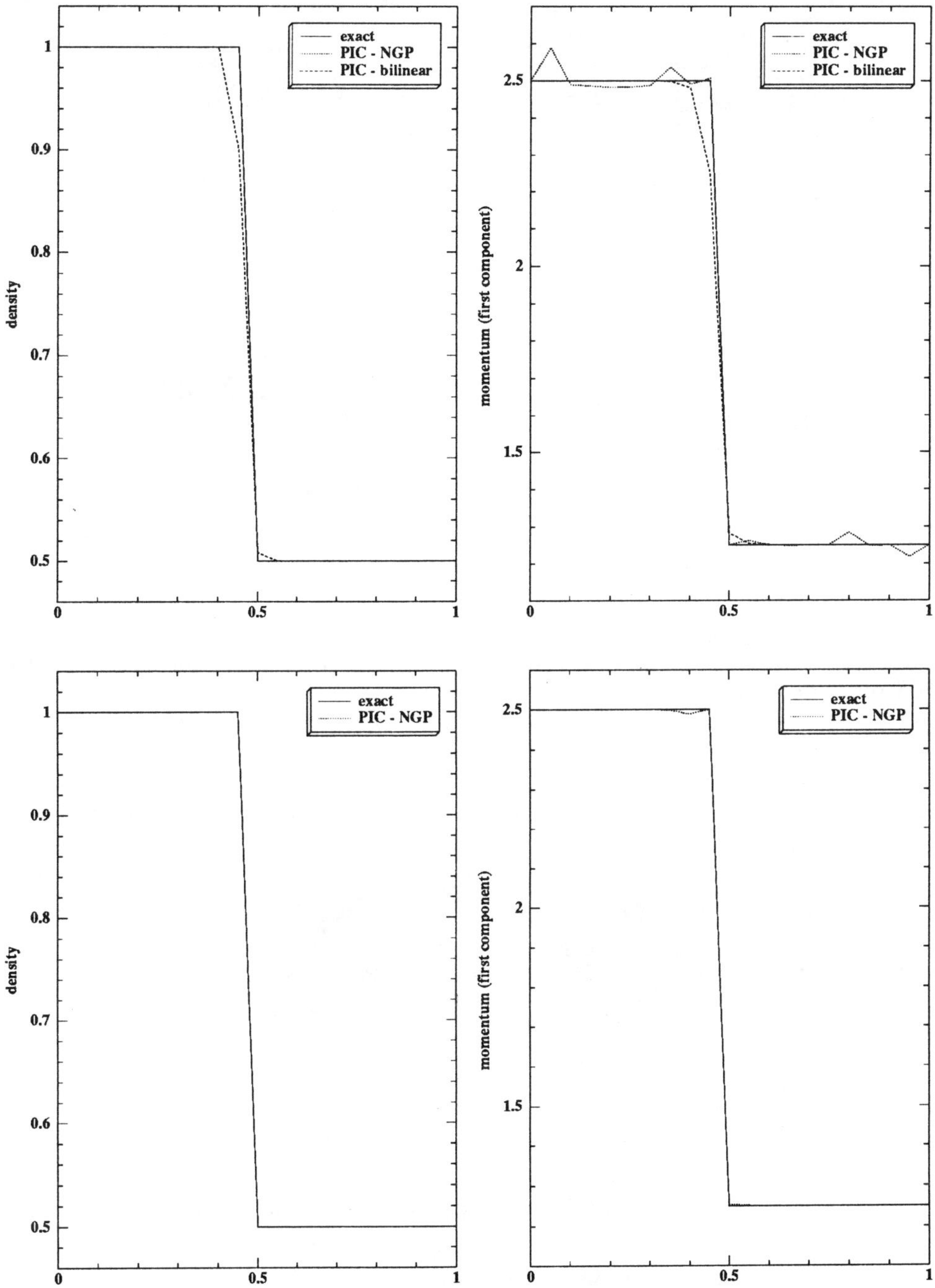

Fig. 5. Test 3: density and velocity profiles for the PIC method using NGP or bilinear interpolation with $CFL = .25$ (top pictures) and $CFL = .1$ (bottom pictures)

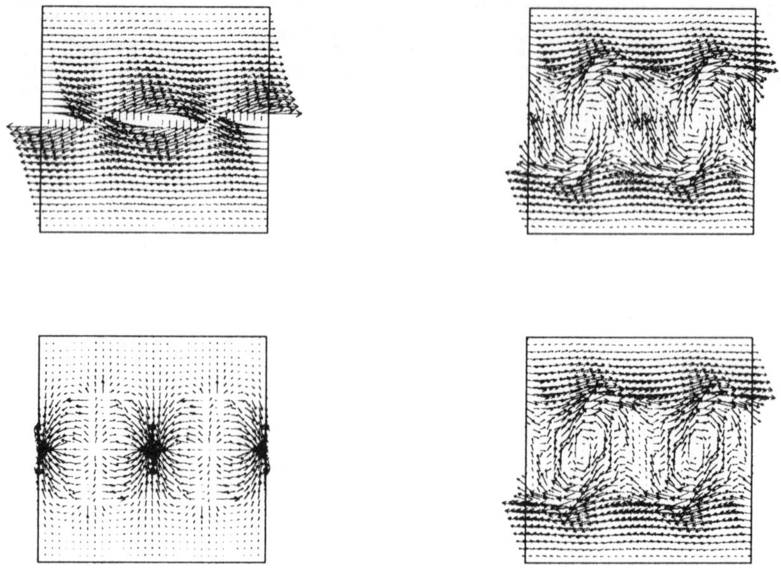

Fig. 6. Kelvin-Helmoltz instability: velocity profiles at times .0013 (top-left) and .134 (top-right) and potential (bottom-left) vs rotational (bottom-right) component of the velocity at $t = .134$

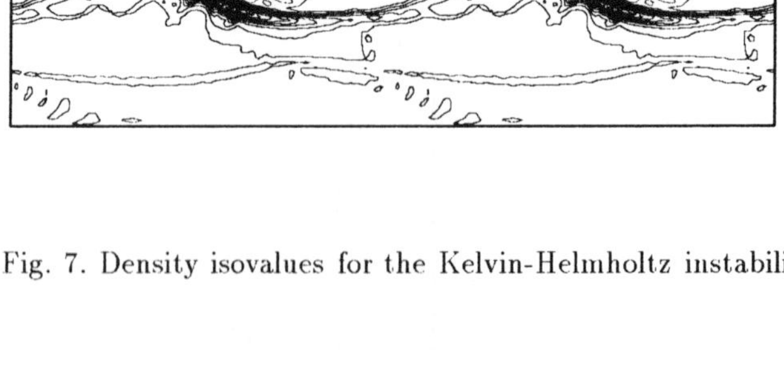

Fig. 7. Density isovalues for the Kelvin-Helmholtz instability

References

[1] J.U. Brackbill, H.M. Ruppel, Flip: A method for adaptively zoned, Particle-in-cell calculations of fluid flows in two dimensions, J. Comp. Phys. **65**, 314-343 (1986).

[2] O. Buneman, Compressible flow simulation using Hamilton's equations and Clebsch-type vortex parameters, in *Seventh International Conference on Numerical Methods in Fluid Dynamics*, Stanford, Springer Verlag ed. (1980).

[3] O. Buneman, Ideal gas dynamics in Hamiltonian form with benifit for numerical schemes, Abstract CH4, Bull. APS **24**, 1139 (1979).

[4] J.P. Christiansen, Numerical simulation of hydrodynamics by the method of point vortices, J. Comp. Phys. **13**, 363-379 (1973).

[5] G.-H. Cottet, Convergence of a Vortex In Cell method for the two-dimensional Euler equations, Math. Comp., **49**, 407-425, 1987.

[6] B. Engquist, S. Osher, Upwind difference schemes for systems of conservation laws-potential flow calcualtions, MRC Technical Report 2186, 1981.

[7] P. Glaister, A Riemann solver for barotropic flow, J. Comp. Phys., **93**, 477-480, 1991.

[8] H. Harlow, A. Amsden, Slip instability, Phys. of Fluids **7** , 327-334 (1964).

[9] F. Harlow, The particle-in-cell computing method for the fluid dynamics, Methods in Comp. Phys. **3** (1964).

[10] R.W. Hockney, Comp. Phys. Comm. **2** (1971).

[11] S. Huberson and A. Jollès, C. R. Acad. Sci., **309**, Série II, 445-448, Paris, 1989.

[12] M. Louaked, Une méthode P.I.C (particle-in-cell) pour les équations d'Euler compressibles, Rapport interne 90016, Lab. Anal. Num. 1990.

[13] S. Osher, J. Sethian, Fronts propagating with curvature-dependent speed : Algorithms based on Hamilton-Jacobi formulations, J. Comp. Phys. **79**, 12-49 (1988).

[14] S. Osher, C. Shu, High order essentially non-oscillatory schemes for Hamilton-Jacobi equations, Icase Report 90-13.

[15] S. Osher, F. Salomon, Upwind difference schemes for hyperbolic systems of conservation laws, Math. Comp. **38**, (1982), 339-374.

[16] R. Woodward, Simulation of the Kelvin-Helmholtz instability of a supersonic slip surface with the piecewise-parabolic method (P.P.M.) in *Numerical methods for the Euler equations of fluid dynamics*, INRIA workshop, 439-508, SIAM, ed., Philadelphia (1985).